Quadratic Formula (2.2)

If $ax^2 + bx + c = 0$, $a \neq 0$, then

$$x = \frac{-b \pm \sqrt{b^2 - 4ac}}{2a}$$

Variation (2.6)

$y = kx$ y varies directly as x
$y = k/x$ y varies inversely as x
$y = kuv$ y varies jointly as u and v

Distance Formula (3.1)

$$d[(x_1, y_1),(x_2, y_2)] = \sqrt{(x_2 - x_1)^2 + (y_2 - y_1)^2}$$

Lines (3.3)

1. Slope $m = \dfrac{y_2 - y_1}{x_2 - x_1}$
2. Point-slope formula
   $(y - y_0) = m(x - x_0)$
3. Slope-intercept formula
   $y = mx + b$
4. Vertical lines
   $x = a$
5. General formula
   $Ax + By + C = 0$
6. Parallel lines $m_1 = m_2$.
7. Perpendicular lines $m_1 m_2 = -1$

Parabola with vertex at (x_0, y_0)
(3.4 & 3.6)

$y = a(x - x_0)^2 + y_0$
    or
$x = A(y - y_0)^2 + x_0$

Circle of radius r centered at (x_0, y_0)
(3.2 & 3.6)

$(x - x_0)^2 + (y - y_0)^2 = r^2$

Ellipses and Hyperbolas centered at (h,k)
(3.5 & 3.6)

Ellipse: $\dfrac{(x - h)^2}{a^2} + \dfrac{(y - k)^2}{b^2} = 1$

Hyperbola: $\dfrac{(x - h)^2}{a^2} - \dfrac{(y - k)^2}{b^2} = 1$
    or
$\dfrac{(y - k)^2}{b^2} - \dfrac{(x - h)^2}{a^2} = 1$

Logarithms (6.2, 6.3, & 6.5)

$u = \log_b v$ if and only if $v = b^u$
         ($v > 0$ and $0 < b \neq 1$)

$\log_b uv = \log_b u + \log_b v$

Logarithms (6.2, 6.3, & 6.5)(continued)

$\log_b(u/v) = \log_b u - \log_b v$
$\log_b u^r = r \log_b u$
$\log_b 1 = 0$
$\log_b(1/u) = -\log_b u$

$$\log_b u = \frac{\log_a u}{\log_a b}$$

Arithmetic Progressions (8.2 & 8.3)

$a_n = a_{n-1} + d = a_1 + (n - 1)d$

$$a_1 + \cdots + a_n = \frac{n}{2}[a_1 + a_n]$$

Geometric Progressions (8.2 & 8.3)

$a_n = r \cdot a_{n-1} = a_1 \cdot r^{n-1}$

$$\sum_{j=0}^{n-1} ar^j = a\left(\frac{1 - r^n}{1 - r}\right) r \neq 1$$

Summation Symbol (8.3 & 8.4)

$$\sum_{j=k}^{l} (a_j + b_j) = \sum_{j=k}^{l} a_j + \sum_{j=k}^{l} b_j$$

$$\sum_{j=k}^{l} (Ca_j) = C \sum_{j=k}^{l} a_j$$

Factorial (8.4)

$n! = 1 \cdot 2 \cdot \ldots \cdot n$

Binomial Theorem (8.4)

$$(a + b)^n = \sum_{l=0}^{n} \frac{n!}{l!(n - l)!} a^{n-l}b^l$$

Permutations and Combinations (8.5)

$${}_nP_r = \frac{n!}{(n - r)!}$$

$${}_nC_r = \frac{n!}{(n - r)!r!}$$

Probability (8.6)

1. $P(E) = \dfrac{\text{number of ways in which } E \text{ could occur}}{\text{total number of possible outcomes}}$
2. $P(E_1 \text{ or } E_2) = P(E_1) + P(E_2) - P(E_1 \text{ and } E_2)$
3. $P(\sim E) = 1 - P(E)$
4. Independent events
   $P(E_1 \text{ and } \cdots \text{ and } E_n) = P(E_1) \cdot P(E_2) \cdots P(E_n)$
5. Mutually exclusive events
   $P(E_1 \text{ or } \cdots \text{ or } E_n) = P(E_1) + P(E_2) + \cdots + P(E_n)$

College Algebra

College Algebra

Ralph C. Steinlage

Professor of Mathematics
University of Dayton

WEST PUBLISHING COMPANY
St. Paul   New York   Los Angeles   San Francisco

Cover Photograph: Patrick R. Siegrist

Production Coordination: Editing, Design & Production, Inc.

Copyright © 1984 by West Publishing Co.
　　　　　　　　　　50 West Kellogg Boulevard
　　　　　　　　　　P.O. Box 3526
　　　　　　　　　　St. Paul, Minnesota 55165

Printed in the United States of America

Library of Congress Cataloging in Publication Data
Steinlage, Ralph C.
　College algebra.

　Includes index.
　1. Algebra.　　I. Title.
QA154.2.S738　　1984　　　512.9　　　83-19709
ISBN 0-314-77816-0

Contents

Preface

This text is designed to be used in first-year algebra courses at the college or university level. Topics are presented in an intuitive manner so that students can more quickly gain the confidence needed for further progress. Nevertheless, care is taken so that mathematical statements are precise and clear.

Chapter 1 begins with a review of the Real Number System, including negative numbers, fractions, order, and absolute value. The basic development of algebra including exponents, radicals, factoring, polynomials, and rational expressions follows. Complex numbers are introduced in preparation for the discussion of complex roots obtained by the quadratic formula (Chapter 2). Depending on the background of the class, Chapter 1, either partially or in its entirety, may be treated as an optional review.

Chapter 2 is a thorough study of equations and inequalities, and the techniques introduced are used throughout the text. A separate section (2.4) describes a systematic approach to the solution of word problems.

Chapter 3 introduces the student to graphing in a Cartesian coordinate system. Circles, lines, and parabolas are studied extensively in the first part of the chapter. The remainder of the chapter treats the other conic sections: ellipses and hyperbolas. This latter material can be postponed or simply omitted if the instructor chooses. No later topics depend on this material.

Chapter 4 introduces the concept of "function." The techniques of modifying a known graph by translation, reflection, and scaling are emphasized in the general approach to graphing functions.

Polynomial and rational functions and their graphs are discussed in detail in Chapter 5. The chapter begins with an analysis of the roots of a polynomial and methods of finding these roots. These methods include synthetic division, Descartes' rule of signs, the rational root theorem, and successive approximations using a calculator. The chapter concludes with the graphing of polynomial and rational functions. In graphing these functions, emphasis is placed on the importance of roots and on the graphing techniques developed in Chapter 4. This entire chapter is independent of the remainder of the text. It may be postponed or omitted with no loss of continuity.

Chapter 6 is devoted to the uses of exponential and logarithmic functions, such as in the description of exponential growth, compound interest, and carbon dating. Many questions in these areas require finding a logarithm. Thus, even though their use as a computational tool has been usurped by inexpensive calculators, logarithms continue to be important.

Matrices are introduced in Chapter 7 as a means to simplify the solution of a system of equations. A section on matrix algebra is included for those who desire it, but this section (7.9) may be omitted with no loss of continuity. Systems of inequalities are also included in Chapter 7, and these lead naturally into linear programming, provided here as a very practical and powerful application of basic mathematics. No topics later in the text depend on either of these sections.

Finally, Chapter 8 introduces the topics of discrete mathematics: mathematical induction, progressions, the binomial theorem, permutations, combinations, and probability. The emphasis in this chapter is on a structured approach to problems rather than memorization of formulas.

FEATURES

Instructional Aids

1. *Maintaining Student Interest:* Introductions provided for each chapter pave the way into the text and enliven the material in the chapter. Historical perspectives are included to give the student an appreciation for the development of mathematics.

2. *Emphasis on Graphing:* An intuitive approach is taken to the development of most topics. Whenever possible, graphs are used to illustrate the solution of a problem. Many of the exercises also emphasize graphs and graphing techniques.

Study Aids

1. *Exercises:* There are more than 3400 exercises in the text. The exercises sets are graded from easy to challenging, and extensive supplementary exercise sets are included at the end of each chapter. These may be used for additional assignments or as a source of test problems. Answers to all odd-numbered exercises are provided at the end of the book.

2. *Rules Boxes:* Throughout the text, rules and formulas are displayed in boxes, and examples of their use are provided. In many cases, numerical examples appear alongside the rules in these boxes. Thus, these rules stand out and are easy for the student to locate and use.

3. *Listing of Steps:* Whenever several steps are involved in a given procedure, each step is identified and listed so that the student can readily reference steps when needed.

4. *Emphasis on Orderly Procedures in Long Problems:* Solving large systems of equations can sometimes be frustrating. This is especially true when a minor arithmetic mistake has been made in the middle of a problem and a maze of computations must be checked. A record-keeping procedure is introduced to simplify the task of retracing one's steps in order to find an error in solving a system of equations or evaluating a large-order determinant.

5. *Chapter Reviews:* Reviews summarize important terms and concepts, rules and formulas, and techniques that were introduced in the chapter. In most cases, a simple illustration or example appears alongside the concept, rule, or technique.

Accuracy

In order to guarantee the accuracy of the answers provided, all of the exercises and examples have been checked by an independent problems solver.

Applications

Numerous real-life applications are included throughout the text as examples and as exercises. These applications introduce and reinforce the concepts being studied and serve to motivate student whose primary interest is in a discipline other than mathematics. Application topics include medicine, nuclear decay, pH, noise levels, the Richter scale, the effect of size or scale on the strength of an animal, pollution, carbon dating, and many others. Most of the exercise sets include applications problems.

Calculators

The text recognizes and encourages the use of calculators without becoming a "calculator text." The use and operation of a calculator are explained in general terms (without reference to a specific model). "Calculator Comments" boxes describe calculator usage and warn students of erroneous answers that could be obtained by haphazardly using a calculator. Students are encouraged to use calculators rather than tables for computations involving exponents and logarithms. For those who desire it, however, an optional section (6.4) detailing the use of tables is included.

Supplements

1. A students' *Solutions Manual* containing complete worked solutions for representative odd-numbered exercises is available for purchase.

2. An *Instructor's Manual* contains answers to all the even-numbered exercises, three sample tests for each chapter, and three sample final examinations. Answer keys with solutions are also provided for these tests and examinations.

FLEXIBILITY

This text has been written so that it can be used in a variety of courses. A skeleton course is obtained by covering the sections and chapters listed in the first column below. This set of material should be augmented with topics selected from the second column. The topics selected will depend on the nature of the course for which the book is used. Each of the chapters or sections listed in the second column may be covered at the point indicated by the broken arrow, postponed for coverage at a later time if desired, or simply omitted. None of these topics is required in order to cover a later topic in the book. For example, the conic sections (sections 3.5 and 3.6) are covered last in some courses, while in others they may not be covered at all.

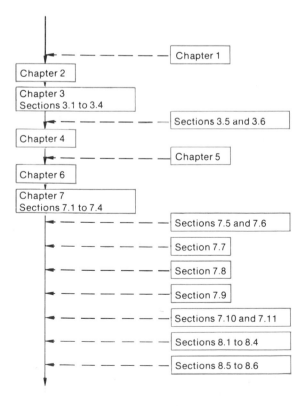

ACKNOWLEDGMENTS

I wish to thank the people who reviewed the manuscript in its various stages of development. Their advice and comments regarding style, inclusion of topics, emphasis on certain topics, and order of presentation were especially helpful. Many pertinent suggestions relating to both the details and the overall organization of the manuscript were obtained from these reviews. The list of reviewers includes the following:

Gerald Bradley—The Claremont Colleges
Mary Jane Causey—University of Mississippi
Frank Gilfeather—University of Nebraska
Paul Hutchens—St. Louis Community College at Florissant Valley
James Lightbourne—West Virginia University
Charles Luning—Sam Houston State University
Anthony Peresini—University of Illinois
Gilbert Perez—North Harris Community College
Samuel Rankin—West Virginia University
Burton Rodin—University of California at San Diego
John Spellman—Southwest Texas State University
Robert B. Thompson—Lane Community College
Arnold Wendt—Western Illinois University
Henry Zatzkis—New Jersey Institute of Technology

A special note of appreciation is also due Mr. Peter Marshall, mathematics editor at West Publishing Company, who coordinated and guided the project from its inception, through the development and reviewing stages, and into production of the final manuscript.

Finally, I wish to thank Gloria Langer of Aurora, Colorado, who checked every exercise and example in the book.

Ralph C. Steinlage
January, 1984

College Algebra

1 Basic Algebra

Preceding the most primitive attempts to record history, numbers and the counting process developed in a natural way to communicate essential ideas such as the size of a herd of animals. With the development of number systems and our dependence on numbers and their properties, numbers are today an absolute necessity for commerce and communication. For example, our monetary system is based on numbers and the arithmetic operations of addition, subtraction, multiplication, and division; space flights would not be possible without algebra, trigonometry, calculus, and so on.

Scientists express the laws of nature in terms of letters that can represent different numbers in different situations. Examples include the distance formula $d = rt$ and the mass-energy equation $E = mc^2$ developed by Albert Einstein (1879–1955). In **algebra,** letters and symbols are used to represent numbers and to state rules by indicating that a property holds for all numbers. Since letters are simply names for numbers, the rules of arithmetic apply. We will review the elementary operations of arithmetic as they apply to the system of real numbers.

1.1 THE REAL NUMBER SYSTEM

The counting numbers 1, 2, 3, . . . are called **natural numbers.** A number such as 21, which is the product of two or more natural numbers, is called a **composite number;** the numbers whose product is the composite number are called **factors** of the composite number. For example, we say that 21 can be **factored** into 3 times 7. A natural number whose only factors are 1 and itself is called a **prime number.** For example, 2, 3, 5, 7, 11, 13, 17, 19 are prime numbers, whereas $4 = 2 \cdot 2$, $9 = 3 \cdot 3$, and $12 = 3 \cdot 4$ are composite

1

numbers. Every natural number can be factored into a product of primes. For instance, $12 = 2 \cdot 2 \cdot 3$.

Factoring can facilitate finding the **least common multiple (LCM)** and **greatest common divisor (GCD)** of several numbers. For instance, we write

$$60 = 2 \cdot 2 \cdot 3 \cdot 5$$

and

$$84 = 2 \cdot 2 \cdot 3 \cdot 7$$

The factors 2, 2, and 3 divide both 60 and 84; thus, the GCD of 60 and 84 is $2 \cdot 2 \cdot 3 = 12$. On the other hand, any multiple of 60 and 84 must have factors of 2, 2, 3, 5, and 7. The LCM of 60 and 84 is $2 \cdot 2 \cdot 3 \cdot 5 \cdot 7 = 420$. These techniques will be adapted to algebraic expressions later.

The **integers** are the natural numbers together with their negatives and 0:

$$\ldots, -3, -2, -1, 0, 1, 2, 3, \ldots$$

HISTORICAL PERSPECTIVE

One of the best-known theorems in mathematics is the **Pythagorean theorem** which states that *the sum of the squares of the legs of a right triangle equals the square of the hypotenuse.* (The product of a number a with itself is denoted a^2 [read "a squared"]. Thus, 7 squared is $7^2 = 7 \cdot 7 = 49$.)

The so-called Pythagorean school of mathematics, science, and philosophy was so well disciplined and organized that it survived for over 100 years after the death of Pythagoras (ca. 575–500 B.C.). All the works of the school were attributed to Pythagoras. However, there is no direct evidence that Pythagoras, or even the Pythagorean school, proved the Pythagorean theorem. In fact, the Babylonians knew of the "Pythagorean property" more than 1000 years before the time of Pythagoras, but they were unable to prove it.

The Pythagorean school evolved into a cult. Philosophy and mathematics became so intertwined that its members believed everything depended on the whole numbers. This is perhaps the only example in history of a religion based in mathematics. Their belief in the omnipotence of the whole numbers encompassed the fractional or rational numbers also; 1/3 could be considered as being obtained by dividing a whole into three equal parts.

Figure 1–1   The Length $\sqrt{2}$

The Pythagorean theorem indicates the existence of a line segment whose length is $\sqrt{2}$; (the symbol $\sqrt{a}$ is used to denote a number that when multiplied by itself yields a; it is called the **square root** of a—thus, $\sqrt{2} \cdot \sqrt{2} = 2$.) Here the $\sqrt{2}$ is the hypotenuse of a right triangle with legs of length 1 as shown in Figure 1–1. One of the greatest contributions of the Pythagoreans to the development of mathematics was their discovery that $\sqrt{2}$ could not be a rational number; it cannot be written as a fraction or ratio of integers m/n. Such numbers are said to be **irrational.** However, this discovery shook the Pythagorean school to its very core. It rendered invalid their basic belief that "everything depends on the whole numbers." Legend has it that Hippasus (ca. 500 B.C.) was set adrift at sea for revealing to outsiders the irrationality of $\sqrt{2}$.

The minus sign ($-$) is used to denote "below zero" or "negative." For instance, $-10°$ represents $10°$ below zero. Death Valley has an elevation of -282 feet; it is 282 feet below sea level.

A number that can be expressed as a ratio of integers m/n with $n \neq 0$ is called a **rational number.** Every integer is also a rational number since $m = m/1$. Rational numbers are used to denote "fractional parts" or "fractions." For instance, $2/7$ is used to indicate two pieces of something that has been divided into seven equal parts. In a fraction a/b, a is called the **numerator** and b is called the **denominator.**

The real number line is used to describe numbers like $\sqrt{2}$, which arise from elementary geometry as lengths of line segments. Choosing a unit of length and marking an origin O on some line l, we assign to each point P on l a number equal to the length of the segment $\overline{OP}$. Points to the right of the origin are assigned positive numbers, and points to the left are assigned negative numbers. The number zero is assigned to the origin O. Each such number is called a **real number.** The line is called the **real number line.** The locations of several numbers on the real line are illustrated in Figure 1–2.

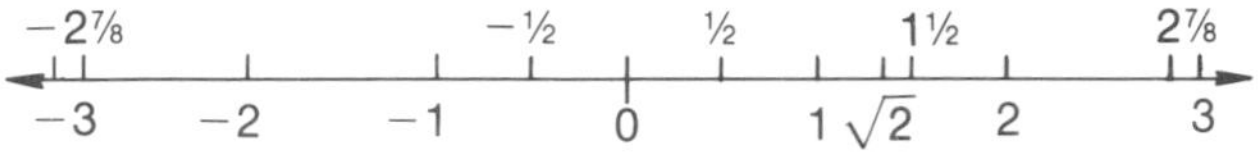

Figure 1–2   **The Real Number Line**

One of the fundamental properties of the real number line is that it contains within itself not only the square roots, cube roots, and so on of each of its positive numbers but also the so-called transcendental numbers, such as π, which are more difficult to describe. Later we shall describe the extension of the real numbers to a larger system, the complex numbers, which will include the square roots of all negative numbers as well.

The two basic arithmetic operations in the real number system are **addition** ($+$) and **multiplication** ($\cdot$). These operations are commutative, associative, and distributive.

Some Properties of Real Numbers

	Property	Example
Commutative	$a + b = b + a$ $ab = ba$	$5 + 4 = 4 + 5 = 9$ $5 \cdot 4 = 4 \cdot 5 = 20$
Associative	$a + (b + c) = (a + b) + c$ $a \cdot (b \cdot c) = (a \cdot b) \cdot c$	$5 + (4 + 3) = (5 + 4) + 3 = 12$ $5 \cdot (4 \cdot 3) = (5 \cdot 4) \cdot 3 = 60$
Distributive	$a \cdot (b + c) = ab + ac$ $(a + b) \cdot c = ac + bc$	$5 \cdot (4 + 3) = 5 \cdot 4 + 5 \cdot 3 = 35$ $(5 + 4) \cdot 3 = 5 \cdot 3 + 4 \cdot 3 = 27$

Subtraction ($-$) and **division** ($\div$) are in a sense "inverses" of addition and multiplication. For instance, for any two numbers a and b, the **difference** $a - b$ is that number c, which when added to b yields a.

$$a - b = c \quad \text{means} \quad a = b + c$$

For example, $5 - 3 = 2$ since $5 = 3 + 2$. Retail salesclerks generally subtract in exactly this fashion. For rather than formally subtract the purchase price from the amount tendered, the clerk simply returns the excess amount by counting from the purchase price up to the tendered amount. The rules governing subtraction and negative numbers are summarized as follows.

Rules for Negative Numbers

Rule	Example
1. $a + (-b) = a - b$	**1.** A loss is a negative profit. Thus, a profit of \$10,000 plus a loss of \$2000 yields a net profit of \$8000: $\$10,000 + (-\$2000) = \$8000 = \$10,000 - \$2000$
2. $a - (-b) = a + b$	**2.** $5280 - (-282) = 5280 + 282 = 5562$
3. $-(a + b) = -a - b$	**3.** $-(2 + 3) = -5 = -2 - 3$
4. $a(-b) = -(ab) = (-a)b$ In particular $(-b) = (-1)b$	**4.** $\left.\begin{array}{l} 2(-3) = (-3) + (-3) = -6 \\ (-2)(3) = (-2) + (-2) + (-2) = -6 \end{array}\right\} = -(2 \cdot 3)$ $(-1)b = -(1 \cdot b) = -b$
5. $-(-a) = a$	**5.** $-(-2) = 2$; the opposite of a \$2 loss is a \$2 gain.
6. $(-a)(-b) = ab$	**6.** $(-2)(-7) = -[2(-7)] = -[-14] = 14 = 2 \cdot 7$

Division of a real number a by $b \neq 0$ (denoted a/b or $a \div b$) is that number c, which when multiplied by b yields a; c is called the **quotient** of a and b. For instance, $8/4 = 2$ since $8 = 4 \cdot 2$.

$$\frac{a}{b} = c \qquad \text{means} \qquad a = bc, \qquad (b \neq 0)$$

Division by 0 is never permitted. For $a = 0 \cdot c$ cannot hold unless $a = 0$. This indicates that, at best, only 0 could be divided by 0. On the other hand, we could write $0 = 0 \cdot c$ for any number c. Thus, $0/0$ would be ambiguous. Hence, we *never* divide by 0.

The rules governing division and rational expressions are summarized in the following table.

Rule 1 is called the **cancellation law.** Factoring both numerator and denominator of a fraction and eliminating or "canceling" common factors yield a fraction equal to the original. The numerator and denominator of the resulting fraction will share no common factors other than ± 1. Such a fraction is said to

Rules for Fractions

Rule	Example

1. $\dfrac{k \cdot m}{k \cdot n} = \dfrac{m}{n}$

1. $\dfrac{3 \cdot 1}{3 \cdot 2} = \dfrac{3}{6} = \dfrac{1}{2}$

2. $\dfrac{a}{b} + \dfrac{c}{d} = \dfrac{ad + bc}{bd}$

2. $\dfrac{1}{2} + \dfrac{4}{3} = \dfrac{1}{2} \cdot \dfrac{3}{3} + \dfrac{2}{2} \cdot \dfrac{4}{3} = \dfrac{1 \cdot 3 + 2 \cdot 4}{2 \cdot 3} = \dfrac{11}{6}$

3. $\dfrac{a}{b} - \dfrac{c}{d} = \dfrac{ad - bc}{bd}$

3. $\dfrac{4}{3} - \dfrac{1}{2} = \dfrac{4}{3} \cdot \dfrac{2}{2} - \dfrac{3}{3} \cdot \dfrac{1}{2} = \dfrac{4 \cdot 2 - 3 \cdot 1}{3 \cdot 2} = \dfrac{5}{6}$

4. $-\dfrac{a}{b} = \dfrac{-a}{b} = \dfrac{a}{-b}$

4. $-\left(\dfrac{2}{3}\right) = \dfrac{-2}{3} = \dfrac{2}{-3}$ since when any of these is added to 2/3, the result is 0

5. $\dfrac{a}{b} \cdot \dfrac{c}{d} = \dfrac{a \cdot c}{b \cdot d}$

5. $\dfrac{1}{5} \cdot \dfrac{2}{3} = \dfrac{2}{15}$

6. $\dfrac{\frac{a}{b}}{\frac{c}{d}} = \dfrac{a}{b} \cdot \dfrac{d}{c}$

6. $\dfrac{\frac{2}{3}}{\frac{5}{7}} = \dfrac{\frac{2}{3} \cdot \frac{7}{5}}{\frac{5}{7} \cdot \frac{7}{5}} = \dfrac{\frac{2}{3} \cdot \frac{7}{5}}{1} = \dfrac{2}{3} \cdot \dfrac{7}{5}$

be written in **lowest terms.** For instance, 30/105 is reduced to lowest terms as follows:

$$\frac{30}{105} = \frac{2 \cdot \cancel{3} \cdot \cancel{5}}{\cancel{3} \cdot \cancel{5} \cdot 7} = \frac{2}{7}$$

Addition and subtraction of fractions is complicated by the fact that we must compare only like quantities. Although Rules 2 and 3 bypass this requirement, they can result in some very large numbers, especially in the denominator. But only a common denominator is really needed.

DEFINITION

The *least common denominator* or LCD of a collection of fractions is the LCM of the denominators of these fractions.

EXAMPLE 1   Find 11/36 − 7/30 using the LCD.

Solution

$$\frac{11}{36} - \frac{7}{30} = \frac{11}{6 \cdot 6} - \frac{7}{5 \cdot 6} = \frac{5 \cdot 11}{5 \cdot 6 \cdot 6} - \frac{7 \cdot 6}{5 \cdot 6 \cdot 6}$$

$$= \frac{55 - 42}{5 \cdot 6 \cdot 6} = \frac{13}{180}$$

Decimals

Each real number has a decimal representation. The long-division process can be used to find the decimal representation for any rational number such as 13/11 as follows.

$$
\begin{array}{r}
1.\stackrel{\frown}{1\ 8}\ \stackrel{\frown}{1\ 8}\ .\ .\ . \\
11\overline{)1\ 3\ .\ 0\ \ 0\ \ 0\ \ 0\ \ 0\ \ 0\ .\ .\ .} \\
\underline{1\ 1} \\
②\ 0 \\
\underline{1\ 1} \\
⑨\ 0 \\
\underline{8\ 8} \\
②\ 0 \\
\underline{1\ 1} \\
⑨\ 0 \\
\underline{8\ 8} \\
② \\
\end{array}
$$

$$\frac{13}{11} = 1.\stackrel{\frown}{18}\stackrel{\frown}{18}\stackrel{\frown}{18}\ .\ .\ .$$

The expansion for 13/11 is a **repeating decimal;** the $\stackrel{\frown}{18}$ block repeats itself continuously without stopping. In fact, every rational number either has a repeating decimal representation or terminates after a finite number of places (e.g., $1/4 = 0.25$). For once we have progressed to the stage of the long-division process where only zeros are being brought down, a remainder must eventually be repeated. This happens because each remainder must be less than the divisor; hence, only a finite number of remainders is possible. From then on the steps repeat those that followed the first appearance of the remainder.

Conversely, every repeating or terminating decimal expansion represents a rational number. For example, finding the rational number represented by $0.246246246\ .\ .\ .$ is illustrated as follows:

Let

$$x = 0.246246246\ .\ .\ .$$

Then

$$
\begin{aligned}
1000x &= 246.246246246\ .\ .\ . \\
x &= 0.246246246\ .\ .\ . \\
\hline
999x &= 246
\end{aligned}
$$

$$x = \frac{246}{999} = \frac{82}{333}$$

that is,

$$0.246246\ .\ .\ . = \frac{82}{333}$$

If the repeating block had been two units long or four units long, we would have multiplied the original number by 100 or 10,000, respectively, to line up the decimal points appropriately as in the above illustration. Thus, the irrational numbers are precisely those that have nonterminating, nonrepeating decimal expansions.

   In practice, decimal numbers are oftentimes "rounded off" to the degree of accuracy desired in a given calculation. For instance, 13/11 = 1.1818 . . . can be rounded off to the nearest thousandth as 1.182 (since it is closer to 1.182 than to 1.181).

   A decimal expansion that terminates with a 5 is midway between the two choices for rounding off the number. We could "round up" or "round down" in these cases. Current accepted practice is to round such a number so that the last digit of the rounded off number is even. This prevents prejudicing large collections of data either upward or downward.

Percent

Interest rates, profits, losses, and so on are frequently expressed as a *percent* (%) of some figure such as cost or investment. The symbol "%" or the word "percent" means "per centum" or "per hundred." Thus, for example, if you pay 20 percent of your income in taxes, this means that $20 of every $100 you earn is paid in taxes. Your tax is 20/100 of what you earn. To say 20 percent is just another way of saying 20/100 or 0.20. Similarly,

$$5\% = \frac{5}{100} = 0.05$$

$$67\% = \frac{67}{100} = 0.67$$

$$250\% = \frac{250}{100} = 2.50$$

EXAMPLE 2   If you miss 7 problems on a test consisting of 28 problems, what percent of the questions did you answer correctly?

Solution   Twenty-one (21) of the questions were answered correctly. Thus,

$$\frac{21}{28} = \frac{3}{4} = \frac{3 \cdot 25}{4 \cdot 25} = \frac{75}{100} = 75\%$$

of the answers were correct.   ■

EXAMPLE 3   A stereo is advertised as being on sale at 30 percent off.

   a.   If the original selling price is $400, what is its sale price?

   b.   If the sale price of another stereo is $350, what was its original price?

Solution   **a.**   If the original price was $400, the discount is 30 percent of $400 or

$$(0.30) \cdot 400 = \$120$$

Hence, the sale price is

$$\$400 - \$120 = \$280$$

b. On the other hand, if the sale price is \$350 after a discount of 30 percent, this \$350 represents 70 percent ($= 100\% - 30\%$) of the original price P; that is,

$$(0.70)P = 350$$
$$P = \frac{350}{0.7} = \$500$$

■

Hierarchy of Operations

In an expression involving more than one operation, such as $2 \cdot 3 + 4$, parentheses are sometimes used to indicate that certain operations are to be performed before others. The order of computation can make a difference. For example,

$$(2 \cdot 3) + 4 = 10$$
$$2 \cdot (3 + 4) = 14$$

To prevent ambiguities, mathematicians have adopted the following conventions.

---

1. Expressions in parentheses are to be treated as a single number that should be evaluated before being combined with numbers outside the parentheses.

2. Multiplications and divisions are performed before additions and subtractions and in the order in which they appear from left to right.

3. Then additions and subtractions are done in the order in which they appear from left to right.

EXAMPLE 4 Evaluate the following expressions.

a. $4 \div 2 - 4 \cdot 2 + 7$

b. $4 \div (2 - 4) \cdot 2 + 7$

Solution **a.** According to the hierarchy of operations, we have

$$4 \div 2 - 4 \cdot 2 + 7 = (4 \div 2) - (4 \cdot 2) + 7$$
$$= 2 - 8 + 7$$
$$= 1$$

b. Note that the division operation precedes multiplication when reading from left to right. Thus,

$$4 \div (2 - 4) \cdot 2 + 7 = 4 \div (-2) \cdot 2 + 7$$
$$= (-2) \cdot 2 + 7$$
$$= -4 + 7$$
$$= 3 \qquad \blacksquare$$

CALCULATOR COMMENTS

One must be careful when solving a problem that requires more than a single operation on a calculator. Some of the simpler calculators perform the operations in the order in which they are entered. On the other hand, scientific calculators are programmed to respect the hierarchy of operations as described earlier and will perform multiplication before addition. Some calculators also provide parentheses to enable the user to impose a desired operational order. But without parentheses one must be careful to enter the problem in such a way that the correct answer results. Read the owner's manual carefully, and *know your calculator. Be careful when borrowing one.*

Example	Result from a scientific calculator	Possible result from a simple calculator	Obtain correct result by entering as
$2 + 3 \cdot 4$	$2 + 12 = 14$	$(2 + 3) \cdot 4 = 20$	$3 \cdot 4 + 2 = 12 + 2 = 14$
$4\frac{1}{9} = 4 + 1 \div 9$	$4.111111111\ldots$	$(4 + 1) \div 9 \approx 0.555555\ldots$	$1 \div 9 + 4 \approx 4.1111\ldots$

EXERCISES 1.1

Factor the following into primes.

1. 24
2. 187
3. 396
4. 391
5. 819
6. 3850

Find the LCM and GCD of the following sets of numbers.

7. 45, 75
8. 68, 738
9. 129, 215
10. 76, 1425
11. 6, 15, 21
12. 36, 48, 60

Locate each of the following points on the number line.

13. 7.3
14. 9.45
15. −3.2
16. −4.65
17. 0.666 . . .
18. 0.4999 . . .

Reduce each of the following to lowest terms.

19. $\dfrac{15}{27}$
20. $\dfrac{-21}{28}$
21. $\dfrac{57}{95}$
22. $\dfrac{15}{-12}$
23. $\dfrac{63}{72}$
24. $\dfrac{-30}{231}$
25. $\dfrac{-660}{450}$
26. $\dfrac{2310}{5005}$

Express each of the following as a single integer.

27. $-(-[-6])$
28. $-(2 - 7)$
29. $(-5) - (-7)$
30. $(-5) - (-3)$
31. $(-8)(-6)$
32. $(-2)(-4)(-3)$
33. $3(-11) + 4(-2)(-1) - 13(-2)$
34. $-(-2)[4 + 6 \cdot (-3)]$
35. $2 - 3 \cdot 5 - 6 \cdot 5 - 2 - 7$
36. $[2 - 3(5 - 6)][5 - (2 - 7)]$
37. $-5 \cdot 6 - (-10) + 3(-10) - 8$
38. $-5[6 - (-10) + 3(-10 - 8)]$
39. $10 + 6 \div 2 + 6 \div 2 + 2 \cdot 3$
40. $(10 + 6) \div (2 + 6) \div 2 + 2 \cdot 3$
41. $(10 + 6) \div [(2 + 6) \div (2 + 2)] \cdot 3$
42. $10 + [6 \div 2 + 6 \div 2 + 2] \cdot 3$

Perform the following operations and express the answers in lowest terms.

43. $\dfrac{1}{4} + \dfrac{3}{8}$
44. $\dfrac{1}{3} + \dfrac{1}{6}$
45. $\dfrac{1}{2} - \dfrac{1}{3}$
46. $\dfrac{5}{7} - \dfrac{5}{9}$
47. $\dfrac{3}{4} - \left(\dfrac{1}{2} + \dfrac{1}{6}\right)$
48. $\left(\dfrac{3}{4} - \dfrac{1}{2}\right) + \dfrac{1}{6}$

49. $\dfrac{3}{7} + \dfrac{4}{9} + \dfrac{5}{21}$

50. $\dfrac{3}{7} + \dfrac{-4}{9} - \dfrac{5}{21}$

51. $\dfrac{10}{14} \cdot \dfrac{3}{-8} \cdot \dfrac{35}{50}$

52. $2\dfrac{1}{16} \cdot 3\dfrac{1}{3} \cdot \left(-1\dfrac{1}{5}\right)$

53. $3\dfrac{1}{3} \cdot 4\dfrac{1}{2} \cdot 5\dfrac{1}{5}$

54. $\dfrac{\dfrac{9}{16}}{\dfrac{21}{32}}$

55. $\dfrac{2}{3} \div \dfrac{7}{3} \cdot \dfrac{7}{18}$

56. $\dfrac{5}{6} \div 4\dfrac{4}{9}$

57. $\left(\dfrac{3}{4} - \dfrac{1}{3}\right) \div \left(\dfrac{2}{3} - \dfrac{1}{2}\right)$

58. $\dfrac{\dfrac{1}{2} + \dfrac{1}{3}}{\dfrac{2}{3} + \dfrac{1}{6}}$

59. $\left(2\dfrac{1}{3} + 1\dfrac{1}{6}\right) \div \left(4\dfrac{1}{2} - 3\dfrac{1}{4}\right)$

60. $\left(1\dfrac{1}{4} - 2\dfrac{1}{3}\right) \div \left[\left(\dfrac{3}{4} \cdot \dfrac{1}{2}\right) - \dfrac{1}{8}\right]$

61. $\dfrac{3}{\dfrac{3}{5} + \dfrac{2}{4 - \dfrac{2}{3}}}$

62. $\dfrac{\dfrac{1}{5}}{7 - \dfrac{\dfrac{1}{6}}{\dfrac{1}{2} + \dfrac{2}{3}}}$

Find a repeating decimal expansion for each of the given rational numbers.

63. $\dfrac{6}{7}$

64. $\dfrac{4}{9}$

65. $\dfrac{23}{11}$

66. $\dfrac{9}{13}$

67. $\dfrac{17}{24}$

68. $\dfrac{27}{41}$

Write each of the following in the form m/n.

69. $0.4999\ldots$

70. $1.010101\ldots$

71. $5.1272727\ldots$

72. $3.27327\ldots$

73. $672.3456456\ldots$

74. $414.201420\ldots$

75. Round off to the nearest thousandth each of the numbers in Exercises 63–74.

"Mental magicians" are adept at exploiting the commutative, associative, and distributive laws. For example, a quick mental calculation of $6 \cdot 97$ can be accomplished as follows:

$$6 \cdot 97 = 6(100 - 3) = 600 - 18 = 582$$

Another use of these properties is illustrated in computing $18 \cdot 25$:

$$18 \cdot 25 = 9 \cdot 2 \cdot 25 = 9 \cdot 50 = 450$$

Exploit the commutative, associative, and distributive laws to evaluate the following mentally.

76. $5 \cdot 297$

77. $19 \cdot 40$

78. $7 \cdot 69$

79. $24 \cdot 7$

80. $99 \cdot 98$

81. $8 \cdot 19$

Fill in the missing forms of the given numbers.

	Fraction	Decimal	%
82.		0.2	
83.			500
84.	$\dfrac{3}{20}$		
85.			$\dfrac{3}{10}$
86.		0.035	
87.	$\dfrac{5}{4}$		

Determine the missing numbers

88. 6% of 27 is _____.

89. _____% of 60 is 45.

90. 25% of _____ is 17.

91. _____% of 20 is 8.

92. 5% of _____ is 25.

93. 0.03% of 600 is _____.

94. When adding a column of numbers, we generally add from top to bottom. Then we check our work by adding from bottom to top. Which property or properties of the real number system are we using?

95. Is subtraction commutative? Associative?

96. Is division commutative? Associative?

97. Does division distribute over addition? That is to say, is
$$a \div (b + c) = (a \div b) + (a \div c)?$$

98. Basic Building Blocks are sold by Creative Toys in rectangular packages of 24 square blocks. What different shapes could the company use for the packages?

99. Two hardware companies produce bolts of the same quality. Acme Fasteners packages 32 bolts per box; Better Bolts puts 40 in each box. If neither company will sell partial boxes, what is the smallest order that both companies could fill?

100. The highest and lowest points in the 48 continental United States lie within 85 miles of one another. Both are in California. The highest point is Mount Whitney at 14,494 feet; the lowest is, of course, Death Valley at −282 feet. What is the change in altitude in these 85 miles?

101. At the previous fill-up, the odometer on Bernice's car registered 26,739.2. At the time of this fill-up requiring 18.7 gallons, it registers 26,986.8. How many miles per gallon did the car get?

102. Fixed expenses at Cinema X are $289 per day. If 62 patrons pay $1.50 each for the early show and 120 patrons pay $3.75 each for the evening show, how much profit is made?

103. Some supermarkets have introduced a "unit-pricing" concept. If a 16-ounce loaf of bread costs 83¢, what is its unit price or price per ounce? If a 20-ounce loaf costs $1.05, which is the better buy?

104. The Intellectual Quotient or IQ is defined as 100 times mental age divided by chronological age:

$$IQ = 100\ \frac{\text{mental age}}{\text{chronological age}}$$

a. Johnny is 10 years old and has the mental capabilities of a 12-year-old. What is his IQ?

b. Joan is 15 years old and has an IQ of 115. What is her mental age?

105. Paving a certain portion of highway requires 15,000 cubic yards of concrete. If a concrete truck can haul $6\frac{2}{3}$ cubic yards, how many truckloads are needed?

106. If a hen and a half can lay an egg and a half in a day and a half, how many eggs will 12 hens lay in 12 days?

107. Patio Covers, Inc. advertises a Sizzling Summer Sale offering 15 percent off the regular price. If the original price is $310.00, what is the sale price? If the sale price is $340, what was the original price?

108. A TV is advertised at $175 off the list price of $575. What percent savings does this discount represent.

109. Ms. Adams purchases an investment property for $100,000. To cover her expenses she must realize 12 percent on her investment each year. If there are four apartments in this property, what is the minimum monthly rent she must charge to cover her expenses.

110. Sudsy Sam buys automatic washers for $400 and then marks up the selling price by 50 percent. They are moving very slowly so he discounts them 40 percent from the selling price, figuring to make at least a 10 percent profit. Did Sam really make a profit?

1.2 ORDER AND ABSOLUTE VALUE

Order

The so-called inequality symbols are used to facilitate the comparison of numbers. Say, for instance, we want to express the fact that a particular business may go bankrupt if its debts exceed the value of its assets. We use the symbols < and > (which indicate **less than** and **greater than,** respectively) to describe the relationship between debts and assets. We might write 3 < 7, −3 < −2, and −1 < 2, as illustrated in Figure 1–3.

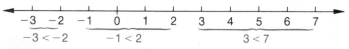

Figure 1–3

Numbers greater than zero are called **positive** and those less than zero are called **negative.** If $a < b$, then a positive number must be added to a in order to obtain b; that is, $b - a > 0$. It should be clear that $a < b$ if and only if a lies to the left of b on the real number line.

The following statements are equivalent:

1. a is less than b.

2. $a < b$.

3. $b - a > 0$.

4. a lies to the left of b on the real number line.

An expression involving $<$ or $>$ is called an **inequality**. The following rules govern operations on inequalities.

Rules for Inequalities

Rule	Example
1. $a < b$ if and only if $b - a > 0$.	1. $2 < 5$ and $5 - 2 = 3 > 0$
2. If $a < b$ and $b < c$, then $a < c$.	2.
3. If $a < b$, then $(a + c) < (b + c)$.	3.
4. If $a < b$, then $-a > -b$.	4. $2 < 5$ and $-2 > -5$

5. If $a < b$ and
  a. $x = 0$, then $ax = bx$
  b. $x > 0$, then $ax < bx$
  c. $x < 0$, then $ax > bx$
	5. $3 < 5$ and
	  a. $3 \cdot 0 = 5 \cdot 0 = 0$
	  b. $3 \cdot 2 < 5 \cdot 2$ since $6 < 10$
	  c. $3(-2) > 5(-2)$ since $-6 > -10$

6. If $0 < a < b$, then $0 < \dfrac{1}{b} < \dfrac{1}{a}$.	6. Larger denominators make smaller fractions. $0 < 2 < 5$ and $0 < \dfrac{1}{5} < \dfrac{1}{2}$

The symbols $\leq$ and $\geq$ are read *"less than or equal to"* and *"greater than or equal to,"* respectively. Thus,

---

$a \leq b$	means	$a < b$ or $a = b$
$a \geq b$	means	$a > b$ or $a = b$

Rules 1 through 6 also apply to $\leq$ and $\geq$. For example, rule 5(c) becomes: if $a \leq b$ and $x \leq 0$, then $ax \geq bx$. For if either $a = b$ or $x = 0$, it follows that $ax = bx$.

Intervals

Certain collections or sets of real numbers are defined in terms of the order or inequality symbols $<$ and $>$ as shown in Figure 1–4. The symbols $\{x : a < x < b\}$

are read "the set of those x's such that $a < x < b$." These types of sets are called **intervals;** a and b are **endpoints** of these intervals. Note that a parenthesis, "(" or ")," indicates that the endpoint is not in the interval, whereas a square bracket, "[" or "]," indicates that the endpoint is in the interval. Intervals that include their endpoints are called **closed intervals;** those that include neither endpoint are called **open intervals.** Intervals of the form $(a, b]$ or $[a, b)$ are called **half-open intervals** (see Fig. 1–4).

$$(a, b) \quad = \{x : a < x < b\}$$

$$[a, b) \quad = \{x : a \leq x < b\}$$

$$(a, b] \quad = \{x : a < x \leq b\}$$

$$[a, b] \quad = \{x : a \leq x \leq b\}$$

$$(-\infty, b) = \{x : x < b\}$$

$$(-\infty, b] = \{x : x \leq b\}$$

$$(a, \infty) \quad = \{x : a < x\}$$

$$[a, \infty) \quad = \{x : a \leq x\}$$

$$(-\infty, \infty) = \{x : x \text{ is a real number}\}$$

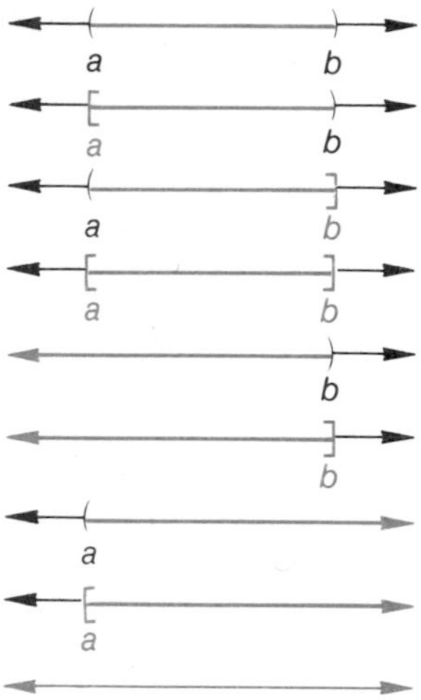

Figure 1–4

The symbols $-\infty$ and ∞ are not regarded as numbers. They are merely place holders used to describe intervals that are unbounded on the left or on the right, respectively.

EXAMPLE 1   Sketch the following intervals.

    a. $(2, 4)$ **b.** $(-2, 3]$ **c.** $[1, 5]$

Solution   Figure 1–5 shows the intervals.

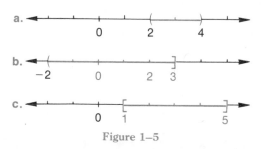

Figure 1–5

Absolute Value   The **absolute value** of a number is its distance from zero on the number line. For example, the absolute value of 7 is 7; the absolute value of -7 is also 7. Formally we define the absolute value of a number x, denoted $|x|$, as follows:

DEFINITION

$$|x| = \begin{cases} x & \text{if } x \geq 0 \\ -x & \text{if } x < 0 \end{cases}$$

(*Note:* $-x > 0$ when $x < 0$.)

Many students are confused by the statement that $|x| = -x$ when $x < 0$. However, note that

$$|-7| = 7 = -(-7)$$

the absolute value of -7 is the negative of -7.

There are many situations in which the absolute value of a number is more important than the number itself. For example, our familiar alternating current (ac) electrical system operates by alternating the direction in which the current flows. In a 60-ampere (amp) circuit, the current alternates between $+60$ and -60 amps. The shock felt by someone touching an electrical wire depends on the numerical size of the current (say 60 amps or 0.00006 amps); its sign ($\pm$) at any given instant is relatively unimportant.

The following rules govern operations on absolute values.

Rules for Operations on Absolute Value

Rule	Example
1. $|a| \geq 0$	1. $|-3| = 3 \geq 0$
2. $|-a| = |a|$	2. $|-7| = |7| = 7$
3. $|x - y| = |y - x|$	3. $|5 - 3| = |2| = 2$ $|3 - 5| = |-2| = 2$

Rule	Example
4. $|x| < a$ if and only if $-a < x < a$	4.

Rule	Example
$|x| > a$ if and only if $x > a$ or $x < -a$	
5. $|ab| = |a||b|$	5. $|(-2) \cdot 7| = |-2| \cdot 7$ since $|-14| = 2 \cdot 7$
6. $\left|\dfrac{a}{b}\right| = \dfrac{|a|}{|b|}$	6. $\left|\dfrac{8}{-4}\right| = \dfrac{|8|}{|-4|}$ since $\left| -2 \right| = \dfrac{8}{4}$
7. $|a + b| \leq |a| + |b|$	7. $|5 + 3| = 8 = |5| + |3|$ $|5 + (-3)| = 2 < |5| + |-3|$
8. $|x - y| \leq |x - z| + |z - y|$	8. $|7 - 4| < |7 - 2| + |2 - 4|$ since $3 < 5 + 2$ $|7 - 4| = |7 - 5| + |5 - 4|$ since $3 = 2 + 1$

Note in the sketch associated with Rule 3 that $|x - y|$ is a measure of the distance between x and y on the real number line. Rule 8 then says that the direct distance from x to y is less than the distance found by detouring through some third point z. Actually Rules 7 and 8 are the same rules; this can be seen by letting $a = x - z$ and $b = z - y$ to relate the two rules. Each of these rules is called the *triangle inequality*. Note in Rule 8 that $|x - y| = |x - z| + |z - y|$ only when z is between x and y; in Rule 7, $|a + b| = |a| + |b|$ only when a and b have the same sign.

EXAMPLE 2 Mr. Hofmann wants to determine the area of a meeting hall so that he can order the correct amount of floor covering. He measures the length of the floor to be 79 feet and the width to be 62 feet, correct to the nearest foot. How accurate is his area computation?

Solution Since the measurements are given to the nearest foot, the length l and the width w satisfy the following:

$$l = 79 + x \qquad |x| \leq \frac{1}{2}$$

$$w = 62 + y \qquad |y| \leq \frac{1}{2}$$

The actual area is

$$A = lw = (79 + x)(62 + y)$$
$$= 79 \cdot 62 + 79 \cdot y + x \cdot 62 + x \cdot y$$

Comparing this with his estimate of the area as $79 \cdot 62 = 4898$ square feet, we see a possible error E of

$$E = 79 \cdot y + x \cdot 62 + x \cdot y$$

Note that

$$|E| = |79 \cdot y + x \cdot 62 + x \cdot y|$$
$$\leq 79 \cdot |y| + |x| \cdot 62 + |x| \cdot |y|$$
$$\leq 79 \cdot \left(\frac{1}{2}\right) + \left(\frac{1}{2}\right) \cdot 62 + \left(\frac{1}{2}\right) \cdot \left(\frac{1}{2}\right)$$
$$= 39.5 + 31 + 0.25$$
$$= 70.75$$

His area computation is accurate to within 70.75 square feet. ■

EXERCISES 1.2

Express an order relation ($<$, $=$, $>$) between the following numbers.

1. 2, 7   2 < 7

3. -2, -6

5. 2, -6

2. 16, 5

4. 6, -2

6. -5, -4 $<$

7. -100, 17

8. 5, 5 $>$

Express an order relation between ax and bx when a and b are as in Exercises 1–8 and x is given here for each pair respectively.

9. -1

10. 0

11. 2

12. 6

13. -2

14. -3

15. Write the absolute value of each of the numbers in Exercises 9–14.

16. For each of the pairs a, b in Exercise 1–8, verify that $|a + b| \le |a| + |b|$ and that equality holds if and only if a and b have the same sign.

Sketch the following intervals on the real number line.

17. $(-\infty, 2)$

18. $[3, 4]$

19. $(-2, 0]$

20. $(-\infty, 4]$

21. $(4, 6)$

22. $[-1, 5)$

23. $(3, \infty)$

24. $[1, \infty)$

25. State the rules for inequalities in terms of $\le$.

Express an order relation between the following rational numbers. A bar indicates a repeating block of digits.

26. $\dfrac{3}{5}, \dfrac{7}{11}$

27. $\dfrac{7}{11}, \dfrac{19}{37}$

28. $\dfrac{17}{21}, \dfrac{23}{29}$

29. $\dfrac{6}{7}, \dfrac{54}{93}$

30. $\dfrac{13}{25}, \dfrac{10}{21}$

31. $0.\overline{023}, 0.0\overline{23}$

32. $2.3\overline{7}, 2.3\overline{7}$

33. $7.3\overline{54}, 7.35\overline{4}$

Arrange the following sets of numbers in order from lowest to highest.

34. $\dfrac{4}{7}, \dfrac{6}{11}, \dfrac{7}{13}, \dfrac{8}{15}, \dfrac{9}{17}, \dfrac{10}{19}$

35. $4.\overline{87}, 4.8\overline{7}, 4.8, 4.87, 4.\overline{8}$

36. $\overline{2.367}, 2.\overline{367}, 2.36\overline{7}, 2.36\overline{7}, 2.367$

Find a rational number between the following pairs of numbers.

37. $\dfrac{3}{5}, \dfrac{7}{11}$

38. $\overline{4.5}, \overline{4.6}$

39. $7.3\overline{54}, 7.35\overline{4}$

40. If Bill owns 2/5 of his company's stock and Janice owns 3/7 of her company's stock, who owns the larger share?

41. Potato Ruffles come in several different sized packages. A 12-ounce package sells for $1.59, 10 ounces for $1.29, and $7\frac{1}{2}$ ounces for $1.09. Which is the best buy?

42. Generic Label corn sells in 370-gram packages for 59¢, and Yellow Valley corn comes in 454-gram packages for 75¢. Which is more expensive?

43. Super Suds detergent in the economy size 5-pound package is priced at $5.23. The giant size $3\frac{1}{2}$ pound box sells for $3.59. The large size 24-ounce package is $1.59. Which package is the best buy?

44. If the base of a triangle is 10 feet and its area is between 60 and 70 square feet, describe its set of possible altitudes as an interval.

45. Gasoline purchases are usually expressed to the nearest tenth of a gallon. Odometer readings are given to the nearest tenth of a mile. The previous odometer reading on Eileen's car was 23,468.7. She now purchases 17.3 gallons at an odometer reading of 23,684.3. Indicate her possible miles-per-gallon ratings in terms of inequalities. You may wish to use a hand calculator.

46. In Example 2, how accurately can Mr. Hofmann determine the length of molding needed to trim the perimeter of the floor?

47. The dimensions of a rectangle are measured as 10 centimeters accurate to within 0.01 centimeter and 5 centimeters accurate to within 0.002 centimeters. Its area is then estimated as 50 square centimeters. How accurate is this estimate?

48. If $x = 2.3$ to within 0.05 and $y = 4.7$ to within 0.03, how accurately can xy and $x + y$ be determined?

1.3 INTEGRAL EXPONENTS

For each real number x and each positive integer n, the nth power of x is defined as the product of x with itself n times:

DEFINITION

$$x^n = \underbrace{x \cdot \cdots \cdot x}_{n \text{ times}}$$

In particular,

$$x^1 = x,$$
$$x^2 = x \cdot x,$$
$$2^3 = 2 \cdot 2 \cdot 2 = 8.$$

In the notation x^n, x is called the **base** and n is called the **exponent** or **power**.

Rules for working with positive integer exponents follow. Note that no rules are given regarding sums and differences. Such relationships are impossible. For example,

$$1^2 + 2^2 \neq 3^2 \qquad \text{since} \qquad 1 + 4 \neq 9,$$

and

$$2^2 + 2^3 \neq 2^5 \qquad \text{since} \qquad 4 + 8 \neq 32.$$

Rules for (Positive) Integer Exponents

Rule	Example
1. $x^m x^n = x^{m+n}$	1. $x^3 x^2 = (x \cdot x \cdot x)(x \cdot x) = x^5$
2. $(x^m)^n = x^{mn}$	2. $(x^2)^3 = x^2 \cdot x^2 \cdot x^2 = x^6$
3. a. $(xy)^n = x^n y^n$	3. a. $(2 \cdot 3)^2 = (2 \cdot 3)(2 \cdot 3) = (2 \cdot 2)(3 \cdot 3)$ $= 2^2 \cdot 3^2$
b. $\left(\dfrac{x}{y}\right)^n = \dfrac{x^n}{y^n}$	b. $\left(\dfrac{4}{2}\right)^3 = \dfrac{4}{2} \cdot \dfrac{4}{2} \cdot \dfrac{4}{2} = \dfrac{4 \cdot 4 \cdot 4}{2 \cdot 2 \cdot 2} = \dfrac{4^3}{2^3}$
4. $\dfrac{x^m}{x^n} = \begin{cases} x^{m-n} & \text{if } m > n \\ 1 & \text{if } m = n \\ \dfrac{1}{x^{n-m}} & \text{if } n > m \end{cases}$	4. $\dfrac{x^3}{x^2} = \dfrac{x \cdot x \cdot x}{x \cdot x} = x = x^{3-2}$ $\dfrac{x^2}{x^4} = \dfrac{x \cdot x}{x \cdot x \cdot x \cdot x} = \dfrac{1}{x^2} = \dfrac{1}{x^{4-2}}$

EXAMPLE 1 Use the preceding rules for (positive) integer exponents to simplify each of the following. Recall that parentheses indicate a certain order of operation.

a. $\dfrac{(-2)^5}{(-2)^3}$

b. $\dfrac{3x^2 y^3 z}{2x^2 y z^2}$

c. $\dfrac{(k^2 m^2)^3}{(k^3 m)^2}$

d. $[(p^2 q)^3 (q p^3)^2]^4$

Solution a. $\dfrac{(-2)^5}{(-2)^3} = (-2)^{5-3} = (-2)^2 = 4$

b. $\dfrac{3x^2 y^3 z}{2x^2 y z^2} = \dfrac{3}{2} \cdot \dfrac{x^2}{x^2} \cdot \dfrac{y^3}{y} \cdot \dfrac{z}{z^2} = \dfrac{3}{2} \cdot 1 \cdot y^2 \cdot \dfrac{1}{z} = \dfrac{3y^2}{2z}$

c. $\dfrac{(k^2 m^2)^3}{(k^3 m)^2} = \dfrac{(k^2)^3 (m^2)^3}{(k^3)^2 m^2} = \dfrac{k^6 m^6}{k^6 m^2} = 1 \cdot m^{6-2} = m^4$

d. $[(p^2 q)^3 (q p^3)^2]^4 = [(p^6 q^3)(q^2 p^6)]^4 = [p^{12} q^5]^4 = p^{48} q^{20}$ ■

EXAMPLE 2   The population of the United States has been doubling approximately every 60 years. If this growth rate is sustained and the current population is estimated at 250 million people, what population can be projected for 300 years from now?

Solution   In 300 years the population will have doubled 5 times ($60 \cdot 5 = 300$). Thus, the expected population will be

$$(250,000,000) \cdot 2^5 = 8,000,000,000$$

or 8 *billion* people. ∎

To define x^0 in a manner consistent with the properties of exponents, we must have

$$x^m \cdot x^0 = x^{m+0} = x^m$$

and

$$\frac{x^m}{x^0} = x^{m-0} = x^m$$

That is, multiplying or dividing by x^0 must be equivalent to multiplying or dividing by 1. Therefore, we have the following:

DEFINITION
$$x^0 = 1 \qquad \text{for any real number } x \neq 0$$

$0^0 = $ indeterminant

However, 0^0 presents difficulties that we will not discuss here.

The laws of exponents can also be extended to negative exponents. Then we must have, for example,

$$x^5 x^{-2} = x^{5+(-2)} = x^{5-2} = x^3 = \frac{x^5}{x^2}$$

Thus, multiplication by x^{-2} should be equivalent to dividing by x^2. Consequently, x^{-n} is defined as follows for any integer $n > 0$.

DEFINITION
$$x^{-n} = \frac{1}{x^n} \qquad \text{for any real number } x \neq 0$$

Using these definitions of x^0 and x^{-n}, we absorb Rule 4 for exponents into Rule 1: $x^m x^n = x^{m+n}$. Specifically,

$$\frac{x^m}{x^n} = x^m x^{-n} = x^{m+(-n)} = x^{m-n}$$

$$= \begin{cases} x^{m-n} & \text{if } m > n \\ x^0 = 1 & \text{if } n = m \\ x^{-(n-m)} = \dfrac{1}{x^{n-m}} & \text{if } n > m \end{cases}$$

Furthermore, *all the rules for positive integer exponents mentioned earlier hold for negative integers as well.* For instance, $(x^m)^{-n} = x^{m(-n)}$ since

$$(x^m)^{-n} = \frac{1}{(x^m)^n} = \frac{1}{x^{mn}} = x^{-mn} = x^{m(-n)}$$

We have defined $x^{-2} = 1/x^2$; that is, *a negative exponent in the numerator is equivalent to a positive exponent in the denominator.* The opposite is also true: *A negative exponent in the denominator is equivalent to a positive exponent in the numerator.* For example,

$$\frac{1}{x^{-10}} = \frac{1}{\dfrac{1}{x^{10}}} = x^{10}$$

Thus, $x^{-n} = 1/x^n$ regardless of whether n is positive, negative, or zero.

EXAMPLE 3 Simplify the following and express the answer with nonnegative exponents.

a. $\dfrac{x^4}{x^{10}}$ b. $(x^{-2})^{-3}$ c. $\dfrac{12a^{-3}b^4}{4ab^{-2}}$

Solution a. $\dfrac{x^4}{x^{10}} = x^{4-10} = x^{-6} = \dfrac{1}{x^6}$

b. $(x^{-2})^{-3} = x^6$
An alternate solution is

$$(x^{-2})^{-3} = \frac{1}{(x^{-2})^3} = \frac{1}{\left(\dfrac{1}{x^2}\right)^3} = \frac{1}{\dfrac{1}{x^6}} = x^6$$

c. $\dfrac{12a^{-3}b^4}{4ab^{-2}} = \dfrac{12b^4b^2}{4aa^3} = \dfrac{3b^6}{a^4}$

◼

Scientific Notation Scientists often use very small or very large numbers. For example, the frequency of some X rays is $300{,}000{,}000{,}000{,}000{,}000{,}000$ cycles per second; the charge on an electron is 0.0000000000000000001602 coulombs. To facilitate working with such numbers, scientists express them as some number times a power of 10. Multiplication or division by 10 moves the decimal point to the right or left, respectively. We can write $300{,}000{,}000{,}000{,}000{,}000{,}000 =$

$3(10^{20})$ and we can write $0.0000000000000000001602 = 1.602(10^{-19})$. Other examples of large or small numbers follow.

Frequency of TV waves	$3(10^8)$ cycles per second
Threshold of human hearing	10^{-2} watts per square meter or 0 decibels
Height of Mt. Everest	$2.52(10^6)$ centimeters

Any number expressed in the form

$$b \cdot 10^k, \quad 1 \le b < 10,$$

is said to be written in **scientific notation.** Scientific notation enables us to compare numbers very quickly by looking at the respective powers of 10. Calculations involving such numbers are also simplified by using the laws of exponents on the powers of 10.

CALCULATOR COMMENTS

Many of the better calculators are programmed to work in scientific notation. In fact, some will even convert automatically to scientific notation when a certain number of digits are introduced into a calculation. But even the simplest calculators can be used to work with scientific notation. The powers of 10 can be manipulated by hand, using the laws of exponents; the calculator need only be used to do the "hard arithmetic." For instance,

$$\frac{(0.000064)(637,000,000)}{(0.00000562)}$$
$$= \frac{(6.4)(10^{-5})(6.37)(10^8)}{(5.62)(10^{-6})}$$
$$= \frac{(6.4)(6.37)}{5.62}10^{-5}10^810^6$$
$$= \frac{(6.4)(6.37)}{5.62}10^9 \quad \text{——Use a calculator.}$$
$$\approx (7.254)10^9$$
$$= 7,254,000,000$$

EXERCISES 1.3

Simplify the following, eliminating parentheses and negative exponents, if any.

1. $3^2 \cdot 3^3$

2. $3^{-2}(-3)^3$

3. $2^3 \cdot 2^4$

4. $3^2 \cdot 2^4$

5. $3^{-2}2^{-4}$

6. $\dfrac{5^5}{5^4}$

7. $\dfrac{5^5}{(-5)^{-4}}$

8. $\dfrac{(-2)^5}{(-2)^3}$

9. $\dfrac{-2^5}{(-2)^3}$

10. $(2^3)^{-2}$

11. $(-2)^2(-2)^3$

12. $(-2^2)(-2^3)$

13. $(-2^3)^2$

14. $(-2^2)^3$

15. $[(-2)^3]^2$

16. $[(-2)^2]^3$

17. y^5y^{12}

18. $t^{-7} \cdot t^5$

19. $x^{13} \cdot x^{17}$

20. $u^{-13}u^{-4}$

21. $\dfrac{a^{10}}{a^5}$

22. $\dfrac{b^{12}}{b^3}$

23. $\dfrac{xy^3}{x^2y}$

24. $\dfrac{u^2v}{uv^2}$

25. $\dfrac{(x^{-2}y)}{xy^{-2}}$

26. $(3x^2y)(2y^3)$

27. $(2xy^2z)(6x^3yz^2)$

28. $(-2ax^{-2}b)(4a^{-1}b^{-2}x^{-2})$

29. $(7a^3b)(-2a^{-7}b)$

30. $(t^2)^3$

31. $(x^6)^{-2}$

32. $(kL^2m^3)^4$

33. $(xy^2z)^3(x^2yz^2)^2$

34. $(x^2y^3z^5)^4(z^3x^2y)^3$

35. $27x^{-2}(u^{-73}v^{2900}w^{163})^0$

36. $(2x^{-2}y)^{-3}$

37. $(a^{-3}b^{-2})^{-3}$

38. $(xy^2z)^{-3}(x^{-2}yz^2)^{-2}$

39. $\dfrac{(x^2y^{-4})^{-3}}{(x^{-3}y^{-2})^2}$

40. $[(ab^2c)^2(c^2ba^3)^4]^3$

41. $[(x^{-2}y)^3(y^{-1}x^3)]^{-4}$

42. $\dfrac{(14s^{-2}t^{-3})}{(7s^{-4}t^{-4})}$

43. $\dfrac{(xy^2z)^{-3}}{(x^{-2}yz^2)^{-2}}$

44. $\dfrac{(q^2r^3)^4}{(qr^5)^6}$

45. $\left(\dfrac{2s^2t^3}{4st^2}\right)^3$

46. $\left(\dfrac{3xy^2}{2z}\right)^3 \cdot \left(\dfrac{4x^2z}{y^4}\right)^2$

47. $\dfrac{(r^2s^{-2})^{-3}}{(r^{-3}s)^2}$

48. $[(2x^{-2}y)^{-3}(xy^2z^{-3})^{-4}]^2$

49. $[(ab^{-2}c^{-1})^{-2}(c^{-2}ba^3)^4]^{-3}$

50. $\left[\dfrac{(3^{-2}rs^{-4})^2}{(2^2r^{-2}s^2)^3}\right]^{-2}$

51. $\left[\dfrac{(x^2y^{-4}z^3)^{-2}}{(x^{-5}y^3z^2)^{-1}}\right]^{-2}$

52. $3x^{-1} - 2y^{-2}$

53. $(ab)^2(2a^{-1} + b^{-1})$

54. $\dfrac{2}{t^{-1}} + \dfrac{3t^{-1}}{t^{-2}}$

55. $\dfrac{p^{-1} + q^{-1}}{(p + q)^{-1}}$

Verify the following relationships.

56. $2^2 + 3^2 \neq 5^2$

57. $3^3 - 2^3 \neq (3 - 2)^3$

58. $2^2 - 3^2 \neq (-1)^2$

59. $3^2 + 3^2 \neq 6^2$

60. $2^2 + 2^3 \neq 2^5$

61. $(2 + 3)^{-2} \neq 2^{-2} + 3^{-2}$

62. $(x + y)^{-1} \neq x^{-1} + y^{-1}$

63. A googol is a 1 followed by 100 zeros, and a googolplex is a 1 followed by a googol of zeros. Express these numbers as powers of 10.

Express each of the following in decimal notation.

64. $5.5(10^{15})$, the surface of the earth in square feet.

65. $9.11(10^{-31})$, the mass of an electron in kilograms.

66. $1.67(10^{-27})$, the mass of a proton in kilograms.

67. $3(10^{10})$, the speed of light in centimeters per second.

Express each of the following in scientific notation.

68. 26

69. 2463

70. 230,000,000,000

71. 143.17

72. 0.023

73. 0.00132

74. 0.000341

75. 0.0000000000000712

76. The number of seconds in a life span of 70 years.

77. The surface area of the earth in acres and in hectares (1 acre = 43,560 square feet, 1 hectare = 10,000 square meters). See Exercise 64 and use 1 meter ≈ 39.37 inches.

Perform the following calculations by first converting the numbers to scientific notation. Express your answer in scientific notation and in terms of decimals.

78. $\dfrac{(2000)(0.00009)}{0.03}$

79. $\dfrac{(300.000)(0.0000006)}{(0.0000000015)(40,000)}$

80. $\dfrac{(250,000,000)(63,000)}{(0.0007)(0.09)(500)}$

81. $\dfrac{(0.000004)(600,000)}{(0.004)(0.003)}$

82. $\dfrac{(76,300,000,000)(0.0293)(0.0000000479)}{(0.00000543)(167,000)}$

83. Integrated circuits (ICs) have caused a revolution in the computer and calculator industries. The number of components that can be put into an integrated circuit has doubled each year since

the transistor was developed in 1959. If the single component transistor is considered a 1959-grade IC, how many components can presently be incorporated into an IC?

84. Wavelengths and atomic and molecular dimensions are frequently given in angstroms. One angstrom equals 10^{-8} centimeters. In 1977 the manufacturing of microelectronic chips required circuit elements having a width of only 2500 angstroms. If straight line circuits 2500 angstroms wide are etched on a chip with buffers 2500 angstroms wide between circuits, how many circuits can be etched on a chip 2 centimeters square?

85. The population of the world has been doubling approximately every 35 years. Estimating the current world population at 4 billion, what will the population be in 280 years? What would the result be if the doubling time were 70 years?

86. If an insect doubles its weight every 48 hours and now weighs 2 milligrams, how much will it weigh in 30 days? In 60 days?

87. The gross national product (GNP) of the United States has recently been growing at a rate so that it triples each 15 years. If the GNP was $2.5 trillion in 1980, what is it expected to be in the year 2025? What was it in 1950? In 1965?

88. Scientists tell us that light rays travel at the rate of 186,000 miles per second. Celestial distances are so great that astronomers use a light-year as the unit of measurement; 1 light-year is the distance traveled by a ray of light in 1 year.
   a. Express the number of miles in 1 light-year in scientific notation.
   b. A furlong is 1/8 mile; a fortnight is 14 days. Express the speed of light in furlongs per fortnight.

89. If the sun suddenly burned out, we would not know it for 8 minutes. How far is the sun from us? (See Exercise 88.)

90. Our solar system is located in a galaxy known as the Milky Way. The diameter of the Milky Way is 100,000 light-years. Its thickness is 10,000 light-years. Express these dimensions as miles in scientific notation.

91. It is possible to view certain events today that happened a very long time ago. For example, astronomers view events in the Andromeda Galaxy as they happened 750,000 years ago. How far away is the Andromeda Galaxy in miles? Write your answer in scientific notation.

1.4 RADICALS AND FRACTIONAL EXPONENTS

Biophysicists tell us that the walking speed of an animal is related to the square root of the length of its legs. (With tongue in cheek, one might be tempted to say that the walking speed is related to the number of feet per leg.) With legs L feet long, an animal might be capable of walking $7\sqrt{L}/3$ miles per hour. According to this, a man with 36-inch legs should be capable of walking $7\sqrt{3}/3 \approx 4$ miles per hour. Square roots (and other roots) occur in other areas of science and engineering as well as in biophysics. In this section, we shall consider sums, differences, products, and quotients of expressions involving roots and relate roots to fractional exponents.

To preserve the "power-of-a-power" rule for positive rational exponents, we must require that

$$(x^{1/n})^n = x^{n/n} = x^1 = x$$

that is, $x^{1/n}$ must be a number that, when multiplied by itself n times, yields x. Such a number is called an nth root of x. (If $n = 2$ or 3, we say "square root" or "cube root," respectively.)

Every positive real number has two square roots. On the other hand, negative numbers have no real square roots since $a^2 \geq 0$ for every real number a.

The same comments hold regarding nth roots whenever n is even. For instance, 16 has two real fourth roots: 2 and -2. But -8 has no real square roots or fourth roots; -8 does have precisely one real cube root: -2. In fact, every real number has precisely one real nth root for each odd integer $n > 0$.

The notations $x^{1/n}$ and $\sqrt[n]{x}$ are used interchangeably to denote the nth root of x (the positive root if there are two roots). In particular, $x^{1/2}$ or $\sqrt[2]{x}$ is used to denote the positive square root of a nonnegative number x. We generally write $\sqrt{}$ instead of $\sqrt[2]{}$, but the index n must be used for other roots such as the cube root of x, denoted $\sqrt[3]{x}$. The symbol $\sqrt{}$ is called a **radical.**

DEFINITION

$$x^{1/n} = \sqrt[n]{x}$$

If n is even and

$$x \geq 0, \quad \text{then } \sqrt[n]{x} \text{ denotes the positive } n\text{th root of } x$$
$$x < 0, \quad \sqrt[n]{x} \text{ is undefined}$$

If n is odd, $\sqrt[n]{x}$ denotes the unique nth root of x.

For example,

$$\textbf{a.} \quad 36^{1/2} = \sqrt{36} = 6$$

$$\textbf{b.} \quad (-8)^{1/3} = \sqrt[3]{-8} = -2$$

$$\textbf{c.} \quad (-25)^{1/2} = \sqrt{-25} \text{ is not defined.}$$

Writing roots and radicals in terms of fractional exponents, we obtain the following properties.

$$\textbf{a.} \quad \sqrt[n]{x}\,\sqrt[n]{y} = \sqrt[n]{xy} \text{ since } (x^{1/n}y^{1/n}) = (xy)^{1/n},$$

$$\textbf{b.} \quad \frac{\sqrt[n]{x}}{\sqrt[n]{y}} = \sqrt[n]{\frac{x}{y}} \text{ since } \frac{x^{1/n}}{y^{1/n}} = \left(\frac{x}{y}\right)^{1/n}, \text{ and}$$

$$\textbf{c.} \quad \sqrt[k]{\sqrt[n]{x}} = \sqrt[kn]{x} \text{ since } (x^{1/n})^{1/k} = x^{1/nk}.$$

We now define $x^{m/n}$ whenever $\sqrt[n]{x} = x^{1/n}$ exists.

DEFINITION

$$x^{m/n} = (x^{1/n})^m = (x^m)^{1/n} \qquad \text{when } \sqrt[n]{x} \text{ exists}$$
$$= (\sqrt[n]{x})^m = \sqrt[n]{x^m}$$

For example,

$$(-8)^{2/3} = [(-8)^{1/3}]^2 = (\sqrt[3]{-8})^2 = (-2)^2 = 4$$
$$(-8)^{2/3} = [(-8)^2]^{1/3} = (64)^{1/3} = \sqrt[3]{64} = 4$$

CAUTION: *Be careful when simplifying rational exponents.* For instance, $(-1)^{2/6} \neq (-1)^{1/3}$ since $(-1)^{2/6} = (\sqrt[6]{-1})^2$, which does not exist ($\sqrt[6]{-1}$ is not a real number); yet $(-1)^{1/3} = \sqrt[3]{-1} = -1$.

As usual we define $x^{-m/n} = 1/x^{m/n}$. With these definitions it can be shown that the previously derived properties of exponents hold for rational exponents as well. The basic properties of exponents and radicals that include all the others are listed here for reference.

Rules for Exponents and Radicals

Rule	Example
1. $x^a x^b = x^{a+b}$	1. $2^2 2^3 = 2^5$ since $4 \cdot 8 = 32$
2. $(x^a)^b = x^{ab}$	2. $(2^3)^2 = 2^6$ since $8^2 = 64 = 2^6$
3. $(xy)^a = x^a y^a$	3. $(2 \cdot 3)^2 = 2^2 3^2$ since $36 = 4 \cdot 9$
4. $x^0 = 1$	4. $10^0 = 1$
5. $x^{-a} = \dfrac{1}{x^a}$	5. $2^{-2} = \dfrac{1}{2^2}$
6. $\sqrt[n]{x^n} = \begin{cases} \|x\| & \text{if } n \text{ is even} \\ x & \text{if } n \text{ is odd.} \end{cases}$	6. $\begin{cases} \sqrt[4]{(-2)^4} = \sqrt[4]{16} = 2 = \|-2\| \\ \sqrt[3]{(-2)^3} = \sqrt[3]{-8} = -2 \end{cases}$
7. $\begin{cases} \sqrt[n]{a} \cdot \sqrt[n]{b} = \sqrt[n]{ab} \\ \dfrac{\sqrt[n]{a}}{\sqrt[n]{b}} = \sqrt[n]{\dfrac{a}{b}} \end{cases}$	7. $\begin{cases} \sqrt{4}\,\sqrt{9} = 2 \cdot 3 = 6 = \sqrt{36} = \sqrt{4 \cdot 9} \\ \dfrac{\sqrt{100}}{\sqrt{4}} = \dfrac{10}{2} = 5 = \sqrt{25} = \sqrt{\dfrac{100}{4}} \end{cases}$
8. $\sqrt[k]{\sqrt[n]{x}} = \sqrt[kn]{x}$	8. $\sqrt[3]{\sqrt[2]{64}} = \sqrt[3]{8} = 2 = \sqrt[6]{64}$
9. Only like terms can be collected	9. $3\sqrt{2} + 5\sqrt{2} = 8\sqrt{2}$ but $\sqrt{2} + \sqrt{3} \neq \sqrt{5}$

We should emphasize Rule 6 when n is even. Thus

$$\sqrt{x^2} = |x|$$

since the radical denotes the *positive* root when n is even. In particular

$$\sqrt{(-3)^2} = \sqrt{9} = |-3|$$

EXAMPLE 1   Simplify the following radicals.

    a. $\sqrt{64x^2y^4z^5}$

    b. $\sqrt[3]{-27x^3y^6z^4}$

    c. $\sqrt[3]{8}\,\sqrt[4]{y^{12}}\,\sqrt{x^3}$

Solution   **a.** $\sqrt{64x^2y^4z^5} = \sqrt{64}\,\sqrt{x^2}\,\sqrt{y^4}\,\sqrt{z^4}\,\sqrt{z}$
                        $= 8|x|y^2z^2\,\sqrt{z}$

b. $\sqrt[3]{-27x^3y^6z^4} = \sqrt[3]{-27}\ \sqrt[3]{x^3}\ \sqrt[3]{y^6}\ \sqrt[3]{z^3}\ \sqrt[3]{z}$

$$= -3xy^2z\sqrt[3]{z}$$

c. $\sqrt[3]{8\sqrt[4]{y^{12}}\ \sqrt{x^3}} = \sqrt[3]{8}\ \sqrt[3]{\sqrt[4]{y^{12}}}\ \sqrt[3]{\sqrt{x^3}} = 2\sqrt[12]{y^{12}}\ \sqrt[6]{x^3}$

$$= 2|y|x^{3/6} = 2|y|\sqrt{x}$$

EXAMPLE 2 Simplify the following expressions.

a. $3\sqrt{2} + 4\sqrt{8} - \sqrt{32}$

b. $\dfrac{\sqrt[4]{32x^3} - \sqrt[4]{8x^5}}{\sqrt[4]{2x}}$

c. $(4\sqrt{5} - \sqrt{6})(2\sqrt{5} + 3\sqrt{6})$

Solution **a.** $3\sqrt{2} + 4\sqrt{8} - \sqrt{32} = 3\sqrt{2} + 4\sqrt{4}\ \sqrt{2} - \sqrt{16}\ \sqrt{2}$

$$= 3\sqrt{2} + 8\sqrt{2} - 4\sqrt{2}$$
$$= 7\sqrt{2}$$

b. $\dfrac{\sqrt[4]{32x^3} - \sqrt[4]{8x^5}}{\sqrt[4]{2x}} = \dfrac{\sqrt[4]{32x^3}}{\sqrt[4]{2x}} - \dfrac{\sqrt[4]{8x^5}}{\sqrt[4]{2x}} = \sqrt[4]{\dfrac{32x^3}{2x}} - \sqrt[4]{\dfrac{8x^5}{2x}}$

$$= \sqrt[4]{16x^2} - \sqrt[4]{4x^4} = \sqrt[4]{16}\ \sqrt[4]{x^2} - \sqrt[4]{4}\ \sqrt[4]{x^4}$$
$$= 2\sqrt{\sqrt{x^2}} - \sqrt{\sqrt{4}x} = 2\sqrt{x} - \sqrt{2x}.$$

c. When multiplying lengthy expressions involving radicals, it is helpful to keep track of the work by computing the product in the displayed form illustrated below. However, like terms must be listed in the same column even though they may not be generated in the same order each time. For example, the product $(4\sqrt{5} - \sqrt{6})(2\sqrt{5} + 3\sqrt{6})$ could be displayed as follows.

$$
\begin{array}{r}
4\sqrt{5} - \sqrt{6} \\
2\sqrt{5} + 3\sqrt{6} \\
\hline
8\cdot 5 - 2\sqrt{5}\ \sqrt{6} \\
-3\cdot 6 + 12\sqrt{5}\ \sqrt{6} \\
\hline
22 + 10\sqrt{30}
\end{array}
$$

EXAMPLE 3 Simplify the following.

a. $4^{3/2}$

b. $\left(-\dfrac{1}{8}\right)^{-2/3}$

c. $(-0.0000128)^{1/7}$

d. $(25a^{-4}b^6)^{-3/2}$

Solution **a.** $4^{3/2} = (\sqrt{4})^3 = 2^3 = 8$

b. $\left(-\dfrac{1}{8}\right)^{-2/3} = \left(\sqrt[3]{-\dfrac{1}{8}}\right)^{-2}$

$= \left(-\dfrac{1}{2}\right)^{-2}$

$= \dfrac{1}{\left(-\dfrac{1}{2}\right)^2}$

$= \dfrac{1}{\dfrac{1}{4}} = 4$

c. $(-0.0000128)^{1/7} = [(-128) \cdot (10^{-7})]^{1/7}$

$= (-128)^{1/7}(10^{-1})$

$= -2 \cdot 10^{-1}$

$= -0.2$

d. $(25a^{-4}b^6)^{-3/2} = (25)^{-3/2}(a^{-4})^{-3/2}(b^6)^{-3/2}$

$= (\sqrt{25})^{-3}(\sqrt{a^{-4}})^{-3}(\sqrt{b^6})^{-3}$

$= 5^{-3}(a^{-2})^{-3} \mid b^3 \mid^{-3}$

$= \dfrac{a^6}{125 \mid b \mid^9}$

EXAMPLE 4 A certain bacteria colony doubles its population every hour. If 10,000 bacteria are present at noon, how many are present at 3:30 P.M.?

Solution The population n hours after noon is $10,000 \cdot 2^n$. Assuming the population grows steadily rather than by sudden large increases each hour, we may also assume that the population $3\frac{1}{2}$ hours after noon is

$$10,000(2^{3\frac{1}{2}}) = 10,000(2^3)(2^{1/2})$$
$$= 10,000 \cdot 8 \cdot \sqrt{2}$$
$$= 80,000\sqrt{2}$$
$$\approx 113,140$$

EXERCISES 1.4

In each of the following, evaluate and/or simplify the given expressions. Write the answers in terms of radicals or positive exponents.

1. $\sqrt{49}$

2. $\sqrt{625}$

3. $\sqrt{72}$

4. $\sqrt{675}$

5. $\sqrt[3]{125}$

6. $\sqrt[6]{64}$

7. $\sqrt[5]{-32}$

8. $\sqrt[4]{625}$

9. $\sqrt{8} \cdot \sqrt{2}$

10. $\dfrac{\sqrt{8}}{\sqrt{2}}$

11. $\dfrac{\sqrt{2}}{\sqrt[3]{8}}$

12. $\dfrac{\sqrt{125}}{\sqrt{5}}$

13. $\sqrt{3} \cdot \sqrt{27}$

14. $\sqrt[3]{27} \cdot \sqrt{9}$

15. $\dfrac{\sqrt[3]{81}}{\sqrt[3]{3}}$

16. $\dfrac{\sqrt[4]{16}}{\sqrt{4}}$

17. $\sqrt[4]{32} \sqrt[4]{8}$

18. $\sqrt{400 + 225}$

19. $\sqrt[3]{32\sqrt{4}}$

20. $\sqrt{400} + \sqrt{225}$

21. $\sqrt{\sqrt[3]{729}}$

22. $9^{5/2}$

23. $4^{3/2}$

24. $(-8)^{5/3}$

25. $(-27)^{-1/3}$

26. $(64)^{-5/6}$

27. $(27)^{-2/3}$

28. $(16)^{3/4}$

29. $(-27)^{2/3}$

30. $(-32)^{-6/5}$

31. $\left(\dfrac{1}{27}\right)^{-2/3}$

32. $\left(-\dfrac{1}{32}\right)^{-2/5}$

33. $\left(\dfrac{1}{81}\right)^{-3/4}$

34. $\left(\dfrac{1}{81}\right)^{1/4}$

35. $\sqrt[3]{\dfrac{(270,000,000)(0.000008)}{1.25}}$

36. $\left[\dfrac{1600(0.0025)}{900}\right]^{-3/2}$

37. $\sqrt[4]{81r^8s^4}$

38. $\dfrac{\sqrt{3x^3y^3z}}{\sqrt{27xy^5z^3}}$

39. $\sqrt[3]{16a^4b}\ \sqrt[3]{ab^4}$

40. $\dfrac{\sqrt[4]{256u^{10}v^{20}}}{\sqrt{4u^3v^2}}$

41. $\sqrt{125\sqrt{25x^4y^6}}$

42. $\sqrt{256\sqrt[3]{x^2}}$

43. $(-2x^{3/2}y^{2/5})(4x^{3/2}y^{8/5})$

44. $(3r^{-3/2}s^{1/4})(r^{-1/2}s^{1/4})$

45. $(u^{1/3}v^{2/3}w^{5/3})^6$

46. $(x^{-1/3}y^{2/3}z^{-5/3})^6$

47. $(-x^{1/2}y^{-2/3})(3x^{-3/2}y^{2/3})$

48. $(x^{1/8}y^{1/4}z^2)^2$

49. $(p^{-1/8}q^{1/4}r^{-2})^{-2}$

50. $(s^{2/3}t^{-5/12}u^{-1/6})^{-3}$

51. $(a^{27}b^9z^3)^{2/3}$

52. $(a^8b^{-4}c^2)^{-3/2}$

53. $(x^{5/3}y^{-5/12}z^{5/6})^{3/5}$

54. $(x^{2/3}y^{-5/12}z^{-1/6})^{-3}$

55. $(x^{1/4}y^{3/8}x^{3/16})^{16}$

56. $\dfrac{(a^{-2/3}b^{1/3})^3}{(a^{1/4}b^{-1/8})^4}$

57. $\dfrac{(u^3v^4)^{-3}}{(v^4w^{-2}u^2)^{-2}}$

58. $\left(\dfrac{25a^{-4}b^{-6}c^2}{16x^{-2}y^6z^{-2}}\right)^{-3/2}\cdot\left(\dfrac{2a^2b^{-1}c^{-3}}{x^2y^{-2}z^3}\right)^{-1}$

59. $5\sqrt{27}+\sqrt{75}-\sqrt{147}$

60. $\sqrt[3]{-576}-\sqrt[3]{243}-\sqrt[3]{72}$

61. $3\sqrt{8}-\sqrt[3]{81}-\sqrt{128}+\sqrt[3]{375}$

62. $\sqrt[4]{576}-\sqrt{150}+\sqrt[6]{216}$

63. $(2\sqrt{5}+3\sqrt{3})(2\sqrt{5}-3\sqrt{3})$

64. $(8\sqrt{6}+\sqrt{15})(\sqrt{6}-2\sqrt{15})$

65. $(\sqrt{5}+\sqrt{20})^2$

66. $(\sqrt{2}+\sqrt{3}+\sqrt{5})(\sqrt{2}-\sqrt{3}-\sqrt{5})$

67. $(3\sqrt{6}-\sqrt{2}+\sqrt{3})(2\sqrt{6}+\sqrt{2}+\sqrt{3})$

68. $(\sqrt[3]{9}+\sqrt[3]{4})(\sqrt[3]{3}+\sqrt[3]{2})$

69. $(\sqrt[3]{9}+\sqrt[3]{6}+\sqrt[3]{4})(\sqrt[3]{3}-\sqrt[3]{2})$

70. $(\sqrt[3]{25}-\sqrt[3]{10}+\sqrt[3]{4})(\sqrt[3]{5}+\sqrt[3]{2})$

71. $\dfrac{\sqrt{18}+\sqrt{32}}{\sqrt{2}}$

72. $\dfrac{4\sqrt{18}-2\sqrt{12}}{\sqrt{3}}$

Verify the following statements.

73. $\sqrt[3]{1+8}\neq\sqrt[3]{1}+\sqrt[3]{8}$

74. $(1+3)^{1/2}\neq 1^{1/2}+3^{1/2}$

75. $\sqrt{2^2+2^2}\neq 2+2$

76. If an animal with legs L feet long can walk $7\sqrt{L}/3$ miles per hour and Elmo the giraffe has legs 84 inches long, how fast can he walk?

77. The number of strides per minute taken by a running animal with legs L feet long is approximately $99/\sqrt{L}$. Approximately how many strides per minute are taken by Elmo the giraffe (Exercise 76) when he runs? How many are taken by Wiener the dog if his legs are 4 inches long?

78. The velocity of a wave on a banjo string is given by $\sqrt{T/\rho}$, where T is the tension and ρ is the mass per unit length. If T is given in newtons N and ρ is given in kilograms per meter kg/m, the velocity is obtained in meters per second. Find the velocity if $T=48N$, $\rho=0.012$ kg/m.

79. If the gross national product was \$2.5 trillion in 1980 and triples every 15 years, what would you expect it to be in the year 2000? What would it have been in 1970?

80. If the U.S. population sustains its doubling time of 60 years and the 1980 population was estimated at 222 million people, what will the population be in the year 2000?

81. The doubling time of a certain bacteria culture is 2 hours. If the culture starts with 5000 bacteria, how many will there be $5\frac{1}{2}$ hours later?

82. If the doubling time of a bacteria culture is 3 hours and there are 1000 present now, how many were there 1 hour ago?

83. The temperature in a blast furnace is 2000°C. Cooling chambers are designed to permit molten metal to lose one quarter of its temperature (in °C) in 20 minutes. What will the temperature of the metal be after 45 minutes in the cooling chamber?

84. The rate at which crude oil flows from a tank depends on the weight or amount of oil remaining in the tank. If 1/5 of the oil flows out every hour and 50,000 imperial gallons remain in the tank after 3 hours, how much will be in the tank after 10 hours?

85. An insect doubles its weight every 48 hours and now weighs 4 milligrams.
    a. How much will it weigh in 12 hours?
    b. How much will it weigh in 3 days?

86. The half-life of radium is approximately 1600 years; that is, the quantity of radium in an ore sample decomposes to half its original amount in 1600 years. Two thousand units of radium are now present in a certain sample.

a. How many units will be present 800 years from now?

b. How many units will be present 400 years from now?

c. How many units were present 800 years ago?

d. How many units were present 400 years ago?

1.5 EXPANDING AND FACTORING

Letters or symbols that stand for numbers in an algebraic expression are called **variables.** For example, in the expression $2x + 3$, x is a variable. We can substitute many numbers for x. If $x = 2$, the expression has a value of $2 \cdot 2 + 3 = 7$; if $x = -1$, its value is $2(-1) + 3 = 1$.

Since the letters in an algebraic expression represent numbers, we shall manipulate them in precisely the same way that we manipulate the real numbers. For example, the distributive law is used in the simple addition or subtraction of like terms:

$$3x + 2x = (3 + 2)x = 5x$$

To add or subtract expressions involving unlike terms, we also use the associative and commutative laws since these permit the rearrangement of terms in an expansion.

EXAMPLE 1   Find the sum and difference of the given algebraic expressions.

$$2x^3 + 3x^2 + 4x - 1 \qquad 6x^3 - x + 5$$

Solution   Use the commutative and associative laws to collect like terms in the sum as follows.

$$(2x^3 + 3x^2 + 4x - 1) + (6x^3 - x + 5)$$
$$= (2x^3 + 6x^3) + 3x^2 + (4x - x) - 1 + 5$$
$$= 8x^3 + 3x^2 + 3x + 4$$

Next, we subtract the second expression from the first.

$(2x^3 + 3x^2 + 4x - 1) - (6x^3 - x + 5)$
$\quad = (2x^3 + 3x^2 + 4x - 1) + [(-1)(6x^3 - x + 5)]$

(property of subtraction)

$\quad = (2x^3 + 3x^2 + 4x - 1) + (-6x^3 + x - 5)$

(distributive law)

$\quad = (2x^3 - 6x^3) + 3x^2 + (4x + x) - 1 - 5$

(collecting like terms)

$\quad = -4x^3 + 3x^2 + 5x - 6$ ■

We see in the preceding example that algebraic expressions are added or subtracted by merely adding or subtracting like terms. This shortcut is incorporated in the method of displaying the work in the familiar manner that is used for addition and subtraction of numbers. This so-called displayed form is illustrated in Example 2.

EXAMPLE 2   Rework Example 1 by writing one expression below the other in column formation.

Solution   The addition is displayed as follows.

$$
\begin{array}{r}
2x^3 + 3x^2 + 4x - 1 \\
+\,6x^3 \qquad\quad -\ x + 5 \\
\hline
8x^3 + 3x^2 + 3x + 4
\end{array}
$$

To subtract we must remember that $-(a + b) = -a - b$. The subtraction is then displayed in two steps.

$$
\begin{array}{r}
2x^3 + 3x^2 + 4x - 1 \\
-\,(6x^3 \qquad\quad -\ x + 5)
\end{array}
=
\begin{array}{r}
2x^3 + 3x^2 + 4x - 1 \\
-\,6x^3 \qquad\quad +\ x - 5 \\
\hline
-4x^3 + 3x^2 + 5x - 6
\end{array}
$$

When using the displayed form of subtraction, we generally do the transition mentally and immediately write the second step.   ■

Multiplication of algebraic expressions depends on the distributive law and the laws of exponents. A simple addition problem remains after applying these laws. The "displayed form" of multiplication is an excellent computational tool.

EXAMPLE 3   Find the product of $(3x^2 + x - 1)$ and $(2x^2 - 2x + 3)$.

Solution   We can use the distributive law to write

$$(3x^2 + x - 1)(2x^2 - 2x + 3)$$
$$= (3x^2 + x - 1)(2x^2) + (3x^2 + x - 1)(-2x) + (3x^2 + x - 1)(3)$$

Use the distributive law again:

$$= (6x^4 + 2x^3 - 2x^2) + (-6x^3 - 2x^2 + 2x) + (9x^2 + 3x - 3)$$

Collect like terms:

$$= 6x^4 - 4x^3 + 5x^2 + 5x - 3$$

It is more convenient to display the work as follows:

$$
\begin{array}{l}
3x^2 +\ x - 1 \\
2x^2 - 2x + 3 \\
\hline
6x^4 + 2x^3 - 2x^2 \\
\quad\ \ -\,6x^3 - 2x^2 + 2x \\
\quad\qquad\quad\ +\,9x^2 + 3x - 3 \\
\hline
6x^4 - 4x^3 + 5x^2 + 5x - 3
\end{array}
\qquad
\boxed{
\begin{array}{l}
= (3x^2 + x - 1)(2x^2) \\
= (3x^2 + x - 1)(-2x) \\
= (3x^2 + x - 1)(3)
\end{array}
}
$$

■

The following is a list of product–factor relationships occurring so often in mathematics that each should be memorized. Each can be verified by simply carrying out the indicated multiplication. Their verifications are left as exercises.

COMMON EXPANSIONS

1. $(a \pm b)^2 = a^2 \pm 2ab + b^2$

2. $a^2 - b^2 = (a + b)(a - b)$

3. $(a \pm b)^3 = a^3 \pm 3a^2b + 3ab^2 \pm b^3$

4. $a^3 + b^3 = (a + b)(a^2 - ab + b^2)$

5. $a^3 - b^3 = (a - b)(a^2 + ab + b^2)$

Formulas 1 and 3 are each really two formulas in one. In any such double formula the upper signs stay together and the lower signs stay together. Thus,

$$(a + b)^2 = a^2 + 2ab + b^2$$
$$(a - b)^2 = a^2 - 2ab + b^2$$

EXAMPLE 4   Find the following products by using formulas (1) through (5) in the list of common expansions.

    a. $(2x^2 + 3y)^2$

    b. $(3r - 2s)(3r + 2s)$

    c. $(2u^2 - 3v)^3$

Solution   **a.**   We use formula (1) with $a = 2x^2$ and $b = 3y$.

$$(2x^2 + 3y)^2 = (2x^2)^2 + 2(2x^2)(3y) + (3y)^2$$
$$= 4x^4 + 12x^2y + 9y^2$$

    b.   This product fits formula (2) with $a = 3r$ and $b = 2s$.

$$(3r - 2s)(3r + 2s) = (3r)^2 - (2s)^2$$
$$= 9r^2 - 4s^2$$

    c.   We recognize this as the cube of a difference given by formula (3) with $a = 2u^2$ and $b = 3v$.

$$(2u^2 - 3v)^3 = (2u^2)^3 - 3(2u^2)^2(3v) + 3(2u^2)(3v)^2 - (3v)^3$$
$$= 8u^6 - 36u^4v + 54u^2v^2 - 27v^3 \qquad \blacksquare$$

Using these expansion formulas or the distributive law (p. 3) in reverse is called **factoring** since this entails finding the factors of an expression.

EXAMPLE 5   Factor the following expressions.

    a. $16x^2 + 8xy + 24x$

    b. $25 - u^2$

    c. $w^3 - 27$

Solution   **a.**   Note that $8x$ is a factor of each term in this expression. We use the distributive law in reverse to factor out $8x$

$$16x^2 + 8xy + 24x = 8x(2x + y + 3)$$

b.   Applying formula (2) of the common expansions, we have

$$\begin{aligned}
25 - u^2 &= 5^2 - u^2 \\
&= (5 + u)(5 - u)
\end{aligned}$$

c.   Since $27 = 3^3$, we use formula (5) of the common expansions

$$\begin{aligned}
w^3 - 27 &= w^3 - 3^3 \\
&= (w - 3)(w^2 + 3w + 9)
\end{aligned}$$ ■

In Example 5(a), the factor $8x$ is called a **common factor** of the terms $16x^2$, $8xy$, and $24x$. When we use the distributive law in reverse as in Example 5(a), we are factoring out a common factor.

More involved applications of the common expansions are given in Example 6.

EXAMPLE 6   Factor the following expressions.

a. $25x^2 - 20xy + 4y^2$

b. $81r^4 - 16s^4$

c. $27r^6 + 8s^3$

Solution   **a.** $\begin{aligned}[t] 25x^2 - 20xy + 4y^2 &= (5x)^2 - 2(5x)(2y) + (2y)^2 \\ &= (5x - 2y)^2 \end{aligned}$

b. $\begin{aligned}[t] 81r^4 - 16s^4 &= (9r^2 + 4s^2)(9r^2 - 4s^2) \\ &= (9r^2 + 4s^2)(3r + 2s)(3r - 2s) \end{aligned}$

c. $\begin{aligned}[t] 27r^6 + 8s^3 &= (3r^2)^3 + (2s)^3 \\ &= (3r^2 + 2s)[(3r^2)^2 - (3r^2)(2s) + (2s)^2] \\ &= (3r^2 + 2s)(9r^4 - 6r^2s + 4s^2) \end{aligned}$ ■

Factoring by Completing the Square

Certain expressions resemble the "perfect square" forms $a^2 \pm 2ab + b^2$ but they don't match it exactly. In this case we adjust the offending term to fit the form and then balance the expression by adding or subtracting a term so that the result is equal to the original expression. This process is called *completing the square*. Sometimes, the result can then be factored as illustrated in Example 7.

EXAMPLE 7   Factor the following expressions by first completing the square.

a. $9x^4 + 15x^2y^2 + 16y^4$

b. $a^2 - 8a + 7$

c. $3x^2 - 4x + 1$

Solution **a.** The expression $9x^4 + 15x^2y^2 + 16y^4$ resembles the form $a^2 + 2ab + b^2$ except that $15x^2y^2 \neq 2ab$. To match the form, the middle term should be $2(3x^2)(4y^2) = 24x^2y^2$. Thus, we write

$$\begin{aligned}
9x^4 + 15x^2y^2 + 16y^4 &= (9x^4 + 24x^2y^2 + 16y^4) - 9x^2y^2 \\
&= (3x^2 + 4y^2)^2 - 9x^2y^2 \\
&= (3x^2 + 4y^2 + 3xy)(3x^2 + 4y^2 - 3xy)
\end{aligned}$$

b. In this situation, it is more convenient to adjust the last term rather than the middle term. Observe that $a^2 - 8a + 16 = (a - 4)^2$ and write

$$\begin{aligned}
a^2 - 8a + 7 &= (a^2 - 8a + 16) - 9 \\
&= (a - 4)^2 - 9 \\
&= [(a - 4) + 3][(a - 4) - 3] \\
&= (a - 1)(a - 7)
\end{aligned}$$

c. Here we adjust the first term.

$$\begin{aligned}
3x^2 - 4x + 1 &= (4x^2 - 4x + 1) - x^2 \\
&= (2x - 1)^2 - x^2 \\
&= (2x - 1 - x)(2x - 1 + x) \\
&= (x - 1)(3x - 1)
\end{aligned}$$

Scissors Factoring

Many expressions do not fit any of the forms discussed so far. To factor them we must rely on experience and ingenuity. An analysis of the manner in which products are formed sometimes helps. For instance, in Figure 1–6 we have the "scissors" formula for the product $(ax + b)(cx + d)$, so called because the coefficient of x is found by multiplying the coefficients at opposite ends of the cross or "scissors."

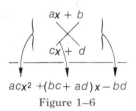

$$acx^2 + (bc + ad)x - bd$$

Figure 1–6

The scissors formula is important as a factoring aid. To begin the factoring process, we note that ac is the coefficient of x^2 and bd is the constant term.

EXAMPLE 8 Factor each of the following.

a. $u^2 - 5u + 4$

b. $z^2 + 4z - 5$

c. $2t^2 + 7t + 6$

d. $10x^2 - 13x - 3$

Solution   **a.**   Since the coefficient of u^2 is 1, we use the scissors formula with $a = c = 1$. Then

$$(u^2 - 5u + 4) = (u + b)(u + d)$$

with $bd = 4$ and $(b + d) = -5$. Thus, b and d have the same sign since $bd > 0$; in fact, both must be negative since the coefficient of the middle term is negative. The factors of 4 are $1 \cdot 4$ and $2 \cdot 2$. Since $1 + 4 = 5$ and $2 + 2 \neq 5$, the only possibility is

$$u^2 - 5u + 4 = (u - 4)(u - 1).$$

b.   Again, write

$$z^2 + 4z - 5 = (z + b)(z + d)$$

with $bd = -5$ and $(b + d) = 4$. This time b and d must have opposite signs since $bd < 0$; the positive term must be larger since $b + d > 0$. Since the only factors of 5 are $1 \cdot 5$, we have

$$z^2 + 4z - 5 = (z + 5)(z - 1).$$

c.   We shall use the scissors formula and try to write

$$(2t^2 + 7t + 6) = (2t + b)(t + d)$$

with $bd = 6$. Thus, b and d have the same sign since $bd > 0$. In fact, both must be positive since the coefficient of the middle term $(=7)$ is positive. The factors of 6 are $1 \cdot 6$ and $2 \cdot 3$. Trial and error leads to only one successful combination:

$$(2t^2 + 7t + 6) = (2t + 3)(t + 2)$$

d.   Using the scissors formula we factor $10x^2 - 13x - 3$ as follows:

$$(10x^2 - 13x - 3) = \begin{cases} (10x + b)(x + d) \\ \text{or} \\ (2x + b)(5x + d) \end{cases}$$

with $bd = -3$. Thus, b and d have opposite signs, and numerical values of 1 and 3 suggest themselves. After trying the various possibilities, we find that

$$(10x^2 - 13x - 3) = (2x - 3)(5x + 1)$$

is the only possible factorization.   ■

EXAMPLE 9   The La-Zee Lady Chair Company finds that a webbing machine generates a monthly profit of $60 + 37t - 2t^2$ thousand dollars t months after its purchase. How long does the machine remain profitable?

Solution   The machine ceases to produce a profit when $60 + 37t - 2t^2$ is reduced to 0. This expression for the profit factors into

$$60 + 37t - 2t^2 = (20 - t)(3 + 2t)$$

and equals 0 if and only if one of its factors is 0. This clearly happens when $t = 20$. We only consider $t \geq 0$ since t measures operating time since the machine was purchased. Thus, $(3 + 2t) > 0$ for all $t \geq 0$ and $(20 - t) > 0$ for $0 \leq t < 20$. The machine ceases to be profitable after 20 months. ∎

Factoring by Grouping

When more than three terms appear in an expression to be factored, it sometimes helps to partition the expression into two or more smaller groups. This technique is known as **factoring by grouping.**

EXAMPLE 10   Factor the following expression.

a. $6x^2 + 3xy + 8x + 4y$

b. $x^2 - y^2 + 2x^2y + 2xy^2$

c. $3x - 2y + xy - 6$

Solution   a.   Group the first two and the last two terms and then factor as follows:

$$\begin{aligned} 6x^2 + 3xy + 8x + 4y &= (6x^2 + 3xy) + (8x + 4y) \\ &= 3x(2x + y) + 4(2x + y) \\ &= (3x + 4)(2x + y) \end{aligned}$$

b. $\begin{aligned} x^2 - y^2 + 2x^2y + 2xy^2 &= (x^2 - y^2) + (2x^2y + 2xy^2) \\ &= (x - y)(x + y) + 2xy(x + y) \\ &= (x - y + 2xy)(x + y) \end{aligned}$

c.   After rearranging terms, we can again group and factor as follows.

$$\begin{aligned} 3x - 2y + xy - 6 &= 3x + xy - 2y - 6 \\ &= x(3 + y) - 2(y + 3) \\ &= (x - 2)(y + 3) \end{aligned}$$ ∎

EXERCISES 1.5

Perform the indicated operations, collecting any like terms.

1. $(2x^2 + 3x - 2) + (4x^2 + 3x - 6)$

2. $(10y^5 + 3y^3 - 2y + 4)$
$- (5y^4 - 3y^3 + 2y^2 - 4y - 2)$

3. $(6a^3 - 2a^2 - a + 4)$
$+ (3a^4 - 2a^3 - a^2 + a + 5)$

4. $(3u^4 - 10u^3 + 2u + 1) + (2u^3 - 3u^2 + 4)$

5. $(14t^7 - 20t^6 + 4t^2 - 2t + 8)$
$- (17t^6 - 12t^5 + 5t^3 - 2t - 8)$

6. $(6b^4 - 4b^3 + 3b - 4)$
$- (5b^4 - 4b^3 + 2b^2 - 3)$

7. $(4y + 2)(3y - 1)$ 8. $(x + y)(3y - 2x)$

9. $(t - 2)(t - 1)(t + 3)$

10. $(z^2 + z - 1)(z + 3)$

11. $(2s^3 + 3s - 2)(s^2 + 4)$

12. $(r^2 - r + 1)(r^2 + 2r + 5)$

13.   Verify the formulas in the list of common expansions.

Use the common expansions to find the following.

14. $(x + 4)^2$ **15.** $(4a - 2b)^2$

16. $(r - 3)^3$ **17.** $(3x + 5y)^3$

18. $(2p - 3q)^3$

19. $(r - s + t)(r - s - t)$

20. $(x - y + z)(x + y - z)$

21. $(u + 2)(u - 2)(u^2 + 4)$

Factor the following as far as possible.

22. $8a + 4$ **23.** $25 - 50x$

24. $6y^2 + 3y - 9ay$

25. $4r^3s^2 - 2rs^8 + 10r^4s^4 - 12r^2s^2$

26. $r^2 - 25$ **27.** $36 - s^2$

28. $256 - w^4$ **29.** $36x^2 - 49y^2$

30. $w^3 - 27$ **31.** $8 - u^3$

32. $s^3 + 8$ **33.** $64 + v^3$

34. $8t^3 + 27u^3$ **35.** $y^2 - 9y + 20$

36. $t^2 - 14t + 24$ **37.** $p^2 - p - 20$

38. $q^2 - 18qr + 72r^2$

39. $4a^2 - 16a + 15$

40. $24b^2 - 6b - 18$

41. $4r^2 - 28rs + 49s^2$

42. $25u^2 + 60uv + 36v^2$

43. $a^4 - 2a^2 + 1$ **44.** $w^4 + 8w^2 - 9$

45. $r^4 - 7r^2 - 18$ **46.** $u^2 - 2uv - 8v^2$

47. $16x^8 - 72x^4 + 81$

48. $8a^3 + 36a^2b + 54ab^2 + 27b^3$

49. $8r^3 + 12r^2s + 6rs^2 + s^3$

50. $27 - 54x + 36x^2 - 8x^3$

51. $125p^6 - 64q^6$

Factor by completing the square.

52. $x^4 + x^2y^2 + y^4$ **53.** $u^8 + 4$

54. $r^4 - 3r^2s^2 + s^4$ **55.** $256 - 17t^2 + t^4$

56. $w^2 + 4w - 21$ **57.** $p^2 - 6p + 5$

Factor the following by grouping.

58. $2ar + 3br + 4as + 6bs$

59. $x^2y^2 - 4y^2 + x^2 - 4$

60. $x^5 - x^3 + x^2 - 1$

61. $x^3 - y^3 + x^2 - y^2$

62. $9r^2u^2 - r^2v^2 - 9s^2u^2 + s^2v^2$

63. $x^2y^2 + 6xy^2 + 5y^2 - 9x^2 - 54x - 45$

64. The formula $a^2 - b^2 = (a + b)(a - b)$ can be used to perform seemingly complicated multiplications very quickly. For instance, $51 \cdot 49 = (50 + 1)(50 - 1) = 50^2 - 1^2 = 2500 - 1 = 2499$. Use this technique to evaluate the following.

a. $17 \cdot 13$ **b.** $75 \cdot 65$ **c.** $23 \cdot 37$

65. A sporting goods manufacturer makes downhill skis, cross country skis, and water skis. Each pair of downhill skis sells for \$120 and costs \$78 to manufacture and distribute. Cross country skis sell for \$75 and cost \$43 to manufacture and distribute. The water skis cost \$55, but they are sold for \$95. If x downhill skis, y cross country skis, and z water skis are sold, express (a) the total sales receipts, (b) the total expenses, and (c) the total net profit algebraically.

66. The altitude of a scientific weather rocket t seconds after it is fired is given by the polynomial $-16t^2 + 56t + 120$.

a. How high is the rocket 3 seconds after firing?

b. Factor the expression to determine when the rocket hits the ground.

1.6 POLYNOMIALS AND RATIONAL EXPRESSIONS

Polynomials

In the last section we worked with expressions of the form

$$a_n x^n + a_{n-1} x^{n-1} + \cdots + a_1 x + a_0$$

where $a_n, \ldots, a_0$ are fixed constants and x is a variable. These expressions are known as **polynomials** in one variable. The largest exponent is called the **degree** of the polynomial. For example,

$$x^3 - 3x^2 + 1 \text{ is a polynomial of degree 3,}$$
$$4x^2 + 6x - 2 \text{ is a polynomial of degree 2,}$$
$$5x^{100} + 2x^{17} \text{ is a polynomial of degree 100.}$$

A constant polynomial A is said to have degree 0 since $A = Ax^0$.

A polynomial with one term is called a **monomial.** A polynomial with two terms is called a **binomial.** One with three terms is called a **trinomial.** Thus,

$6x^2$ is a monomial,
$6x^2 + 2$ is a binomial,
$6x^2 + 3x + 2$ is a trinomial.

The polynomial identically 0 is called the "zero polynomial."
Expressions of the form

$$a_{n,k}x^n y^k + \cdots + a_{m,l}x^m y^l$$

are known as **polynomials in two variables.** Polynomials in three or more variables are defined similarly. The sum of the exponents in any term of a polynomial is the degree of that term; the highest degree of any of its terms is the **degree of the polynomial.** For instance,

$$3x^2 y + 2xy + 4x + 2y - 10$$

has degree 3, the degree of the term $3x^2 y$. The polynomial

$$29x^{100}y^{10} - 46y^{60}x^3 + 20$$

has degree 110, the degree of the term $29x^{100}y^{10}$. The polynomial

$$6x^2 yz^2 + 6xy - 3z$$

has degree 5, the degree of the term $6x^2 yz^2$.

Rational Expressions

In this section we shall be concerned with **quotients of polynomials.** Just as a quotient of two integers is called a rational number, a quotient of two polynomials will be called a **rational expression.** Examples of rational expressions include:

$$\frac{3x^2 + 2x - 4}{x^2}$$

$$\frac{2x^3 - 3x + 10}{3x^2 + 2}$$

$$\frac{2xy - 3}{4x^2 + 3xy^3 - 4y^2}$$

Least common multiples (LCMs) apply to polynomials, too. For example, the LCM of $(x + 1)^2(x - 1)$ and $(x + 1)(x - 1)^3(x + 2)$ is $(x + 1)^2(x - 1)^3(x + 2)$. As with numbers, factoring polynomials facilitates finding the LCM. The least common denominator (LCD) of two or more rational expressions is the LCM of their respective denominators.

In Section 1.1 we listed the rules for algebraic operations on fractions. Since rational expressions represent numbers, the same rules must apply. Thus, in order to combine rational expressions, we use the familiar laws governing fractions.

COMBINING RATIONAL EXPRESSIONS

1. To add or subtract, first find the LCD, then write each in terms of this LCD, and finally add or subtract numerators.

2. To multiply, multiply numerators and multiply denominators separately.

3. To divide, invert and multiply.

4. Finally, simplify the results by cancelling like factors in the numerator and denominator.

The following example illustrates these procedures.

EXAMPLE 1 Perform the indicated operations and simplify the results.

a. $\dfrac{2x - 8}{x^2 - 1} + \dfrac{3}{x - 1} + \dfrac{5x^2}{x + 1}$

b. $\dfrac{x^2 + 2x + 4}{x + 3} \cdot \dfrac{x^2 + 6x + 9}{x - 4}$

c. $\dfrac{x^2 - 4}{x^2 + 5x + 6} \div \dfrac{x - 2}{x^2 + 7x + 12}$

Solution a. Since $x^2 - 1 = (x - 1)(x + 1)$, the least common denominator is clearly $(x - 1)(x + 1)$. Hence, we write each fraction in terms of the LCD, add the resulting numerators, and simplify as follows.

$$\frac{2x - 8}{x^2 - 1} + \frac{3}{x - 1} + \frac{5x^2}{x + 1}$$

$$= \frac{2x - 8}{(x - 1)(x + 1)} + \frac{3(x + 1)}{(x - 1)(x + 1)} + \frac{5x^2(x - 1)}{(x + 1)(x - 1)}$$

$$= \frac{(2x - 8) + (3x + 3) + (5x^3 - 5x^2)}{(x + 1)(x - 1)}$$

$$= \frac{5x^3 - 5x^2 + 5x - 5}{(x + 1)(x - 1)} = \frac{5x^2(x - 1) + 5(x - 1)}{(x + 1)(x - 1)}$$

$$= \frac{(5x^2 + 5)(x - 1)}{(x + 1)(x - 1)}$$

$$= \frac{5(x^2 + 1)}{x + 1}$$

b. We simply multiply the numerators and multiply the denominators as with any fraction. The result is then simplified by canceling like terms in numerator and denominator as follows.

$$\frac{x^2 + 2x + 4}{x + 3} \cdot \frac{x^2 + 6x + 9}{x - 4}$$

$$= \frac{(x^2 + 2x + 4)(x^2 + 6x + 9)}{(x + 3)(x - 4)}$$

$$= \frac{(x^2 + 2x + 4)(\cancel{x + 3})(x + 3)}{(\cancel{x + 3})(x - 4)}$$

$$= \frac{x^3 + 5x^2 + 10x + 12}{x - 4}$$

c. As with ordinary division of fractions, we "invert and multiply," then proceed as in solution (b). Note the factorization of polynomials in the numerator and denominator; this factorization enables us to see the common factors that can be canceled.

$$\frac{x^2 - 4}{x^2 + 5x + 6} \div \frac{x - 2}{x^2 + 7x + 12}$$

$$= \frac{x^2 - 4}{x^2 + 5x + 6} \cdot \frac{x^2 + 7x + 12}{x - 2}$$

$$= \frac{(\cancel{x + 2})(\cancel{x - 2})(\cancel{x + 3})(x + 4)}{(\cancel{x + 2})(\cancel{x + 3})(\cancel{x - 2})}$$

$$= x + 4 \qquad \blacksquare$$

Certain relationships in medicine and optics involve rational expressions, the numerators and denominators of which are also rational expressions. Just as complex fractions reduce to a simple fraction, these complex rational expressions will reduce to a simple rational expression.

EXAMPLE 2   Simplify the following complex rational expression.

$$\frac{x + \dfrac{1}{x} + 2}{x - \dfrac{1}{x}}$$

Solution   Two solutions are shown to illustrate the various techniques that may be used. In the first solution the terms in the numerator are combined into a single rational expression. After doing the same in the denominator, we carry out the indicated division by inverting and multiplying:

$$\frac{x + \dfrac{1}{x} + 2}{x - \dfrac{1}{x}} = \frac{\left(\dfrac{x^2 + 1 + 2x}{x}\right)}{\left(\dfrac{x^2 - 1}{x}\right)}$$

$$= \frac{x^2 + 1 + 2x}{\cancel{x}} \cdot \frac{\cancel{x}}{x^2 - 1}$$

$$= \frac{(x + 1)^{\cancel{2}}}{\cancel{(x + 1)}(x - 1)}$$

$$= \frac{x + 1}{x - 1}$$

An alternate solution is found by simply multiplying numerator and denominator by x.

$$\frac{x + \dfrac{1}{x} + 2}{x - \dfrac{1}{x}} = \frac{x^2 + 1 + 2x}{x^2 - 1}$$

$$= \frac{(x + 1)^{\cancel{2}}}{\cancel{(x + 1)}(x - 1)}$$

$$= \frac{x + 1}{x - 1}$$ ■

EXAMPLE 3   The circulatory system in the human body is a maze of blood vessels connected in various ways. The blood vessels illustrated in Figure 1–7 are said to be connected in parallel. Each vessel offers a certain resistance to the flow of blood. Diseases like arteriosclerosis increase this resistance and either reduce the flow of blood, increase the blood pressure, or both. If the resistance of v_1 is x and the resistance of v_2 is y, the net resistance of the parallel network of Figure 1–7 is $\dfrac{1}{1/x + 1/y}$.

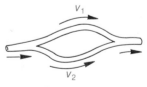

Figure 1–7

a.   Simplify the expression for the resistance of a parallel network.

b.   To see how pressure is reduced in parallel networks, let the resistance of v_1 be 10 and in v_2 be 5. Compute the resistance of the parallel network.

Solution   **a.** $\dfrac{1}{\dfrac{1}{x} + \dfrac{1}{y}} = \dfrac{1}{\dfrac{y + x}{xy}} = \dfrac{xy}{x + y}$

b.   Letting $x = 10$, $y = 5$, we have

$$\frac{1}{\dfrac{1}{x} + \dfrac{1}{y}} = \frac{xy}{x + y} = \frac{10 \cdot 5}{10 + 5} = \frac{50}{15} = 3\frac{1}{3}$$

The resulting resistance of the parallel network is less than that in either of the two separate branches.   ■

Polynomial Long Division

We have seen that some rational expressions can be simplified rather easily by factoring and applying the rules for manipulating fractions and the laws of exponents. However, other expressions are not so readily simplified because their factors are not readily apparent. For instance, it is not immediately obvious that

$$\frac{x^5 + 6x^4 - x^3 - 14x^2 + 18x + 8}{x^3 - 3x + 4}$$

can be simplified. For this kind of problem, the familiar procedure for long division of numbers can be adapted and used to advantage. In this particular problem we have

$$
\begin{array}{r}
x^2 + 6x + 2 \\
x^3 - 3x + 4 \overline{)x^5 + 6x^4 - x^3 - 14x^2 + 18x + 8} \\
\underline{x^5 - 3x^3 + 4x^2} \\
6x^4 + 2x^3 - 18x^2 \\
\underline{6x^4 - 18x^2 + 24x} \\
2x^3 - 6x + 8 \\
\underline{2x^3 - 6x + 8} \\
0
\end{array}
$$

Thus,

$$\frac{x^5 + 6x^4 - x^3 - 14x^2 + 18x + 8}{x^3 - 3x + 4} = x^2 + 6x + 2.$$

LONG DIVISION OF POLYNOMIALS

1. Always write the polynomials with terms in order of *decreasing powers*. Then at each stage, only the first terms of the expressions need be examined.

2. Leave room for any missing powers of x since they may appear in the intermediate stages of the long division.

3. Just as in long division of numbers, a remainder is sometimes left. The remainder will be a polynomial whose degree is less than that of the divisor.

The long division can be checked by verifying that

$$(x^3 - 3x + 4)(x^2 + 6x + 2)$$
$$= x^5 + 6x^4 - x^3 - 14x^2 + 18x + 8.$$

EXAMPLE 4   Use the long-division algorithm on the indicated quotient.

$$\frac{6x^3 - x^4 - 14x - 8}{3x + x^2 - 4}$$

Solution   After rearranging the terms in decreasing powers of x and leaving space for an x^2 term, we can carry out the long division.

$$
\begin{array}{r}
-x^2 + 9x - 31 \\
x^2 + 3x - 4 \overline{)\,-x^4 + 6x^3 + 0x^2 - 14x - 8} \\
-x^4 - 3x^3 + 4x^2 \\
\hline
9x^3 - 4x^2 - 14x \\
9x^3 + 27x^2 - 36x \\
\hline
-31x^2 + 22x - 8 \\
-31x^2 - 93x + 124 \\
\hline
115x - 132
\end{array}
$$

Note the remainder of $115x - 132$. As with long division of numbers, we can write

$$\frac{-x^4 + 6x^3 - 14x - 8}{x^2 + 3x - 4} = -x^2 + 9x - 31 + \frac{115x - 132}{x^2 + 3x - 4}$$

Combining the terms on the right side of the equality according to the laws of fractions will verify the relationship.   ■

Rationalizing Denominators

When an expression involves radicals, it is sometimes preferable to rewrite the expression in a form with no radicals in the denominator. Frequently, multiplying by an appropriate factor will remove such radicals. Both numerator and denominator must be multiplied by the same factor to avoid changing the value of the given expression. Removing the radicals in a denominator by this process is known as *rationalizing*.

EXAMPLE 5   Simplify the following expressions and rationalize the denominator in each case.

a. $\dfrac{\sqrt{5}}{\sqrt{2}}$ **b.** $\dfrac{\sqrt[3]{5} + \sqrt[3]{2}}{\sqrt[3]{10}}$ **c.** $\dfrac{5}{\sqrt{2} + \sqrt{3}}$ **d.** $\dfrac{\sqrt[3]{xy}}{\sqrt[3]{x} + \sqrt[3]{y}}$

Solution   **a.**   Multiply numerator and denominator by $\sqrt{2}$ to obtain

$$\frac{\sqrt{5}}{\sqrt{2}} = \frac{\sqrt{5}\,\sqrt{2}}{\sqrt{2}\,\sqrt{2}} = \frac{\sqrt{10}}{2}$$

b. To eliminate $\sqrt[3]{10}$ in the denominator, multiply by $\sqrt[3]{10}$ twice more. The expression then simplifies as follows.

$$\frac{\sqrt[3]{5} + \sqrt[3]{2}}{\sqrt[3]{10}} = \frac{(\sqrt[3]{5} + \sqrt[3]{2})\,\sqrt[3]{10}\,\sqrt[3]{10}}{\sqrt[3]{10}\,\sqrt[3]{10}\,\sqrt[3]{10}}$$

$$= \frac{\sqrt[3]{500} + \sqrt[3]{200}}{10}$$

$$= \frac{\sqrt[3]{125}\,\sqrt[3]{4} + \sqrt[3]{8}\,\sqrt[3]{25}}{10}$$

$$= \frac{5\sqrt[3]{4} + 2\sqrt[3]{25}}{10}$$

c. To eliminate $\sqrt{2} + \sqrt{3}$, multiply numerator and denominator by $\sqrt{2} - \sqrt{3}$ to obtain

$$\frac{5}{\sqrt{2} + \sqrt{3}} = \frac{5}{\sqrt{2} + \sqrt{3}} \cdot \frac{\sqrt{2} - \sqrt{3}}{\sqrt{2} - \sqrt{3}}$$

$$= \frac{5\sqrt{2} - 5\sqrt{3}}{2 - 3}$$

$$= 5\sqrt{3} - 5\sqrt{2}$$

d. The formula for factoring the sum of two cubes is useful here.

$$\frac{\sqrt[3]{xy}}{\sqrt[3]{x} + \sqrt[3]{y}} = \frac{\sqrt[3]{xy}}{\sqrt[3]{x} + \sqrt[3]{y}} \cdot \frac{(\sqrt[3]{x})^2 - \sqrt[3]{x}\,\sqrt[3]{y} + (\sqrt[3]{y})^2}{(\sqrt[3]{x})^2 - \sqrt[3]{x}\,\sqrt[3]{y} + (\sqrt[3]{y})^2}$$

$$= \frac{x^{1/3}y^{1/3}(x^{2/3} - x^{1/3}y^{1/3} + y^{2/3})}{x + y}$$

$$= \frac{xy^{1/3} - x^{2/3}y^{2/3} + x^{1/3}y}{x + y}$$

$$= \frac{x\sqrt[3]{y} - \sqrt[3]{x^2y^2} + y\sqrt[3]{x}}{x + y} \qquad\blacksquare$$

Expressions of the form $(\sqrt{a} + \sqrt{b})$ and $(\sqrt{a} - \sqrt{b})$ are said to be **conjugates** of one another. To rationalize a denominator of this type, multiply by its conjugate as in Example 5(c). The denominator then becomes $(a - b)$; the radicals may remain in the numerator, however.

Sometimes, rationalizing the denominator requires several steps; for such problems, experience and ingenuity are valuable assets.

EXAMPLE 6 Rationalize the denominator in the following expressions.

a. $\dfrac{\sqrt{2}}{\sqrt{2} + \sqrt{3} + \sqrt{7}}$

b. $\dfrac{1}{\sqrt[4]{a} - \sqrt[4]{b}}$

Solution   **a.**   Write the denomintor in the form $(\sqrt{2} + \sqrt{3}) + \sqrt{7}$ and eliminate the $\sqrt{7}$ as in Example 5(c).

$$\frac{\sqrt{2}}{\sqrt{2} + \sqrt{3} + \sqrt{7}}$$

$$= \frac{\sqrt{2}}{(\sqrt{2} + \sqrt{3}) + \sqrt{7}} \cdot \frac{(\sqrt{2} + \sqrt{3}) - \sqrt{7}}{(\sqrt{2} + \sqrt{3}) - \sqrt{7}}$$

$$= \frac{2 + \sqrt{6} - \sqrt{14}}{(\sqrt{2} + \sqrt{3})^2 - 7} = \frac{2 + \sqrt{6} - \sqrt{14}}{2(\sqrt{6} - 1)}$$

The $\sqrt{6}$ can also be removed if we multiply numerator and denominator by $\sqrt{6} + 1$. Hence, we have

$$\frac{\sqrt{2}}{\sqrt{2} + \sqrt{3} + \sqrt{7}} = \frac{2 + \sqrt{6} - \sqrt{14}}{2(\sqrt{6} - 1)}$$

$$= \frac{2 + \sqrt{6} - \sqrt{14}}{2(\sqrt{6} - 1)} \cdot \frac{\sqrt{6} + 1}{\sqrt{6} + 1}$$

$$= \frac{8 + 3\sqrt{6} - 2\sqrt{21} - \sqrt{14}}{10}$$

b.   Note that $(\sqrt[4]{x})^2 = x^{2/4} = x^{1/2} = \sqrt{x}$. Hence, this expression can be rationalized as follows.

$$\frac{1}{\sqrt[4]{a} - \sqrt[4]{b}} = \frac{1}{\sqrt[4]{a} - \sqrt[4]{b}} \cdot \frac{\sqrt[4]{a} + \sqrt[4]{b}}{\sqrt[4]{a} + \sqrt[4]{b}}$$

$$= \frac{\sqrt[4]{a} + \sqrt[4]{b}}{\sqrt{a} - \sqrt{b}}$$

$$= \frac{\sqrt[4]{a} + \sqrt[4]{b}}{\sqrt{a} - \sqrt{b}} \cdot \frac{\sqrt{a} + \sqrt{b}}{\sqrt{a} + \sqrt{b}}$$

$$= \frac{a^{3/4} + a^{1/2}b^{1/4} + a^{1/4}b^{1/2} + b^{3/4}}{a - b}$$

◼

EXERCISES 1.6

Simplify the given expressions without resorting to long division.

1. $\dfrac{2x^3 - 6x^2 + 2x}{2x}$

2. $\dfrac{3y^4 + 9y^3 - 12y^2}{3y^2}$

3. $\dfrac{u^2 + 5u - 14}{u - 2}$

4. $\dfrac{2v^2 + 13v - 7}{2v - 1}$

5. $\dfrac{z^2 + 2z - 3}{z + 3}$

6. $\dfrac{6r^2 - 5r - 6}{2r - 3}$

In the following exercises, perform the indicated operations and simplify the results.

7. $\dfrac{r}{2 - r} - \dfrac{r}{r - 2}$

8. $\dfrac{2u + v}{2v - u} + \dfrac{u + 2v}{v - 2u}$

9. $\dfrac{1}{x - 3} - \dfrac{x + 2}{x^2 - 3x}$

10. $\dfrac{2a}{a - 2} - \dfrac{a^2 - 5}{a^2 - 4a + 4} - 1$

11. $\dfrac{12p - 6q}{3p^2 - 3q^2} + \dfrac{5}{p + q} - \dfrac{1}{p - q}$

12. $\dfrac{3s + 4}{s^2 - 16} - \dfrac{s + 2}{s^2 - 5s + 4} + \dfrac{1}{s + 4}$

13. $\dfrac{3w^3}{w^3 + 27} - \dfrac{w}{w + 3} + \dfrac{2w^2}{w^2 - 3w + 9}$

14. $\dfrac{2t + 1}{4t^2 - 4t + 1} + \dfrac{1 - 18t}{16t^4 - 8t^2 + 1}$
$\qquad\qquad - \dfrac{3 + t}{4t^2 + 4t + 1}$

15. $(t + 3)\left(\dfrac{1}{t} - \dfrac{1}{3}\right)$ **16.** $(x - y)\left(\dfrac{1}{x} + \dfrac{1}{y}\right)$

17. $\dfrac{a^3 - b^3}{ab} \cdot \dfrac{a^2 b^3}{(a^2 + ab + b^2)}$

18. $\dfrac{u^2 + 4u + 4}{4 - u^2} \cdot \dfrac{u - 2}{u^2 + 2u}$

19. $\dfrac{v^2 - 25}{v^2 - 16} \cdot \dfrac{v^2 + v - 12}{v^2 + 2v - 15}$

20. $\dfrac{r^3 - 8}{r^3 + 8} \cdot \dfrac{(r - 1)(r^2 - 2r + 4)}{(r + 1)(r^2 + 2r + 4)}$

21. $\dfrac{u^2 + 5u + 6}{u^2 - 9} \div \dfrac{2 + u}{3 - u}$

22. $\dfrac{1 - 4w^2}{2w^2 - 3w + 1} \div \dfrac{4w^2 + 8w + 3}{2w^2 - 5w + 3}$

23. $\dfrac{x^2 - 25}{x^2 - 9} \div \dfrac{x^2 - x - 20}{x^2 + 7x + 12}$

24. $\dfrac{t^2 - 7t + 10}{t^2 - 6t - 16} \div \dfrac{t^2 - 11t + 18}{t^2 - 9t + 8}$

25. $\dfrac{x^3}{x^2 - xy + y^2} \cdot \dfrac{x}{x^2 - y^2} \div \dfrac{x^4}{x^3 + y^3}$

26. $\dfrac{18b^2 - 57b - 10}{18b^2 + 33b + 5} \div \dfrac{12b^2 - 47b + 45}{12b^2 + 5b - 72}$
$\qquad\qquad \cdot \dfrac{6b^2 - 11b - 35}{6b^2 - 5b - 56}$

27. $\dfrac{a^2 - 2a}{2a + 4} \div \dfrac{a^2 - 4}{a^2 + 4a + 4} \div \dfrac{a^2 - 4a + 4}{2a^2 + 4a}$

28. $\left(\dfrac{u^2 - u - 2}{u + 1}\right) \div \left(\dfrac{u^2 - u - 20}{u^2 + 2u - 8} \div \dfrac{u^2 - 25}{u^2 + 5u}\right)$

29. $\dfrac{s - \dfrac{9}{s}}{s - 3}$

30. $\left(1 - \dfrac{y^2}{4x^2}\right) \div \left(\dfrac{y}{2x} - 1\right)$

31. $\dfrac{3a}{3 - \dfrac{2}{a}} - \dfrac{9a^2 - \dfrac{2}{a} - 1}{9a + \dfrac{4}{a} - 12}$

32. $x - \dfrac{x}{x - \dfrac{1}{x}}$

33. $\dfrac{1 + \dfrac{1}{1 - \dfrac{1}{w}}}{1 - \dfrac{3}{1 - \dfrac{1}{w}}}$

34. $\dfrac{r + \dfrac{1}{1 + \dfrac{1}{1 + r}}}{r - \dfrac{1}{1 - \dfrac{5}{1 - r}}}$

Use long division to find:

35. $\dfrac{12a^4 + 25a^2 + 42}{2a^2 + 3a + 4}$ **36.** $\dfrac{v^7 + 1}{v + 1}$

37. $\dfrac{2y^3 + 15y - 3y^2 - 1}{y^2 - 5}$

38. $\dfrac{7t^2 + t - t^3 - 3}{t + 3}$

39. $(x^4 + 2x^3 - 13x^2 - 14x + 24) \div (x + 2)$

40. $\dfrac{3s^3 - 2s + s^6 - 1}{s - 2}$

Simplify each of the following and rationalize all denominators in your answers.

41. $\dfrac{\sqrt{2} - \sqrt{3}}{\sqrt{2} + \sqrt{3}}$ **42.** $\dfrac{2\sqrt{3} - \sqrt{8}}{\sqrt{3} - \sqrt{2}}$

43. $\dfrac{5\sqrt{2} - 2\sqrt{5}}{\sqrt{5} + 2\sqrt{2}}$ **44.** $\dfrac{2\sqrt{6} + \sqrt{8}}{\sqrt{2} - 2\sqrt{3}}$

45. $\dfrac{2\sqrt{2}}{1 + \sqrt{3} - \sqrt{2}}$ **46.** $\dfrac{\sqrt{2}}{2 + \sqrt{2} + \sqrt{10}}$

47. $\dfrac{3\sqrt{8}}{\sqrt{3} + \sqrt{2} + \sqrt{5}}$ **48.** $\dfrac{4\sqrt{3}}{\sqrt{3} - \sqrt{2} - \sqrt{5}}$

49. $\dfrac{3\sqrt[3]{2}}{1 + \sqrt[3]{2}}$ **50.** $\dfrac{\sqrt[3]{12}}{\sqrt[3]{3} - \sqrt[3]{2}}$

51. $\dfrac{2}{\sqrt[4]{8} - \sqrt[4]{2}}$ **52.** $\dfrac{7\sqrt[3]{800}}{\sqrt[3]{4} - \sqrt[3]{10} + \sqrt[3]{25}}$

53. $\dfrac{\sqrt[3]{3} + \sqrt[3]{7}}{\sqrt[3]{9} + \sqrt[3]{21} + \sqrt[3]{49}}$

54. $\dfrac{\sqrt{u}}{\sqrt{u} + 2}$

55. $\dfrac{\sqrt{a} - \sqrt{b}}{\sqrt{a} + \sqrt{b}}$ **56.** $\dfrac{y^2(x - \sqrt{x^2 - y^2})}{x + \sqrt{x^2 - y^2}}$

57. $\dfrac{\sqrt{r + 1} + \sqrt{r - 1}}{\sqrt{r + 1} - \sqrt{r - 1}}$

58. $\dfrac{\sqrt{uv}}{\sqrt{u} + \sqrt{v} - \sqrt{u + v}}$

59. $\dfrac{(v^2 + w^2)\sqrt[3]{vw}}{(\sqrt[3]{v^4} - \sqrt[3]{v^2 w^2} + \sqrt[3]{w^4})}$

60. $\dfrac{1}{p^{2/3} + \sqrt[3]{pq} + q^{2/3}}$

61. $\dfrac{\sqrt{2\sqrt{2}y^3} + \sqrt{8y^5}}{\sqrt[4]{2y^2}}$

62. $\dfrac{x(1 - y^2)}{\sqrt[4]{x^2} - \sqrt[4]{x^2y^2}}$

63. Jane's Daughter's Wax Company finds that its manufacturing costs for N quarts of wax are $\dfrac{(2N^2 + 7N + 6)}{(N + 2)}$ cents. The Environmental Protection Agency (EPA) and OSHA then require additional manufacturing safeguards and procedures that cost $\dfrac{(N + 2)}{3}$ cents for N quarts. Find an expression for the total cost of manufacturing N quarts and simplify this expression.

64. The yield of a certain plot of ground is generally $\dfrac{8}{(x + 4)}$ hundred bushels, where x is the number of rainless days in the growing season. An invasion of feasting beetles reduces the yield by $\dfrac{(2x + 3)}{[(x + 3)(x + 4)]}$ hundred bushels. What is the resulting yield?

65. The force exerted by two objects on one another is given by $6.67(10^{-11})m_1m_2/r^2$, where m_1, m_2 are the respective masses of the objects in kilograms, and r is the distance between them in meters. Let three masses of 10 kilograms each be located as illustrated. Calculate the net force acting on each of the three masses.

$\overset{10 \text{ meters}}{\underset{m_1 \qquad m_2}{\rule{0pt}{0pt}}} \qquad \overset{20 \text{ meters}}{\underset{\qquad\qquad m_3}{\rule{0pt}{0pt}}}$

10 meters   20 meters
m_1 m_2 m_3

66. In working with cameras and projectors, the distance of the image from the lens is given by the expression

$$\dfrac{1}{\dfrac{1}{x} - \dfrac{1}{y}}$$

where x is the focal length of the lens and y is the distance of the object or slide from the lens.

a. Simplify this expression.

b. Amy wishes to snap a picture of her current boyfriend standing 3 meters away. The lens on her camera has a focal length of 50 millimeters. How far must the lens be positioned from the film in order to focus the picture?

c. Later Amy wishes to view the slide she took. She uses a projector with a fixed lens having a 10-centimeter focal length. The slides are positioned 10.2 centimeters behind the lens. Where should she place the screen?

67. In the network of blood vessels shown in the figure, let x, y, and z denote the various resistances.

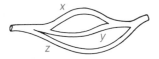

a. Describe the net resistance as a rational expression. (See Example 3.)

b. Find the net resistance if $x = 10$, $y = 10$, and $z = 20$.

68. The manufacturer of Green-Grass Pellets was recently found guilty of false advertising and required by the Federal Trade Commission to spend $20 million on commercial messages disclaiming earlier advertising. The manufacturer's advertising agency claims that the number of products sold is determined by

$$\dfrac{\sqrt{x} + \sqrt{y} + \sqrt{z}}{\sqrt{u} + \sqrt{v}}$$

where x = amount of money spent on television commercials, y = amount of money spent on radio commercials, z = amount of money spent on advertising in print, u = amount of money spent by the government contradicting the manufacturer's claims, and v = amount of money the manufacturer spends on disclaimer commercials. Determine the value of this expression if x = $50 million, y = $32 million, z = $18 million, u = $20 million, and v = $20 million.

1.7 COMPLEX NUMBERS

In the first section of this chapter we reviewed the operations on the real number system. At an early age, we became familiar with the "natural" or counting numbers and the fractions or rational numbers. The real numbers are more extensive than these, encompassing irrational numbers such as π and e as well. Every positive real number has two real square roots, but no negative real number has a real square root. This happens because

$$\sqrt{a} = b \qquad \text{means} \qquad a = b^2 > 0$$

for every real number b. We shall now extend the real numbers to complex numbers in order to generate square roots of negative numbers. Although the complex numbers may seem to be purely artificial mathematical constructions, they are indeed very useful in describing the behavior of certain electronic devices.

Our purpose is to extend the real numbers to a larger number system in which square roots of negative numbers exist. If we agree to write

$$\sqrt{-a} = \sqrt{-1}\,\sqrt{a}$$

for every $a > 0$, we see that by introducing a symbol i for $\sqrt{-1}$, we shall effectively introduce square roots for all negative numbers. Thus, for example,

$$\sqrt{-1} = i$$
$$\sqrt{-49} = \sqrt{-1}\,\sqrt{49} = 7i$$
$$\sqrt{-25} = \sqrt{-1}\,\sqrt{25} = 5i$$
$$\sqrt{-2} = \sqrt{-1}\,\sqrt{2} = \sqrt{2}i.$$

Just as with positive real numbers, every negative real number $-x$ has two square roots given by $\pm\sqrt{-x}$ or $\pm i\sqrt{x}$. The notation $\sqrt{-x}$ denotes the **principal root** $i\sqrt{x}$.

In the real number system we must deal with numbers of the form $1 + \sqrt{2}$ and $7 - 3\sqrt{2}$; similarly, introduction of the number or symbol i means that we must consider numbers of the form $1 + i$ and $7 - 3i$. If we consider

HISTORICAL PERSPECTIVE

Just as the Pythagoreans had difficulty accepting numbers like $\sqrt{2}$, which were not rational, mathematicians of the seventeenth century (and many students today) found it difficult to accept numbers of the form $\sqrt{-4}$, which could not be real numbers. Descartes (1596–1650) called these numbers "imaginary," thereby giving them a permanent stigma. However, they are no more imaginary than any other abstraction such as the concept of "one" or "two."

two such numbers $a + bi$, $c + di$ and add or multiply these according to the usual rules of algebra, letting $i^2 = -1$ since $\sqrt{-1} = i$, we obtain the following.

$$(a + bi) + (c + di) = a + c + bi + di$$
$$= (a + c) + (b + d)i$$
$$(a + bi)(c + di) = ac + bdi^2 + bci + adi$$
$$= (ac - bd) + (bc + ad)i$$

We are now in a position to introduce the system of complex numbers.

DEFINITION

The system of **complex numbers** consists of numbers of the form

$$a + bi \qquad a, b \text{ real numbers}$$

with the operations of addition and multiplication defined as follows:

$$(a + bi) + (c + di) = (a + c) + (b + d)i$$
$$(a + bi) \cdot (c + di) = (ac - bd) + (ad + bc)i$$

The number a is called the **real part** and b the **imaginary part** of the complex number $a + bi$.

It can be shown that the commutative, associative, and distributive laws hold for complex numbers and that additive and multiplicative inverses exist. Subtraction and division are defined in the usual manner:

$$a - b = a + (-b)$$
$$a \div b = a \cdot b^{-1}$$

where $-b$, b^{-1} are the additive and multiplicative inverses of b, respectively.

Furthermore, it can be verified from the preceding rules that $(0 + 1i)^2 = -1$; that is, $i^2 = -1$ (see Exercise 31). It also follows that $(c + 0i)(a + bi) = ca + cbi$ (see Exercise 32); that is,

$$c(a + bi) = ca + cbi$$

Thus,

$$3(6 - 7i) = 18 - 21i$$

and

$$2(4 + 3i) = 8 + 6i$$

> The number i satisfies
>
> $$i^2 = -1$$
>
> To work with complex numbers $(a + bi)$, we apply the usual algebraic rules of combination, replacing i^2 with -1 whenever it appears.

It then follows that

$$i^3 = i^2i \qquad \text{and} \qquad i^4 = (i^2)^2$$
$$= -i \qquad\qquad\qquad\quad = (-1)^2$$
$$= 1$$

Caution: *The familiar rule $\sqrt{ab} = \sqrt{a}\,\sqrt{b}$ for positive real numbers does not necessarily hold for negative real numbers a and b.* For instance

$$\sqrt{-4}\,\sqrt{-9} = 2i \cdot 3i = -6$$

but

$$\sqrt{(-4)(-9)} = \sqrt{36} = 6$$

EXAMPLE 1 Simplify each of the following into the form $a + bi$

   a. $(5 + 3i) + (2 + 4i)$

   b. $(5 + 3i) - (2 + 4i)$

   c. $(5 + 3i)(2 + 4i)$

   d. $(2 + 4i)^2$

   e. $2(5 + 3i) - 3(2 - 4i)$

   f. $2i(5 + 3i) + 3i^3(2 - 4i)$

   g. $i^{27} - i^{36}$

Solution **a.** $(5 + 3i) + (2 + 4i) = 7 + 7i$

   b. $(5 + 3i) - (2 + 4i) = 5 + 3i - 2 - 4i$
$$= 3 - i$$

   c. $(5 + 3i)(2 + 4i) = (10 - 12) + (20 + 6)i$
$$= -2 + 26i$$

   d. $(2 + 4i)(2 + 4i) = (4 - 16) + (8 + 8)i$
$$= -12 + 16i$$

   e. $2(5 + 3i) - 3(2 - 4i) = 10 + 6i - 6 + 12i$
$$= 4 + 18i$$

f. $2i(5 + 3i) + 3i^3(2 - 4i) = 10i + 6i^2 + 6i^3 - 12i^4$
$$= 10i - 6 + 6i(i^2) - 12 \cdot 1$$
$$= 10i - 6 - 6i - 12$$
$$= -18 + 4i$$

g. $i^{27} - i^{36} = i^{24}i^3 - (i^4)^9$
$$= (i^4)^6i^3 - 1$$
$$= i^3 - 1$$
$$= (i^2)i - 1$$
$$= -1 - i$$

DEFINITION

Two complex numbers are **equal** if their real and imaginary parts are equal; that is,

$$(a + bi) = (c + di) \quad \text{means} \quad a = c \quad \text{and} \quad b = d$$

Thus,

$$a + bi = 7 + 3i \quad \text{means} \quad a = 7, \quad b = 3$$

Division

Division of complex numbers is accomplished in the same way that fractions with denominators of the form $a + b\sqrt{c}$ were simplified. Since the symbol $\sqrt{-1}$ now makes sense, we can simplify an expression of the form

$$\frac{2 + \sqrt{-1}}{3 + 4\sqrt{-1}}$$

by "rationalizing the denominator." The numerator and denominator must be multiplied by the **conjugate** of the denominator: $3 - 4\sqrt{-1}$.

$$\frac{2 + \sqrt{-1}}{3 + 4\sqrt{-1}} \cdot \frac{3 - 4\sqrt{-1}}{3 - 4\sqrt{-1}}$$

$$= \frac{6 - 5\sqrt{-1} - 4(-1)}{9 - 16(-1)} = \frac{10 - 5\sqrt{-1}}{25}$$

$$= \frac{2}{5} - \frac{1}{5}i$$

DEFINITION

The **conjugate** of a complex number $z = a + bi$ is the complex number $\bar{z} = a - bi$.

For example,

$$\overline{1 + 3i} = 1 - 3i$$
$$\overline{1 - 3i} = 1 - (-3)i = 1 + 3i$$

Note that the product of a complex number and its conjugate is always real and nonnegative. For

$$(a + bi)\overline{(a + bi)} = (a + bi)(a - bi)$$
$$= a^2 - (bi)^2$$
$$= a^2 - b^2 i^2$$
$$= a^2 + b^2 \geq 0$$

since a and b are both real numbers. Thus,

$$z\bar{z} \geq 0 \qquad \text{for all complex numbers } z.$$

It is this property that makes division by complex numbers possible.

To divide one complex number by another, write the indicated division as a "fraction" and multiply both numerator and denominator by the conjugate of the denominator. The result is a new fraction with a positive real denominator. Then divide both parts of the numerator by the real denominator.

EXAMPLE 2 Simplify the following into the form $a + bi$.

a. $\dfrac{2}{3i}$

b. $\dfrac{1 + \sqrt{2}i}{5 - \sqrt{2}i}$

c. $\dfrac{2 - i}{(1 + i)(3 + i)}$

Solution a. $\dfrac{2}{3i} = \dfrac{2}{3i} \cdot \dfrac{(-3i)}{(-3i)} = \dfrac{-6i}{9} = \dfrac{-2}{3}i$

b. $\dfrac{1 + \sqrt{2}i}{5 - \sqrt{2}i} = \dfrac{1 + \sqrt{2}i}{5 - \sqrt{2}i} \cdot \dfrac{5 + \sqrt{2}i}{5 + \sqrt{2}i}$

$$= \dfrac{5 + \sqrt{2}i + 5\sqrt{2}i - 2}{25 + 2} = \dfrac{3 + 6\sqrt{2}i}{27}$$

$$= \dfrac{1}{9} + \dfrac{2\sqrt{2}}{9}i$$

$$\textbf{c.} \quad \frac{2 - i}{(1 + i)(3 + i)} = \frac{2 - i}{3 - 1 + 4i} = \frac{2 - i}{2 + 4i}$$

$$= \frac{2 - i}{2 + 4i} \cdot \frac{2 - 4i}{2 - 4i}$$

$$= \frac{4 + 4 - 10i}{4 + 16} = \frac{8 - 10i}{20}$$

$$= \frac{2}{5} - \frac{1}{2}i$$

EXERCISES 1.7

Simplify each of the following into the form $a + bi$.

1. $(2 + i) + (1 - 3i)$
2. $(5 + 12i) + (4 + 3i)$
3. $(3 + i) + (2 - 5i)$
4. $(2 + 3i) + (4 - 5i)$
5. $(2 + 3i) - (-2 - 4i)$
6. $(1 + 3i) - (-2 + 5i)$
7. $(3 - 2i) - (5 - 4i)$
8. $2(2 + 5i) + 3i(3 - 2i)$
9. $i^2(10 + i) - 3(2 - 4i)$
10. $i(4 + 9i) - 2i^2(1 + 3i)$
11. $2i^3(3 + 2i) - 2i(3 + 2i)$
12. $(1 - 2i)^2 i^3$
13. $3i(1 + i) - 2i^3(1 - i)$
14. $(1 - i)(2 + i)$ 15. $(1 - 4i)(3 - i)$
16. $(2 - i)(1 - 3i)$ 17. $(3 - 4i)(3 + 4i)$
18. $(1 + 2i)(1 + i)i^2$
19. $(2 + i)^2$
20. $(1 - \sqrt{3}i)^2$ 21. $(1 + 2i)^3$
22. $(1 + 2i)^2(1 - i)^3$
23. $(2 - i)^{30}(i^2 + 1)$
24. i^{39} 25. i^{632}
26. $i^{57} - i^{26}$ 27. $i^{17} + i^{12} - i^3$
28. $3i^{23} - 2i^{42} + 3i^3$
29. $-2i^{63} - (2i)^3 + 2i^3$
30. $4i^{27} + 3i^7 - (5i)^2$
31. Show that when $(0 + i)^2$ is expanded according to the rules given on page 47, we obtain $(0 + i)^2 = -1 + 0i$; that is, $i^2 = -1$.
32. Show that when the real number c is written as a complex number $c + 0i$, we have $(c + 0i)(a + bi) = ca + cbi$; that is, $c(a + bi) = ca + cbi$.

Find $\bar{z}$ for each of the following.

33. $2 - 5i$ 34. $3 + 4i$
35. $1 + \sqrt{2}i$ 36. $-3 - 2i$

37. $-5 - 3i$ 38. $4 - 3i$
39. $-2 + 6i$ 40. $-6 - 3i$
41. $\sqrt{-4} + \sqrt{-9}i$ 42. $\sqrt{-25} - \sqrt{16}i^3$

Write each of the following in the form $a + bi$.

43. $\dfrac{-2}{1 + 3i}$ 44. $\dfrac{3i}{1 + i}$

45. $\dfrac{-3i^3}{2 - i}$ 46. $\dfrac{2 - i}{1 - 2i}$

47. $\dfrac{-2 + \sqrt{3}i}{2 - \sqrt{3}i}$ 48. $\dfrac{3 - 4i}{1 + \sqrt{2}i}$

49. $\dfrac{2 - 3i}{3 + 2i}$ 50. $\dfrac{3 + 2i}{2 - 3i}$

51. $\dfrac{1 + i}{1 - i}$ 52. $\dfrac{4 - 3i}{3 + 4i}$

53. $\dfrac{5 + 3i}{2 - i}$ 54. $\dfrac{1 + i}{(2 + i)^2}$

55. $\dfrac{1 + i}{(3 + i)(2 - i)}$ 56. $\dfrac{3 + i}{2 - i} + \dfrac{2 + i}{1 - i}$

57. $\dfrac{-1 + 3i}{1 - 2i} + \dfrac{i - 1}{i + 1}$

Verify the following statements.

58. $\overline{z_1 + z_2} = \overline{z_1} + \overline{z_2}$
59. $\overline{z_1 z_2} = \overline{z_1}\,\overline{z_2}$
60. $\bar{z} = z$ if and only if z is a real number.
61. $\overline{\bar{z}} = z$

62. $\overline{\left(\dfrac{z_1}{z_2}\right)} = \dfrac{\overline{z_1}}{\overline{z_2}}$

63. $z + \bar{z}$ is always a real number.
64. $z + \bar{z} = 0$ if and only if the real part of z is 0.
65. $\overline{z_1 - z_2} = \overline{z_1} - \overline{z_2}$
66. $\overline{az} = a\bar{z}$ for a a real number.

1.8 CHAPTER REVIEW

TERMS AND CONCEPTS

- Natural Numbers — 1, 2, 3, . . .
  1. Composite number — $6 = 2 \cdot 3$
  2. Prime number — 2, 3, 5, 7, 11, 13, . . .
  3. Factor — 1, 2, 3, 6 are factors of 6
  4. LCM — LCM $(4,6) = 12$
  5. GCD — GCD $(4,6) = 2$
- Integers — . . . , $-2, -1, 0, 1, 2, . . .$
- Rational Number — $\dfrac{m}{n}, n \neq 0$
- Irrational Number — $\sqrt{2}, \pi$
- Real Number — Includes all of the above
- Complex Number — $a + bi, i^2 = -1$
- Decimal Representation and Percent — $67\% = \dfrac{67}{100} = .67$

RULES AND FORMULAS

- Negative Numbers
  1. $a + (-b) = a - b$ $\$10,000 + (-\$2000) = \$8000 = \$10,000 - \$2000$
  2. $a - (-b) = a + b$ $5280 - (-282) = 5280 + 282 = 5562$
  3. $-(a + b) = -a - b$ $-(2 + 3) = -5 = -2 - 3$
  4. $a(-b) = -(ab) = (-a)b$ $2(-3) = -(2 \cdot 3) = (-2)(3) = -6$
  5. $-(-a) = a$ $-(-2) = 2$
  6. $(-a)(-b) = ab$ $(-2)(-7) = 14 = 2 \cdot 7$
- Rational Numbers
  1. $\dfrac{k \cdot m}{k \cdot n} = \dfrac{m}{n}$ $\dfrac{3}{6} = \dfrac{3 \cdot 1}{3 \cdot 2} = \dfrac{1}{2}$
  2. $\dfrac{a}{b} + \dfrac{c}{d} = \dfrac{ad + bc}{bd}$ $\dfrac{1}{2} + \dfrac{4}{3} = \dfrac{1 \cdot 3 + 2 \cdot 4}{2 \cdot 3} = \dfrac{11}{6}$
  3. $\dfrac{a}{b} - \dfrac{c}{d} = \dfrac{ad - bc}{bd}$ $\dfrac{4}{3} - \dfrac{1}{2} = \dfrac{4 \cdot 2 - 3 \cdot 1}{3 \cdot 2} = \dfrac{5}{6}$
  4. $-\dfrac{a}{b} = \dfrac{-a}{b} = \dfrac{a}{-b}$ $-\left(\dfrac{2}{3}\right) = \dfrac{-2}{3} = \dfrac{2}{-3}$
  5. $\dfrac{a}{b} \cdot \dfrac{c}{d} = \dfrac{a \cdot c}{b \cdot d}$ $\dfrac{1}{5} \cdot \dfrac{2}{3} = \dfrac{2}{15}$
  6. $\dfrac{\dfrac{a}{b}}{\dfrac{c}{d}} = \dfrac{a}{b} \cdot \dfrac{d}{c}$ $\dfrac{\dfrac{2}{3}}{\dfrac{5}{7}} = \dfrac{2}{3} \cdot \dfrac{7}{5}$
- Absolute Value and Order
  1. $|x| = \begin{cases} x & \text{if } x \geq 0 \\ -x & \text{if } x < 0 \end{cases}$ $|-7| = 7 = -(-7)$
  2. $\begin{cases} |x| < a \text{ if and only if } -a < x < a \\ |x| > a \text{ if and only if } x > a \text{ or } x < -a \end{cases}$

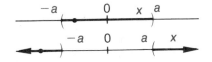

3. $|a + b| \leq |a| + |b|$

$$\begin{cases} |5 + 3| = 8 = |5| + |3| \\ |5 + (-3)| = 2 < |5| + |-3| \end{cases}$$

• Exponents and Radicals

1. $x^a x^b = x^{a+b}$

$2^2 2^3 = 2^5$ since $4 \cdot 8 = 32$

2. $(x^a)^b = x^{ab}$

$(2^3)^2 = 2^6$ since $8^2 = 64 = 2^6$

3. $(xy)^a = x^a y^a$

$(2 \cdot 3)^2 = 2^2 3^2$ since $36 = 4 \cdot 9$

4. $x^0 = 1$

$10^0 = 1$

5. $x^{-a} = \dfrac{1}{x^a}$

$2^{-2} = \dfrac{1}{2^2}$

6. $x^{1/n} = \sqrt[n]{x}$

$4^{1/2} = \sqrt{4} = 2$

7. $\sqrt[n]{x^n} = \begin{cases} |x| \text{ if } n \text{ is even} \\ x \text{ if } n \text{ is odd.} \end{cases}$

$\begin{cases} \sqrt[4]{(-2)^4} = \sqrt[4]{16} = 2 = |-2| \\ \sqrt[3]{(-2)^3} = \sqrt[3]{-8} = -2 \end{cases}$

8. $\sqrt[n]{a} \cdot \sqrt[n]{b} = \sqrt[n]{ab}$

$\sqrt{4}\,\sqrt{9} = 2 \cdot 3 = 6 = \sqrt{36} = \sqrt{4 \cdot 9}$

9. $\dfrac{\sqrt[n]{a}}{\sqrt[n]{b}} = \sqrt[n]{\dfrac{a}{b}}$

$\dfrac{\sqrt{100}}{\sqrt{4}} = \dfrac{10}{2} = 5 = \sqrt{25} = \sqrt{\dfrac{100}{4}}$

10. $\sqrt[k]{\sqrt[n]{x}} = \sqrt[kn]{x}$

$\sqrt[3]{\sqrt[2]{64}} = \sqrt[3]{8} = 2 = \sqrt[6]{64}$

11. Only like terms can be collected.

$3\sqrt{2} + 5\sqrt{2} = 8\sqrt{2}$ but
$\sqrt{2} + \sqrt{3} \neq \sqrt{5}$

• Common Expansions

1. $(a \pm b)^2 = a^2 \pm 2ab + b^2$

$(x - 1)^2 = x^2 - 2x + 1$

2. $a^2 - b^2 = (a + b)(a - b)$

$x^2 - 1 = (x + 1)(x - 1)$

3. $(a \pm b)^3 = a^3 \pm 3a^2 b + 3ab^2 \pm b^3$

$(x - 1)^3 = x^3 - 3x^2 + 3x - 1$

4. $a^3 + b^3 = (a + b)(a^2 - ab + b^2)$

$x^3 + 1 = (x + 1)(x^2 - x + 1)$

5. $a^3 - b^3 = (a - b)(a^2 + ab + b^2)$

$x^3 - 1 = (x - 1)(x^2 + x + 1)$

• Hierarchy of Operations

1. Expressions in parentheses are to be treated as a single number, which should be evaluated before being combined with numbers outside the parentheses.

$(8 + 2) \div 2 + 6 \div 3 \cdot 2$
$= 10 \div 2 + 6 \div 3 \cdot 2$

2. Multiplications and divisions are performed before additions and subtractions and in the order in which they appear from left to right.

$= 5 + 2 \cdot 2$

3. Then additions and subtractions are done in the order in which they appear from left to right.

$= 5 + 4$
$= 9$

TECHNIQUES

• Factoring

1. Common expansions in reverse

$x^2 + 6xy + 9y^2 = (x + 3y)^2$

2. Completing the square

$x^2 + 6x - 7 = (x^2 + 6x + 9) - 16$
$\qquad\qquad = (x + 3)^2 - 4^2$
$\qquad\qquad = (x + 3 + 4)(x + 3 - 4)$
$\qquad\qquad = (x + 7)(x - 1)$

3. Scissors factoring

$2x^2 + x - 3 = (2x + 3)(x - 1)$

    4. Grouping and taking out common factors

$$2ax + 3bx - 2ay - 3by$$
$$= (2a + 3b)x - (2a + 3b)y$$
$$= (2a + 3b)(x - y)$$

- Long Division of Polynomials
  1. Always write the polynomials with terms in order of *decreasing powers*. Then at each stage, only the first terms of the expressions need be examined.
  2. Leave room for any missing powers of x since they may appear in the intermediate stages of the long division.
  3. Just as in long division of numbers, a remainder is sometimes left. The remainder will be a polynomial whose degree is less than that of the divisor.

$$
\begin{array}{r}
x - 1 \\
x + 1 \overline{)x^2 \qquad + 1} \\
\underline{x^2 + x} \\
-x \\
\underline{-x - 1} \\
2 = R
\end{array}
$$

- Combining Rational Expressions
  1. To add or subtract, first find the LCD, then write each in terms of this LCD, and finally add or subtract numerators.

$$\frac{3}{x} - \frac{2}{x + 1} = \frac{3(x + 1)}{x(x + 1)} - \frac{2x}{x(x + 1)} = \frac{x + 3}{x(x + 1)}$$

  2. To multiply, multiply numerators and multiply denominators separately.

$$\frac{3}{x} \cdot \frac{2}{x + 1} = \frac{6}{x(x + 1)}$$

  3. To divide, invert and multiply.

$$\frac{3}{x} \div \frac{6}{x} = \frac{3}{x} \cdot \frac{x}{6} = \frac{3x}{6x}$$

  4. Simplify the results by cancelling like factors in the numerator and denominator.

$$\frac{3x}{6x} = \frac{3x}{2 \cdot 3x} = \frac{1}{2}$$

- Division by a Complex Number or by Radicals, Rationalizing
  1. Complex number

$$\frac{1}{a + bi} = \frac{1}{a + bi} \cdot \frac{a - bi}{a - bi} = \frac{a - bi}{a^2 + b^2}$$

  2. Radicals

$$\frac{1}{\sqrt{3} - \sqrt{2}} = \frac{1}{\sqrt{3} - \sqrt{2}} \cdot \frac{\sqrt{3} + \sqrt{2}}{\sqrt{3} + \sqrt{2}}$$
$$= \frac{\sqrt{3} + \sqrt{2}}{3 - 2} = \sqrt{3} + \sqrt{2}$$

1.9 SUPPLEMENTARY EXERCISES

Factor the following into primes.

    1. 76   **2.** 341   **3.** 5280

Find the LCM and GCF of the following sets of numbers.

    4. 68, 76   **5.** 210, 1425   **6.** 18, 24, 40

Reduce each of the following to lowest terms.

    7. $\dfrac{21}{66}$ **8.** $\dfrac{96}{108}$ **9.** $\dfrac{-360}{144}$

Express each of the following as a single integer.

    10. $2 + (-5)$ **11.** $(-3)(5)$

    12. $8 - 2 \cdot 4 + (-10) - 7(6 - 8)$

    13. $7 - 2[8 + 2 - (-9)] + 6$

    14. $\left(\dfrac{1}{2}\right)[4 + 8(6 - 22)(-1)]$

    15. $-2 + 3 \cdot 4 - 3 \div 6 - 8$

    16. $-2 + 3 \cdot (4 - 3) \div (6 - 8)$

17. $(-2 + 3) \cdot 4 - (3 \div 6 - 8)$

Perform the indicated operations and express the answer in lowest terms.

18. $\dfrac{1}{4} + \dfrac{1}{3}$

19. $\dfrac{2}{3} - \dfrac{1}{5} + 3\dfrac{1}{2}$

20. $\dfrac{1}{2} + \dfrac{1}{3} - \dfrac{1}{4}$

21. $\dfrac{25}{28} \cdot \dfrac{14}{15} \cdot \dfrac{36}{10}$

22. $\dfrac{7}{9} \div \dfrac{-4}{15}$ **23.** $2\dfrac{1}{3} \div 4\dfrac{1}{2}$

24. $\left(\dfrac{1}{2} - \dfrac{2}{3}\right) \div \left(\dfrac{1}{3} - \dfrac{3}{4}\right)$

25. $\dfrac{7 + \dfrac{1}{6}}{\dfrac{2}{3} + \dfrac{5}{14}}$

Locate each of the following points on the number line.

26. 1.25   **27.** -3.63 **28.** $2.333\ldots$

Sketch the following intervals on the real number line.

29. $(-3, 6]$ **30.** $[2, 5]$ **31.** $(1, \infty)$

Find a repeating decimal expansion for each of the given rational numbers.

32. $\dfrac{5}{7}$ **33.** $\dfrac{8}{9}$ **34.** $\dfrac{37}{13}$

Write each of the following in the form m/n.

35. $1.23434\ldots$ **36.** $2.506506\ldots$

37. $15.20652065\ldots$

38. Round off each of the numbers in Exercises 32–37 to the nearest thousandth.

39. Determine the missing numbers.

   a. 1000% of 2 is _____

   b. $\dfrac{1}{4}$% of _____ is 5.

   c. _____% of 2 is 6

Express an order relation ($<$, $=$, $>$) between the following numbers. A bar denotes a repeating block of digits.

40. $-4, -10$ **41.** $\dfrac{2}{3}, \dfrac{3}{4}, \dfrac{5}{7}, \dfrac{7}{11}$

42. $0.\overline{204},\ 0.2\overline{04}$

43. $13.24\overline{5},\ 13.2\overline{45},\ 13.\overline{245}$

44. $\dfrac{2}{3}, \dfrac{262}{393}$

Insert a rational number between the following.

45. 13.87 and 13.8

46. $\dfrac{5}{7}$ and $\dfrac{7}{11}$

Simplify the given expressions. Write the answers in terms of radicals or positive exponents if necessary. Rationalize any denominators in your answers.

47. $(2^3)^2$ **48.** $\dfrac{(-3)^2}{(-3)^3}$

49. $\left(\dfrac{1}{16}\right)^{1/4}$ **50.** $\left(\dfrac{1}{32}\right)^{-3/5}$

51. $\sqrt{16a^2}$ **52.** $(2^{-2})^3$

53. $\dfrac{10^{20}}{10^{-4}}$ **54.** $(81)^{3/4}$

55. $(-8)^{2/3}$ **56.** $(25)^{-1/2}$

57. $\sqrt{144 + 25}$

58. $\sqrt[7]{0.0000128} \cdot \sqrt[5]{0.00243}$

59. $\dfrac{\sqrt{6} + \sqrt{3}}{\sqrt{6} - \sqrt{3}}$

60. $\sqrt[3]{54} + 3\sqrt[3]{24} + 2\sqrt[3]{250} - 5\sqrt[3]{-81}$

61. $(\sqrt[3]{9} - \sqrt[3]{6} + \sqrt[3]{4})(\sqrt[3]{2} + \sqrt[3]{3})$

62. $\sqrt[4]{64x^2y^6}$ **63.** $(t^3)^2$

64. $\dfrac{x^{10}}{x^4}$ **65.** $s^4 s^{-16}$

66. $(3x^{-2}y)(2y^{-3})$ **67.** $(2x^2)^3$

68. $(ab^2c^3)^4$ **69.** $[(x^2y)^3(yx^3)^2]^4$

70. $\dfrac{(x^2y^2)^3}{(x^3y)^2}$ **71.** $\dfrac{x^{-10}}{x^{-3}}$

72. $\dfrac{xy^{-3}}{x^{-2}y}$

73. $(2x^{-1}y^2z^{-1})(6x^3y^{-1}z^2)$

74. $(ab^2c^{-3})^{-4}$

75. $(uv^2w)^{-3}(u^{-2}vw^2)^{-2}$

76. $\left[\dfrac{(a^{-2}b^3)^{-4}}{(ab^{-5})^3}\right]^2$

77. $\dfrac{(x - y)^{-2}}{x^{-2} - y^{-2}}$ **78.** $\sqrt[3]{125x^6y^{12}}$

79. $\sqrt{8a^{10}b^2} \cdot \sqrt{2a^4b^6}$

80. $(a^{27}b^9z^3)^{2/3}$

81. $\dfrac{(a^{-3/2}y^{-1/4})^{-2}}{(y^{-2/3}z^{-5/6})^{-3}}$

82. $\left(\dfrac{25a^{-4}b^{-6}c^2}{16x^{-2}y^6z^{-2}}\right)^{-3/2} \cdot \left(\dfrac{2a^2b^{-1}c^{-3}}{x^2y^{-2}z^3}\right)^{-1}$

83. $(x^{3/4}y^{-3/8}z^{9/24})^{4/3}$

84. $\dfrac{x^2 - y^2}{\sqrt{x} + \sqrt{y}}$

Verify the following

85. $2^3 + 4^3 \neq 6^3$

86. $1^{-3} + 2^{-3} \neq 3^{-3}$

87. $2^{-2} - 3^{-2} \neq (-1)^{-2}$

88. $(x + y)^{-2} \neq x^{-2} + y^{-2}$

Express each of the following in decimal notation.

89. $6.5(10^9)$ **90.** $4.2(10^{-6})$

91. $6.63(10^{-34})$

Express each of the following in scientific notation.

92. $16,700,000$ **93.** 0.2372

94. 0.0000000019 **95.** 26.0014

Perform the following calculations by first converting the numbers to scientific notation. Express your answers in scientific notation and in terms of decimals.

96. $\dfrac{(0.000004)(1,200,000)}{0.0008}$

97. $\dfrac{(26,200,000,000)(54,300)}{(0.000357)}$

Perform the indicated operations, collecting like terms, if any.

98. $(r^2 + 3r - 2) + (3r^2 - 2r + 1)$

99. $(3r - 5)(2r + 7)$

100. $(2w + 3)(2w - 3)(4w^2 + 9)$

101. $\dfrac{3x^2}{x^3 - 8} - \dfrac{1}{x - 2} - \dfrac{3x}{x^2 + 2x + 4}$

102. $\dfrac{w^2 - 16}{w^2 - 9} \cdot \dfrac{w^2 + 5w + 6}{w^2 + 6w + 8}$

103. $\dfrac{y^2 + 14y - 15}{y^2 + 4y - 5} \div \dfrac{y^2 + 12y - 45}{y^2 + 6y - 27}$

104. $x - \dfrac{x^2}{x - \dfrac{x}{y - 2}}$

Factor the following as far as possible.

105. $16 + 12a$ **106.** $x^2 - 5x + 6$

107. $x^2 - x - 20$ **108.** $x^2 - 10x + 21$

109. $81 - a^4$ **110.** $25a^2 - 36b^2$

111. $x^8 - 4$ **112.** $64x^6 - y^6$

113. $4x^2 + 12xy + 9y^2$

114. $x^2 - 18xy + 81y^2$

115. $625 + 25z^2 + z^4$

116. $uv + 2v - 3u - 6$

117. $8 - 36y + 54y^2 - 27y^3$

118. $x^3 + 6x^2y + 12xy + 8y^3$

119. $4x^2y^2 - 4x^2 - y^4 + y^2$

Use long division to find:

120. $\dfrac{u^8 - 1}{u - 1}$ **121.** $\dfrac{4r^3 + 7r + 5}{1 + 2r}$

Simplify each of the following into the form $a + bi$.

122. $(3 + 2i) + (2 - 4i)$

123. $(1 + 2i) - (3 - i)$

124. $(4 + i)(2 - 4i)$

125. $3(2 - i) + 4i(2 + i)$

126. $(3 + i)^3$ **127.** i^{57}

128. $i^{54} + i^{27}$ **129.** $2i^{58} + 3i^{62} - 5i^2$

130. $\dfrac{3}{2 + i}$ **131.** $\dfrac{5 - 2i}{1 + 3i}$

132. $\dfrac{-1 - i}{-1 + i}$ **133.** $\dfrac{1 + i}{2 + i} + \dfrac{1 - i}{3 - i}$

Find $\bar{z}$ for each of the following.

134. $7 - 3i$ **135.** $6 + 5i$

136. $\sqrt{-2} + \sqrt{-3}i$ **137.** $\sqrt{-16} + \sqrt{25}i^3$

138. The Soft-Eze Tissue Company distributes its paper to various retailers. Some vendors run specials of six rolls for $1.00. Others need more money to cover overhead expenses and prefer to sell four for $1.00. What is the maximum number of rolls the company should place in a package so that neither vendor need sell broken packages?

139. The highest U.S. city is Leadville, Colorado, with an elevation of 10,200 feet. The lowest U.S. town is Calipatria, California, which lies 183 feet below sea level. How much higher is Leadville than Calipatria?

140. An arena with an advertised seating capacity of 13,257 is filled except for three rows of 63 seats each. How many people are in attendance?

141. A dress pattern requires $2\frac{2}{3}$ yards of cotton and $1\frac{1}{8}$ yards of velvet. If cotton costs $5 per yard and velvet costs $12 per yard, what is the cost of the material needed to make the dress?

142. If $x = 9.73$ to within 0.01 and $y = 4.52$ to within 0.002, to what accuracy can $x + y$ and xy be determined?

143. Proxima Centauri is the star nearest to us other than our sun. When astronomers view Proxima Centauri by telescope, they see it as it was 4 years, 4 months ago. How far is Proxima Centauri from earth? (See Exercise 88, Sect. 1.3.)

144. A certain microorganism causes the human little finger to twitch. The number of people it has infected quintuples every week. If one person initially becomes infected, how many will have twitchy little fingers in 25 days? In 365 days?

145. If the half-life of radium is 1600 years and at a given time there are 100 units present, how

1.9 SUPPLEMENTARY EXERCISES • 57

many units are present 6400 years later? How many units were present 1600 years earlier?

146. If a ball is dropped from a height of 500 feet and returns to 80 percent of its original height after each bounce, how high does it bounce after hitting the pavement for the third time?

147. The "doubling time" of a bacteria culture is 2 hours. If the culture starts with 500 bacteria, how many will there be 6 hours later?

148. If the doubling time of a bacteria culture is

known to be 3 hours and there are 1000 bacteria present now, how many were there 6 hours ago?

149. The enrollment in a given school district dropped from 12,000 to 8500. What was the percentage of the decrease?

150. In the school district of Exercise 149, the enrollment increased by 40 percent the following year. Did the new enrollment reach the level of enrollment before the decline?

2 Equations and Algebraic Inequalities

An **equation** is the indicated equality of two quantities or expressions such as

$$x + 5 = 0$$
$$\text{or} \quad t^2 + 3t + 7 = 10$$

The letters in an equation are simply placeholders for numbers; various numbers may be substituted in place of the letters. In Chapter 1 we called these letters **variables.**

Substituting certain numbers into an equation may yield a true statement; substituting others may not. The equation's **solution** consists of those numbers that, when assigned to the variables, make the equation true. Such numbers are said to *satisfy* the equation. Since the values that satisfy an equation are not generally known beforehand, the variables are also called **unknowns.**

We can perform various algebraic manipulations on an equation until the solution is apparent. But to maintain the equality represented by the original equation, whatever is done to one side of the equation must be done to the other. This is analogous to a balanced seesaw: if the weight on one side is increased or decreased by a given amount, the other side must be changed by the same amount in order to maintain the balance. At the end of this chapter, we will apply the techniques and skills for solving equations to inequalities.

2.1 FIRST-DEGREE EQUATIONS IN ONE VARIABLE

Certain equations, such as $3x - 2x = x$, are true for all values of x. Such equations are called **identities.** Others, such as $x + 5 = 0$, are true only for certain values of x. These are called **conditional equations.** In this chapter, we shall be concerned with conditional equations in one variable.

If both sides of an equation are polynomials, the larger degree of the two polynomials is called the degree of the equation. Thus, $x + 5 = 0$ is an equation of degree 1; $t^2 + 3t + 7 = 10$ is an equation of degree 2. The numbers (not variables) in an equation are called **constants** or **coefficients.** For example, 3 is the coefficient of t in $t^2 + 3t + 7$. The solution(s) of an equation are also called **root(s)** of that equation.

The various techniques we will discuss for solving equations are all based on the following rule.

> If the same operations are performed on equal quantities, then the results must be equal.

In other words, whatever you do to one side of an equation, you must do to the other.

SOLVING FIRST-DEGREE EQUATIONS IN ONE VARIABLE

> 1. Simplify the algebraic expressions.
>
> 2. Isolate the variable on one side of the equation by
>   a. adding or subtracting terms
>   b. multiplying or dividing by a constant.

EXAMPLE 1   Solve the following equation.

$$7y + 6 = y + 6$$

Solution

$$
\begin{aligned}
7y + 6 &= y + 6 \\
7y &= y && \text{(Subtract 6 from both sides.)} \\
6y &= 0 && \text{(Subtract } y \text{ from both sides.)} \\
y &= 0 && \text{(Divide both sides by 6.)}
\end{aligned}
$$

Thus, $y = 0$ is the only solution. The possibility of y being 0 prevented us from dividing both sides of the equation in step 2 by y. Had we attempted this division, we would have obtained the impossible statement $1 = 7$. ■

EXAMPLE 2   Solve the following equation.

$$3(x - 2) - 4(x - 3) = 8$$

Solution   We first remove parentheses and collect like terms. Then we proceed as in Example 1 to isolate x on one side of the equation.

$$3(x - 2) - 4(x - 3) = 8$$

$$3x - 6 - 4x + 12 = 8 \qquad \text{(Remove parentheses.)}$$

$$-x + 6 = 8 \qquad \text{(Collect terms.)}$$

$$-x = 2 \qquad \text{(Subtract 6 from both sides.)}$$

$$x = -2 \qquad \text{(Multiply both sides by } -1.$$

The solution to the equation is $x = -2$. ∎

EXAMPLE 3 Solve the following equation.

$$\frac{2x + 3}{5} - 10 = \frac{4 - 3x}{2}$$

Solution Our first task is to simplify the left side of this equation. We can do this and simultaneously eliminate the fractions if we multiply both sides by the LCD of the two fractions.

$$10 \cdot \left(\frac{2x + 3}{5} - 10\right) = \left(\frac{4 - 3x}{2}\right) \cdot 10 \qquad \text{(Multiply both sides by 10.)}$$

$$4x + 6 - 100 = 20 - 15x \qquad \text{(Remove parentheses.)}$$

$$4x - 94 = 20 - 15x \qquad \text{(Collect like terms.)}$$

$$19x - 94 = 20 \qquad \text{(Add } 15x \text{ to both sides.)}$$

$$19x = 114 \qquad \text{(Add 94 to both sides.)}$$

$$x = \frac{114}{19} \qquad \text{(Divide both sides by 19.)}$$

$$x = 6$$ ∎

Adding or subtracting the same term on both sides of an equation is sometimes called transposition; in effect, the given term is **transposed** to the other side of the equation, and its sign is changed. For instance, in the preceding examples we could write

1. $7y + 6 = y + 6$

$$7y = y + 6 - 6$$
$$7y = y$$

$$7y - y = 0$$
$$6y = 0$$

2. $-x + 6 = 8$

$$-x = 8 - 6$$
$$-x = 2$$

3. $4x - 94 = 20 - 15x$

$$19x - 94 = 20$$

$$19x = 114$$

EXAMPLE 4 The sales of thermal underwear in a region are related to the average winter temperature of the region: Low temperatures give high sales volume and vice versa. A simplified expression exhibiting this behavior is $S = [150 - 2(t + 7)]$.

a.   If S represents sales volume in hundred case lots, and t temperature in degrees Farenheit, at what temperature do sales drop to 0?

b.   If the manufacturing plant can produce only 200 lots annually, at what temperature is the capacity of the plant reached?

c.   Solve the given equation for t in terms of S.

Solution   **a.**   The company sells no thermal underwear when

$$150 - 2(t + 7) = 0$$

Solving this equation for t, we obtain

$$150 - 2t - 14 = 0$$
$$136 - 2t = 0$$
$$-2t = -136$$
$$t = 68$$

In regions with an average winter temperature of 68°F or above there would be no sales.

b.   On the other hand, the capacity of the plant is reached when 200 lots are sold annually:

$$150 - 2(t + 7) = 200$$
$$-2(t + 7) = 50$$
$$t + 7 = -25$$
$$t = -32$$

In a region with an average winter temperature of -32°F, the plant would be operating at full capacity.

c.   Solving for t in terms of S, we have

$$S = 150 - 2(t + 7)$$
$$2(t + 7) = 150 - S$$
$$t + 7 = \frac{150 - S}{2}$$
$$= 75 - \frac{S}{2}$$
$$t = 68 - \frac{S}{2}$$

$\blacksquare$

EXERCISES   2.1

Solve the following equations.

1. $4x - 5 = 2x + 5$

2. $5 - 7z = 5z - 19$

3. $6 - 2t = 4t + 12$

4. $\frac{1}{5}u - 3 = 9 + \frac{1}{2}u$

5. $\frac{1}{3}v + 4 = 7 - \frac{1}{2}v$

6. $5y - 3y = 32 - 2y$

7. $6z + 4 = 10z - 20$

8. $2 + 7t + 3 = 9t + 6 + 4t$

9. $4u + 6u = 2u$

10. $3u + 21 = 0$

11. $4t - 3 + 2t = (8t + 5) - (2t + 8)$

12. $2(u - 1) + 3(2 - u) = (4 - u)$

13. $4x - 6 = 3x + 10$

14. $7(q - 2) = 6(q + 3)$

15. $4(x + 7) + 3(8 - 2x) = 0$

16. $3(3 - 2w) = 2(w - 5) + 3$
17. $4(p - 1) = 5 - 3(4p + 3)$
18. $(3s + 2) - (6s - 3) + 1 = 0$
19. $7(2 - y) - 26 = 2(8 - 7y)$
20. $4(3z - 2) + 1 = 6(1 + z) + 5$
21. $(4t + 3)3 - 4(1 - t) = 5$
22. $3(2x + 1) - 2(3x + 1) = 1$
23. $3(s + 3) + (s - 2) = 2(2s + 4) - 1$
24. $3(2x + 1) + 2(3x + 1) = 17$
25. $(2 - y) + 4(2y - 1) = 12$
26. $(r + 6) - (5 - 2r) + 2 = 0$
27. $4(2v + 3) - 2 = 5v + 4$
28. $\dfrac{3p + 4}{2} - 6 = \dfrac{1 - 3p}{5}$
29. $\dfrac{1 + u}{3} - \dfrac{u}{2} = \dfrac{u + 2}{7}$
30. $\dfrac{x + 600}{3} - \dfrac{x}{2} = 100$
31. $\dfrac{2 - w}{3} + \dfrac{w - 2}{3} = 0$
32. $\dfrac{1 + 2x}{2} + \dfrac{2 - x}{4} = \dfrac{3x + 2}{6} + \dfrac{6x + 8}{12}$
33. $\dfrac{2w - 1}{3} + 2 = \dfrac{10 + 3w}{2}$
34. $\dfrac{q + 1}{2} = \dfrac{2q - 1}{5}$ 35. $\dfrac{r + 3}{4} = \dfrac{4 - r}{3}$
36. $\dfrac{2s - 5}{3} + \dfrac{s - 4}{8} = \dfrac{-7}{12}$
37. $\dfrac{2 - x}{7} = \dfrac{x}{2} - \dfrac{x + 1}{3}$
38. $v - 2 = \dfrac{v + 2}{6} - \dfrac{v - 7}{3}$
39. $\dfrac{w + 1}{6} + \dfrac{w - 1}{4} + \dfrac{w - 2}{3} + 3 = 0$
40. $\dfrac{3 + 2s}{3} + \dfrac{1 - 3s}{2} + 1 = 0$
41. $z - 3i = 5i - z$
42. $z + 2i - 4 = 3iz - 2$
43. $z + 3i + 2 = 4iz - 6 + i$
44. $(z + 2i)^2 = (z + 3i)(z - 3i)$
45. $(z - 3i - 2)(z + 2i) = (z + i + 1)(z - i)$
46. $(z + i + 2)(z - i - 2) = (z + 3i)(z - 2i)$
47. $3x + 5iy = 15$ 48. $2x + 5iy = 15i$
49. $2x + 10i = 10 + 2yi$
50. $-2x + 21yi = 8 - 3i$
51. The formula for converting Celsius temperatures to Farenheit is $F = \dfrac{9}{5}C + 32$.

   a. Convert 77°F to Celsius.

   b. At what temperature do the Celsius and Farenheit temperature scales agree?
   c. Express the Celsius temperature C in terms of F.

52. The distance d traveled by a moving object equals its speed r times the length of time t it travels: $d = rt$.
   a. If Beth travels 2100 miles in 6 hours to visit her mother, how fast does the plane fly?
   b. Express r in terms of d and t.

53. The interest I earned on an investment equals the principal P invested multiplied by the interest rate r and the elapsed time t: $I = Prt$.
   a. If Mary earns $21 in 6 months at a rate of 6%, how much money has she invested?
   b. Express each of the variables P, r, t in terms of the others and I.

54. The total resistance in the parallel network of blood vessels pictured here is 5. If one of the branches has a resistance of 6, find the resistance of the other. (*Hint:* $\dfrac{1}{\dfrac{1}{x} + \dfrac{1}{6}} = 5$; see Example 3, Section 1.6.)

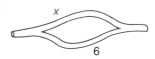

55. The formula $\dfrac{1}{x} + \dfrac{1}{y} = \dfrac{1}{r}$ arises in the study of resistance in a parallel network of blood vessels or electrical circuits. Solve this expression for y in terms of x and r.

56. The circumference C and area A of a circle of radius r are given by $C = 2\pi r$ and $A = \pi r^2$. Solve for r in terms of
   a. C b. A

57. The surface area S and volume V of a sphere of radius r are given by $S = 4\pi r^2$ and $V = \dfrac{4}{3}\pi r^3$.
   Solve for r in terms of
   a. S b. V

58. The volume V of a right circular cone of base radius r and altitude h is given by $V = \dfrac{1}{3}\pi r^2 h$.
   Solve this expression for
   a. r b. h

59. The area A and perimeter P of a rectangle hav-

ing length l and width w are given by $A = lw$ and $P = 2(l + w)$. Solve each of these expressions for

a. l **b.** w

60. Einstein's mass-energy equation is $E = mc^2$. Solve this expression for

a. m **b.** c

2.2 QUADRATIC EQUATIONS

A second-degree equation in one variable is called a **quadratic equation.** For example,

$$x^2 - x - 12 = 0$$
$$2t^2 + 3t - 2 = 7t - 3$$
$$u^2 - 5u + 4 = 2u^2 - 3u + 2$$

are all quadratic equations. Quadratic equations have the general form

$$ax^2 + bx + c = 0 \qquad a \neq 0$$

or can be reduced to this form.

Solution by Factoring

This method relies on the following property of the real number system.

> The product of two real numbers is 0 if and only if at least one of the factors is 0:
>
> $$uv = 0 \quad \text{if and only if} \quad u = 0 \text{ or } v = 0.$$
>
> If the quadratic equation can be factored into the form
>
> $$(ax + b)(cx + d) = 0$$
>
> then x satisfies the equation if and only if one of these factors is 0.

EXAMPLE 1 Solve the following quadratic equation.

$$x^2 - x - 12 = 0$$

Solution Factor the quadratic expression by the scissors method of Chapter 1 to obtain

$$(x - 4)(x + 3) = 0$$

This product is 0 if and only if

$$x - 4 = 0 \qquad \text{or} \qquad x + 3 = 0.$$

Hence, $x = 4$ and $x = -3$ are the solutions to the equation. ∎

EXAMPLE 2   Solve the following equation.

$$4u^2 - 12u + 9 = 0$$

Solution   This quadratic expression is a perfect square: $(a - b)^2 = a^2 - 2ab + b^2$ with $a = 2u$ and $b = 3$. The solution is obtained as follows:

$$4u^2 - 12u + 9 = 0$$
$$(2u - 3)^2 = 0$$
$$2u - 3 = 0$$
$$u = \frac{3}{2}$$

In this example there is only one solution since the original expression was a perfect square; its factors were identical.   ■

Completing the Square

A variation on the factoring method involves completing the square (see Section 1.5) and then taking the square root of both sides of the equation.

To solve a quadratic equation $ax^2 + bx + c = 0$ $(a \neq 0)$ by completing the square:

1. Transpose the constant term to the right side of the $=$ sign and divide through by a if necessary to put the equation in the form

$$x^2 + Bx = C$$

2. Add $(B/2)^2$ to both sides of the equation to obtain the form

$$\left(x + \frac{B}{2}\right)^2 = D.$$

3. Take the square root of both sides.

4. Isolate x on one side of the $=$ sign.

EXAMPLE 3   Solve the following quadratic equation by completing the square.

$$x^2 - 4x - 5 = 0$$

Solution

$$x^2 - 4x = 5$$
$$x^2 - 4x + 4 = 5 + 4 \qquad \text{(Complete the square.)}$$
$$(x - 2)^2 = 9$$
$$x - 2 = \pm 3 \qquad \text{(Take the square root}$$
$$x = 2 \pm 3 \qquad \text{of both sides.)}$$

The solutions are $x = 2 + 3 = 5$
and $x = 2 - 3 = -1$.   ■

EXAMPLE 4 Complete the square to solve the equation.

$$2x^2 + x - 3 = 0$$

Solution

$$2x^2 + x = 3$$

$$x^2 + \frac{1}{2}x = \frac{3}{2}$$

$$x^2 + \frac{1}{2}x + \left(\frac{1}{4}\right)^2 = \frac{3}{2} + \frac{1}{16}$$

$$\left(x + \frac{1}{4}\right)^2 = \frac{25}{16}$$

$$x + \frac{1}{4} = \pm\frac{5}{4}$$

$$x = -\frac{1}{4} \pm \frac{5}{4}$$

$$x = \frac{4}{4}, \ -\frac{6}{4}$$

$$x = 1, \ -\frac{3}{2}$$

■

The Quadratic Formula

The quadratic formula enables us to solve any quadratic equation in one variable. It can be derived by completing the square in a general quadratic equation:

$$ax^2 + bx + c = 0$$

$$x^2 + \frac{b}{a}x + \frac{c}{a} = 0$$

$$x^2 + \frac{b}{a}x = -\frac{c}{a}$$

$$x^2 + \frac{b}{a}x + \left(\frac{b}{2a}\right)^2 = -\frac{c}{a} + \frac{b^2}{4a^2}$$

$$\left(x + \frac{b}{2a}\right)^2 = \frac{b^2 - 4ac}{4a^2}$$

Take the square root of both sides:

$$x + \frac{b}{2a} = \pm\frac{\sqrt{b^2 - 4ac}}{2|a|}$$

Since we are interested in both roots (those corresponding to the positive and negative signs on the right side), we can replace $|a|$ with a. Thus, we obtain the **quadratic formula**. (See box on next page.)

The number $b^2 - 4ac$ is called the **discriminant** of the quadratic equation. It indicates the number and form of the roots:

The roots of a quadratic equation with real coefficients

$$ax^2 + bx + c = 0, \qquad a \neq 0$$

are given by

$$x = \frac{-b \pm \sqrt{b^2 - 4ac}}{2a}$$

---

1. If $b^2 - 4ac > 0$, there are two distinct real roots.

2. If $b^2 - 4ac = 0$, there is precisely one real root.

3. If $b^2 - 4ac < 0$, there are two complex roots. These roots are complex conjugates of one another; they are:

$$x_1 = \frac{-b}{2a} + i\,\frac{\sqrt{|b^2 - 4ac|}}{2a}$$

and

$$x_2 = \frac{-b}{2a} - i\,\frac{\sqrt{|b^2 - 4ac|}}{2a}$$

EXAMPLE 5 Solve the following equation by the quadratic formula.

$$2x^2 + 5x - 3 = 0$$

Solution We use $a = 2$, $b = 5$, and $c = -3$ in the quadratic formula to find

$$
\begin{aligned}
x &= \frac{-5 \pm \sqrt{5^2 - 4 \cdot 2 \cdot (-3)}}{2 \cdot 2} \\
&= \frac{-5 \pm \sqrt{25 + 24}}{4} \\
&= \frac{-5 \pm 7}{4} \\
&= \frac{-5 + 7}{4} \quad \text{or} \quad \frac{-5 - 7}{4} \\
&= \frac{1}{2}, \; -3
\end{aligned}
$$

The solutions are $x = 1/2$ and $x = -3$.

EXAMPLE 6   Solve the following quadratic equation.

$$x^2 - 2x + 2 = 0$$

Solution   By the quadratic formula

$$x = \frac{-(-2) \pm \sqrt{(-2)^2 - 4 \cdot 1 \cdot 2}}{2 \cdot 1}$$
$$= \frac{2 \pm \sqrt{4 - 8}}{2}$$
$$= \frac{2 \pm \sqrt{-4}}{2}$$
$$= \frac{2 \pm 2i}{2}$$
$$= 1 \pm i$$

The roots are $x = 1 + i$ and $x = 1 - i$.

EXAMPLE 7   Physicists tell us that the altitude h in feet of a projectile t seconds after firing is

$$h = -16t^2 + v_u t + h_0$$

where v_u is the *upward* component of its initial velocity in feet per second and h_0 is the altitude in feet from which it is fired. A rocket is launched from a hilltop 2400 feet above the desert with an initial upward velocity of 400 feet per second. When will it land on the desert? (See Fig. 2–1.)   ■

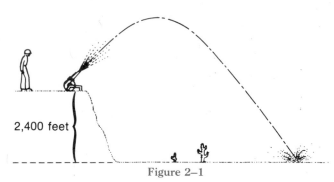

2,400 feet

Figure 2–1

Solution   The altitude of the rocket t seconds after launching is

$$-16t^2 + 400t + 2400$$

When it lands, its altitude is zero. To determine the time when its altitude is zero, we must solve the quadratic equation

$$h = -16t^2 + 400t + 2400 = 0$$

Simplifying and applying the quadratic formula, we find

$$-16(t^2 - 25t - 150) = 0$$
$$t^2 - 25t - 150 = 0$$
$$t = \frac{25 \pm \sqrt{625 + 600}}{2}$$
$$= \frac{25 \pm \sqrt{1225}}{2}$$
$$= \frac{25 \pm 35}{2}$$
$$= \frac{60}{2}, \frac{-10}{2}$$
$$= 30, -5$$

The solution $t = -5$ represents the time 5 seconds *before* launching and is irrelevant. Thus, the rocket should land on the desert 30 seconds after it is fired. ∎

EXERCISES 2.2

Solve the following equations by factoring.

1. $x^2 - 4 = 0$
2. $u^2 - 9 = 0$
3. $x^2 + 4 = 0$
4. $u^2 + 9 = 0$
5. $s^2 - 5s + 6 = 0$
6. $4y^2 - 4y + 1 = 0$
7. $z^2 + 4z + 4 = 0$
8. $4x^2 - 4x - 15 = 0$
9. $v^2 - v - 2 = 0$
10. $w^2 + w - 20 = 0$
11. $6x^2 + 5x - 6 = 0$
12. $4y^2 + 3y - 1 = 0$
13. $9z^2 + 12z + 4 = 0$
14. $3s^2 - s - 10 = 0$
15. $2t^2 + 9t - 5 = 0$
16. $4u^2 + 11u - 3 = 0$
17. $25v^2 - 10v + 1 = 0$
18. $3y^2 = 4y + 4$
19. $2u^2 - 7u - 15 = 0$
20. $3z^2 + 5z + 2 = 0$

Solve the following by completing the square.

21. $y^2 + 4y - 21 = 0$
22. $x^2 + 2x - 3 = 0$
23. $p^2 + 18p - 88 = 0$
24. $t^2 - 8t + 13 = 0$
25. $w^2 + 4w + 2 = 0$
26. $3s^2 + 12s - 15 = 0$
27. $2q^2 + 4q - 1 = 0$
28. $2u^2 - 6u + 1 = 0$
29. $z^2 - z - 72 = 0$
30. $r^2 + 3r + 5 = 0$
31. $r^2 - 4r - 12 = 0$
32. $u^2 + 4u - 2 = 0$
33. $2x^2 - 4x - 3 = 0$
34. $5p^2 - 10p + 3 = 0$
35. $4y^2 - 4y + 1 = 0$
36. $x^2 + 3x - 4 = 0$
37. $4v^2 - 4v - 3 = 0$
38. $3s^2 - s - 10 = 0$

Use the quadratic formula to solve the following equations.

39. $x^2 + 3x - 4 = 0$
40. $y^2 - 3y - 4 = 0$
41. $9u^2 - 3u - 2 = 0$
42. $2z^2 + 9z - 5 = 0$
43. $4u^2 - 12u + 9 = 0$
44. $25v^2 + 10v + 1 = 0$
45. $3p^2 + 5p + 2 = 0$
46. $x^2 - 2x + 2 = 0$

47. $y^2 - y + 1 = 0$
48. $z^2 - 4z + 9 = 0$
49. $r^2 - 6r + 34 = 0$
50. $4w^2 + 11w - 3 = 0$
51. $3z^2 = 4z + 4$
52. $(x - 1)(x - 2) + x = 0$
53. $5w^2 - 3w - 14 = 0$
54. $s^2 - 2s + 5 = 0$
55. $t^2 - 2t + 4 = 0$
56. $u^2 - 2u + 3 = 0$
57. $v^2 + 2v + 2 = 0$
58. $2w^2 - 2w + 1 = 0$
59. $p^2 + 8p + 25 = 0$
60. $6v^2 + v - 2 = 0$
61. $r^2 - 6r + 16 = 0$
62. $4y^2 + 3y - 1 = 0$
63. $3s^2 - 14s + 8 = 0$
64. $q^2 - 2q + 10 = 0$
65. $2x^2 - 2\sqrt{2}x + 3 = 0$
66. $y^2 + y + 1 = 0$
67. $p^2 + 2p + 5 = 0$
68. $u^2 + 4u + 5 = 0$
69. $w^2 + 4w + 13 = 0$
70. $2z^2 + 4z + 3 = 0$
71. $q^2 - 2q + 2 = 0$
72. $v^2 + 2v + 6 = 0$
73. $t^2 + 6t + 10 = 0$

Without solving the quadratic equation, determine the number of real roots in each of the following.

74. $2x^2 - 5x - 9 = 0$
75. $8y^2 - 30y + 27 = 0$
76. $3z^2 - 4z + 2 = 0$
77. $r^2 - 8r + 16 = 0$
78. $3s^2 - 9s + 4 = 0$
79. $4t^2 - 20t + 25 = 0$
80. $u^2 - 2u - 2 = 0$
81. $8v^2 - 14v - 15 = 0$
82. $x^2 - x + 1 = 0$
83. $2y^2 - 17y + 36 = 0$

Solve the following quadratic equations with complex coefficients by using the quadratic formula.

84. $2z^2 + 2iz - 13 = 0$
85. $2iz^2 - 5z - 3i = 0$
86. $3z^2 - 5iz + 2 = 0$
87. $-2iz^2 + 2z + 13i = 0$
88. $7z^2 - 4iz + 3 = 0$

89. The Rescue Service wishes to send supplies by projectile to a mountaintop 3200 feet above their location. The projectile is fired with an initial upward velocity of 480 feet per second. When do the supplies land (from above) on the mountaintop? (See Example 7.)

90. A plane flying at an altitude of 1000 feet over the ocean ejects a cannister downward with a velocity of 120 feet per second. How long does it take the cannister to reach the ocean? (See Example 7.)

91. It is found that the concentration of a pollutant in a river decreases steadily as one moves downstream from the polluting source. **a.** If the concentration of the pollutant x kilometers from the source is $(10,000 - 3x - x^2/20)$ parts per million, how far must one be from the source to find water containing only 365 parts per million of the pollutant? **b.** 9.8 parts per million?

92. Suppose that a lake covering x square acres can accommodate $(10,000x^2 + 45,000x)$ fish of various sizes and varieties. How many acres are then required to support one million fish?

93. For various diseases, the amount of serum injected in humans is computed in terms of body weight. If the proper dosage for a person weighing w kilograms is $(w^2 + 100w)$ milligrams, what size person requires an injection of 11.9 grams?

94. Limpin' Lou's sporting goods store finds that sales receipts depend on the size of the showroom. Suppose that the store can sell $(100T^2 + 10T + 194)$ dollars worth of sporting goods per day with a showroom space of T thousand square feet. How large should the showroom be in order to sell \$2000 worth each day?

95. If running at a speed of s miles per hour burns $(5s^2 + 20s + 25)$ calories per hour, how fast should one run in order to burn 505 calories in one hour?

96. Suppose that Flo's Fashions can sell $(596 - 4d - d^2)$ coats if the markup on each coat is \$$d$. **a.** What should the markup be if Flo has 200 coats to sell? **b.** What if she has 375 coats to sell?

2.3 EQUATIONS SOLVABLE WITH QUADRATIC TECHNIQUES

Certain other kinds of equations can also be solved by using the techniques for solving quadratic equations. For instance, in an equation of the form

$$ax^{2k} + bx^k + c = 0,$$

substitution of $u = x^k$ results in a quadratic equation. Such a substitution is illustrated in Examples 1 and 2 with k an integer and k a rational number, respectively.

EXAMPLE 1 Solve the following equation.

$$x^4 - 2x^2 - 3 = 0$$

Solution This is a quadratic equation in x^2. Setting $u = x^2$, this equation becomes

$$u^2 - 2u - 3 = 0$$

Solving for u, we obtain

$$u = \frac{2 \pm \sqrt{(-2)^2 - 4 \cdot 1 \cdot (-3)}}{2}$$
$$= \frac{2 \pm \sqrt{16}}{2}$$
$$= \frac{2 \pm 4}{2}$$
$$= 3, -1$$

But $u = x^2$, so that $x = \pm\sqrt{u}$; thus,

$$x = \pm\sqrt{3}, \quad \pm i$$

are the roots of the equation. ■

EXAMPLE 2 Solve the following equation.

$$x + \sqrt{x} - 6 = 0$$

Solution We set $u = \sqrt{x}$ and obtain

$$u^2 + u - 6 = 0$$

By the quadratic formula,

$$u = \frac{-1 \pm \sqrt{1 - 4(-6)}}{2}$$
$$= \frac{-1 \pm 5}{2}$$
$$= -3, 2$$

But the symbol $\sqrt{x}$ stands for the positive square root of x. Thus, $\sqrt{x} = -3$ is impossible. Hence, the only solution is

$$\sqrt{x} = 2$$

or

$$x = 4$$ ∎

Equations Involving Radicals

Another type of problem that can lead to a quadratic equation involves radicals. Squaring both sides of the equation one or more times will eliminate the radicals.

EXAMPLE 3   Solve the following equation.

$$\sqrt{x + 1} + \sqrt{2x - 5} = 3$$

Solution   First, we square both sides of the equation and simplify as follows.

$$(x + 1) + 2\sqrt{x + 1} \sqrt{2x - 5} + (2x - 5) = 9$$
$$2\sqrt{x + 1} \sqrt{2x - 5} = 13 - 3x$$

Squaring again eliminates the remaining radicals:

$$4(x + 1)(2x - 5) = 13^2 - 2 \cdot 13 \cdot 3x + 9x^2$$
$$8x^2 - 12x - 20 = 169 - 78x + 9x^2$$
$$0 = x^2 - 66x + 189$$
$$= (x - 3)(x - 63)$$

Thus, we have *apparent* solutions $x = 3, 63$. But 63 is clearly too large to be a solution since $\sqrt{63 + 1} = 8 > 3$. The apparent root $x = 63$ is a root of the equation twice squared but not of the original equation. Such a root is said to be extraneous and must be rejected. ∎

> Multiplying an equation by an expression containing the variable (as when the equation is squared) can introduce extraneous (i.e., false) roots.

If x satisfies the original equation, x will also satisfy the equation that results from multiplying by any expression. Hence, in the preceding example we can say only that if x is a solution, then $x = 3$ or $x = 63$; that is, 3 and 63 are the only *possible* solutions. Checking the original equation shows us that $x = 3$ is the only solution.

The introduction of extraneous roots is illustrated very vividly by modifying Example 3 slightly as in Example 4.

EXAMPLE 4   Solve the following equation.

$$\sqrt{x + 1} + \sqrt{2x - 5} = -3$$

Solution   There are *no solutions* because the sum of two positive numbers cannot be negative. However, if we square the equation, we generate the same algebra as in Example 3 and obtain the two extraneous solutions, which we would have to reject. ∎

**Equations
Involving
Fractions**

To solve an equation involving rational expressions (fractions)

1. Multiply through by the LCD of the fractions in the equation.

2. Simplify and solve the resulting equation.

3. Check the solutions in the original equation

Whenever the LCD involves the variable we must check to see that the apparent solutions satisfy the original equation; we multiply by an expression involving the variable in this case.

EXAMPLE 5   Solve the following equation.

$$\frac{x - 1}{x + 2} + \frac{x + 7}{x + 4} = 0$$

Solution   Multiplying through by the LCD $= (x + 2)(x + 4)$, we obtain

$$(x - 1)(x + 4) + (x + 7)(x + 2) = 0$$
$$(x^2 + 3x - 4) + (x^2 + 9x + 14) = 0$$
$$2x^2 + 12x + 10 = 0$$
$$x^2 + 6x + 5 = 0$$
$$(x + 1)(x + 5) = 0$$
$$x = -1, -5$$

Checking $x = -1$, we have

$$\frac{(-1) - 1}{(-1) + 2} + \frac{(-1) + 7}{(-1) + 4} = \frac{-2}{1} + \frac{6}{3} = -2 + 2 = 0.$$

Similarly, $x = -5$ can be shown to satisfy the original equation. The solutions are $x = -1$ and $x = -5$.

Since we multiplied both sides of the original equation by $(x + 2)(x + 4)$, we *must* check to see that our apparent solutions satisfy the original equation; we have multiplied by an expression involving the variable. For instance, had we found $x = -2$ as a possible solution, we would have had to reject it because checking the original problem would have entailed division by zero. ∎

EXERCISES 2.3

Find all real solutions to the following equations.

1. $x^4 - 6x^2 + 5 = 0$
2. $u^4 + 4u^2 - 12 = 0$
3. $2w^4 + 15w^2 - 8 = 0$
4. $y^4 + y^2 - 2 = 0$
5. $z^4 - 8z^2 + 7 = 0$
6. $w^4 - 5w^2 + 4 = 0$
7. $r^6 - 6r^3 + 5 = 0$
8. $s^6 + 4s^3 - 12 = 0$
9. $t^6 + 6t^3 + 9 = 0$
10. $9p^6 - 12p^3 + 4 = 0$
11. $t^6 + 3t^3 - 10 = 0$
12. $z^6 + 2z^3 - 15 = 0$
13. $q^8 - 6q^4 + 8 = 0$
14. $x^8 + 10x^4 - 24 = 0$
15. $3v^8 - 5v^4 + 2 = 0$
16. $6y^8 - 5y^4 - 6 = 0$
17. $z^{10} - z^5 - 6 = 0$
18. $3s^{10} + 8s^5 - 3 = 0$
19. $y^{12} - y^6 - 6 = 0$
20. $4t^{16} + 4t^8 - 15 = 0$
21. $2w - 5\sqrt{w} = 3$
22. $x - 5\sqrt{x} + 6 = 0$
23. $w - \sqrt{w} - 2 = 0$
24. $t + 3\sqrt{t} - 10 = 0$
25. $2p + 3\sqrt{p} - 2 = 0$
26. $3q - 7\sqrt{q} - 6 = 0$
27. $4r - 4\sqrt{r} + 1 = 0$
28. $v - \sqrt{v} - 6 = 0$

29. $\sqrt{r} + \sqrt[4]{r} - 20 = 0$
30. $\sqrt[3]{s} + \sqrt[6]{s} - 6 = 0$
31. $\sqrt{z - 4} = 4 - z$
32. $4 + \sqrt{x + 2} = x$
33. $\sqrt{r + 2} = r - 10$
34. $\sqrt{4 - s} = 2 - s$
35. $\sqrt{2u + 9} = u - 3$
36. $\sqrt{3w + 1} = w - 3$
37. $5 + \sqrt{y - 3} = y$
38. $\sqrt{3s - 2} = 2 + \sqrt{s}$
39. $\sqrt{3w + 4} + 6 - \sqrt{w} = 0$
40. $\sqrt{r + 20} = \sqrt{r + 4} + 2$
41. $2 + \sqrt{q - 4} = \sqrt{2q - 1}$
42. $\sqrt{3x - 8} - \sqrt{5x - 11} = -1$
43. $\sqrt{w + 3} - \sqrt{11w + 58} + 5 = 0$
44. $\sqrt{p + 6} + \sqrt{p + 3} = \sqrt{5 - 2p}$
45. $\sqrt{v + 1} + \sqrt{v - 2} = \sqrt{2v + 3}$
46. $\frac{3}{t^2} - \frac{11}{t} - 20 = 0$

47. $\frac{4}{u^4} - \frac{17}{u^2} + 4 = 0$
48. $\frac{6}{r^2} - \frac{5}{r} - 6 = 0$
49. $\frac{9}{z^4} - \frac{10}{z^2} + 1 = 0$
50. $\frac{w - 1}{w + 1} + \frac{w + 1}{w - 1} = 0$
51. $\frac{v + 2}{v - 1} = \frac{v - 3}{2v - 2}$
52. $\frac{s + 3}{s - 1} + \frac{s + 1}{s - 8} = 0$

53. $\dfrac{3}{2y + 7} - 2y - 5 = 0$

54. $5 - \dfrac{24}{x + 3} = \dfrac{1}{x - 2}$

55. $\dfrac{3}{2x + 1} - 3 = \dfrac{2x}{3x - 1}$

56. $\dfrac{4}{y + 2} + \dfrac{2y}{y - 1} - 5 = 0$

57. $\dfrac{2}{z - 3} + 1 = \dfrac{6}{z - 8}$

58. $\dfrac{1}{3p - 4} + 3p - 2 = 0$

59. $\dfrac{45q}{3q - 4} - 1 = \dfrac{40q}{2q - 3}$

60. $\dfrac{2u}{3u + 2} - 3 = \dfrac{5u}{1 - 2u}$

61. $\dfrac{2}{r^2 - 3} + 1 = \dfrac{6}{r^2 - 8}$

62. $\dfrac{v^2 + 2}{v^2 - 3} = \dfrac{3v^2 - 6}{2v^2 - 7}$

63. $\dfrac{w^3 - 3}{w^3 + 2} + \dfrac{w^3 + 1}{w^3 + 6} + 4 = 0$

64. $\dfrac{3}{2 - 3z} + \dfrac{1}{z - 1} + \dfrac{-1}{3z - 2} = 0$

65. According to Poiseuille's law, the rate at which fluid flows through an unobstructed pipe is given by $Q = kr^4$, where r is the radius of the pipe and k is a constant related to the viscosity of the fluid. An obstruction generates turbulence and decreases the flow. A certain blood vessel has fatty deposits that reduce the flow to

$$Q = k(r^4 - ar^2) \text{ cubic centimeters} \\ \text{per second}$$

where r is given in centimeters and $k = 120$, $a = 1/20$ for blood. What radius must the blood vessel have in order to handle 6 cubic centimeters a second?

66. The cost of manufacturing a calculator increases directly as its complexity. On the other hand, more sophisticated units generate a larger market, thereby lowering the manufacturing cost per unit. Suppose that a calculator with n functions can profitably be sold at a price of $\$(2n - 6\sqrt{n})$ each. How many functions can be expected in a calculator selling for $56?

67. According to the theory of relativity, the energy E, momentum p, and mass m of a particle are related to the speed of light c as follows:

$$E^2 = p^2 c^2 + m^2 c^4.$$

Express c explicitly in terms of the other quantities.

2.4 WORD PROBLEMS

Before a real-life problem can be solved mathematically, it must be translated into mathematical terms. Developing the ability to do this requires time and practice at increasing levels of difficulty. Solving word problems helps to develop this ability. Students find the solution of word problems challenging, mainly because they haven't had much practice at solving them. As with most new skills, practice is the key to success. The steps outlined in the box on the next page are essential for solving word problems.

EXAMPLE 1 Mr. Pleiman sells an investment property and, after paying off the mortgage, places the proceeds in certificates of deposit (CDs). The maximum current interest rate is with a 1-year certificate paying 12 percent interest. He can lock in a 10 percent rate for $2\frac{1}{2}$ years. Thinking that interest rates may go down, he splits his money, depositing three times as much in the 12 percent account as in the 10 percent account. The yearly return on these deposits is $8280. How much did he realize from the sale of his property? How much did he invest in each account?

SOLVING WORD PROBLEMS

1. *Understand the problem.* Read the problem through quickly to pick up its general thrust. Then read it through again very carefully and completely, possibly several times. Make a sketch if it will help to understand the problem.

2. *Identify the unknown quantities* that are to be determined.

3. *Name these quantities.* Use a variable such as x or y. Try to minimize the number of variables used. If you have drawn a sketch, label these quantities on the sketch.

4. *Express all known relationships mathematically.* Do so in terms of the variable names x, y, and so on.

5. *Reread the problem.* Check the relationships you have written down. Has all the information supplied in the problem been incorporated into these relationships?

6. *Solve the resulting equations.*

7. *Interpret your results* in terms of the original problem.

8. *Check your solutions against the conditions of the problem.*

Solution Let

$$x = \text{amount invested at 10\%}$$
$$3x = \text{amount invested at 12\%}$$
$$4x = \text{amount realized from sale of the house}$$
$$(0.10)x = \text{yearly interest from the 10\% investment}$$
$$(0.12)3x = \text{yearly interest from the 12\% investment}$$

For each year's income, we have the following equation:

$$\begin{pmatrix} \text{interest from} \\ \text{10\% account} \end{pmatrix} + \begin{pmatrix} \text{interest from} \\ \text{12\% account} \end{pmatrix} = (\text{total interest})$$
$$(0.10)x \qquad + \qquad (0.12)3x \qquad = 8280.00$$

We then solve for the unknowns:

$$(0.10)x + (0.36)x = 8280.00$$
$$10x + 36x = 828,000$$
$$46x = 828,000$$
$$x = 18,000$$
$$3x = 54,000$$
$$4x = 72,000$$

Mr. Pleiman realized $72,000 from the sale of his property. He then deposited $18,000 at 10 percent interest and $54,000 at 12 percent interest. ■

FUNKY WINKERBEAN   Tom Batiuk

EXAMPLE 2 The Strong and Weak Acid Company (SWACO) stocks two solutions of its product. Its strong solution is 80 percent acid; its weak solution is only 20 percent acid. A potential customer must have a 30 percent solution in order to clean his machinery without causing any damage. Develop a formula for obtaining a 30 percent acid solution by mixing the 80 and 20 percent solutions. Do so by determining the amount of 80 percent solution that must be added to 10 liters of the 20 percent solution in order to obtain a 30 percent solution.

Solution Let

x = number of liters of 80 percent solution that must be added to 10 liters of 20 percent solution to yield a 30 percent solution.

The total amount of acid in each container is shown in Figure 2–2.

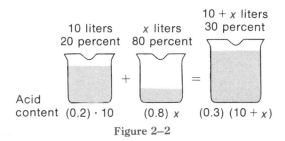

Figure 2–2

To find x, we must solve the equation

$$(0.2) \cdot 10 + (0.8)x = (0.3)(10 + x)$$
$$2 + 0.8x = 3 + 0.3x$$
$$0.5x = 1$$
$$5x = 10$$
$$x = 2$$

Thus, 2 liters of 80 percent solution must be added to 10 liters of 20 percent solution to yield a 30 percent solution. The formula for 30 percent solution is two parts 80 percent solution with ten parts 20 percent solution or "strong and weak acids are mixed in the ratio 1 to 5." ■

EXAMPLE 3 The Vacation Realty Company plans to sell shares in a luxury resort condominium. Each share will entitle the shareholder to 1-week's use of the condominium each year. The company already has 50 prospective buyers. Knowing that not everyone will want to return to this particular spot every year, the company advises the prospective buyers to allow the project to be oversold. By allowing 25 additional shares to be sold, the original investors can save $1200 each. What is the total sale price of the condominium? What is the cost per share if 50 or 75 shares are sold, respectively?

Solution   Let

$$P = \text{total sales price of the condominium}$$
$$\frac{P}{50} = \text{cost per share with 50 investors}$$
$$\frac{P}{75} = \text{cost per share with 75 investors}$$

With 75 investors, the cost per share is $1200 less than with 50 investors, which gives us the equation

$$\begin{pmatrix} \text{cost with 50} \\ \text{investors} \end{pmatrix} - \$1200 = \begin{pmatrix} \text{cost with 75} \\ \text{investors} \end{pmatrix}$$
$$\frac{P}{50} - 1200 = \frac{P}{75}$$

This equation is solved for the unknowns as follows.

$$150 \cdot \left(\frac{P}{50} - 1200 \right) = \frac{P}{75} \cdot 150$$
$$3P - 180,000 = 2P$$
$$P = 180,000$$
$$\frac{P}{50} = \frac{180,000}{50} = 3600$$
$$\frac{P}{75} = \frac{180,000}{75} = 2400$$

The condominium costs $180,000; the costs per share are $3600 or $2400 depending on whether 50 or 75 shares are sold.   ■

EXAMPLE 4   Economists say that a given commodity is in equilibrium when supply equals demand. Suppose that when corn sells for $1.50 per bushel, there is a demand for 100,000 metric tons and a supply of only 50,000 metric tons. Also, suppose that each x¢ increase in the price per bushel stimulates an additional production of $5x^2$ metric tons and decreases the demand by $750x$ metric tons. What is the equilibrium price for corn?

Solution   At a price of $1.50 + x¢ per bushel,

$$\text{Production} = 50,000 + 5x^2$$
$$\text{Demand} \ \ = 100,000 - 750x$$

Equating these two for equilibrium:

$$50,000 + 5x^2 = 100,000 - 750x$$
$$5x^2 + 750x - 50,000 = 0$$
$$x^2 + 150x - 10,000 = 0$$
$$(x + 200)(x - 50) = 0$$
$$x = -200, 50$$

The solution $x = -200$ represents a decrease of $2.00 in the price of corn to $-50¢$. Such a solution is meaningless. The solution $x = 50$ makes sense. An equilibrium price for corn is $1.50 + x = 1.50 + 50¢ = 2.00$ per bushel. ■

The supply and demand relationship is actually much more complicated than that given in Example 4. In this and other examples and exercises, simplified expressions are used so that the problem can be worked in this course.

EXAMPLE 5 The $60,000 cost of a charter flight must be shared equally by those who take the trip. A certain number of travelers have already expressed an interest. If 100 additional customers sign up, the cost to each of the original patrons will be reduced by $100. How many have already signed up?

Solution Let

$$x = \text{number of people already signed up}$$
$$\frac{60,000}{x} = \text{cost to each patron}$$
$$x + 100 = \text{total number of people if 100 more people sign up}$$
$$\frac{60,000}{x + 100} = \text{cost to each patron if 100 more people sign up}$$

Thus, we obtain the equation

$$\text{(original cost)} - \$100 = \text{(cost if 100 extra sign up)}$$
$$\frac{60,000}{x} - 100 = \frac{60,000}{x + 100}$$
$$x(x + 100)\left[\frac{60,000}{x} - 100\right] = \left[\frac{60,000}{x + 100}\right] \cdot x(x + 100)$$
$$60,000x + 6,000,000 - 100x^2 - 10,000x = 60,000x$$
$$-100x^2 - 10,000x + 6,000,000 = 0$$
$$x^2 + 100x - 60,000 = 0$$
$$(x + 300)(x - 200) = 0$$
$$x = -300, 200$$

Thus, 200 people have already signed up. ■

EXAMPLE 6 On an outing, Cindy's Explorer troop rowed upstream 60 miles and returned. The entire trip took 40 hours. If the stream is flowing at the rate of 2 miles per hour, how fast can Cindy's troop row in still water? How much more time was spent rowing upstream than downstream?

Solution   Let

$$x = \text{rate in miles per hour the troop can row in still water}$$
$$x + 2 = \text{rate they travel downstream}$$
$$x - 2 = \text{rate they travel upstream}$$

Since

$$\text{distance} = \text{rate} \cdot \text{time}$$
$$d = rt$$

we have

$$t = \frac{d}{r}$$

Thus, for the 60-mile trip, we have

$$\frac{60}{x + 2} = \text{length of time for the return trip}$$

$$\frac{60}{x - 2} = \text{length of time for the upstream trip}$$

We then obtain the equation:

$$(\text{time upstream}) + (\text{time downstream}) = (\text{total time})$$
$$\frac{60}{x - 2} + \frac{60}{x + 2} = 40$$
$$(x - 2)(x + 2)\left[\frac{60}{x - 2} + \frac{60}{x + 2}\right] = 40(x - 2)(x + 2)$$
$$60x + 120 + 60x - 120 = 40x^2 - 160$$
$$-40x^2 + 120x + 160 = 0$$
$$x^2 - 3x - 4 = 0$$
$$(x - 4)(x + 1) = 0$$
$$x = 4, \; -1$$

Cindy's troop can row 4 miles per hour in still water. They traveled 2 ($= x - 2$) miles per hour upstream and 6 miles per hour ($= x + 2$) downstream. The upstream trip took them 30 hours [$= 60/(x - 2)$] compared with 10 hours [$= 60/(x + 2)$] for the downstream trip.   ◼

EXAMPLE 7   A painter knows that her assistant can paint only half as much in a day as she can. She is planning to fire the assistant, but she must finish painting the houses in a certain real estate development by a given date or forfeit a bond. In the past 5 days they have painted three houses working together. How long would it take each woman working alone to paint a house?

Solution   Let

$$t = \text{number of days required for the painter to paint a house}$$
$$2t = \text{number of days required for the assistant to paint a house}$$
$$\frac{1}{t} = \text{number of houses painted in 1 day by the painter}$$
$$\frac{1}{2t} = \text{number of houses painted in 1 day by the assistant}$$
$$\frac{1}{t} + \frac{1}{2t} = \text{number of houses painted in 1 day by the two working together}$$

Since they can paint three houses in 5 days working together, this yields the equation:

$$\left(\frac{1}{t} + \frac{1}{2t}\right)5 = 3$$
$$\frac{3}{2t} \cdot 5 = 3$$
$$\frac{5}{2t} = 1$$
$$5 = 2t$$
$$2\frac{1}{2} = t$$

The painter can paint a house in $2\frac{1}{2}$ days while her assistant takes 5 days.

EXERCISES 2.4

1. The manager of a bicycle shop knows she can sell 400 Reflecto handlebars if she charges $8 each. Suppose that for each 50¢ increase in price she loses 20 sales. At what selling price does she lose all sales?

2. If a television set sells for $150 and the markup is 30 percent of the selling price, how much did the dealer pay for the set?

3. If a television set sells for $156 and the markup is 30 percent of the dealer's cost, how much did he pay for the set?

4. On a recent visit to Las Vegas, Sandy doubled her money at the roulette wheel and then spent $300. She then tripled her money at the craps table and spent $540. Later she quadrupled her money at blackjack and spent $720. If she went home with $480 in her pocket, how much did she have when she arrived in Las Vegas?

5. If Todd spends one third of his money, loses two thirds of the remainder, and is left with $10, with how much did he start?

6. The manager of a toy store paid $80 for some scooters. She was unable to sell three of them but made a profit of $6 on each of the others for a total return of $80. How many did she purchase?

7. Suppose that when the price of apples increases by $1 per bushel, the grower sells one less bushel than before for the same income of $30. What are the respective prices?

8. Suppose that when fish oil extract sells for $50 per gallon, there is a demand for 50,000 gallons per week and a supply of 80,000 gallons per week. Also, suppose that if the price decreases $x per gallon, the demand increases $200x^2$ gallons while the supply decreases $1000x$ gallons. What is the equilibrium price for fish oil extract?

9. The Zany auto parts store can sell 2000 "Pedal Power" units if the price is $8.00 per unit. Suppose that each increase in price of 25x¢ decreases sales by $5x^2$ units. At what price do sales drop off to zero?

10. Find the dimensions of a rectangular garden plot if it has an area of 600 square feet and can be surrounded with 100 feet of fence.

11. Dale is given instructions to construct an open metal box by cutting identical pieces from the four corners of a rectangular piece of metal. The box is to be 4 centimeters high and have a volume of 160 cubic centimeters. Its base is to be a rectangle 3 centimeters longer than it is wide. With what size piece of sheet metal should he begin?

12. Enclosing a certain kennel having an area of 1350 square meters requires 150 meters of fence. What are the dimensions of the kennel?

13. Mr. Rindler wishes to plant a border of flowers around the outside of his garden. He plans to use two-thirds of his total available area for produce. How wide should his border be if his available space is a rectangle 50 feet by 60 feet?

14. The Great Hall in Nature Museum consists of a 10-foot wide walkway above an open display area. The open area is 50 percent longer than it is wide. What are the dimensions of the museum if 800 square yards of carpet are required to carpet the entire first floor together with the walkway?

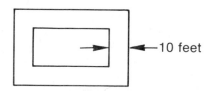

15. Two blood vessels connected in parallel have a net resistance of 2. The resistance of one is three units more than the resistance of the other. What are the individual resistances of the

blood vessels? *Hint:* With resistances of r_1 and r_2, the net resistance is $\dfrac{1}{\dfrac{1}{r_1} + \dfrac{1}{r_2}}$.

16. Distance at sea is measured in nautical miles. Two ocean-going vessels leave port at noon traveling in perpendicular directions in the Atlantic Ocean. One travels 10 nautical miles per hour (knots) faster than the other. If the boats are 50 nautical miles apart at 1 P.M., find their respective rates of speed.

17. Suppose that an animal with legs L feet long walks at the rate of $7\sqrt{L}/3$ miles per hour and that Black Beauty's legs are exactly 3 feet longer than Rover's. If they are separated by 14 miles and walk toward each other, they will meet in 2 hours. How long are their legs?

18. Fred and his mother drive separate cars to a destination 150 miles away. Fred's shoes are heavier than his mother's so he drives 10 miles per hour faster than his mother and arrives 1/2 hour sooner. How fast did each drive?

19. An outdoors enthusiast takes a boat 40 miles upstream and then back again in 15 hours. If the stream is flowing at 2 miles per hour, find the speed of the boat in still water. How long did each part of the trip take?

20. Making a round trip from Dayton to Chicago, a pilot faces a 30-mile-per-hour head wind one way and a 30-mile-per-hour tail wind on the return trip. The return trip takes 45 minutes less than the initial journey. **a.** If the distance between the two cities is 350 miles, what is the air-speed? **b.** How long did each part of the trip take?

21. Bill and Jane are moving to another city. Being a two-car family, they must each drive a car to their new home. Bill drives at 50 miles per hour but Jane drives 10 miles per hour faster. It takes Bill 1 hour longer than Jane to make the trip. Find their respective travel times and the distance between the cities.

22. A plane travels for 4 hours at a steady speed with a tail wind. On the return trip it must face the wind, thereby reducing its speed 100 miles per hour and increasing its time by 1 hour. **a.** Find the respective speeds for both the initial and return trips. **b.** What is the one-way distance involved?

23. Dan bid the labor for a remodeling job at $336. However, the job took 4 hours longer than he estimated. Consequently, he earned $2 an hour less than he had planned. What was the estimated time?

24. Ms. Ranly repaired a broken antique chair for $240. She earned 50¢ less per hour than she planned because the work took 2 hours longer than expected. In what amount of time had she planned to fix the chair?

25. Mary recently moved to a colder climate and must strengthen the antifreeze in her car. Her radiator is now filled to capacity with 12 liters of a 40 percent solution. How much must be drained and replaced with pure antifreeze to obtain a 65 percent solution?

26. Thirty quarts of 20 percent medicinal saline solution is to be strengthened to 50 percent by replacing some of it with 80 percent solution. How much must be replaced?

27. The rangers at the Fish and Wildlife Service wish to spray the perimeter of Lake Mosquito in order to protect the campers and cabin dwellers from mosquitos. Last year they sprayed a solution consisting of 10 percent pesticide, which was ineffective. Because last year's price was so good they overstocked the 10 percent solution by 42 barrels. This year they have decided to use a 25 percent solution but wish to use their leftover 10 percent solution. A 60 percent solution is available. How many barrels of 60 percent solution must be added to their 42 barrels of 10 percent solution to get a 25 percent solution?

28. Jan has $6000 in a savings account. She withdraws $1000 to loan to a co-worker at twice the bank interest rate. Find the respective rates if the annual interest earned is $280.

29. Roger has invested $6000 in a CD at 5 percent. He later found a better opportunity offering 8 percent. If he is now realizing 7 percent on his total investment, how much additional money did he invest at 8 percent?

30. Chuck has deposits of $12,000 and $9000 in separate CDs. The $9000 earns a 1/2 percent higher interest rate but $120 per year less than the $12,000. Find the respective interest rates.

31. A class decides to buy a calculator for $160. Had there been eight fewer members in the class, the cost per student would have risen by

only $1. How many students are in the class?

32. A group of 10 investors plans to construct an office building, but the cost is very high. If they can find five additional investors their individual investment will be reduced by $10,000. How much will the building cost assuming each must invest the same amount?

33. If Mary can paint a room in 2 hours and Janice can do so in 3 hours, how long will it take them to paint a room working together? Assume that they do not get in each other's way or otherwise distract one another. Thus, the answer is *not* 5 hours.

34. Dan can build a garage in 14 days less time than Dale can. They complete a garage in 24 days when they work together. How long does it take each to build a garage working alone?

35. Todd can paint a barn in 6 days less than it takes Doug. Working together they paint the barn in 4 days. How long would it take each to paint the barn working alone?

36. A large pipe can fill a pool in 39 hours less than a small pipe can. If both pipes are running, it takes 40 hours to fill the pool. How long would it take each pipe alone to fill the pool?

37. At the poultry plant, Gert can clean 500 turkeys in half a day (4 hours) while Gail can do only 500 turkeys in a full 8-hour day. Working together, how long would it take them to clean 1000 turkeys?

38. An oil mixture worth $1.17 a quart is to be made by adding oil worth 72¢ a quart to 100 quarts of oil priced at $1.35 a quart. How much must be added?

39. A magician asks one of his spectators to
    a. Think of a number
    b. Double it
    c. Add 4
    d. Divide by 2
    e. Subtract the original number
    The magician then states that the result is 2. Explain. (*Hint:* Let x denote the original number and then list the result of each action as it is taken.)

40. To conceal the trick of Exercise 39 from his audience, the magician instructs his next subject to perform the same steps except that some other number is used in part (c). He knows that the final result will be half the number added in part (c). Explain.

Some mentalists like to give the impression that they have memorized the calendar and are capable of performing feats of lightning addition. Typical stunts are described in the exercises that follow. Each of three participants from the audience is asked to enclose a three-by-three block of nine dates in a square as illustrated here. Explain each of the stunts described in the following three exercises.

13	14	15
20	21	22
27	28	29

41. The first participant is requested to add the numbers in the four corners of the square and state the result. The mentalist then recites all nine dates enclosed.

42. The next participant is asked to add the numbers in any two opposite corners of the square and state the result. The mentalist can then list all entries in the three-by-three block without even knowing which two corners were chosen!

43. The third participant is asked to indicate the smallest number in his chosen block. The mentalist immediately gives the total of all nine numbers in the block.

The following are variations of ancient puzzles. Find the solution to each.

44. Three honest women are to share equally in a gift from their mother. Confident of their honesty, the mother stacks the money on a table for them to pick up at their convenience. They arrive at separate times. Each thinks she is the first to arrive and takes only 1/3 of the money that is on the table when she arrives. After they have what they consider to be their respective shares, the mother finds $800 left on the table. What was the total of her original gift?

45. A hungry man meets two friends with three sandwiches and five sandwiches, respectively. After seeing his plight, they agree to share the food equally three ways. The hungry man then pays his friends a total of 80¢. How much money did each friend receive?

46. A messenger is sent from the rear to the front of a column of soldiers 3 kilometers long. **a.** If the column is marching forward at 5 kilometers per hour and the messenger travels at 25 kilo-

meters per hour, how long does it take the messenger to deliver the message and return? **b.** What would be the corresponding time for a messenger delivering a message from front to rear and returning?

47. On the gravestone of Diophantus is a riddle that purports to indicate his age. The message is that he spent
   a. One sixth of his life as a child
   b. One twelfth of his life as a youth
   c. One seventh of his life as an unmarried adult.
   He had a son who lived to be half as old as Diophantus. The son was born 5 years after Diophantus was married and died 4 years before Diophantus did. How old was Diophantus?

48. In a given room half the people are Caucasian, one fourth are black, one seventh are Chicano, and three are Oriental. How many people are in the room?

49. On the statue of Pallas Athene in Greece is engraved a message claiming that of all the gold used to make the statue
   a. Half came from Kariseus
   b. One eighth came from Thespian
   c. One tenth came from Solon
   d. One twentieth came from Themison
   e. Nine units came from Aristodikos
   How many units of gold are in the statue?

The following are "puzzle" problems that some people find interesting.

50. Two numbers differ by 3. Three times the larger number is 7 less than four times the smaller number. Find the numbers.

51. Find two consecutive even integers the squares of which differ by 28.

52. Find three consecutive integers such that the sum of the first two is 2 more than the last.

53. Find three consecutive odd integers such that the product of the first two is 20 less than the product of the last two.

54. If a square is enlarged to a rectangle by lengthening one pair of sides 2 centimeters and the other pair of sides three centimeters, the area is increased by 36 square centimeters. Find the dimensions of the original square.

55. The border around a square picture is 2 inches wide and has an area of 56 square inches. Find the dimensions of the picture.

56. Find two consecutive even integers the product of which is 168.

57. Find two consecutive even integers, the product of which is 528.

58. Find an integer that is 5 more than 24 times its reciprocal.

59. Find the two positive integers such that one is 6 more than the other and their reciprocals sum to 1/4.

2.5 PROPORTIONALITY; VARIATION

We noted earlier that the circumference and area of a circle of radius r are given as $C = 2\pi r$ and $A = \pi r^2$, respectively. We say that the circumference is directly proportional to r and the area is directly proportional to r^2. The concept of *proportionality* or *variation* is formalized as follows:

DEFINITIONS

If there is a constant $k \neq 0$ and an $r > 0$ such that one of the following relationships holds, then k is called a *constant of proportionality* and the terminology of proportion and variation is used as indicated.

1. $y = kx^r$ y *varies directly* as the rth power of x, or y is *directly proportional* to x^r.

2. $y = \dfrac{k}{x^r}$ y *varies inversely* as the rth power of x, or y is *inversely proportional* to x^r.

3. $y = kuv$ y *varies jointly* as u and v, or y is directly proportional to u and v.

4. $y = \dfrac{k}{uv}$ y is inversely proportional to u and v.

The reader should now be able to expand on the preceding notions of proportionality. For instance, to say that y is directly proportional to x^3 and w^2 and inversely proportional to $\sqrt[3]{u}$ and $v^{3/2}$ means that there is a constant $k \neq 0$ such that

$$y = k\frac{x^3 w^2}{\sqrt[3]{u}\, v^{3/2}}.$$

In many attempts to develop a mathematical model for a physical phenomenon, the investigator has some idea regarding direct and inverse variation among variables. The task then is to determine the constant of proportionality k. This can be done by obtaining corresponding values of the variables and solving the proportionality equations for k. The procedure is illustrated in the following examples.

EXAMPLE 1   The volume of a sphere is directly proportional to the cube of its radius. If the volume is 36π when the radius is 3, write the volume as a function of radius.

Solution   There is some constant $k \neq 0$ such that

$$V = kr^3.$$

When $r = 3$, $V = 36\pi$

$$36\pi = k \cdot 3^3 = 27k$$
$$k = \frac{36\pi}{27} = \frac{4\pi}{3}$$

Thus,

$$V = \frac{4\pi}{3}r^3.$$

■

EXAMPLE 2   If y varies directly as x^2 and inversely as z and if $y = 50$ when $x = 5$ and $z = 1$, write y in terms of x and z.

Solution   We have

$$y = k\frac{x^2}{z}$$

for some constant $k \neq 0$. The given conditions tell us that

$$50 = k\frac{5^2}{1}$$
$$= 25k$$

That is,

$$k = 2$$

so that

$$y = \frac{2x^2}{z}.$$

■

The proportionality relationship alone sometimes gives information that can be used even if we do not know the constant of proportionality. This situation is illustrated in the following examples.

EXAMPLE 3   Suppose that the efficiency of the liver is inversely proportional to the square of the amount of alcohol in the blood stream. What effect does doubling the alcohol content have on the liver's efficiency?

Solution   Letting E denote the efficiency of the liver and a denote the alcohol content, we have

$$E = \frac{k}{a^2}$$

for some constant k. Doubling a, the efficiency is reduced to

$$E^* = \frac{k}{(2a)^2} = \frac{k}{4a^2} = \frac{1}{4}\frac{k}{a^2}$$
$$= \frac{1}{4}E$$

Doubling the alcohol content reduces the efficiency of the liver to 1/4 its original value. ∎

EXAMPLE 4 Let y vary directly as u^2 and v and inversely as $\sqrt{w}$. If u is tripled, v is halved, and w is quadrupled, what happens to y?

Solution We can write

$$y = \frac{ku^2v}{\sqrt{w}}$$

However, if u is tripled, v is halved, and w is quadrupled, the new value y^* is found as follows:

$$y^* = \frac{k(3u)^2\left(\dfrac{v}{2}\right)}{\sqrt{4w}}$$

$$= \frac{k9u^2\left(\dfrac{v}{2}\right)}{2\sqrt{w}}$$

$$= \frac{k9u^2v}{4\sqrt{w}}$$

$$= \frac{9}{4}\frac{ku^2v}{\sqrt{w}}$$

$$= \frac{9}{4}y$$

that is, y is multiplied by 9/4 when the given changes are effected on u, v, and w. ∎

EXERCISES 2.5

1. Let y vary directly as x.
    a. If x is tripled, what happens to y?

    b. Write y in terms of x if $y = 2$ when $x = 10$.
    c. Find y when $x = 30$.

2. Let P vary inversely as q^2.
   a. If q is doubled, what happens to P?
   b. Write P in terms of q if $P = 3$ when $q = 3$.
   c. Find P when $q = 1$.

3. Let M vary directly as v^3.
   a. If v is halved, what happens to M?
   b. Write M in terms of v if $M = 500$ when $v = 5$.
   c. Find M when $v = 4$.

4. Let y vary directly as x and inversely as z^2 and w^3.
   a. If x, w, and z are all doubled, what happens to y?
   b. Write y in terms of x, w, and z if $y = 4$ when $x = 144$, $w = 2$, and $z = 3$.
   c. Under the conditions of (b), find y when $x = 162$, $z = 2$, and $w = 3$.

5. Let s be directly proportional to t^2 and inversely proportional to $\sqrt{u}$ and v^2.
   a. If t is doubled, u is multiplied by 9, and v is quadrupled, what happens to s?
   b. Write s in terms of t, u, and v if $s = 6$ when $t = 10$, $u = 4$, and $v = 5$.
   c. Under the conditions of (b), find s when $t = 2$, $u = 9$, and $v = 2$.

6. Let r vary directly as s^2 and inversely as h and l.
   a. If s, h, and l are all tripled, what happens to r?
   b. Write r in terms of s, h, and l if $r = 5$ when $s = 10$, $h = 2$, and $l = 10$.
   c. Under the conditions of (b), find r when $s = 20$, $h = 40$, and $l = 8$.

7. Let y be directly proportional to w and $\sqrt{x}$ and inversely proportional to z^3.
   a. If x is quadrupled, w is tripled, and z is halved, what happens to y?
   b. Write y in terms of w, x, and z if $y = 80$ when $x = 4$, $w = 128$, and $z = 4$.
   c. Under the conditions of (b), find y when $x = 25$, $w = 243$, and $z = 3$.

8. At a fixed velocity the distance traveled by a moving particle is directly proportional to the time of travel. Write distance in terms of time if the particle travels 100 feet in 20 seconds.

9. The time required for an automobile to travel a fixed distance is inversely proportional to its speed. Express the time t in terms of speed r if $t = 3$ seconds when $r = 80$ feet per second.

10. The intensity of illumination that an object receives from a light source varies inversely as the square of its distance from the source. If the distance from the source is tripled, how is the intensity of illumination affected?

11. The force of attraction between two objects varies directly as their masses and inversely as the square of the distance between them. If one of the masses is halved while the other is tripled and the distance between them shrinks to 1/3 its original value, how is the force affected?

12. The kinetic energy of a moving object is directly proportional to its mass and the square of its velocity. If the mass is quadrupled, what adjustment in velocity must be made in order to maintain the same kinetic energy?

13. The load that can be safely carried by a steel beam is directly proportional to its width and the square of its depth and inversely proportional to its length.
   a. If the initial construction plans called for beams 20 feet long, 5 inches wide, and 12 inches high to support a load of 30 tons, express the safe load in terms of width, depth, and length (all in inches).
   b. If for some reason a supporting wall is eliminated so that the span must now be 80 feet rather than 20 feet, how much wider must the beams be in order to support the same weight, assuming the depth remains the same? How much deeper must the beams be in order to support the same weight, assuming the width remains the same?
   c. If steel is sold by weight and weight is proportional to width, depth, and length, is it better for the contractor to buy wider or deeper beams in order to support the 30 tons over the longer span of 80 feet?

14. The electrical current carried by a wire is directly proportional to the applied voltage and inversely proportional to the resistance of the wire. If the voltage is doubled and the resistance is halved, how is the current flow affected?

15. The resistance of a wire is directly proportional to its length and inversely proportional to the square of its cross-sectional radius. If the diameter of the wire is tripled, how much longer can the length be without increasing the resistance?

16. **a.** Using the information given in the preceding two problems, indicate how the current flow, i, varies with respect to the length, L, of a wire, its radius, r, and the applied voltage, v.

   b. If 20 amps current must flow through 660 feet of wire and 220 volts suffice if the wire has a radius = 1/16 inch, how large a wire must be used if only 110 volts are available?

17. In physics, Hooke's Law says that the force exerted by a spring is directly proportional to the amount that it is stretched. If a force of 100 pounds is required to stretch the spring 4 inches, how many pounds of force are required to stretch the spring 25 inches?

18. The time it takes for a falling object to reach the ground is directly proportional to the square root of the altitude from which it falls. If it takes 2 seconds for an object to fall 64 feet, how long would it take the object to fall 400 feet?

19. According to Charles' Law in physics, the absolute temperature of an enclosed gas varies directly as the pressure and volume.

   a. How does the volume vary with respect to the temperature and pressure?

   b. How does the pressure vary with respect to the volume and temperature?

   c. If the absolute temperature remains the same and the volume is halved, how is the pressure affected?

20. If the circumference of a circle is doubled, how is the area affected? (See Exer. 56, Sect. 2.1.)

21. If the perimeter of a square is tripled, how is the area affected? (See Exer. 59, Sect. 2.1.)

22. If the circumference of a great circle on a sphere is tripled, how is its surface area affected? (See Exer. 57, Sect. 2.1.)

23. If the altitude of a triangle is doubled and its base is decreased to 1/3 its original length, how is its area affected?

24. Certain science-fiction movies depict giant insects wreaking havoc on the human population. These movies are more fiction than science. a little mathematics should convince us that such giant insects would be so weak they wouldn't even be able to move their own limbs. Recall that the volume of a sphere of radius r is $4\pi r^3/3$ and the area of a circle of radius r is πr^2. It is reasonably accurate to say that the volume, and hence, the weight, of an animal or insect is directly proportional to the cube of its height or length; the cross-sectional area of its muscles varies directly as the square of its height or length. The weight the animal can lift is directly proportional to the cross-sectional area of its muscles. The strength of an animal is defined as

$$\text{Strength} = \frac{\text{weight the animal can lift}}{\text{weight of the animal}}$$

   a. Show that the strength s of an animal is inversely proportional to its height or length h: $s = k/h$ for some constant k; the constant k will depend on the type of animal.

   b. An average-sized ant 1 centimeter long can lift about three times its own weight; how strong would a human-sized ant 180 centimeters long be?

   c. An average-sized man 180 centimeters tall can lift about half his body weight; how strong would an ant-sized man 1 centimeter tall be?

   d. Whose body structure is inherently stronger, the human's or the ant's?

25. The rate at which an animal metabolizes energy is directly proportional to its surface area, which, in turn, varies directly as the square of its height or length. The oxygen for this metabolism is transported from the lungs to individual cells by the bloodstream. Thus, the rate of metabolism is directly proportional to the availability of oxygen, which, in turn, varies directly as the volume of blood pumped by each heartbeat and the rate at which the heart beats. The volume of the heart is directly proportional to the cube of the height or length of the animal.

   a. Show that the heart rate r of an animal is inversely proportional to its length h: $r = k/h$ for some constant k, depending on the type of animal.

   b. Compare the heart rates of
      1. An 18-inch infant
      2. A 2-year-old who is 3 feet tall
      3. An adult 6 feet tall

2.6 ALGEBRAIC INEQUALITIES WITH ONE VARIABLE

Since algebraic expressions represent numbers, the order and absolute value rules discussed in Section 1.2 also hold for operations on such expressions.

Expressions of the form

$$2x^2 + 3x < 5$$
$$3y + 10 \le 0$$
$$5z + 20z^3 > 16z^2$$

are called algebraic inequalities. The solution to an inequality consists of all values that, when assigned to the variable, make the inequality true. The following table reviews the rules for operations on inequalities.

Rule	Example
1. $a < b$ if and only if $b - a > 0$.	1. $2 < 5$ and $5 - 2 = 3 > 0$
2. If $a < b$ and $b < c$, then $a < c$.	2. $\xleftarrow{\qquad\quad a \qquad\qquad b \quad c \qquad}\rightarrow$
3. If $a < b$, then $(a + c) < (b + c)$.	3. $\xleftarrow{\quad a \quad a+c \qquad b \quad b+c \quad}\rightarrow$
4. If $a < b$, then $-a > -b$.	4. $\xleftarrow{\quad -5 \quad -2 \quad 0 \quad 2 \quad 5 \quad}\rightarrow$ $2 < 5$ and $-2 > -5$
5. If $a < b$ and **a.** $x = 0$, then $ax = bx$ **b.** $x > 0$, then $ax < bx$ **c.** $x < 0$, then $ax > bx$	5. $3 < 5$ and **a.** $3 \cdot 0 = 5 \cdot 0 = 0$ **b.** $3 \cdot 2 < 5 \cdot 2$ since $6 < 10$ **c.** $3 \cdot (-2) > 5 \cdot (-2)$ since $-6 > -10$
6. If $0 < a < b$, then $0 < \dfrac{1}{b} < \dfrac{1}{a}$.	6. Larger denominators make smaller fractions. $0 < 2 < 5$ and $0 < \dfrac{1}{5} < \dfrac{1}{2}$.
7. The product of two positive or two negative numbers is positive.	7. $2 \cdot 3 = 6$ $(-2)(-3) = 6$
8. The product of a negative and a positive number is negative.	8. $2(-3) = -6$

Linear Inequalities

In this section we shall restrict our attention to inequalities that can be solved by analyzing expressions of the form $ax + b$. These are called **linear inequalities.** As with first-degree equations in one variable, we work to isolate the variable on one side of the inequality.

EXAMPLE 1 Solve the inequality

$$10 - x > 2x - 2$$

and sketch the solution on the real number line.

SOLVING LINEAR INEQUALITIES

> 1. Simplify the algebraic expressions.
>
> 2. Isolate the variable on one side of the inequality by
>
>   a. adding or subtracting terms.
>
>   b. multiplying or dividing by a constant. Note that multiplication or division by a negative number reverses the inequality.

Solution

$$10 - x > 2x - 2$$
$$10 - 3x > -2 \qquad \text{(Subtract } 2x \text{ from both sides.)}$$
$$-3x > -12 \qquad \text{(Subtract 10 from both sides.)}$$
$$3x < 12 \qquad \text{(Reverse the inequality.)}$$
$$x < 4 \qquad \text{(Divide both sides by 3.)}$$

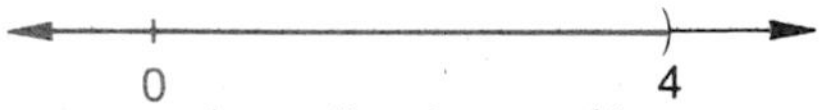

$$(-\infty, 4) = \{x : x < 4\}$$

Figure 2–3

The solution consists of the set of all $x < 4$ as shown in Figure 2–3. Recall from Section 1.2 that such a set is an interval, denoted $(-\infty, 4)$. Also recall that the parenthesis next to the 4 indicates that 4 is not included in the set. Had 4 been included, we would have used a square bracket "]" and written $(-\infty, 4] = \{x : x \leq 4\}$. ∎

EXAMPLE 2 The manager of the Campus Bookstore can sell 4000 copies of "Passing Tests Made Easy" if he charges $8 per copy. For each 50¢ increase in price, he loses 100 sales. If each book costs the store $5, express the profits in terms of the selling price x. At what selling prices does the store make a profit?

Solution Let

$$x = \text{sales price in dollars}$$
$$x - 8 = \text{increase in price over \$8}$$
$$2(x - 8) = \text{number of 50¢ increases over \$8}$$
$$2(x - 8)100 = \text{number of sales lost}$$
$$4000 - 2(x - 8)100 = \text{resulting number of sales}$$
$$x - 5 = \text{profit per book}$$
$$[4000 - 2(x - 8)100](x - 5) = \text{total profit}$$

The bookstore makes a profit when

$$[4000 - 2(x - 8)100](x - 5) > 0$$

Since no profit is made if the book sells for no more than the store's purchase price, the manager will clearly set $x > 5$. Thus, he makes a profit when

$$4000 - 2(x - 8)100 > 0$$

since $(x - 5) > 0$ when $x > 5$. Solving this inequality, we have

$$4000 - 2(x - 8)100 > 0$$
$$4000 > 200(x - 8)$$
$$20 > x - 8$$
$$28 > x$$

The store makes a profit if the book sells for more than $5 and less than $28.

■

Multiple Inequalities

Problems often arise that require two or more inequalities to be satisfied simultaneously. Such multiple inequalities are treated as separate inequalities that are solved separately. Only values of the variable that satisfy each of the separate inequalities will satisfy the multiple inequality.

EXAMPLE 3   Solve the following double inequality.

$$2 - x < 3x + 2 \leq x + 4$$

Solution   We consider the two inequalities indicated. These are solved separately as follows.

The first inequality:	The second inequality:
$2 - x < 3x + 2$	$3x + 2 \leq x + 4$
$-x < 3x$	$3x \leq x + 2$
$0 < 4x$	$2x \leq 2$
$0 < x$	$x \leq 1$

Both inequalities are satisifed when $0 < x \leq 1$. The solution consists of all x's in the interval $(0, 1]$ as illustrated in Figure 2–4.

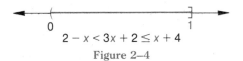

$$2 - x < 3x + 2 \leq x + 4$$

Figure 2–4

■

Higher-Order Inequalities

1. Express the inequality as a product of linear factors (if possible) with 0 on the other side of the inequality symbol.

2. Find the points in which these factors become 0; the factor will be negative on one side of this 0-point and positive on the other.

3. Collect this information into a "sign diagram" (see Figures 2–5 and 2–7).

4. An odd number of negative factors are required to make the product negative. Otherwise the product is positive (except at the 0-points).

EXAMPLE 4   Solve the following inequality.

$$(2u - 6)(3u + 12) < 0$$

Solution   Since this inequality is a product of linear factors, the two factors are analyzed separately in order to determine the intervals on which they are positive and negative.

Observe that	Similarly
$2u - 6 < 0$	$2u - 6 > 0$
when	when
$2u < 6$	$u > 3$
$u < 3$	

Also note that	Consequently
$3u + 12 < 0$	$3u + 12 > 0$
when	when
$3u < -12$	$u > -4$
$u < -4$	

We note that the signs of the given factors may change at $u = 3$ and at $u = -4$. It is convenient to indicate the signs of $(2u - 6)$ and $(3u + 12)$ on the intervals of the real line determined by $u = 3$ and $u = -4$ in Figure 2–5.

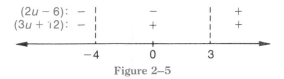

Figure 2–5

In order for the product $(2u - 6)(3u + 12)$ to be negative, exactly one of the factors must be negative while the other is positive. We see from Figure 2–5 that this happens precisely when $-4 < u < 3$. The solution set is the interval $(-4, 3)$ as shown in Figure 2–6.

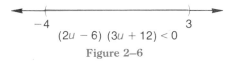

$(2u - 6)(3u + 12) < 0$

Figure 2–6

EXAMPLE 5   Solve the following inequality.

$$(x^2 - 7x + 12)(2x + 10) \le 0$$

Solution   To solve this inequality, we factor the quadratic expression to obtain

$$(x - 4)(x - 3)(2x + 10) \leq 0$$

Analyzing each factor separately, we obtain the sign diagram shown in Figure 2–7. Since we must have an odd number of negative factors, the solution consists of those x's satisfying $x \leq -5$ or $3 \leq x \leq 4$ as indicated in the figure.

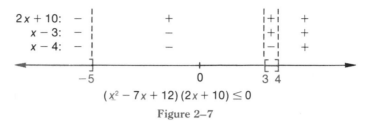

$$(x^2 - 7x + 12)(2x + 10) \leq 0$$

Figure 2–7

Fractional Inequalities

Although "cross-multiplying" is sometimes a quick procedure for solving fractional equations, this procedure should be avoided with inequalities. For multiplication by an expression involving the variable may reverse the inequality (if the multiplier is negative). Rather than consider additional cases (positive vs. negative multipliers), it is best to proceed as follows.

INEQUALITIES INVOLVING FRACTIONS

1. Bring all terms to one side.

2. Combine the fractions.

3. Simplify and factor.

4. Analyze the individual factors and construct a sign diagram.

EXAMPLE 6 Solve the following inequality.

$$\frac{2}{t + 3} \leq \frac{1}{t - 4}$$

Solution

$$\frac{2}{t + 3} - \frac{1}{t - 4} \leq 0$$

$$\frac{2t - 8 - t - 3}{(t + 3)(t - 4)} \leq 0$$

$$\frac{t - 11}{(t + 3)(t - 4)} \leq 0$$

The inequality is satisfied when we have an odd number of negative factors. Constructing the sign diagram (Fig. 2–8), we see that this happens when $t \leq -3$ and when $4 \leq t \leq 11$. But $t = -3$ and $t = 4$ must be discarded since these values would indicate division by 0. Thus, the solution set consists of those t's satisfying $t < -3$ or $4 < t \leq 11$ as illustrated in Figure 2–8.

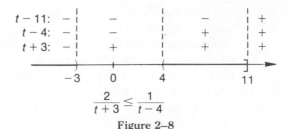

$$\frac{2}{t+3} \le \frac{1}{t-4}$$

Figure 2–8

Blindly cross-multiplying in Example 6 would have given

$$2(t - 4) \le (t + 3)$$
$$2t - 8 \le t + 3$$
$$t \le 11$$

But we saw that not every $t \le 11$ satisfies the inequality. Since multiplication by a negative number reverses an inequality, we *must consider the signs* of $(t + 3)$ and $(t - 4)$ when cross-multiplying. If we were careful in this respect, we could then arrive at the same solution as before.

EXERCISES 2.6 ODD

Solve the following inequalities. Sketch the solution sets on the real number line.

1. $2y - 16 > 0$
2. $4z + 20 \le 0$
3. $3x - 2 < 10$
4. $3r + 15 \le 0$
5. $3s - 9 > 0$
6. $3t - 2 > \dfrac{(10 - 2t)}{8}$
7. $u + 1 < 7u - 2$
8. $4w - 8 \ge w + 1$
9. $3 < 9 - 3z \le 6$
10. $2 \le \dfrac{4 - 2r}{3} \le 8$
11. $-2 < 2y + 6 < 4$
12. $t - 2 \le 3t - 1 \le 2 - 3t$
13. $4u - 8 < u + 1 < 6u + 2$
14. $2w - 4 \ge 2 - 4w > w - 13$
15. $(x - 3)(x + 2) < 0$
16. $(z + 5)(z - 1) > 0$
17. $(2s + 1)(3s - 6) \ge 0$
18. $(2v + 4)(3v - 1)(v - 7) < 0$
19. $(6 - w)(4 + 2w)(w - 5) > 0$
20. $(2x + 8)(3x - 12)(4 - x) \le 0$
21. $(u - 3)(2 - u)(14u + 28) \ge 0$
22. $x^2 - x + 2 \le 4$
23. $2y^2 + 3y - 2 \ge 0$
24. $8r^2 + 5r + 4 < 2r^2 - 5r - 8$
25. $(6t^2 + 6t - 36)(3t - 6) > 0$
26. $(s^2 + s - 2)(s + 3) \le 0$
27. $(v^2 - v - 6)(-4v^2 + 8v - 3) < 0$

28. $\dfrac{1}{w - 2} < 0$
29. $\dfrac{2}{v^2 - v - 2} < 0$
30. $\dfrac{2x + 8}{x^2 - 9} \le 0$
31. $\dfrac{w - 1}{w^2 - w - 2} \ge 0$
32. $\dfrac{2y + 8}{(y^2 - 4)(2y^2 + 5y - 3)} > 0$
33. $\dfrac{2}{x - 3} \le 6$
34. $1 \le \dfrac{2}{y - 4} < 3$
35. $0 < \dfrac{1}{r - 5} \le \dfrac{3}{r + 4}$
36. $-1 < \dfrac{1}{2z + 1} \le 2$
37. $\dfrac{2}{3s + 6} < \dfrac{1}{s - 7} \le 0$
38. $\dfrac{1}{(t - 2)(t - 3)} \le \dfrac{4}{t + 2} \le 2$
39. $\dfrac{1}{(u - 1)(u - 2)} \le \dfrac{2}{(u + 1)(u - 2)}$
$$\le \dfrac{2}{(u + 3)(u - 3)}$$

40. Mary's Florist has fixed costs of $900 each month. For each $100 in retail sales, she has

additional costs of $70. How much business must she have each month to make a profit?

41. A heart patient enters the hospital with a resting heart rate of 100 beats per minute. Her physician wants to decrease this rate medically to between 60 and 80 beats per minute. If 2 milligrams of medication per hour reduces the rate by 5 beats per minute, what should the prescription be?

42. If your grades on the first four tests in this course are 75, 95, 70, and 90, what grade must you receive on the fifth test in order to increase your average to 85 or better?

43. What would be the answer to Exercise 42 if the instructor is willing to drop your lowest grade?

44. A rectangle with one side of length 5 centimeters must have an area greater than 25 square centimeters and less than 75 square centimeters. What lengths are permitted for its other dimension?

45. A triangle with base of length 5 inches must have an area greater than 25 square inches and less than 75 square inches. What altitudes are permitted?

2.7 EQUATIONS AND INEQUALITIES INVOLVING ABSOLUTE VALUES

The fundamental relationships between the absolute value and inequality symbols are summarized in the following table.

Expression	Solution	Illustration
$\lvert u \rvert < a$	$-a < u < a$	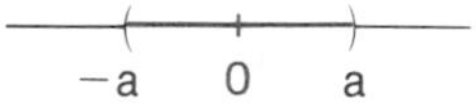 $-a \quad 0 \quad a$
$\lvert u \rvert = a$	$u = \pm a$	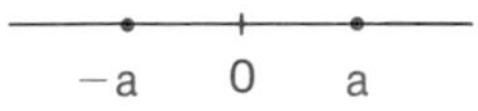 $-a \quad 0 \quad a$
$\lvert u \rvert > a$	$u > a$ or $u < -a$	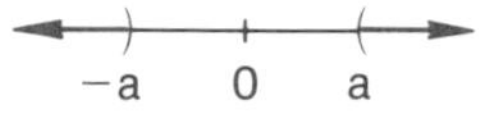 $-a \quad 0 \quad a$

To solve the absolute value inequality $\lvert x - 3 \rvert < 5$, we let $u = x - 3$ and use these properties as follows.

$$\lvert u \rvert < 5 \quad \text{means} \qquad -5 < \quad u \quad < 5$$
$$\lvert x - 3 \rvert < 5 \quad \text{means} \qquad -5 < x - 3 < 5$$
$$-2 < \quad x \quad < 8 \qquad \text{(Add 3 to all parts of the previous inequality.)}$$

See Figure 2–9.

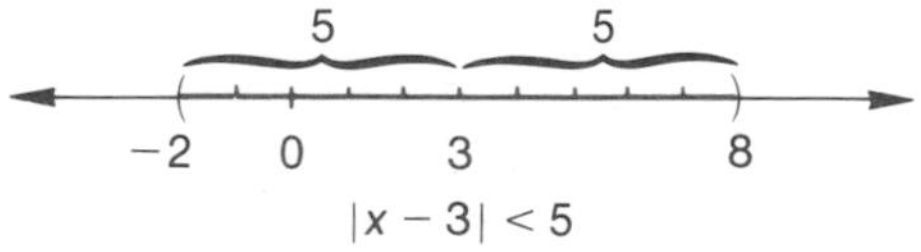

$$\lvert x - 3 \rvert < 5$$

Figure 2–9

EXAMPLE 1   Solve the following inequality.

$$|2x + 1| < 3$$

Solution   We know that $|2x + 1| < 3$ if and only if

$$-3 < 2x + 1 < 3$$
$$-4 < \quad 2x \quad < 2$$
$$-2 < \quad x \quad < 1.$$

The solution set is sketched in Figure 2–10.

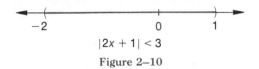

$$|2x + 1| < 3$$

Figure 2–10

EXAMPLE 2   Solve the following equation.

$$|2x - 3| = 5$$

Solution   Observe that $|2x - 3| = 5$ when $2x - 3 = \pm 5$. Each of these equations is solved separately:

$$2x - 3 = 5 \qquad 2x - 3 = -5$$
$$2x = 8 \qquad\qquad 2x = -2$$
$$x = 4 \qquad\qquad x = -1$$

The solutions are $x = 4$ and $x = -1$.

EXAMPLE 3   Solve the following inequality.

$$|3t - 4| \geq 5$$

Solution   Letting $u = 3t - 4$ and using the properties listed at the beginning of this section, we have

$$|u| \geq 5 \quad \text{means} \qquad u \geq 5 \text{ or} \qquad u \leq -5$$
$$|3t - 4| \geq 5 \quad \text{means} \quad 3t - 4 \geq 5 \text{ or } 3t - 4 \leq -5$$

If t satisfies either of these conditions, the original inequality holds. Thus we solve each of these latter inequalities separately.

$$3t - 4 \geq 5 \qquad\qquad 3t - 4 \leq -5$$
$$3t \geq 9 \qquad\qquad\qquad 3t \leq -1$$
$$t \geq 3 \quad \text{or} \qquad\qquad t \leq -\frac{1}{3}$$

The solution set is illustrated in Figure 2–11.

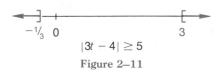

$$|3t - 4| \geq 5$$

Figure 2–11

EXAMPLE 4 Rancher MacDonald must build a corral for his cattle. He plans to use a rectangular canyon 100 feet wide. If he needs an area of approximately 40,000 square feet but will tolerate an error of 1000 square feet, how accurately must his foreman measure the length of the corral before building the fences?

Solution Let x denote the length of the corral. Then $100x$ is its area. The area is to be within 1000 square feet of 40,000:

$$|100x - 40,000| < 1000$$
$$-1000 < 100x - 40,000 < 1000$$
$$39,000 < 100x < 41,000$$
$$390 < x < 410.$$

The length must be between 390 and 410 feet to satisfy Mr. MacDonald's requirements. Figure 2–12 shows that this is equivalent to $|x - 400| < 10$. The foreman should measure the length of the corral as 400 feet; he must be accurate to within 10 feet.

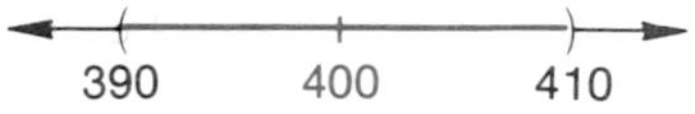

390   400   410

$$|100x - 40,000| < 1000$$

Figure 2–12

EXAMPLE 5 Solve the inequality

$$|x - 4| < |2x - 2|$$

Solution In this problem we must consider separately the two cases determined by the sign of $2x - 2$.

a. $2x - 2 \geq 0$.
In this case $|2x - 2| = 2x - 2$ and the inequality becomes

$$|x - 4| < 2x - 2$$
$$-(2x - 2) < x - 4 < 2x - 2$$
$$-2x + 2 < x - 4 < 2x - 2$$

This is a double inequality of the type studied in the preceding section. Each inequality is solved separately:

$$-2x + 2 < x - 4 \quad \text{and} \quad x - 4 < 2x - 2$$
$$6 < 3x \qquad\qquad -2 < x$$
$$2 < x$$

Don't forget that we are now considering only those values of x for which

$$2x - 2 \geq 0$$
$$x \geq 1$$

All three of these conditions must be satisfied: $x > 2$, $x > -2$, $x \geq 1$. Since $-2 < 1 < 2$, we see that all $x > 2$ are solutions.

b. $2x - 2 \leq 0$. Now $|2x - 2| = -(2x - 2) = 2 - 2x$ and we have

$$|x - 4| < 2 - 2x$$
$$-(2 - 2x) < x - 4 < 2 - 2x$$
$$2x - 2 < x - 4 < 2 - 2x$$

$$2x - 2 < x - 4 \quad \text{and} \quad x - 4 < 2 - 2x$$
$$x < -2 \qquad\qquad 3x < 6$$
$$x < 2$$

We are now considering only those x for which

$$2x - 2 \leq 0$$
$$x \leq 1$$

Again all three conditions must be satisfied: $x < -2$, $x < 2$, $x \leq 1$. Since $-2 < 1 < 2$, all $x < -2$ are solutions.

Combining **a** and **b**, the solution set is sketched in Figure 2–13.

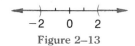

Figure 2–13

EXERCISES 2.7 *ODDS*

Solve the following equations and inequalities. Sketch the solution set on the real number line.

1. $|x + 5| = 7$
2. $|y - 3| = 2$
3. $|2z + 4| = 10$
4. $|3r - 7| = 2$
5. $|2s - 10| = -5$
6. $|t - 3| = 3 - t$
7. $|3u - 9| = 3u - 9$
8. $|3v + 1| \leq 8$
9. $|11 - 3x| \geq 13$
10. $|2w - 5| > 11$
11. $|3y - 12| \geq 12$
12. $|2z + 5| < 3$
13. $|2r - 3| < -2$
14. $|t - 7| \geq 3$
15. $|2u - 5| > 5$
16. $\left|\dfrac{x - 2}{3}\right| < 4$
17. $\left|\dfrac{3q - 1}{2}\right| < 5$
18. $\left|\dfrac{2 - 3p}{4}\right| \geq 3$
19. $\left|\dfrac{3 - x}{4}\right| \geq 2$
20. $|2s + 3| < 9$
21. $|z + 5| < 2$
22. $|2s - 10| \geq 4$
23. $|2v - 3| \leq 3$
24. $|2t + 5| > 1$
25. $|1 - 3t| < 4$

26. $|z + 5| < |z - 3|$
27. $|10 - 3t| > |2t - 5|$
28. $|2v - 3| > |2v + 3|$
29. $\left|\dfrac{3x - 1}{2}\right| \leq \left|\dfrac{3 - x}{4}\right|$
30. $|x^2 - 6x + 9| < (x - 3)$
31. $|x^2 - 4| < |x - 2|$
32. In order for Rare Oak acorns to germinate, the temperature must be within 5° of 27°C. **a.** Write an absolute value inequality that must be satisfied by the temperature T in order for germination to take place. **b.** Express the solution to this inequality as an interval.
33. The Gas-N-Power Utility Company estimates the monthly cost of heating or cooling a medium-sized home to be $3.75 times the difference between 60°F and the average temperature during the month. Ms. Walsh can afford at most $150 per month for heating and cooling.

a. Write an absolute value inequality that must be satisfied by the temperature T if she is to be able to pay her utility bills.

b. In what temperature range can she meet her obligation to the utility company? Express the solution set as an interval.

34. Certain sensitive instruments must be shipped in a special cylinder. The circumference of the cylinder must be approximately 16π centimeters. An error of no more than $\pi/5$ centimeters from this figure will be tolerated. How accurately must the radius of the cylinder be determined? (*Hint:* The circumference of a circle of radius r is $2\pi r$ where π is a constant with value approximately 3.14159.)

2.8 CHAPTER REVIEW

TERMS AND CONCEPTS

• Equations
The indicated equality of two quantities or expressions.
1. Identities — True for all values of the unknown.
2. Conditional equations — True for only certain values of the unknown.
• Variable
Represents a number (whose value is not specified).
• Constant
A fixed or specified number.
• Solution
Values of the variable that make an equation or inequality true.
• Proportionality
1. y is directly proportional to x^r — $y = kx^r$
2. y is inversely proportional to x^r — $y = k/x^r$

RULES AND FORMULAS

• Equations
1. Operations performed on one side of an equation must also be done on the other side.
2. $uv = 0$ if and only if $u = 0$ or $v = 0$.
• Quadratic Formula
If $ax^2 + bx + c = 0$, $a \neq 0$, then
$$x = \frac{-b \pm \sqrt{b^2 - 4ac}}{2a}.$$

SOLUTION TECHNIQUES

• First Degree Equations in One Variable
1. Simplify the algebraic expressions.
2. Isolate the variable on one side of the equation by adding or subtracting terms and multiplying or dividing by a constant.

$$2x - 6 = 0$$
$$2x = 6$$
$$x = 3$$

• Quadratic Equations
1. Factoring

$$x^2 - 4x + 3 = 0$$
$$(x - 3)(x - 1) = 0$$
$$x = 1, 3$$

2. Completing the square

$$x^2 - 2x - 1 = 0$$
$$x^2 - 2x + 1 = 2$$
$$(x - 1)^2 = 2$$
$$x - 1 = \pm \sqrt{2}$$
$$x = 1 \pm \sqrt{2}$$

3. Quadratic Formula

$$x^2 - 2x - 1 = 0$$
$$x = \frac{-(-2) \pm \sqrt{(-2)^2 - 4 \cdot 1 \cdot (-1)}}{2 \cdot 1}$$
$$= \frac{2 \pm 2\sqrt{2}}{2} = 1 \pm \sqrt{2}$$

- Word Problems
  1. Understand the problem; read it several times.
  2. Identify the unknown quantities.
  3. Name these quantities.
  4. Express all known relationships mathematically.
  5. Reread the problem to check your information.
  6. Solve the resulting equations.
  7. Interpret your results.
  8. Check your solutions.

- Linear Inequalities
  1. Simplify the algebraic expressions.
  2. Isolate the variable on one side of the inequality. Note that the multiplication or division by a negative number reverses the inequality.

$$5x - 2 < 2x + 10$$
$$5x < 2x + 12$$
$$3x < 12$$
$$x < 4$$

- Higher Order Inequalities
  1. Express the inequality as a product of linear factors (if possible) with 0 on the other side of the inequality symbol.
  2. Find the points in which these factors become 0; the factor will be negative on one side of the 0-point and positive on the other.
  3. Collect this information into a "sign diagram."
  4. An odd number of negative factors make the product negative. Otherwise, the product is positive (except at the 0-points).

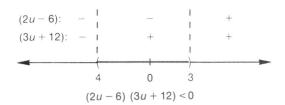

$$(2u - 6)(3u + 12) < 0$$

- Equations Involving Fractions

  1. Multiply through by the LCD.

  2. Solve the resulting equation.

  3. Check your answers in the original equation; you may have introduced extraneous roots.

$$\frac{3}{x + 1} = \frac{2}{x}$$

$$3x = 2(x + 1)$$
$$3x = 2x + 2$$
$$x = 2$$

- Inequalities Involving Fractions

  1. Bring all terms to one side.

  2. Combine the fractions.

  3. Simplify and factor.

  4. Analyze the individual factors and construct a sign diagram.

$$\frac{3}{x + 1} < \frac{2}{x}$$

$$\frac{3}{x + 1} - \frac{2}{x} < 0$$

$$\frac{3x - 2(x + 1)}{x(x + 1)} < 0$$

$$\frac{x - 2}{x(x + 1)} < 0$$

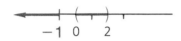

- Equations and Inequalities Involving Absolute Value

  1. $|u| < a$ means $-a < u < a$

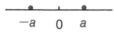

  2. $|u| = a$ means $u = \pm a$

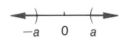

  3. $|u| > a$ means $u > a$ or $u < -a$

$$|2x + 1| < 3$$
$$-3 < 2x + 1 < 3$$
$$-4 < 2x < 2$$
$$-2 < x < 1$$

2.9 SUPPLEMENTARY EXERCISES

Solve the following equations.

1. $3y - 10 = 6y + 20$
2. $3t - 5 = 3 - 5t$
3. $4x - 6 = 3x + 10$
4. $7(q - 2) = 6(q + 3)$
5. $7(w - 2) - 3(5 - w) = 11$
6. $\dfrac{r}{6} - 5 = \dfrac{r}{2} - 7$
7. $\dfrac{v}{3} - 2 = 3 - \dfrac{v}{2}$
8. $\dfrac{3p + 4}{2} - 6 = \dfrac{1 - 3p}{5}$
9. $\dfrac{1 + u}{3} - \dfrac{u}{2} = \dfrac{u - 2}{7}$

10. $\dfrac{z - 3}{2} + \dfrac{z + 3}{4} = \dfrac{8 - z}{3} + 2$
11. $(x + 1) + 3yi = 5 - 6i$
12. $2x + 3i = 8 + yi$
13. $x^2 - 6x + 9 = 0$
14. $y^2 - 10y + 21 = 0$
15. $3s^2 + 19s + 6 = 0$
16. $3r^2 - 8r - 3 = 0$
17. $x^2 + 2x + 2 = 0$
18. $t^2 - 5t + 6 = 0$
19. $r^2 - 4r - 12 = 0$
20. $2x^2 - 5x - 3 = 0$
21. $8 - 14t - 15t^2 = 0$
22. $2u^2 - 7u - 4 = 0$

23. $3t^2 - 10t + 3 = 0$
24. $4y^2 + 3y - 1 = 0$
25. $2x^2 + 2x + 5 = 0$
26. $w^2 + 2w + 10 = 0$
27. $2y^2 - 6y + 9 = 0$
28. $z^2 + 6z + 25 = 0$
29. $r^2 - 4r + 5 = 0$
30. $t^2 - 6t + 10 = 0$
31. $s^2 + 6s + 13 = 0$
32. $x^2 + 6x + 34 = 0$
33. $u^2 + 8u + 25 = 0$
34. $v^2 - 4v + 13 = 0$
35. $u^4 + 4u^2 - 12 = 0$
36. $s^6 + 10s^3 - 24 = 0$
37. $s^8 - 6s^4 + 8 = 0$
38. $v^{10} + 7v^5 + 12 = 0$
39. $4u^{12} - 3u^6 - 1 = 0$
40. $q - \sqrt{q} - 2 = 0$
41. $\sqrt{z} - 3\sqrt[4]{z} + 2 = 0$
42. $3v + 11\sqrt{v} = 4$
43. $\sqrt{2t} - 5\sqrt{t} + \sqrt{18} = 0$
44. $\sqrt{27u} + 5\sqrt{u} = \sqrt{48}$
45. $\sqrt{2x - 1} + \sqrt{x - 1} = 5$
46. $\sqrt{2x - 1} = 2x - 1$
47. $\sqrt{2 - 4u} + 3 = \sqrt{8u + 5}$
48. $\sqrt{5y + 6} - \sqrt{y + 3} = 3$
49. $\sqrt{r - 6} + \sqrt{2r + 2} = \sqrt{3r + 4}$
50. $\dfrac{x - 1}{x + 1} + \dfrac{5 - 2x}{x - 1} = 0$
51. $\dfrac{6}{u + 1} - \dfrac{3}{u + 2} = 1$
52. $\dfrac{3}{t - 2} - \dfrac{5}{2t - 4} = \dfrac{2}{t + 1}$
53. $\dfrac{3 - 2q^2}{q^2 + q - 2} + \dfrac{2q}{q + 2} = \dfrac{5}{q - 1}$
54. $\dfrac{w}{w^2 + w - 2} + \dfrac{2}{1 - w^2} = \dfrac{-2}{w^2 + 3w + 2}$
55. $|z + 4| = 2$ 56. $|u - 5| = -6$
57. $(z + i + 1)^2 = (z + 3)^2$
58. $(z - i + 3)(z + 2i) = (z - 3i)(z + i + 2)$
59. $z^2 - 2iz - 5 = 0$
60. $-iz^2 + 4z + 2i = 0$
61. $3iz^2 + 5z + 2i = 0$

Solve the following inequalities and sketch the solution sets on the real number line.

62. $3r + 15 \leq 0$
63. $7x + 10 \geq 3x - 2$
64. $5v - 2 < 3v - 8$

65. $(z + 5)(z - 1) > 0$
66. $(r - 1)(r + 4) \leq 0$
67. $-2x^2 + 4x + 7 > 2x - 5$
68. $(t + 1)(t - 2)(2t - 8) < 0$
69. $(u^2 - 1)(u^2 + u - 6) \geq 0$
70. $\dfrac{2}{v^2 - v - 2} < 0$
71. $-2 \leq \dfrac{10 + 5s}{-5} \leq 2$
72. $2v - 4 > 2 - 4v \geq v - 8$
73. $0 < \dfrac{1}{r - 3} \leq \dfrac{3}{r - 5}$
74. $|5q + 4| \geq 6$
75. $|p + 4| \geq 2$
76. $|2s + 3| < 9$
77. $|3t + 1| < 4$
78. $|r + 3| < 4$
79. $\left|\dfrac{3 - x}{4}\right| \geq 2$
80. $|20q + 4| \geq |3q - 7|$
81. $|3v - 12| < |2v + 5|$
82. $|3t + 1| < |10 - 2t|$
83. $|p^2 + 4p + 4| < |p + 2|$
84. The area of a square varies directly as the square of its side length; its perimeter varies directly as its side length. If the side length is tripled, what happens to the perimeter? the area?
85. Let r vary directly as p^3 and s and inversely as the square root of q.
    a. If p is doubled, s is halved, and q is quadrupled, what happens to r?
    b. Write r in terms of p, q, and s if $r = 270$ when $p = 3$, $q = 4$, and $s = 2$.
    c. Under the conditions of (b), find r when $p = 1$, $q = 9$, and $s = 12$.
86. Suppose that the walking speed of an animal is directly proportional to the square root of its leg length.
    a. If a fully grown person with 30-inch legs walks at a rate of 4 miles per hour, how fast will a 2-year-old with 15-inch legs walk?
    b. How fast will a horse walk if its legs are 42 inches long?
    c. How fast will a giraffe walk if its legs are 5 feet long?
87. The cost of operating an automobile varies directly as its age and the square root of its

weight. If it costs 20¢ per mile to operate a 2-year-old 4000-pound automobile, how much will it cost to operate a 5-year-old 3000-pound car?

88. Because of fixed operating costs, the profit earned by an electronics supply firm varies directly as the square of the number of units of Super Stereos it sells and inversely as its utility costs. If it traditionally sells 50 percent more units in the winter when its utility costs are double those of the summer, compare its summer and winter profits.

89. To service a backwoods cabin, a utility company must run an electric cable through the forest to a pole along the highway. The perpendicular distance from the cabin to the highway is 200 meters less than the direct distance from the cabin to the pole. By running the cable directly from the cabin to the pole, rather than perpendicular to and along the highway, the company saves 200 meters of wire.
   a. How far is the cabin from the highway?
   b. How much wire is needed if the cable is to be stretched directly from cabin to this nearest pole?

90. Ten thousand dollars is invested in two CDs paying 6 percent and 8 percent interest, respectively. If the annual return on these investments is $660, find the amount invested at each rate. (*Hint:* Let the respective amounts be x and $(10,000 - x)$.)

91. Five small businesses plan to share in the purchase of a small computer. If they encourage three additional companies to join them, their individual investments will be decreased by $1200. How much does the computer cost?

92. Mary bought an order of roses for $1000. She can resell the roses at a profit of $25 per carton. But on delivery, four cartons were inadvertently left standing in the sun until they were ruined. The rest of the order was then sold for a net profit of $200. How many cartons of roses did she order?

93. Members of Friendship Fraternity plan to build a meeting house for $30,000. If they can find five more members, the cost per member will be reduced $500. How many members do they now have?

94. With a 15-mile-per-hour head wind, it takes a plane 1 hour, 40 minutes longer to travel 300 miles than with no wind. What is its airspeed?

95. A boat travels 15 miles upstream and then back again in 2 hours. a. If the stream is flowing at 4 miles per hour, find the speed of the boat in still water. b. How long did each part of the trip take?

96. Two kinds of industrial cleanser worth $1.20 a gallon and $1.80 a gallon, respectively, are mixed. If the mixture is worth $1.40 a gallon, what is the formula for the mixture? (*Hint:* Determine the relative proportions in a 120-gallon mixture.)

97. A farmhand is paid $5000 and a new automobile for labor in a given year. Had he worked only 7 months he would have received only $1000 in addition to his car. How much is the automobile worth?

98. To encourage quality work, the Plastics Molding Company pays Elizabeth 8¢ for each good case she molds but fines her 5¢ for each defective one. After producing 2600 cases her net earnings were zero. How many bad cases did she mold?

99. A master plumber charges $18 an hour for his labor and $12 an hour for his apprentice. On a given job the apprentice showed up 2 hours late for work one day. The total labor charges for this job were $276. How many hours did each spend on this job?

100. Art patched and painted Bill's hot rod for $360. Finishing the job in 12 hours less than estimated, he increased his wage by $1 per hour. What was the estimated time for this job?

101. Mr. Koesters has a field with dimensions 3 furlongs by 4 furlongs. After beginning to sow wild oats around the outside of the field, he decides to put corn in half the field. If he continues to sow oats around the outside until half the field is in oats, what are the dimensions of the inner plot of corn?

102. If a square is transformed into a rectangle by decreasing one pair of sides 1 inch and increasing the other pair of sides 2 inches, the area remains the same. Find the dimension of the original square.

103. A magician asks you to enclose a two-by-two

block on a calendar and state the sum of the four entries. She then states the four entries. Explain.

104. Find two consecutive odd integers, the product of which is 22 more than the square of the first.

105. Find a positive integer that is 14 more than 51 times its reciprocal.

106. Find an integer that is 1 less than 56 times its reciprocal.

3 *Graphs of Equations*

A **graph** is a pictorial representation of a relationship between two or more quantities. Graphs are a very efficient device for communicating information regarding many aspects of daily living. For example, physicians use electrocardiograms to monitor heart function; stockbrokers plot stock prices to predict future trends; and economists plot new car sales as an indicator of economic activity (Fig. 3–1).

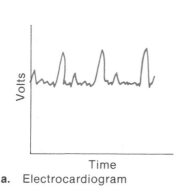

a. Electrocardiogram

b. Stock price

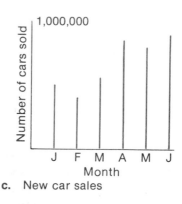

c. New car sales

Figure 3–1

Equations may also be used to describe everyday relationships. For instance, if you travel at a rate of 50 miles per hour, then the distance D traveled in t hours is given by $D = 50t$.

In this chapter we shall be concerned with the graphs of relationships that are expressed as an equation or inequality.

3.1 CARTESIAN COORDINATES; DISTANCE AND CIRCLES

The graph of a relationship between two quantities generally includes a horizontal line and a vertical line on which the sizes of the various quantities are indicated. Accordingly, we shall place two perpendicular lines on a plane and use these in the same fashion. These two lines are called the coordinate lines or **coordinate axes.** The horizontal axis is commonly called the **x-axis** and the vertical axis is called the **y-axis;** however, other letters or symbols that are more appropriate to a given problem are sometimes used to designate the axes. The point O in which these two lines meet is called the **origin.** These lines divide the plane into four sections, called **quadrants.** The quadrants are labeled, I, II, III, and IV as illustrated in Figure 3–2.

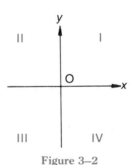

Figure 3–2

Each of the coordinate axes is considered a copy of the real number line. It is customary to assign the positive direction to the right on the x-axis and upward on the y-axis. From each point P in the plane, we drop a perpendicular to each of the coordinate axes. The numbers corresponding to the points at which these perpendiculars meet the respective axes are called, respectively, the **x-coordinate** of P and the **y-coordinate** of P. If a is the x-coordinate of P and b is the y-coordinate of P, we say that P has coordinates (a, b). The context should prevent confusion of the coordinates (a, b) of a point P and the open interval (a, b).

The x-coordinate of a point indicates its distance to the right or left of the y-axis; the y-coordinate gives its distance above or below the x-axis. The coordinates of a number of points are illustrated in Figure 3–3. Note in particular that $(2, -3) \neq (-3, 2)$ since $(2, -3)$ is in Quadrant IV, whereas $(-3, 2)$ is in Quadrant II. Because of the importance of the order in which a and b are written (a, b) is called an **ordered pair.**

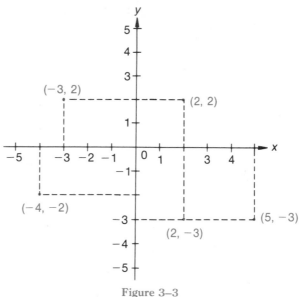

Figure 3–3

We shall see how coordinates are used in three geometric problems: finding the midpoint of a line segment, computing distance, and describing a circle. If P has coordinates (a, b), we shall find it convenient to write $P = (a, b)$. However, if we are working with more than one coordinate system, we must then indicate in which system the coordinates (a, b) apply.

The Midpoint Formula

To obtain the coordinates of the midpoint $M = (x, y)$ of the line segment joining $P = (x_1, y_1)$ and $Q = (x_2, y_2)$, we observe that the x-coordinate of M must be midway between x_1 and x_2 (see Fig. 3–4):

HISTORICAL PERSPECTIVE

This sytem of plotting points is known as the **Cartesian coordinate system** in honor of René Descartes (1596–1650), the French mathematician and philosopher who is credited with its discovery and development. Descartes is said to have been inspired by watching a fly crawling near a corner on a ceiling. A plane on which is imposed a Cartesian coordinate system is called a **Cartesian plane** or **real plane**. Descartes introduced and developed the technique of studying geometry by analyzing coordinates of points and equations of figures. This branch of mathematics is now called **analytic geometry**.

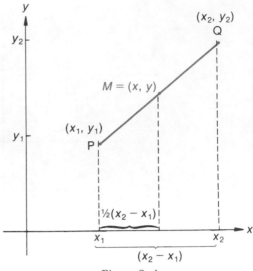

Figure 3–4

$$x = x_1 + \frac{1}{2}(x_2 - x_1)$$

$$= \frac{x_1 + x_2}{2}$$

Similarly,

$$y = \frac{y_1 + y_2}{2}$$

MIDPOINT FORMULA

The midpoint of the line segment joining $P = (x_1, y_1)$ and $Q = (x_2, y_2)$ has coordinates

$$\left(\frac{x_1 + x_2}{2}, \frac{y_1 + y_2}{2} \right)$$

EXAMPLE 1   Find the midpoint of the segment joining the points $(-2, 3)$ and $(6, -5)$.

Solution   The midpoint formula gives the coordinates of the midpoint as

$$\left(\frac{-2 + 6}{2}, \frac{3 - 5}{2} \right) = \left(\frac{4}{2}, \frac{-2}{2} \right) = (2, -1)$$

In Figure 3–5, we see that $(2, -1)$ is indeed the midpoint of the given segment.

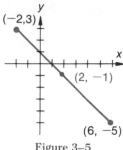

Figure 3–5

**The Distance
Formula**

The distance between two points in a Cartesian plane is determined by their coordinates. If $P = (x_1, y_1)$ and $Q = (x_2, y_2)$ are two points in the plane, let $R = (x_2, y_1)$ as in Figure 3–6. The points P, Q, R then form a triangle with a right angle at R. Recall from Section 1.2 that the distance between two points x_1 and x_2 on the real number line is $|x_1 - x_2|$. If we let $d(A, B)$ denote the distance between two points A and B, the Pythagorean theorem tells us that

$$[d(P, Q)]^2 = [d(P, R)]^2 + [d(R, Q)]^2$$
$$= |x_2 - x_1|^2 + |y_1 - y_1|^2$$

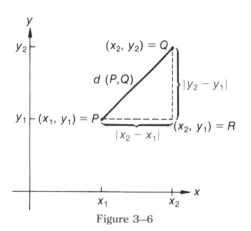

Figure 3–6

Thus, we obtain the distance formula.

DISTANCE FORMULA

The distance between two points $P = (x_1, y_1)$ and $Q = (x_2, y_2)$ in the real plane is given by

$$d(P, Q) = \sqrt{(x_2 - x_1)^2 + (y_2 - y_1)^2}$$

EXAMPLE 2   Find the distance between $(-2, 3)$ and $(4, -1)$.

Solution
$$\begin{aligned}
d[(-2, 3), (4, -1)] &= \sqrt{[4 - (-2)]^2 + [(-1) - 3]^2} \\
&= \sqrt{6^2 + (-4)^2} \\
&= \sqrt{52} \\
&= 2\sqrt{13}
\end{aligned}$$
■

Circles

DEFINITION

A **circle** consists of those points whose distance from a given point A is some fixed constant r. A is called the **center** and r the **radius** of the circle.

If $A = (a, b)$ is the center of a circle of radius r, then $P = (x, y)$ is on this circle if and only if $d(A, P) = r$ (see Fig. 3–7):

$$\sqrt{(x - a)^2 + (y - b)^2} = r$$
$$(x - a)^2 + (y - b)^2 = r^2$$

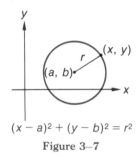

$$(x - a)^2 + (y - b)^2 = r^2$$

Figure 3–7

The circle of radius r centered at (a, b) has equation

$$(x - a)^2 + (y - b)^2 = r^2$$

In particular,

$$x^2 + y^2 = r^2$$

is the circle of radius r centered at the origin.

EXAMPLE 3 Write the equation of the circle of radius 5 centered at (4, 3).

Solution The given circle is just the locus (or set) of points lying 5 units away from (4, 3). According to the preceding formula, the point (x, y) is on this circle if and only if

$$(x - 4)^2 + (y - 3)^2 = 25$$
$$x^2 - 8x + 16 + y^2 - 6y + 9 = 25$$
$$x^2 - 8x + y^2 - 6y = 0$$

EXAMPLE 4 Sketch and write the equation of the circle centered at $(-1, 2)$ that passes through $(-3, 4)$.

Solution If the circle passes through $(-3, 4)$, we can find the radius of the circle as the distance from the center to this point. Thus,

$$\begin{aligned} r &= d[(-1, 2), (-3, 4) \\ &= \sqrt{[-3 - (-1)]^2 + (4 - 2)^2} \\ &= \sqrt{(-3 + 1)^2 + 2^2} \\ &= \sqrt{2^2 + 2^2} \\ &= \sqrt{8} \end{aligned}$$

The equation of the circle is then

$$[x - (-1)]^2 + (y - 2)^2 = 8$$
$$(x + 1)^2 + (y - 2)^2 = 8$$

The sketch is given in Figure 3–8.

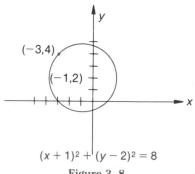

$(x + 1)^2 + (y - 2)^2 = 8$

Figure 3–8

EXAMPLE 5 Describe and sketch the set of points satisfying

$$x^2 - 4x + y^2 + 6y - 12 = 0$$

Solution Rewriting the preceding equation as

$$(x^2 - 4x \quad) + (y^2 + 6y \quad) = 12$$

we complete the squares by adding 4 and 9 to both sides of the equation.

$$(x^2 - 4x + 4) + (y^2 + 6y + 9) = 12 + 4 + 9$$
$$(x - 2)^2 + (y + 3)^2 = 25$$
$$= 5^2$$

Thus, the given equation represents the circle of radius 5 centered at $(2, -3)$ (see Fig. 3–9).

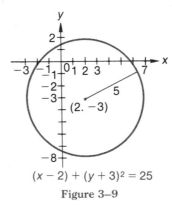

$$(x - 2) + (y + 3)^2 = 25$$

Figure 3–9

By completing the square as in Example 5, it can be shown that any equation of the form

$$\boxed{x^2 + ax + y^2 + by + c = 0}$$

represents either a circle or a single point or is not satisfied by any (x, y). This equation is called the **general equation** of a circle.

EXERCISES 3.1

1. Find the coordinates of each of the points illustrated in Figure A on the following page.

Plot the following points in the real plane.

2. $(1, 2)$
3. $(2, -1)$
4. $(-2, -1)$
5. $(3, -5)$
6. $(0, -1)$
7. $(5, 4)$
8. $(-1, 0)$
9. $(0, 1)$
10. $(10, 2)$
11. $(-3, 2)$
12. $(-2, 1)$
13. $(3, 0)$

14. $(-1, -3)$
15. $(4, 2)$
16. $(-2, 0)$
17. Roadmaps are commonly marked off in sections by a grid. We look for our destination in the index and read, for example, St. Henry, M11. St. Henry can then be found in block M11 as illustrated in the figure. Compile an index for the villages listed on the map in Figure B on the following page.

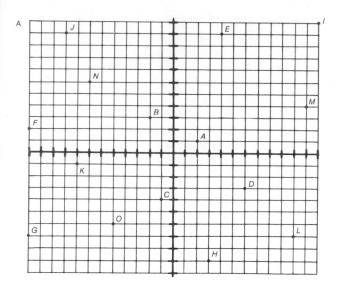

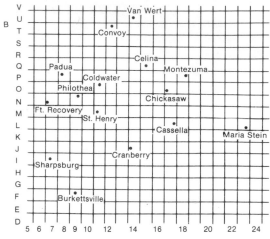

Find the midpoints of the segments joining the following pairs of points.

18. $(-3, 1), (1, 5)$ **19.** $(5, 2), (-1, -2)$

20. $(-3, 0), (-7, 0)$ **21.** $(1, 0), (0, 2)$

22. $(-4, 2), (2, -4)$

23. $(0, -2), (0, 6)$

24. $(3, 2), (1, 10)$

Find the distance between the following pairs of points.

25. $(3, 10), (2, 1)$ **26.** $(-3, 1), (2, 4)$

27. $(-2, 5), (-3, -6)$

28. $(-2, 0), (5, 0)$

29. $(-2, -1), (-1, -2)$

30. $(4, 3), (6, -4)$

31. $(0, -2), (0, -5)$ **32.** $(67, 32), (47, 42)$

33. $(-10, 5), (-10, -3)$

34. $(-10, 5), (6, 5)$

35. $(5, -2), (2, -5)$ **36.** $(-3, 0), (0, -4)$

A partial street naming scheme is to fix a common reference point (such as the intersection of Main St. and Central Ave.) and count "streets" east of this point and "avenues" north of this point. Then, for example, Fifth Street would run north and south and be five blocks east of this reference point.

37. With this street-labeling scheme, find the distance "as the crow flies" between the corner of 59th St. and 7th Ave. and the corner of 64th St. and 19th Ave., if each block is 1/2 mile long and the streets form a rectangular grid.

38. A developer must construct a street running diagonally from the corner of 20th Ave. and 15th St. to the corner of 4th Ave. and 27th St. If the numbered streets and avenues form a rectan-

gular grid and each block is 1/4 mile long, how long is the new street?

39. Show that the points $(-2, 4), (0, -1),$ and $(3, 6)$ are vertices of an isoceles triangle.

40. Show that $(-2, 2), (2, 0), (1, 3),$ and $(-1, -1)$ are vertices of a square.

41. Show that $(-2, 1), (4, 5), (2, 2)$ and $(0, 4)$ are vertices of a parallelogram.

42. Show that the points $(2, 7), (-2, -1), (6, 5)$ are vertices of a right triangle.

Use the distance formula to determine whether the following points fall on a line.

43. $(2, 2), (-1, -4), (7, 12)$

44. $(1, 7), (-2, 2), (2, 9)$

45. $(8, 2), (2, -2), (12, 5)$

46. By finding the midpoints, show that the line segment from $(1, 3)$ to $(-1, -1)$ bisects the segment connecting $(2, 0)$ and $(-2, 2)$.

47. Locate a parallelogram on a convenient coordinate system and show that its diagonals bisect one another.

48. Use a convenient coordinate system to show that the midpoint of the hypotenuse of a right triangle is equidistant from its three vertices.

Sketch and write the equation of the indicated circle.

49. center $(3, 1)$, radius 4

50. center $(4, -1)$, radius 10

51. center $(0, 4)$, radius 7

52. center $(-2, 5)$, radius 5

53. center $(-1, -1)$, radius 1

54. center $(3, 2)$, radius 8

55. center $(2, -2)$, radius 2

56. center $(3, -1)$, radius 9

57. center $(4, 2)$, radius 3

58. center $(3, 0)$, radius 6

59. center $(2, -1)$, radius 4

60. center $(3, 2)$, radius 3

61. center $(-2, 3)$, radius 2

62. center $(-1, -1)$, passing through $(0, -3)$

63. center $(2, -4)$, passing through $(11, -16)$

64. center $(-1, -3)$, passing through $(2, -5)$

65. center $(-2, 6)$, passing through $(2, 9)$

66. center $(2, 4)$, passing through $(5, 8)$

67. center $(1, 5)$, passing through $(6, 17)$

68. center $(3, -2)$, passing through $(4, -4)$

69. $(-2, 0)$ and $(-2, 6)$ are on opposite ends of a diameter

70. $(1, -2)$ and $(7, 6)$ are at opposite ends of a diameter

71. $(-17, -6)$ and $(31, 8)$ are at opposite ends of a diameter

Describe the set of points $P = (x, y)$ satisfying each of the following equations.

72. $x^2 + 2x + y^2 - 4y - 3 = 0$

73. $x^2 - 6x + y^2 + 2y - 15 = 0$

74. $x^2 - 10x + y^2 - 6y + 30 = 0$

75. $x^2 + 4x + y^2 + 8y - 29 = 0$

76. $2x^2 + 4x + 2y^2 + 8y + 8 = 0$

77. $x^2 - 10x + y^2 + 6y + 30 = 0$

78. $x^2 - 4x + y^2 - 8y + 29 = 0$

79. $x^2 - 10x + y^2 + 10y + 25 = 0$

80. $3x^2 + 6x + 3y^2 - 12y + 6 = 0$

81. $x^2 - 6x + y^2 + 2y + 10 = 0$

82. $x^2 - 12x + y^2 - 14y - 15 = 0$

83. $x^2 + 2x + y^2 - 4y - 4 = 0$

84. $x^2 - 4x + y^2 - 8y - 29 = 0$

85. $x^2 - 4x + y^2 + 6y - 3 = 0$

86. $x^2 - 6x + 21 = 8y - y^2$

3.2 GRAPHS

The vertical scale of a graph will now be indicated on a line passing through the zero position of the horizontal scale. In a graph indicating the relationship of energy demand to temperature (Fig. 3–10), we see that daily energy use is 1.5 million kilowatt-hours when the temperature is 70°F. Five million kilowatt-hours are needed when the temperature is 100°F. The graph shows that energy use changes in relation to the temperature. The position of the curve above the x-axis at $x = T$ indicates the daily energy demanded when the temperature is T°F.

The **graph** of any relation between x and y consists of those points (x, y) that satisfy the given relation. The points in which the graph crosses the x-axis are called **x-intercepts**; **y-intercepts** are points in which the graph crosses the y-axis.

PLOTTING POINTS

An elementary approach to graphing a relation is to

1. List corresponding x and y values in a table.

2. Plot these points on a coordinate system.

3. "Connect the dots" when enough points have been plotted to indicate a trend.

Convenient points to plot are those corresponding to $x = 0$ and $y = 0$. These are the y- and x-intercepts of the graph, respectively.

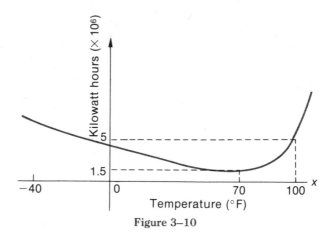

Figure 3–10

EXAMPLE 1 Sketch the graph of the following relation.

$$2x - y + 1 = 0$$

Solution The graph consists of those points (x, y) that satisfy the equation $2x - y + 1 = 0$. Observe that y can be written explicity in terms of x:

$$y = 2x + 1$$

In the table accompanying Figure 3–11, several values for x are listed with corresponding y values. The graph can then be sketched by plotting these points and connecting them as in Figure 3–11.

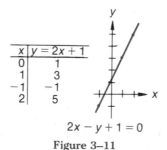

Figure 3–11

We can indeed connect the dots to form a straight line in Example 1. However, we should be careful in this regard because a graph may contain bends and kinks that we may have missed. Later we shall develop analytical methods to assist us in connecting the dots so as to ascertain the general shape of the graph. In the next section, we shall show that the graph in Example 1 is indeed a straight line and that, in fact, the graph of any equation of the form

$$ax + by + c = 0 \qquad a \text{ and } b \text{ not both } 0$$

is a straight line. Consequently, the graph of any such equation can be sketched by plotting just two points and drawing the line that is determined by these two points.

To graph an inequality, we first graph the corresponding equality and then determine how the inequality relates to the equality. This technique is illustrated in Example 2.

EXAMPLE 2   Sketch the graph of the relation

$$2x - y + 1 > 0$$

Solution   This relation can be rewritten as

$$y < 2x + 1$$

In Example 1, the y-altitude was graphed as $y = 2x + 1$. Here, the y-altitude must be less than $2x + 1$; that is, this graph consists of all points below the graph of Example 1, as shown in Figure 3–12.

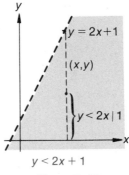

$$y < 2x + 1$$

Figure 3–12

EXAMPLE 3   Sketch the graph of the relation

$$y = x^2$$

Solution   Several points are found by listing in a table those y values corresponding to certain x values. We then plot these points and connect the dots as in Figure 3–13.

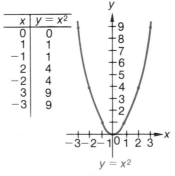

Figure 3–13

A U-shaped graph such as the one illustrated in Figure 3–13 is called a **parabola.**

Information that might otherwise be difficult to find can sometimes be obtained very easily from a graph. For example, the demand for and supply of a given product fluctuate depending on the price the product commands. The economy is in equilibrium when supply and demand are equal. Graphing the supply and demand curves helps to find this equilibrium point as illustrated in the following example.

EXAMPLE 4 A commercial bank finds that when its prime lending rate is 10 percent it has $10 million to lend and loan applications totaling $15 million. An increase of x percent in the prime rate increases the money supply by $\$x^2/5$ million and decreases the loan demand to $\$15/\left(1 + \dfrac{x}{10}\right)$ million. Find the equilibrium interest rate at which supply and demand are equal.

Solution The money supply and loan demand at an interest rate of $(10 + x)$ percent are given, in millions of dollars, by

$$S(x) = 10 + \frac{x^2}{5}$$

and

$$D(x) = \frac{15}{1 + \dfrac{x}{10}}$$

respectively. To solve for x in the equation $S(x) = D(x)$ would be difficult. Thus, we graph the two curves separately; tables of values are included with Figure 3–14.

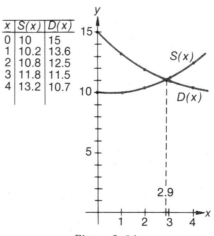

x	S(x)	D(x)
0	10	15
1	10.2	13.6
2	10.8	12.5
3	11.8	11.5
4	13.2	10.7

Figure 3–14

Note that different scales are used on the horizontal and vertical axes in Figure 3–14 in order to enlarge the graph. This enables us to estimate more accurately the equilibrium point x for which $S(x) = D(x)$. In this case, x is approximately 2.9. The bank should set its prime rate at $(10 + 2.9)$ percent = 12.9 percent. ■

EXERCISES 3.2

Sketch the graphs of the following relations. Use either an x-y or a u-v coordinate system as required.

1. $3u + v = 6$
2. $2x - 4y = 8$
3. $3u + v \le 6$
4. $2x - 4y > 8$
5. $y = x$
6. $u = 2v$
7. $y < x$
8. $u \le 2v$
9. $v = |u|$
10. $x^2 = y^2$
11. $v \ge |u|$
12. $x^2 < y^2$
13. $y^2 = 4x^2$
14. $v - 4u^2 = 0$
15. $y^2 > 4x^2$
16. $v > 4u^2$
17. $x = y^2$
18. $x = 2y^2$
19. $x \ge y^2$
20. $x \ge 2y^2$
21. $y = -x^2$
22. $y = -2x^2$
23. $y \ge -x^2$
24. $y \ge -2x^2$
25. $v^2 = |u|$
26. $|u| = 2|v|$
27. $v^2 < |u|$
28. $|u| < 2|v|$
29. $y = |x^3|$
30. $v = u^4$
31. $y \ge |x|^3$
32. $v \le u^4$
33. $|y| < |x|^3$
34. $|v| \le u^4$

To sketch the graphs of the following relations, first replace the inequality symbol with equality and sketch the resulting circle. Then use the distance formula to determine where the points satisfying the inequality are to be plotted.

35. $u^2 + v^2 < 9$
36. $x^2 + y^2 \ge 36$
37. $x^2 - 4x + y^2 + 6y - 12 < 0$
38. $u^2 + 6u \ge 4v - v^2 + 3$
39. $x^2 + y^2 + 30 \le 10x + 6y$
40. $x^2 + 4x + y^2 > 29 - 8y$
41. $x^2 + y^2 < 12x + 14y + 15$

42. When wheat sells for $3.50 per bushel, there is a demand for 150 million metric tons and a supply of only 50 million metric tons. Increasing the price per bushel by $x increases the supply to $\dfrac{(1 + 2x)}{(1 + x)} \times 10^8$ metric tons and decreases the demand by $x^2 \times 10^8$ metric tons. Sketch the supply and demand curves and use these curves to find an equilibrium price for wheat.

43. When eggs sell for 75¢ per dozen, there is a demand for 50,000 cases per week and a supply of 80,000 cases per week. If the price decreases x¢ per dozen, the demand increases $1000x^2$ cases whereas the supply decreases $5000x$ cases. Sketch the supply and demand curves to find an equilibrium price for eggs.

44. Suppose that a lake covering $10x$ acres can support enough fish to regenerate x^2 million fish each year. However, it also attracts fishermen in sufficient numbers to catch $5x$ million fish each year. Sketch the "regeneration" and "catching" curves in order to determine an optimum size for such a lake.

45. The calculator industry has expanded explosively in recent years by compressing the ability to do increasingly complex calculations into smaller and smaller packages. Since 1959, the number of components in an integrated circuit (IC) has doubled each year. However, as the instruments become more complex internally, a higher percentage of the production must be scrapped as defective. When the increase in scrapping costs exceeds the savings achieved by further miniaturization, it is no longer profitable to develop more complex instruments. This phenomenon has led to the manufacturing of electronic chips in "clean" dust-free rooms in orders to decrease the scrapping cost.

Suppose that for a certain calculator costing $100 to produce in 1980, the scrapping costs average $5 per calculator. Its production costs decrease by 25 percent each year thereafter, whereas the scrapping costs increase by 25 percent each year. Use graphs to determine when the *increase* in scrapping costs will exceed the additional savings in production costs.

46. A forest fire starts 50 miles west and 40 miles north of Boondocks. The fire advances at a rate of 2 miles per day in every direction. The rangers direct the local volunteers to dig a fire trench that is to stop the fire after 5 days. Sketch a map indicating the fire trench and the

region to be sacrificed to the forest fire. How close does the fire come to Boondocks?

47. A prisoner escapes from the Placerville jail, but the only means of leaving the area is by foot or horseback. If he can average at most 10 miles per hour, where should the authorities place roadblocks that are operative 5 hours after the escape?

48. Chicago is located at a latitude of 41°52'28" North and a longitude of 87°38'22" West. Dayton's coordinates are 39°45'32" North and 84°11'43" West. Here, 1° represents about 69

miles. If the radar at the Chicago airport has a range of 200 miles while Dayton's has a range of only 150 miles, sketch the area covered by both Dayton's and Chicago's radar units.

49. Sketch the graph consisting of all points equidistant from (2, 4) and (4, 2). Use this distance relationship to find the equation represented by this graph.

50. Sketch the graph consisting of all points whose distance from the x-axis is twice its distance from the y-axis. Use this distance relationship to find the equation represented by this graph.

3.3 LINES AND LINEAR EQUATIONS

When a carpenter describes a roof as having a "5–12 pitch," he means that the roof rises 5 inches for each horizontal advance of 12 inches. He then knows that the roof will rise 5 feet for each horizontal advance of 12 feet; it will rise 10 feet above a horizontal advance of 24 feet, and so on.

In the same way, the vertical **rise** corresponding to a given horizontal advance, called the **run,** indicates the steepness of any nonvertical line. The rise divided by the run is called the **slope** of the line. The similar triangles in Figure 3–15 show that for a given line the ratio slope = rise/run is the same regardless of the location at which this computation is made.

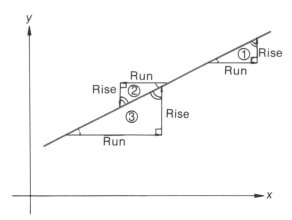

Figure 3–15   **Rise/Run is constant on a given line.**

In triangle 2 of Figure 3–15, we see that the rise and/or run may be negative. If the line falls as we move to the right, a negative rise corresponds to a positive run and the ratio rise/run = slope is negative.

The slope of a line is a measure of its inclination or steepness. A line having 0 slope is horizontal; it does not rise or fall as we move to the right. Lines having small positive or negative slope will rise or fall slowly as we move to the right, whereas lines having a larger slope will be very steep. Figure 3–16 shows

DEFINITION
The slope m of a line l is given by

$$m = \frac{\text{rise}}{\text{run}}$$

lines of different slopes. Note that vertical lines do not have a slope since no horizontal advance can take place on them. This property of no slope is in contrast to horizontal lines, which do have a slope $m = 0$.

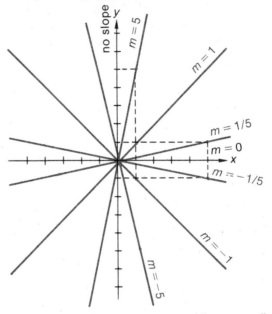

Figure 3–16 **Slope is a measure of "steepness."**

EXAMPLE 1 Sketch the line passing through $(-1, 4)$ and having slope $m = -2$.

Solution

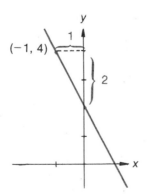

Figure 3–17 **The line through $(-1, 4)$ with slope -2**

Since the slope $m = -2$ is negative, the vertical rise is negative when the horizontal advance is positive; that is, the line falls rather than rises as we move to the right. An advance of one unit to the right causes a vertical drop of two units. Thus, we sketch the line as in Figure 3–17. ∎

Figure 3–18 shows that a vertical line is a set of points with a fixed x-coordinate and arbitrary y-coordinates. A horizontal line is a set of points with a fixed y-coordinate and arbitrary x-coordinates.

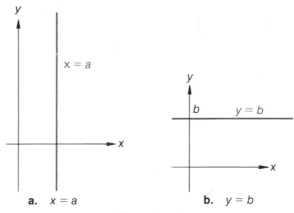

a. $x = a$ **b.** $y = b$

Figure 3–18

---

1. A vertical line with a as its x-intercept has equation $x = a$.

2. A horizontal line with b as its y-intercept has equation $y = b$.

There are several forms for the equation of a nonvertical line. Each of these forms involves the slope of the line.

The Point-Slope Formula

Let (x_0, y_0), (x_1, y_1), and (x_2, y_2) be points on a nonvertical line as in Figure 3–19. The slope of the line, defined as rise/run, is clearly

$$m = \frac{y_2 - y_1}{x_2 - x_1}$$

On the other hand, the slope must be the same regardless of where it is calculated. Thus,

$$m = \frac{y - y_0}{x - x_0}$$

for every point $(x, y) \neq (x_0, y_0)$ on this line. Multiplying this last equation by $(x - x_0)$, we see that *every* point on the line—(x_0, y_0) included—satisfies the equation

$$(y - y_0) = m(x - x_0)$$

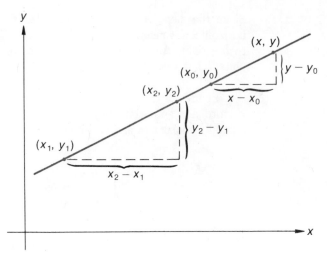

Figure 3–19

Any point not on this line will lie at a different inclination to (x_0, y_0) and hence will not satisfy this equation.

POINT-SLOPE FORMULA

The slope of a nonvertical line may be computed as

$$m = \frac{y_2 - y_1}{x_2 - x_1}$$

for any two distinct points (x_1, y_1), (x_2, y_2) on the line.

The equation of the line passing through (x_0, y_0) with slope m is

$$(y - y_0) = m(x - x_0)$$

EXAMPLE 2   Sketch the line passing through $(2, -3)$ and $(-1, 6)$ and write its equation. Find the points at which it intersects the x- and y-axes, respectively.

Solution   The line can be sketched by simply plotting the two given points as in Figure 3–20. Using the two given points, the slope of the line is found to be

$$m = \frac{-3 - 6}{2 - (-1)} = \frac{-9}{3} = -3$$

Then, using the point-slope formula with the point $(2, -3)$ and $m = -3$, the equation of the line is

$$y - (-3) = -3(x - 2)$$
$$y + 3 = -3x + 6$$
$$y = -3x + 3$$

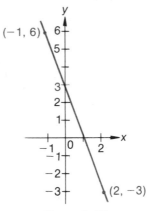

Figure 3–20

This line crosses the x-axis when $y = 0$:

$$0 = -3x + 3$$
$$x = 1$$

It crosses the y-axis when $x = 0$:

$$y = -3 \cdot 0 + 3$$
$$y = 3$$

The Slope-Intercept Formula

An alternate form of the equation of a line is found by rewriting the point-slope formula as

$$y = m(x - x_0) + y_0$$
$$= mx + (y_0 - mx_0)$$

This has the general form

$$y = mx + b$$

When $x = 0$ in this equation, $y = b$. This line $y = mx + b$ crosses the y-axis at $y = b$; b is the y-intercept of this line.

SLOPE-INTERCEPT FORMULA

The equation of a line intersecting the y-axis at $y = b$ with slope m is

$$y = mx + b$$

In particular, when $b = 0$, the lines $y = mx$ pass through the origin. The lines from Figure 3–16 have the equations shown in Figure 3–21.

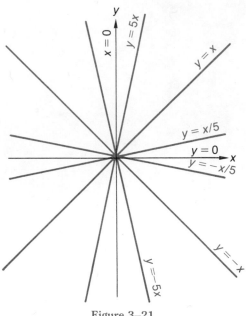

Figure 3–21

EXAMPLE 3   Sketch the line with slope $m = 3$ and y-intercept $b = -6$. Write its equation in slope-intercept form. Find its x-intercept.

Solution   The slope-intercept form of the equation for this line is

$$y = 3x - 6$$

Its x-intercept occurs when $y = 0$:

$$0 = 3x - 6$$
$$x = 2$$

Consequently, the points $(0, -6)$ and $(2, 0)$ are on the line. The line can be sketched by plotting these two points and drawing a line through them.

We could also have moved one unit to the right and three units up from the point $(0, -6)$ to plot the point $(1, -3)$; these two points also determine the line.

Alternatively, we could sketch the line $y = 3x$ and then lower this graph six units. Figure 3–22 shows the graph.   ■

The General Equation of a Line   We have seen that every line has an equation as indicated here:

$$y = mx + b \qquad \text{if the line is not vertical}$$
$$x = a \qquad \text{if the line is vertical}$$

Each of these equations can be rewritten in the general form

$$Ax + By + C = 0 \qquad A, B \text{ not both } 0$$

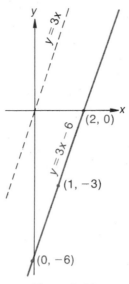

Figure 3–22

Conversely, every equation of this form represents a line. For such an equation can be rewritten in the form

$$y = -\frac{A}{B}x - \frac{C}{B} \quad \text{if } B \neq 0$$

or

$$x = -\frac{C}{A} \quad \text{if } B = 0$$

GENERAL EQUATION OF A LINE

The general form of the equation for a line is

$$Ax + By + C = 0 \qquad A, B \text{ not both } 0$$

Such an equation is called a **linear equation;** if neither A nor B is zero in a linear equation, the equation is said to describe a **linear relationship** between x and y.

EXAMPLE 4   Sketch the line given by the following equation and find its slope and intercepts.

$$6x - 2y + 8 = 0$$

Solution   Since the y-coefficient is not 0, we rewrite the equation as

$$2y = 6x + 8$$
$$y = 3x + 4$$

The line has slope $m = 3$ and y-intercept $b = 4$. Its x-intercept occurs when $y = 0$:

$$0 = 3x + 4$$
$$x = -\frac{4}{3}$$

Figure 3–23 shows the line.

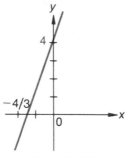

Figure 3–23

EXAMPLE 5 On the Fahrenheit temperature scale, water freezes at 32°F and boils at 212°F at sea level. On the Celsius scale, the freezing and boiling points are 0°C and 100°C, respectively. Assuming a linear relationship between the two scales, express Fahrenheit temperature in terms of Celsius temperature and vice versa. When do the two temperature scales agree?

Solution If we assume a linear relationship, we have

$$F = aC + b$$

where F and C represent Fahrenheit and Celsius temperatures, respectively, a and b are constants to be determined. At the freezing point F = 32 and C = 0:

$$32 = a \cdot 0 + b$$

Thus, $b = 32$ and F = $aC + 32$. At the boiling point F = 212 and C = 100:

$$212 = a \cdot 100 + 32$$
$$180 = a \cdot 100$$
$$a = \frac{180}{100} = \frac{9}{5}$$

Thus,

$$F = \frac{9}{5}C + 32$$

Writing Celsius temperatures in terms of Fahrenheit, we have

$$\frac{9}{5} C = F - 32$$

$$C = \frac{5}{9} (F - 32)$$

The two scales agree when F = C:

$$C = \frac{9}{5} C + 32$$

$$-\frac{4}{5} C = 32$$

$$C = -\frac{5}{4} \cdot 32$$

$$= -40°$$

Thus, $-40°F = -40°C$. Figure 3–24 shows the graph.

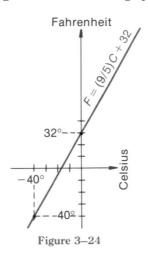

Figure 3–24

Parallel and Perpendicular Lines

Note that the line $y = mx + b$ can be obtained by simply raising or lowering the line $y = mx$. Several lines $y = mx + b$ are sketched in Figure 3–25 for a given m and various values of b. In particular, we note that nonvertical lines are parallel if and only if they have the same slope.

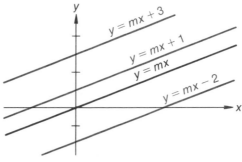

Figure 3–25

It is also possible to state a condition that determines whether two lines are perpendicular (see Exercise 85).

PARALLEL OR PERPENDICULAR LINES

If l_1 and l_2 are two nonvertical lines having equations

$$y = m_1x + b_1$$
$$y = m_2x + b_2$$

respectively,

a. The lines are **parallel** if and only if $m_1 = m_2$.

b. The lines are **perpendicular** if and only if $m_1m_2 = -1$ (equivalently $m_2 = -1/m_1$).

EXAMPLE 6 Write the equations of the lines through $(1, 2)$ that are parallel and perpendicular, respectively, to the line

$$3x + 9y - 18 = 0$$

Solution Solving for y in this equation, we obtain

$$9y = -3x + 18$$
$$y = -\frac{1}{3}x + 2$$

The slope of this line is $-1/3$. The slope of any line parallel to this line is also $-1/3$ (use the point-slope formula); thus, the equation of the line through $(1, 2)$ parallel to this line is

$$(y - 2) = \frac{-1}{3}(x - 1)$$
$$y = 2 - \frac{1}{3}x + \frac{1}{3}$$
$$= \frac{7}{3} - \frac{1}{3}x$$
$$3y = 7 - x$$
$$x + 3y - 7 = 0$$

The slope of any line perpendicular to this line has slope $m = 3$. Hence, the perpendicular line has equation

$$(y - 2) = 3(x - 1)$$
$$y = 2 + 3x - 3$$
$$= 3x - 1$$
$$3x - y - 1 = 0$$

The following table summarizes the forms for linear equations and shows their respective graphs.

Lines	Formula	Sketch
General form	$Ax + By + C = 0$, A, B not both 0	
Slope-intercept form	$y = mx + b$	
Point-slope form	$(y - y_1) = m(x - x_1)$	
Parallel lines	$\left.\begin{array}{l} y = m_1x + b_1 \\ y = m_2x + b_2 \end{array}\right\} m_1 = m_2$	
Perpendicular lines	$\left.\begin{array}{l} y = m_1x + b_1 \\ y = m_2x + b_2 \end{array}\right\} m_1m_2 = -1$	

EXERCISES 3.3

In Exercises 1–8 find the equation of the line with slope m and y-intercept b. Write the equation in general form.

1. $m = 3$, $b = 6$ **2.** $m = -1$ $b = 8$

3. $m = 2$, $b = -12$

4. $m = 6$, $b = 2$

5. $m = -10$, $b = -4$

6. $m = 4/3$, $b = 2/3$

7. $m = -5$, $b = 4$

8. $m = 5$, $b = -20$

In Exercises 9–16 write an equation of the line that

passes through the given point with the indicated slope m. Find the x- and y-intercepts.

9. $(3, 5)$, $m = 1/3$
10. $(-2, 8)$, $m = -10$
11. $(5, -3)$, $m = -1$
12. $(-1, 3)$, $m = -2$
13. $(-3, -1)$, $m = 0$
14. $(3, 2)$, $m = -1/3$
15. $(-2, -5)$, $m = -1/2$
16. $(10, -5)$, $m = 1/5$

In Exercises 17–28 write an equation of the line that passes through the given pair of points.

17. $(-4, 1)$, $(-2, 3)$
18. $(1, 5)$, $(4, 2)$
19. $(-3, 6)$, $(-1, -2)$
20. $(2, -3)$, $(2, 4)$
21. $(4, -2)$, $(-1, 8)$
22. $(1, 2)$, $(3, 6)$
23. $(8, 2)$, $(0, -2)$
24. $(-6, -4)$, $(-3, -2)$
25. $(0, 9)$, $(-5, 4)$
26. $(5, -2)$, $(-2, -2)$
27. $(3, 7)$, $(1, -7)$
28. $(2, -8)$, $(3, 0)$

Graph the following lines. Find their slopes and intercepts.

29. $x - 2y - 3 = 0$ 30. $x + y = 2$
31. $3x - 2y + 12 = 0$
32. $-6x + 4y - 1 = 0$
33. $2x + y - 6 = 0$ 34. $3x = 2$
35. $-4x - 8y + 16 = 0$
36. $2x - 3y + 6 = 0$ 37. $16y = 32$
38. $4x + 7y + 28 = 0$
39. $3x - 6y + 6 = 0$
40. $10x = 100$

In Exercises 41–50 determine whether the following pairs of lines are parallel, perpendicular, or neither.

41. $7x - 3y + 10 = 0$
    $6x + 14y + 5 = 0$
42. $2x - 3y + 4 = 0$
    $-16x + 24y - 2 = 0$
43. $3x - y - 1 = 0$ 44. $x = 10$
    $x - 3y - 1 = 0$ $y = -10$
45. $x - y = 6$ 46. $x = 16$
    $y - x = 3$ $x = -3$
47. $-2x - 2y + 3 = 0$
    $5x - 5y - 3 = 0$

48. $-10x - 5y + 3 = 0$
    $4x + 2y - 10 = 0$
49. $x - 9y + 6 = 0$ 50. $x - 9y - 6 = 0$
    $9x - y - 6 = 0$ $9x + y + 6 = 0$

In each of the following exercises, determine whether the given points fall on a line.

51. $(2, 0)$, $(-6, -6)$, $(6, 3)$
52. $(0, -2)$, $(3, 7)$, $(-2, -8)$
53. $(1, 1)$, $(3, 7)$, $(-2, -6)$
54. $(1, -1)$, $(-2, 5)$, $(5, -9)$
55. $(3, 8)$, $(7, 10)$, $(-9, 2)$
56. $(2, -3)$, $(-2, 5)$, $(5, -5)$

57. Show that $x/a + y/b = 1$, $a \neq 0 \neq b$, is the equation of the line intersecting the x-axis at a and the y-axis at b. This equation is known as the **two-intercept form** of the equation of a line. (*Hint:* Rewrite the equation in general form to observe that it represents a line. Then find its x- and y-intercepts.)

Use Exercise 57 to write the general equation of the line with indicated x- and y-intercepts a and b, respectively. Find the slope of each line.

58. $a = 1$, $b = 2$ 59. $a = 3$, $b = -5$
60. $a = 2$, $b = -4$ 61. $a = -3$, $b = -2$
62. $a = 0$, $b = 2$ 63. $a = 2$, $b = 1$

64. If (x_1, y_1), (x_2, y_2) are any two points on a non-vertical line, establish the **two-point formula** for the equation of the line:

$$\frac{y - y_1}{x - x_1} = \frac{y_2 - y_1}{x_2 - x_1}$$

65. If A and B are both 0 in the linear equation $Ax + By + C = 0$, show that the equation cannot represent a line (*Hint:* Show that the set of points satisfying the equation is either empty or the entire plane depending on the value of C.)

Write the equation of the line through the given point that is parallel to the given line and also for the line that is perpendicular to the given line.

66. $(1, 2)$ $x - y = 6$
67. $(-1, 3)$, $x = 5$
68. $(2, 0)$, $4x - 2y - 6 = 0$
69. $(-1, -2)$, $x + 3y - 2 = 0$
70. $(4, -1)$, $2x - 3y + 4 = 0$
71. $(0, -3)$, $2x + y - 5 = 0$
72. $(2, -11)$, $4x + 7y + 28 = 0$

73. $(6, 2)$, $y = 7$
74. $(100, -1)$, $3x - 9y + 6 = 0$
75. $(-1, 30)$, $x + y = 2$
76. Write the equation of the perpendicular bisector of the segment joining each pair of points given in Exercises 17–28.
77. Write the equation of the line tangent to the circle $x^2 + y^2 = 25$ at the point $(-4, 3)$. (A line is tangent to a circle if it is perpendicular to a line through the center of the circle at the point in which this line meets the circle.)
78. Write the equation of the line tangent to the circle $x^2 + y^2 = 169$ at the point $(5, -12)$. (See Exercise 77.)
79. The Ahlers' house needs a new roof. The cost will depend on the pitch of the roof; a steep roof requires more equipment to keep the workers from falling off. If their house has dimensions shown in the figure, what should they tell the contractor when he requests the pitch of their roof?

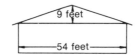

80. For tax purposes, certain property such as real estate, machinery, and equipment can be depreciated in several ways. One such method of depreciation is known as the straight-line method. In this method, the same amount is depreciated each year until the property has been depreciated to $0. If a $400,000 investment property is depreciated over 20 years by using the straight line method, plot the remaining value of the property during each of these 20 years.
81. A car rental agency charges $50 per day for a compact car and an additional 40¢ per mile driven. Write an equation and sketch a graph that indicates the total costs for a one-day rental in terms of the miles driven.
82. An automobile salesperson receives a monthly salary of $400 plus $100 for each car sold. Write an equation and sketch a graph that indicates her monthly income in terms of the number of cars she sells.
83. Each night that a downtown department store is open, the owners must pay $10,000 in addi-

tional salary and utility expenses. The markup on their merchandise equals the price they paid.
   a. Write an equation and sketch a graph that gives profit in terms of total sales for the night.
   b. How much business must they have each night to meet expenses?
   c. If the owners do not feel they can justify remaining open for less than $5000 profit per night, how much business must they have in order to justify the nighttime hours?
84. The RJ Calculator Company developed a new calculator at a cost of $50,000. In mass production it will cost $10 to produce each calculator. Environmental regulations then require an additional investment of $20,000 to bring the production line up to federal standards. Graph the cost of producing x calculators with and without the additional $20,000 investment. With the added investment, how many more units must the company sell before it begins to realize a profit if each unit sells for $15?
85. The formula $m_1 m_2 = -1$ for perpendicular lines is established in this exercise. Let l_1 and l_2 be nonparallel lines neither of which is horizontal or vertical. These lines have nonzero slopes m_1 and m_2, respectively, and intersect in some point $P_0 = (x_0, y_0)$.
   a. Show that the line $x = x_0 + 1$ meets l_1 and l_2 in points

$$P_1 = (x_1, y_1) = (x_0 + 1, y_0 + m_1)$$
$$P_2 = (x_2, y_2) = (x_0 + 1, y_0 + m_2)$$

respectively, as illustrated in the figure.

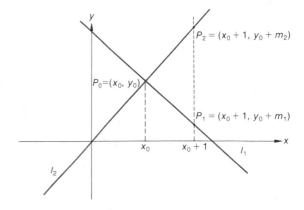

b. Observe that the following statements are equivalent.
  1. l_1 is perpendicular to l_2
  2. $\Delta P_0 P_1 P_2$ is a right triangle with a right angle at P_0

3. $[d(P_1, P_2)]^2 = [d(P_1, P_0)]^2 + [d(P_0, P_2)]^2$, (the Pythagorean formula).

c. Expand statement (3) in terms of the coordinates of P_1, P_2, and P_0 and show that it is equivalent to $m_1 m_2 = -1$.

3.4 PARABOLAS AND QUADRATIC EQUATIONS

The U-shaped curve called a parabola was obtained in Section 3.2 as the graph of the relation $y = x^2$. Parabolas can be used to describe the behavior of objects moving under the influence of gravity. A projectile such as a bullet or flare fired from a gun traces out a parabolic trajectory. Parabolas are also used in many light-focusing devices such as automobile headlights, beacons, and telescopes.

It can be shown (after completing the square) that the graph of any equation of the form

$$y = ax^2 + bx + c, \qquad a \neq 0,$$

or

$$x = Ay^2 + By + C, \qquad A \neq 0,$$

is a parabola. Typical graphs are sketched in Figure 3–26.

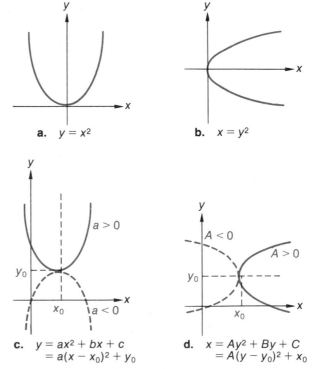

a. $y = x^2$

b. $x = y^2$

c. $y = ax^2 + bx + c$
$= a(x - x_0)^2 + y_0$

d. $x = Ay^2 + By + C$
$= A(y - y_0)^2 + x_0$

Figure 3–26

The point indicated on each of these graphs, at which the parabola "turns around" is called its **vertex**. The line through the vertex that cuts the parabola in half is called its **line of symmetry** or **axis**. In Figure 3–26

c. for $a > 0$

$$y = a(x - x_0)^2 + y_0$$
$$\geq y_0 \quad \text{and} \quad y = y_0 \quad \text{when} \quad x = x_0$$

d. for $A > 0$

$$x = A(y - y_0)^2 + x_0$$
$$\geq x_0 \quad \text{and} \quad x = x_0 \quad \text{when} \quad y = y_0.$$

The cases $a < 0$ and $A < 0$ are similar, but then $y \leq y_0$ and $x \leq x_0$, respectively. Thus, (x_0, y_0) is the vertex in each case.

GRAPHING QUADRATIC EQUATIONS

The graph of a quadratic equation

$$y = ax^2 + bx + c, \quad a \neq 0$$

or

$$x = Ay^2 + By + C, \quad A \neq 0$$

is a parabola. To sketch this graph:

1. Complete the square and write the equation in the alternate form

$$y = a(x - x_0)^2 + y_0$$

or

$$x = A(y - y_0)^2 + x_0$$

2. The vertex of the parabola is at (x_0, y_0).

3. The line of symmetry is

$$x = x_0$$

or

$$y = y_0$$

4. The parabola opens
   upward if $a > 0$; downward if $a < 0$
   to the right if $A > 0$; to the left if $A < 0$

5. Plot additional points to determine how "wide" or "narrow" the parabola is.

6. Determine the y-intercept(s) by setting $x = 0$ and the x-intercept(s) by setting $y = 0$.

EXAMPLE 1 Sketch the graph of the equation

$$y = 2x^2 - 4x + 1$$

Find the vertex, the line of symmetry, and the x- and y-intercepts of the graph.

Solution We can write

$$y = 2(x^2 - 2x \quad) + 1$$

and then complete the square to obtain

$$y = 2(x^2 - 2x + 1) + 1 - 2 \cdot 1$$
$$= 2(x - 1)^2 - 1$$

Since $(x - 1)^2 \geq 0$, the minimum y value occurs when $x = 1$; in this case, $y = -1$. Thus, the vertex of the parabola is the point $(1, -1)$. We know from the coefficient of x^2 that the parabola opens upward.

We must yet determine how wide or narrow the parabola is. We do this by plotting several points on either side of the vertex. Considering $x = 2$ and $x = 0$, we have $y = 2(\pm 1)^2 - 1 = 1$. For $x = 1/2$ and $x = 3/2$, we have $y = 2(\pm 1/2)^2 - 1 = -1/2$. Plotting these points enables us to see the shape of the graph. Its line of symmetry is clearly $x = 1$ (see Fig. 3–27).

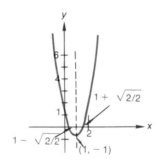

$$y = 2x^2 - 4x + 1 = 2(x - 1)^2 - 1$$

Figure 3–27

The y-intercept occurs when $x = 0$: $y = 1$. The x-intercepts occur when $y = 0$:

$$2x^2 - 4x + 1 = 0$$
$$x = \frac{-(-4) \pm \sqrt{16 - 4 \cdot 2 \cdot 1}}{2 \cdot 2}$$
$$= \frac{4 \pm \sqrt{8}}{4} = \frac{4 \pm 2\sqrt{2}}{4}$$
$$= 1 \pm \frac{\sqrt{2}}{2}$$

EXAMPLE 2   Graph the parabola represented by the following equation. Find its vertex, line of symmetry, and x- and y-intercepts.

$$x = -2y^2 - 8y - 9$$

Solution

$$
\begin{aligned}
x &= -2(y^2 + 4y \quad\quad) - 9 \\
&= -2(y^2 + 4y + 4) - 9 + 8 \\
&= -2(y + 2)^2 - 1
\end{aligned}
$$

The parabola opens to the left; its vertex is $(-1, -2)$, and its line of symmetry is $y = -2$.

At $y = -3$ and $y = -1$, we have $x = -2(\pm 1)^2 - 1 = -3$. At $y = 0$ and $y = -4$, we have $x = -2(\pm 2)^2 - 1 = -9$. These points are indicated in Figure 3–28.

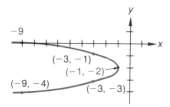

Figure 3–28 $x = -2(y + 2)^2 - 1$

Since the parabola opens to the left and the vertex is at $(-1, -2)$, there can be no y-intercept. The x-intercept(s) occur when $y = 0$: $x = -9$. ■

If we set $x = 0$ and use the quadratic formula to solve for y in Example 2, we find

$$
\begin{aligned}
-2y^2 - 8y - 9 &= 0 \\
y &= \frac{-(-8) \pm \sqrt{(-8)^2 - 4(-2)(-9)}}{2(-2)} \\
&= \frac{8 \pm \sqrt{64 - 72}}{-4} \\
&= \frac{8 \pm \sqrt{-8}}{-4} \\
&= \frac{8 \pm 2\sqrt{2}i}{-4}
\end{aligned}
$$

The roots are complex. This indicates that $x = 0$ for two complex y values but for no real y values. There are no y-intercepts, confirming what we observed in Figure 3–28.

EXAMPLE 3   If the altitude of a rocket t seconds after firing is given by $h = 320t - 16t^2$ feet, find its maximum altitude. When does the rocket attain this maximum altitude? When does it hit the ground?

Solution If we label the horizontal and vertical axes as t and h, respectively, then the graph of the equation $h = 320t - 16t^2$ is clearly a parabola opening downward. We can answer the first two questions by finding the vertex of the parabola. Since

$$
\begin{aligned}
h &= -16t^2 + 320t \\
&= -16(t^2 - 20t \qquad) \\
&= -16(t^2 - 20t + 100) + 1600 \\
&= -16(t - 10)^2 + 1600
\end{aligned}
$$

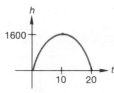

1600

10   20

Figure 3–29

we see that the vertex of the parabola is the point $(10, 1600)$. The maximum altitude of 1600 feet is attained 10 seconds after firing as shown in Figure 3–29. Since $h = 0$ when $t = 0$ and the parabola is symmetric about the line $t = 10$, we see that the rocket hits the ground 20 seconds after firing. ■

EXAMPLE 4 The Solar Oil Company has 1.6 million gallons of gasoline available for a holiday weekend. The usual price is \$1.30 per gallon. For each 10,000 gallons that is held off the market, the selling price of the remainder can be increased by 1¢ per gallon. How much gasoline should the company withhold from the market in order to maximize its revenues?

Solution Let

$$
\begin{aligned}
x &= \text{number of 10,000 gallon units withheld} \\
1,600,000 - 10,000x &= \text{number of gallons sold} \\
130 + x &= \text{cost per gallon in cents}
\end{aligned}
$$

The revenue received is then

$$
\begin{aligned}
y &= (1,600,000 - 10,000x)(130 + x) \\
&= 10,000(160 - x)(130 + x) \\
&= 10,000(20,800 + 30x - x^2) \\
&= -10,000\,[(x^2 - 30x \qquad) - 20,800] \\
&= -10,000\,[(x^2 - 30x + 225) - 21,025] \\
&= -10,000\,[(x - 15)^2 - 21,025] \\
&= -10,000(x - 15)^2 + 2.1025\,(10^8)
\end{aligned}
$$

The graph of this equation is a parabola opening downward with vertex at $(15, 2.1025(10^8))$, as shown in Figure 3–30. The maximum attainable revenue is \$2,102,500. It is obtained by holding back 15 units of 10,000 gallons or 150,000 gallons. The remainder is then sold at \$1.45 per gallon.

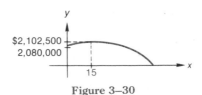

Figure 3–30 ■

Quadratic
Inequalities

Parabolas can also be used to solve quadratic inequalities in one variable. In Section 2.6, solutions to certain quadratic inequalities were found by factoring the quadratic expression and then analyzing the signs of its factors in various regions. The technique introduced here involves only finding the roots of the quadratic equation and looking at the sign of the x^2 coefficient.

SOLVING QUADRATIC INEQUALITIES BY GRAPHING

To solve a quadratic inequality

$$ax^2 + bx + c < 0, \ a \neq 0$$

make a rough sketch of the parabola

$$y = ax^2 + bx + c$$

1. Find the roots of the quadratic equation

    $$ax^2 + bx + c = 0$$

    The parabola crosses the x-axis at these points.

2. Determine from the sign of a whether the parabola opens upward or downward.

3. Roughly sketch the parabola; there is no need even to find the vertex.

4. Determine from this graph the region or regions in which the inequality is satisfied.

This technique applies to $\leq$, $>$, $\geq$ inequalities also.

EXAMPLE 5 Solve the inequality $50x \leq x^2 + 600$.

Solution First we rewrite the inequality as follows:

$$50x \leq x^2 + 600$$
$$-x^2 + 50x - 600 \leq 0$$

The quadratic formula gives the roots of this quadratic as

$$x = \frac{-50 \pm \sqrt{2500 - 2400}}{-2}$$
$$= \frac{-50 \pm 10}{-2}$$
$$= 20, 30$$

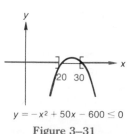

$y = -x^2 + 50x - 600 \leq 0$

Figure 3–31

Furthermore, the graph of $y = -x^2 + 50x - 600$ opens downward. Without even finding the vertex of this parabola, we note that $y \leq 0$ when $x \leq 20$ and when $x \geq 30$ (Fig. 3–31). ∎

EXERCISES 3.4

Graph each of the following parabolas. In each, find the vertex, the line of symmetry, and the intercepts.

1. $y = 1 - x^2$
2. $y = 2x^2 + 1$
3. $x = -3y^2 + 9$
4. $x = -y^2 + 10y$
5. $y = x^2 + 6x$
6. $x = y^2 - 4y + 2$
7. $x = 3y^2 + 6y + 4$
8. $x = 2y^2 - 4y + 6$
9. $y = 2x^2 + 12x + 16$
10. $x = y^2 - 4y + 4$
11. $x = -2y^2 + 4y - 3$
12. $y = x^2 + 6x + 5$
13. $x = -y^2 + 6y - 6$
14. $y = 2x^2 - 4x + 3$
15. $x = -2y^2 + 4y - 2$
16. $y = x^2 - 2x - 1$
17. $x = y^2 + 2y - 5$
18. $x = -2y^2 + 8y - 4$
19. $y = 3x^2 + 6x - 2$
20. $x = 3y^2 + 12y + 4$
21. $x = -y^2 - 2y + 3$
22. $y = -5x^2 + 10x + 4$
23. $y = 3x^2 - 6x + 2$
24. $x = 4y^2 - 4y - 2$

In the following problems, let h be the altitude in feet of a missile t seconds after it is fired. Determine

a. *Its maximum altitude*
b. *The time at which it attains its maximum altitude*
c. *The time at which it hits the ground*

25. $h = 640t - 16t^2$
26. $h = 480t - 16t^2 + 2800$
27. $h = 640t - 16t^2 + 3600$
28. $h = 160t - 16t^2 + 1200$
29. $h = 320t - 16t^2 + 2000$

Solve the following inequalities.

30. $-3x^2 - 8x + 3 \geq 0$
31. $2x^2 + x - 1 \leq 0$
32. $x^2 - 2x + 1 > 0$
33. $4x^2 - 12x + 5 < 0$
34. $x^2 + 5 < 2x$
35. $6x^2 + 4 \leq 14x$
36. $5x^2 > 12 + 4x$
37. $4x^2 - 4x + 1 \geq 0$
38. $2x^2 \geq 2x - 1$
39. $x^2 < 3x - 4$
40. $2x \leq x^2 - 2$
41. $3x^2 + 5x - 8 \leq 2x^2 + 6x + 4$
42. $2x^2 + 8x - 8 \leq x^2 + 2x - 1$

In the following, let $u = x^k$ and sketch a $u - v$ parabola in order to solve the inequalities.

43. $x^6 + 10x^3 - 24 \leq 0$
44. $x^4 - 5x^2 + 4 \geq 0$
45. $x - \sqrt{x} - 6 > 0$
46. $x - 5\sqrt{x} + 4 < 0$

47. The operator of a motorized bicycle business has determined that he can sell 40 bicycles per week when the price is $250 each. For each $25 increase in price, he sells two fewer units per week. How should he adjust his price in order to maximize weekly receipts?

48. The businessman in Exercise 47 sells his deluxe model for $400. At this price he can sell 20 units per week. Each $25 decrease in the sales price nets him two additional sales per week. How should he adjust his price in order to maximize his weekly receipts?

49. The Red-White-Blue Paint Store generally sells 6000 gallons of paint each week at $15 per gallon. For each $1 decrease in the price per gallon, the store will sell an additional 1000 gallons each week. What sale price should the store advertise in order to maximize its income?

50. The Clean-Air Manufacturing Company can produce 4000 pollution control units per week to sell for $100 each. For each additional 100 units produced each week, the sales cost is reduced $1 per unit. At what production level is the total sales cost for the week maximized?

51. The Wonderful Gadget Company expects each salesperson to sell 100 gadgets each month. The seller is paid $10 for each unit sold up to 100 units. For each additional 10 units in a given month, a bonus of $2 per unit is payable on all units sold that month. The units are sold for $50 each. A salesperson who sells too many gadgets could conceivably be paid more than the $50 per unit under this formula. At what point should the company modify its incentive plan in order to maximize its revenues?

52. Suppose that the cost of producing x widgets per day is $\$(2x^2 + 2400)$. These x widgets can be sold for $\$140x$. How many widgets can be produced at a profit?

53. As an incentive to its sales personnel, the Universal Motors Company determines the commission rate by the number of cars sold in a

given month. If a given salesperson sells x cars in a given month, her commission is $\$(80 + x)$ per car. She also earns a base salary of $225 per month. How many cars must she sell in order to earn at least $1650 this month?

3.5 CENTRAL ELLIPSES AND HYPERBOLAS (Optional)

(This section may be postponed with no loss of continuity. No later topics in this text depend on this material.)

An ellipse is a certain type of oval shaped curve. Hyperbolas have two distinct branches and are more difficult to describe geometrically. Orbiting bodies, including stars, planets, and artificial satellites, travel in elliptical paths. Television adventure stories have popularized the concept of locating an object by "triangulation." The hyperbola is the basis for the method of triangulation.

An ellipse or a hyperbola is defined as a locus of points satisfying a given condition; that is, it consists of precisely those points that satisfy the stated condition.

Ellipses

DEFINITION

An **ellipse** is the locus of points, the sum of whose distances from two fixed points remains constant. Each of the two fixed points is called a **focus** of the ellipse.

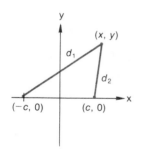

Figure 3–32
$d_1 + d_2$ = constant

To obtain the equation of an ellipse, we set up a coordinate system with the x-axis passing through the two foci. Place the y-axis midway between the foci. These foci then have coordinates $(\pm c, 0)$ for some $c > 0$ as indicated in Figure 3–32. Let (x, y) be an arbitrary point on the ellipse and label d_1, d_2 as in the figure. The condition stated then requires that $d_1 + d_2$ remain constant. It will be convenient to label this constant $2a$.

Using the distance formula to find d_1 and d_2, we find that the defining condition becomes

$$d_1 + d_2 = 2a$$
$$\sqrt{[x - (-c)]^2 + (y - 0)^2} + \sqrt{(x - c)^2 + (y - 0)^2} = 2a$$
$$\sqrt{(x + c)^2 + y^2} = 2a - \sqrt{(x - c)^2 + y^2}$$

Square both sides to obtain

$$x^2 + 2xc + c^2 + y^2 = 4a^2 - 4a\sqrt{(x - c)^2 + y^2} + x^2 - 2xc + c^2 + y^2$$
$$4a\sqrt{(x - c)^2 + y^2} = 4a^2 - 4xc$$
$$a\sqrt{(x - c)^2 + y^2} = a^2 - xc$$

Square both sides again:

$$a^2(x^2 - 2xc + c^2 + y^2) = a^4 - 2a^2xc + x^2c^2$$

Collect terms:

$$x^2(a^2 - c^2) + a^2y^2 = a^2(a^2 - c^2)$$

Since the sum of the lengths of two sides of triangle exceeds the length of the third side, we see in Figure 3–32 that $2a = d_1 + d_2 > 2c$ or $a > c$. Thus, we can let $b^2 = a^2 - c^2 > 0$. The preceding equation then becomes

$$x^2b^2 + y^2a^2 = a^2b^2 \qquad a^2 > 0, \ b^2 > 0$$

Dividing this equation by a^2b^2, we obtain the **standard equation of an ellipse.**

$$\frac{x^2}{a^2} + \frac{y^2}{b^2} = 1$$

A convenient method of sketching an ellipse is to use a pencil, two tacks, and a piece of string. Place the two tacks at the focal points and knot the string so that its length is $2a = d_1 + d_2$ units longer than the distance between the tacks. Let the string guide the pencil and sketch the ellipse as indicated in Figure 3–33.

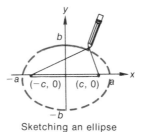

Sketching an ellipse

Figure 3–33

Alternately letting $y = 0$ and $x = 0$ in the equation for the ellipse, we see that it intersects the x- and y-axes at $(\pm a, 0)$ and $(0, \pm b)$, respectively. These points are called the **vertices** of the ellipse. The vertices and foci are related by the equation

$$b^2 = a^2 - c^2$$

or

$$b^2 + c^2 = a^2$$

This relationship is illustrated in Figure 3–34 along with the graph of the ellipse.

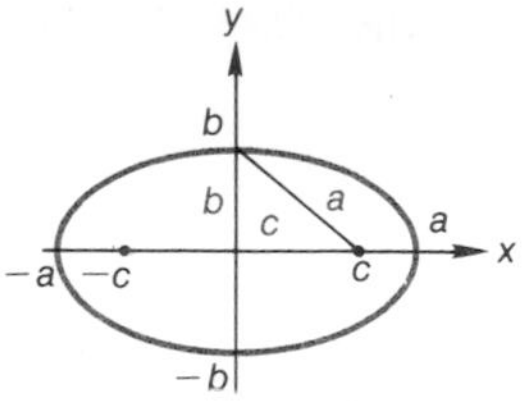

$$a^2 = b^2 + c^2$$

Figure 3–34 $\dfrac{x^2}{a^2} + \dfrac{y^2}{b^2} = 1,\ b > a$

It is clear that the graph of any equation of the form

$$\frac{x^2}{a^2} + \frac{y^2}{b^2} = 1$$

is symmetric about the x-axis, the y-axis, and the origin. Thus, an ellipse has two axes of symmetry. The intersections of these axes of symmetry with the interior of the ellipse are called the **major** and **minor axes,** respectively, depending on which is larger. In Figure 3–34 with $a > b$, the major axis has length $2a$, whereas the length of the minor axis is $2b$. Half the major and minor axes are called the **semimajor** and **semiminor axes,** respectively. Of course, if $a = b$, the ellipse degenerates into a circle.

If $b > a$, the ellipse is elongated vertically. This means that the foci are on the y-axis at points $(0, \pm c)$. The relationship between vertices and foci illustrated in Figure 3–34 can be generally stated as *"use the semimajor axis as the hypotenuse of a right triangle in order to find the foci."* This property is illustrated in Figure 3–35 for an ellipse with $b > a$.

The point in which its major and minor axes intersect is called the **center** of the ellipse. In this section we shall treat only the **central ellipses:** ellipses centered at the origin with the x- and y-axes as the axes of symmetry.

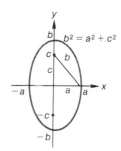

Figure 3–35
$\dfrac{x^2}{a^2} + \dfrac{y^2}{b^2} = 1,\ b > a$

CENTRAL ELLIPSES

Equation: $\dfrac{x^2}{a^2} + \dfrac{y^2}{b^2} = 1$

Vertices: $(\pm a,\ 0)$ and $(0,\ \pm b)$.

Elongation: **1.** Horizontal if $a > b$
 2. Vertical if $b > a$
 3. Reduces to a circle if $a = b$

Foci: $(\pm c,\ 0)$ or $(0,\ \pm c)$ as indicated in Figure 3–36.

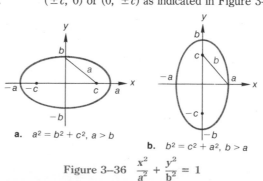

a. $a^2 = b^2 + c^2,\ a > b$

b. $b^2 = c^2 + a^2,\ b > a$

Figure 3–36 $\dfrac{x^2}{a^2} + \dfrac{y^2}{b^2} = 1$

EXAMPLE 1   Sketch the ellipse given by the following equation, and label its vertices and foci.

$$4x^2 + 25y^2 = 100$$

Solution   Dividing by 100, we obtain the standard equation of an ellipse:

$$\frac{x^2}{25} + \frac{y^2}{4} = 1$$

Its vertices are clearly $(\pm 5, 0)$ and $(0, \pm 2)$. Use the semimajor axis $a = 5$ as the hypotenuse of a right triangle and the semiminor axis $b = 2$ as one of its legs. The foci are then found by using the Pythagorean formula on this triangle:

$$b^2 + c^2 = a^2$$
$$4 + c^2 = 25$$
$$c = \sqrt{21}$$

The foci are locted at $(\pm \sqrt{21}, 0)$ (see Fig. 3–37).

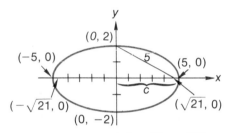

Figure 3–37 $4x^2 + 25y = 100$

EXAMPLE 2   Find the equation and sketch the graph of the central ellipse passing through $(2, 0)$ and $(-1, 3\sqrt{3})$. Label its vertices and foci.

Solution   The vertices of a central ellipse fall on the x- and y-axes. Thus, $(2, 0)$ is a vertex; that is, $a = 2$. The equation then has the form

$$\frac{x^2}{2^2} + \frac{y^2}{b^2} = 1$$

since $(-1, 3\sqrt{3})$ is on the ellipse, it must be the case that

$$\frac{(-1)^2}{4} + \frac{(3\sqrt{3})^2}{b^2} = 1$$

Solving for b^2, we obtain $b^2 = 36$. Thus, the equation of the ellipse is

$$\frac{x^2}{4} + \frac{y^2}{36} = 1$$

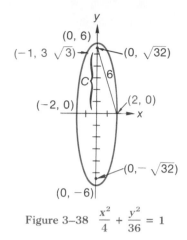

Figure 3–38 $\dfrac{x^2}{4} + \dfrac{y^2}{36} = 1$

Its vertices are $(\pm 2, 0)$ and $(0, \pm 6)$. Using 6 as the hypotenuse of a right triangle having one leg of length 2 as illustrated in Figure 3–38, we find the foci at $(0, \pm\sqrt{32})$:

$$2^2 + c^2 = 6^2$$
$$4 + c^2 = 36$$
$$c = \sqrt{32}$$

Ellipses have the property that a ray of light emanating from one focus is reflected to the other focus. This is the principle behind the so-called "whispering galleries." One of these whispering galleries is the Rotunda of the Capitol Building in Washington, D.C. Two spots are marked on the floor. Two persons, one standing on each of these two spots, can whisper and hear each other quite plainly. The two spots are foci for the elliptical ceiling! (see Fig. 3–39).

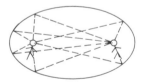

Figure 3–39   Focal properties of an ellipse.

Another use of the focal properties of an ellipse is found in the design of dental and surgery lights. The reflecting surface is elliptical with the light source at one of its foci. The light then becomes concentrated at the other focus of the ellipse. By adjusting the position of the light so that the point of concentration is at the opening of the patient's mouth, several things happen:

1.   The patient is not bothered by glare from the reflector.

2.   A small area inside the patient's mouth is illuminated very brightly.

3.   The dentist's hands do not interfere with the light.

Ellipses are also important in astronomy and celestial mechanics. Johannes Kepler (1571–1630) showed that the orbit of a celestial body is an ellipse with the sun at one focus. Similarly, the moon and orbiting satellites launched from the earth trace out ellipses with the center of the earth at one focus (see Fig. 3–40).

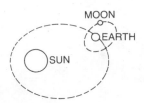

Figure 3–40   Elliptical orbits in space.

Hyperbolas

The locus definition of a hyperbola is very similar to that of an ellipse except that the difference rather than the sum of distances is used.

DEFINITION

A **hyperbola** is the locus of points for which the difference of the distances to two fixed points remains constant. Each of the fixed points is called a **focus** of the hyperbola.

In this context, the "difference" means "larger minus smaller." As with the ellipse, it is convenient to set up a coordinate system with one of the axes passing through the two foci. For the sake of the derivation, let the x-axis pass through the foci and place the y-axis midway between the foci. The foci then have coordinates $(\pm c, 0)$ for some $c > 0$. Let (x, y) be an arbitrary point on the hyperbola and label d_1, d_2 as in Figure 3–41.

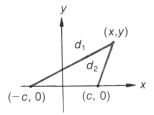

Figure 3–41 $|d_1 - d_2| = \text{constant}$.

The defining condition then requires that $|d_1 - d_2|$ remain constant. As before, label this constant as $2a$. Then

$$|d_1 - d_2| = 2a$$
$$d_1 - d_2 = \pm 2a$$
$$d_1 = d_2 \pm 2a$$

Using the distance formula to find d_1 and d_2, we can rewrite this equation as

$$\sqrt{[x - (-c)]^2 + (y - 0)^2} = \sqrt{(x - c)^2 + (y - 0)^2} \pm 2a \qquad (1)$$

As with the ellipse, repeated squaring operations reduce this equation to

$$x^2(a^2 - c^2) + a^2y^2 = a^2(a^2 - c^2) \tag{2}$$

Since the shortest distance between two points is a straight line, we see in Figure 3–41 that

$$d_2 + 2c > d_1$$
$$2c > d_1 - d_2 = 2a$$

Thus, $c > a$, and we can set $b^2 = c^2 - a^2 > 0$. Equation (2) then becomes

$$-b^2x^2 + a^2y^2 = -a^2b^2, \qquad a^2 > 0, \, b^2 > 0$$

Dividing by $-a^2b^2$, we obtain the **standard equation of a hyperbola with horizontal axis:**

$$\boxed{\dfrac{x^2}{a^2} - \dfrac{y^2}{b^2} = 1} \tag{3}$$

Again, note the symmetry with respect to the x-axis, the y-axis, and the origin. Thus, in graphing this equation, we shall at first concern ourselves only with the first quadrant; that is, consider only $x \geq 0$ and $y \geq 0$. Note that there can be no y-intercept; setting $x = 0$ yields the impossible statement $-y^2/b^2 = 1$. Setting $y = 0$, we obtain the x-intercepts $x = \pm a$. Equation (3) can be rewritten as

$$\dfrac{x^2}{a^2} = 1 + \dfrac{y^2}{b^2} \tag{4}$$

In Equation (4) we observe that

$$\dfrac{x^2}{a^2} \geq 1$$

or

$$x^2 \geq a^2$$

Hence, $x \geq a$ in the first quadrant. We could also solve Equation (3) for y in terms of x to obtain

$$\dfrac{y^2}{b^2} = \dfrac{x^2}{a^2} - 1 = \dfrac{x^2 - a^2}{a^2}$$
$$y^2 = \dfrac{b^2}{a^2}(x^2 - a^2) = \dfrac{b^2}{a^2}x^2\left(1 - \dfrac{a^2}{x^2}\right)$$

Hence, in the first quadrant

$$y = \frac{b}{a}x \sqrt{1 - \frac{a^2}{x^2}} < \frac{b}{a}x \tag{5}$$

Now as x gets very large,

$$1 - \frac{a^2}{x^2} \approx 1$$

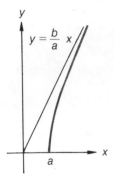

$y = \frac{b}{a} x$

a

Figure 3–42

Equation (5) then indicates that for large x,

$$y \approx \frac{b}{a}x \quad \text{and} \quad y < \frac{b}{a}x$$

Thus, y approaches $(b/a)x$ from below as x gets large. Plotting additional points if necessary, we see finally that the graph of Equation (3) in the first quadrant is as given in Figure 3–42.

Using the symmetry as previously mentioned, we can then complete the graph as illustrated in Figure 3–43. It is called a hyperbola. For a hyperbola, the coordinates $(\pm c, 0)$ of the focus are related to a and b by

$$b^2 = c^2 - a^2$$

or

$$c^2 = a^2 + b^2$$

This relationship is also illustrated in Figure 3–43.

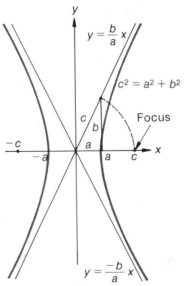

Figure 3–43 $\dfrac{x^2}{a^2} - \dfrac{y^2}{b^2} = 1$

The lines $y = \pm(b/a)x$ are called **asymptotes** of the hyperbola. The points $(\pm a, 0)$ are its **vertices**. The point halfway between the vertices is called its **center**. As with the ellipses, we shall be concerned only with **central hyperbolas**: those with the origin as center and with the x- and y-axes as axes of symmetry.

Placing the foci on the y-axis at $(0, \pm c)$ simply interchanges the roles of x and y in the preceding derivation. Thus, the **standard equation of a hyperbola with vertical axis** is

$$\frac{y^2}{b^2} - \frac{x^2}{a^2} = 1 \qquad (6)$$

The focus is related to a and b in the same way as before:

$$c^2 = a^2 + b^2$$

Expressing y in terms of x in Equation (6), we have

$$\frac{y^2}{b^2} = 1 + \frac{x^2}{a^2} \geq 1 \qquad (7)$$

Thus,

$$y \geq b$$

in the first quadrant; there is no x intercept. Continuing from Equation (7), we see that

$$\frac{y^2}{b^2} = \frac{a^2 + x^2}{a^2} = \frac{x^2}{a^2}\left(1 + \frac{a^2}{x^2}\right)$$

Thus,

$$y = \frac{bx}{a}\sqrt{1 + \frac{a^2}{x^2}} > \frac{bx}{a}$$

in the first quadrant. For large x,

$$y \approx \frac{bx}{a} \quad \text{and} \quad y > \frac{bx}{a}$$

that is, y approaches bx/a from above as x gets large. The graph of Equation (6) is seen to be another hyperbola with the same asymptotes as Equation (3). The vertices are now $(0, \pm b)$ and the hyperbola opens vertically. It is sketched and its foci are indicated in Figure 3–44. The hyperbolas given by Equations (3) and (6) are said to be **conjugate** to one another.

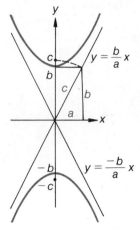

Figure 3–44 $\dfrac{y^2}{b^2} - \dfrac{x^2}{a^2} = 1$

HYPERBOLAS

The equations

$$\frac{x^2}{a^2} - \frac{y^2}{b^2} = 1 \qquad \frac{y^2}{b^2} - \frac{x^2}{a^2} = 1$$

represent conjugate hyperbolas. Each can be sketched by constructing a rectangle centered at the origin with horizontal sides of length $2a$ and vertical sides of length $2b$. The distance of the foci from the origin equals half the length of the diagonal of this rectangle:

$$c^2 = a^2 + b^2$$

The diagonals of the rectangle are asymptotes for each of these conjugate hyperbolas. The vertices occur at the intersection of the respective axes with the rectangle constructed as indicated (see Fig. 3–45 on next page).

EXAMPLE 3   Sketch the conjugate hyperbolas. Label the vertices and foci.

a. $\dfrac{x^2}{4} - \dfrac{y^2}{9} = 1$

b. $\dfrac{y^2}{9} - \dfrac{x^2}{4} = 1$

Solution   In each of these hyperbolas, $a = 2$ and $b = 3$. Thus, the asymptotes are $y = \pm (3/2)x$. The vertices of the first hyperbola are $(\pm 2, 0)$; the vertices of the second are $(0, \pm 3)$. We sketch the rectangle centered at the origin and passing

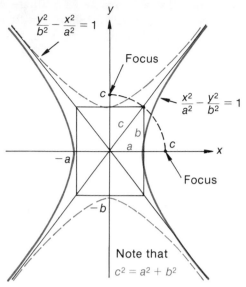

Figure 3–45

through these vertices. The diagonals of this rectangle are the asymptotes of both hyperbolas. It is then a simple matter to sketch the hyperbolas. The foci can be determined from the relationship $c^2 = 2^2 + 3^2 = 13$ (see Fig. 3–46).

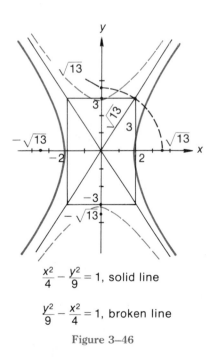

$\dfrac{x^2}{4} - \dfrac{y^2}{9} = 1$, solid line

$\dfrac{y^2}{9} - \dfrac{x^2}{4} = 1$, broken line

Figure 3–46

EXAMPLE 4   Find the equation and sketch the graph of the central hyperbola with vertices at $(\pm 5, 0)$ and the line $y = 2x$ as an asymptote.

Solution   Note that $a = 5$ since $(\pm 5, 0)$ are vertices. The slopes of the asymptote lines are $\pm(b/a) = \pm b/5$. Since $y = 2x$ is an asymptote, we then have $b/5 = 2$ or $b = 10$. Since the vertices are on the x-axis, the equation of the hyperbola must then be

$$\frac{x^2}{25} - \frac{y^2}{100} = 1$$

It is graphed in Figure 3–47.

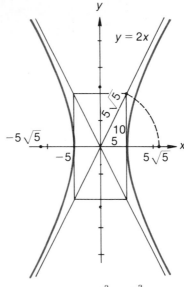

Figure 3–47 $\dfrac{x^2}{25} - \dfrac{y^2}{100} = 1$

EXAMPLE 5   Find the equation and sketch the graph of the central hyperbola passing through $(1, 0)$ and $(2, 2\sqrt{3}\,)$.

Solution   Since this is a central hyperbola, $(1, 0)$ must be a vertex. Thus, the hyperbola opens horizontally and has the form

$$\frac{x^2}{a^2} - \frac{y^2}{b^2} = 1$$

Since $(1, 0)$ is a vertex, $a = 1$; the equation is

$$\frac{x^2}{1} - \frac{y^2}{b^2} = 1$$

or

$$x^2 - \frac{y^2}{b^2} = 1$$

Since $(2, 2\sqrt{3})$ is on the graph, we have

$$2^2 - \frac{(2\sqrt{3})^2}{b^2} = 1$$

$$4 - \frac{12}{b^2} = 1$$

From this we find

$$b^2 = 4$$

The equation of the hyperbola is

$$x^2 - \frac{y^2}{4} = 1$$

See Figure 3–48.

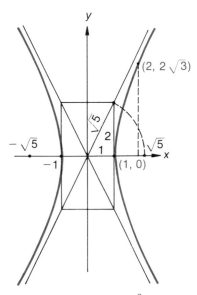

Figure 3–48 $x^2 - \dfrac{y^2}{4} = 1$

EXERCISES   3.5

Sketch the following ellipses; indicate vertices and foci.

1. $\dfrac{x^2}{9} + \dfrac{y^2}{16} = 1$ **2.** $\dfrac{x^2}{16} + \dfrac{y^2}{9} = 1$

3. $\dfrac{x^2}{25} + \dfrac{y^2}{9} = 1$ **4.** $\dfrac{x^2}{25} + \dfrac{y^2}{16} = 1$

5. $x^2 + 9y^2 = 9$ **6.** $9x^2 + y^2 = 9$

7. $4x^2 + 9y^2 = 36$ **8.** $9x^2 + 4y^2 = 36$

9. $25x^2 + 36y^2 = 900$ **10.** $25x^2 + 9y^2 = 225$

11. $16x^2 + 4y^2 = 64$

12. $36x^2 + 81y^2 = 2916$

13. $169x^2 + 25y^2 = 4225$

14. $576x^2 + 625y^2 = 360{,}000$

Find the equation and sketch the central ellipse satisfying the following conditions. Label the vertices and foci.

15. Vertices at $(0, 2)$ and $(-3, 0)$
16. Vertical major axis length of 26, horizontal minor axis length of 24
17. Horizontal major axis of length 10, minor axis of length 6
18. Focus at $(3, 0)$ and horizontal axis of length 10
19. Focus at $(-12, 0)$, vertical axis of length 10
20. Focus at $(0, -\sqrt{5})$ and vertex at $(0, 3)$
21. Focus at $(7, 0)$ and vertex at $(0, -24)$
22. Focus at $(2, 0)$ and minor axis of length 2
23. Focus at $(0, -12)$ and major axis of length 26
24. Vertex at $(4, 0)$, passing through $(2, -3\sqrt{3}/2)$
25. Vertex at $(0, 8)$, passing through $(-\sqrt{3}, 4)$
26. Passing through $(0, 1)$ and $(1, \sqrt{2}/2)$
27. Passing through $(2, 0)$ and $(1, \sqrt{15}/2)$

Sketch the graph of each of the given equations. In each case, write the equation of the conjugate hyperbola and also sketch its graph. Indicate vertices, foci, and asymptotes.

28. $\dfrac{x^2}{16} - \dfrac{y^2}{9} = 1$
29. $\dfrac{y^2}{16} - \dfrac{x^2}{9} = 1$
30. $\dfrac{x^2}{25} - \dfrac{y^2}{144} = 1$
31. $x^2 - y^2 = 9$
32. $9x^2 - y^2 = 9$
33. $x^2 - 4y^2 = 16$
34. $25y^2 - 4x^2 = 100$
35. $144y^2 - 25x^2 = 3600$
36. $16y^2 - 25x^2 = 400$
37. $y^2 - 4x^2 = 12$
38. $4x^2 - 9y^2 = 36$
39. $9x^2 - 4y^2 = 36$
40. $\dfrac{y^2}{49} - \dfrac{x^2}{576} = 1$
41. $64x^2 - 25y^2 = 1600$

Find the equation and sketch the graph of the central hyperbola satisfying the following conditions. Label the vertices, foci, and asymptotes.

42. Vertex $(0, 2)$, asymptote $y = 2x$
43. Vertex $(4, 0)$, asymptote $y = -3x/4$
44. Vertex $(16, 0)$, asymptote $y = 5x/4$
45. Vertex $(0, -1)$, asymptote $y = -3x$
46. Vertex $(0, 2)$, passing through $(1, -3)$
47. Vertex $(3, 0)$, passing through $(-5, 8)$
48. Vertex $(-5, 0)$, passing through $(13, 36)$
49. Vertex $(0, -4)$, passing through $(1, -5)$
50. Vertex $(0, -3)$, focus $(0, 5)$
51. Vertex $(2, 0)$, focus $(-3, 0)$
52. Vertex $(5, 0)$, focus $(13, 0)$
53. Focus $(3, 0)$, asymptote $y = 2x$
54. Focus $(0, 2)$, asymptote $y = x$
55. Focus $(0, -2\sqrt{5})$, asymptote $x = 2y$
56. Sketch and find the equation of the locus of points the sum of whose distances from $(-2, 3)$ and $(4, 3)$ equals 10.
57. Sketch and write the equation of the locus of points, the sum of whose distances from $(-1, 2)$ and $(1, 3)$ is 5.
58. Sketch and write the equation of the locus of points, the difference of whose distances from $(1, 2)$, and $(1, 8)$ is 3.
59. Sketch and write the equation of the locus of points, the difference of whose distances from $(3, -2)$, and $(-5, 4)$ is 8.
60. A race track is in the form of an ellipse with major and minor axes of length 500 and 140 meters, respectively. Posts are driven into the ground at the foci of this ellipse. A rope is looped through Black Beauty's bridle, is tied, and draped over the posts as in Figure 3–33. How long must the rope be if Black Beauty is to be free to run around this track?
61. Bob's backyard is a rectangle 30 meters by 50 meters. He wants to let his dog exercise in the yard, but the neighbors get upset if the dog ventures onto their property. His solution is to drive two stakes into the ground and fasten a chain to both of these stakes after passing it through the dog's collar. Determine where the stakes should be located and how much chain is needed if the dog is to have a maximum exercise area without being able to encroach on neighboring property.
62. Two ranger stations 30 miles apart pick up an SOS broadcast from a hiker who has lost his way. It is determined that Ranger Al receives the signal 1/10,000 of a second later than Ranger Bob.
   a. Sketch a graph indicating the possible locations of the SOS broadcaster. (Radio waves travel at a speed of 186,000 miles per second.)
   b. Indicate how the location of the SOS broadcaster can be pinpointed if a third ranger also picks up the broadcast. (This is an example of "triangulation.")

3.6 THE CONIC SECTIONS AND TRANSLATION
(Optional)

(This section may be postponed with no loss of continuity. No later topics in this text depend on this material.)

You may be surprised to learn that some of the most important curves used by scientists and engineers to describe various natural phenomena can be obtained simply by slicing an ordinary object with which we are all familiar: the (ice cream) cone. These curves, called **conic sections,** are obtained by intersecting a plane and a double cone.

Only five general curves are generated in this way: lines, circles, parabolas, ellipses, and hyperbolas. Figure 3–49 illustrates how these conic sections are generated.

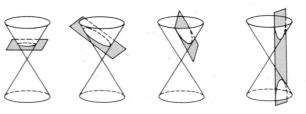

a. Circle **b.** Ellipse **c.** Parabola **d.** Hyperbola

Figure 3–49 The Conic Sections

Each of the conic sections can be defined as a **locus** of points satisfying a given condition; that is, it consists of precisely those points that satisfy the stated condition. For instance,

Circles

> A **circle** is the locus of points whose distance from a given point A is some fixed constant r. A is called the **center** and r the **radius** of the circle.
>
> The distance formula is used to write the equation of the circle of radius r centered at (x_0, y_0) as
>
> $$(x - x_0)^2 + (y - y_0)^2 = r^2$$

Since circles were characterized in this way and studied extensively in Section 3.1, we shall not have any more to say about them in this section.

> A **line** is the locus of points equidistant from two given points.

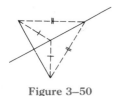

Figure 3–50

This is true because every point on the perpendicular bisector of a line segment is equidistant from the ends of that segment (see Fig. 3–50).

EXAMPLE 1   Write the equation of the line representing the locus of points equidistant from (1, 2) and (3, 8).

Solution   The line described must be the perpendicular bisector of the segment joining (1, 2) and (3, 8). Using the midpoint formula if necessary, we find the midpoint of this segment to be (2, 5). The slope of the line joining the two points is

$$m = \frac{8-2}{3-1} = \frac{6}{2} = 3$$

The slope of the perpendicular bisector is then

$$-\frac{1}{m} = -\frac{1}{3}$$

Thus the equation of the line in question is

$$(y - 5) = -\frac{1}{3}(x - 2)$$

This equation can be simplified into

$$x + 3y = 17$$

See Figure 3–51.

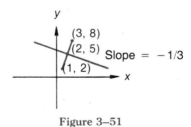

Figure 3–51

Parabolas

> **DEFINITION**
> A **parabola** is the locus of points equidistant from a given line and a given point; the line is called the **directrix,** and the fixed point is called the **focus** of the parabola.

To see why such a locus describes a parabola, set up a coordinate system as follows. Let the *y*-axis be obtained by dropping the perpendicular from the focus to the directrix. Then place the *x*-axis midway between the focus and directrix. In this coordinate system, the focus will have coordinates (0, *d*), and the directrix is the line $y = -d$ for some number d. If (x, y) is on the described locus, and d_1, d_2 are labeled as in Figure 3–52, then $d_1 = d_2$. Thus,

$$\sqrt{(x - 0)^2 + (y - d)^2} = y + d$$

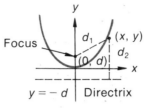

Figure 3–52

Squaring both sides yields

$$x^2 + y^2 - 2dy + d^2 = y^2 + 2dy + d^2$$
$$x^2 = 4dy$$
$$y = \frac{1}{4d} x^2$$

This last equation is the equation of a vertical parabola with its vertex at the origin. See Section 3.4. Similarly, the equation

$$x = \frac{1}{4D} y^2$$

represents a horizontal parabola with vertex at the origin, focus at $(D, 0)$, and directrix at $x = -D$.

The focus of a parabola is very important from a practical viewpoint. The line joining the focus and vertex of a parabola is called its **principal axis.** A light source situated at the focus of a parabola will have its rays reflected parallel to the principal axis of the parabola. This property is used in the design of automobile headlights and beacon lights. The property is used in reverse in designing parabolic-reflector telescopes; all light entering the parabolic reflector is reflected *into* the focus (see Fig. 3–53).

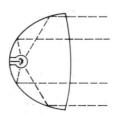 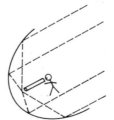

a. Automobile headlight   **b.** Parabolic-reflector telescope

Figure 3–53

EXAMPLE 2   Sketch the graph of $y^2 = -12x$ and find its focus and directrix.

Solution   Writing the equation in the form $x = \dfrac{1}{4D}\, y^2$, we have

$$x = \frac{-1}{12}\, y^2 = \frac{1}{4(-3)}\, y^2$$

Thus, $D = -3$. The graph is a horizontal parabola with vertex at the origin, focus at $(-3,\ 0)$, directrix at $x = -(-3) = 3$, and opening to the left (see Fig. 3–54). ∎

Ellipses and Hyperbolas

These were already characterized in terms of a locus in Section 3.5.

Translation of Axes

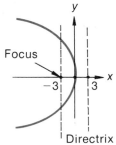

Figure 3–54
$y^2 = -12x$ or
$x = (-1/12)y^2$

Until now we have been concerned only with the central conics. To discuss conics that are not centered at the origin, we must investigate the changes that are effected on an equation when the origin is moved. In this section, we shall only **translate** the origin; that is, move it to a new location in such a way that the axes are not stretched or rotated.

Consider now a plane on which are imposed two coordinate systems parallel to one another and with the same units as illustrated in Figure 3–55. Any point in this plane can then be given coordinates in each of the two coordinate systems. To relate the two coordinate systems, let the origin of the x'-y' system have coordinates $(h,\ k)$ in the x-y system.

Let the coordinates of some point P be $(x',\ y')$ and $(x,\ y)$ in the X'-Y' and X-Y coordinate systems, respectively. It should be clear from Figure 3–55 that

$$x = x' + h$$
$$y = y' + k$$

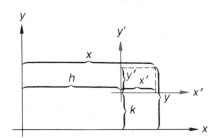

Figure 3–55   Parallel coordinate systems.

Solving for x', y' in terms of x and y, we obtain

$$x' = x - h$$
$$y' = y - k$$

TRANSLATION OF AXES

If we know how to write the equation of a given figure when it is centered at the origin, then the following steps indicate how to write the equation of such a figure when it is centered at (h, k).

1. Introduce a new x'-y' coordinate system with origin at (h, k).

2. Write the equation of the curve in this new coordinate system.

3. Substitute

$$x' = x - h$$
$$y' = y - k$$

into the equation to obtain the equation in the original x-y coordinate system.

We have already seen these relationships in our study of lines and circles and parabolas. Figure 3–56 should illustrate the situation sufficiently. For instance, we should recognize the point-slope formula in Figure 3–56(a).

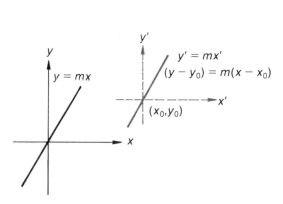

a. A line of slope m through (x_0, y_0)

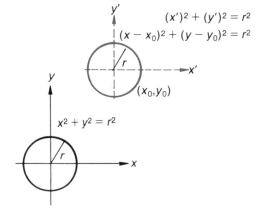

b. A circle of radius r centered at (x_0, y_0)

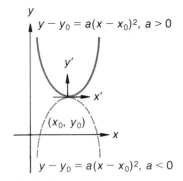

c. Vertical parabolas
$$y' = a(x')^2$$
or
$$y = a(x - x_0)^2 + y_0$$

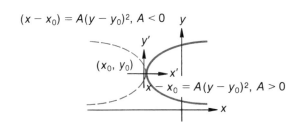

d. Horizontal parabolas
$$x' = A(y')^2$$
or
$$x = A(y - y_0)^2 + x_0$$

Figure 3–56

EXAMPLE 3 Sketch the following parabola; indicate its vertex, focus, directrix, and principal axis.

$$4y = x^2 - 6x + 17$$

Solution First complete the square in order to identify x_h and y_k:

$$
\begin{aligned}
4y &= (x^2 - 6x \qquad) + 17 \\
&= (x^2 - 6x + 9) + 17 - 9 \\
&= (x - 3)^2 + 8 \\
4y - 8 &= (x - 3)^2 \\
4(y - 2) &= (x - 3)^2
\end{aligned}
$$

Thus

$$y - 2 = \frac{1}{4}(x - 3)^2$$

is the equation of the parabola. It opens upward with vertex at $(3, 2)$. In the coordinate system centered at $(3, 2)$, this equation becomes

$$y' = \frac{1}{4}(x')^2$$

If the focus is located d units above the vertex, its equation is

$$y' = \frac{1}{4d}(x')^2$$

Thus, $d = 1$; that is, the focus is one unit above the vertex at $(3, 3)$. The directrix is then one unit below the vertex at $y = 1$. The line $x = 3$ is the principal axis (see Fig. 3–57).

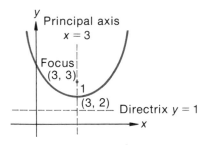

Figure 3–57 $4y = x^2 - 6x + 17$

Ellipses

If an ellipse is centered at (h, k) with semimajor and semiminor axes a and b, it is a simple matter to write its equation in a coordinate system centered at (h, k). The equation is

$$\frac{(x')^2}{a^2} + \frac{(y')^2}{b^2} = 1$$

Substituting the translation conditions into this equation yields

$$\frac{(x - h)^2}{a^2} + \frac{(y - k)^2}{b^2} = 1$$

as the equation of an ellipse centered at (h, k) with semimajor and semiminor axes a and b (see Fig. 3–58).

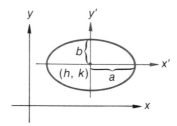

Figure 3–58 $\quad \dfrac{(x - h)^2}{a^2} + \dfrac{(y - k)^2}{b^2} = 1$

EXAMPLE 4 Sketch and find the equation of an ellipse with upper vertex at $(5, 9)$ and rightmost vertex at $(8, 7)$, assuming that its axes of symmetry are parallel to the x- and y-axes, respectively. Find its foci.

Solution The vertices lie on the axes of symmetry. Accordingly, plot the given vertices and draw vertical and horizontal lines through them to find the center of the ellipse at the intersection of these axes of symmetry. Sketching the ellipse, we find its center to be $(5, 7)$; its semimajor and semiminor axes are 3 and 2, respectively. Hence, its equation is

$$\frac{(x - 5)^2}{9} + \frac{(y - 7)^2}{4} = 1$$

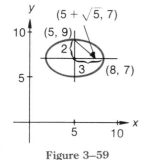

Figure 3–59

For the foci we write

$$3^2 = 2^2 + c^2$$
$$c = \sqrt{5}$$

Thus, the foci are located at $(5 \pm \sqrt{5}, 7)$ (see Fig. 3–59). ■

Hyperbolas For a hyperbola centered at (h, k), we introduce a new coordinate system with origin at (h, k). In this new coordinate system, the equation of the hyperbola is

1. $\dfrac{(x')^2}{a^2} - \dfrac{(y')^2}{b^2} = 1 \qquad$ if the hyperbola opens horizontally or

2. $\dfrac{(y')^2}{b^2} - \dfrac{(x')^2}{a^2} = 1$ if the hyperbola opens vertically

In terms of the original variables, these equations become

1. $\dfrac{(x - h)^2}{a^2} - \dfrac{(y - k)^2}{b^2} = 1$

or

2. $\dfrac{(y - k)^2}{b^2} - \dfrac{(x - h)^2}{a^2} = 1$

Their graphs are shown in Figure 3–60.

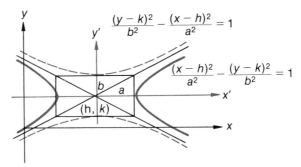

Figure 3–60

EXAMPLE 5   Sketch the graph of the following equation. Find its foci.

$$4x^2 - 9y^2 + 8x + 54y = 113$$

Solution   Manipulate the equation into a more recognizable form.

$$
\begin{aligned}
(4x^2 + 8x \quad) - (9y^2 - 54y \quad) &= 113 \\
4(x^2 + 2x \quad) - 9(y^2 - 6y \quad) &= 113 \\
4(x^2 + 2x + 1) - 9(y^2 - 6y + 9) &= 113 + 4 - 81 \\
4(x + 1)^2 - 9(y - 3)^2 &= 36 \\
\frac{(x + 1)^2}{9} - \frac{(y - 3)^2}{4} &= 1
\end{aligned}
$$

The final form represents a hyperbola opening horizontally and centered at

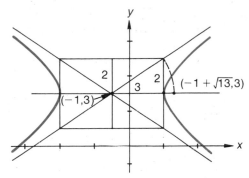

Figure 3-61 $4x^2 - 9y^2 + 8x + 54y = 113$

$(-1, 3)$ with $a = 3$, $b = 2$. It is sketched in Figure 3–61. To find its foci, we write

$$c^2 = 3^2 + 2^2 = 13$$

Thus, the foci are located at $(-1 \pm \sqrt{13}, 3)$.

EXERCISES 3.6

Sketch and write the equation for the locus of points equidistant from each of the following pairs of points

1. $(2, 0)$ and $(0, 4)$ 2. $(2, 3)$ and $(4, 5)$
3. $(3, -1)$ and $(6, 1)$
4. $(4, -2)$ and $(-2, 6)$
5. $(-1, 1)$ and $(-3, 3)$
6. $(-6, -1)$ and $(0, -1)$
7. $(3, -3)$ and $(3, 1)$
8. $(-3, 5)$ and $(-4, -2)$

Sketch and write the equation for the locus of points equidistant from the given point and the given line.

9. $(1, 0)$ and $x = -1$
10. $(-3, 0)$ and $x = 3$
11. $(0, -5)$ and $y = 5$ 12. $(0, 2)$ and $y = 8$
13. $(2, 4)$ and $y = 6$ 14. $(-3, 2)$ and $y = 0$
15. $(5, -7)$ and $x = 9$
16. $(3, 2)$ and $x = -1$

Sketch the graph of each of the following parabolas. Label the vertex, focus, and directrix.

17. $y = x^2/8$ 18. $x^2 = 12y$
19. $x^2 = -20y$ 20. $x^2 = -4y$
21. $4x = y^2$ 22. $y^2 = 16x$
23. $y^2 = -8x$ 24. $y^2 = -20x$

Sketch and write the equations of the following parabolas.

25. Focus at $(-1, 2)$, vertex at $(1, 2)$

26. Focus at $(3, 1)$, directrix $y = 3$
27. Directrix $y = 3$, vertex $(-2, 0)$
28. Vertex $(1, 2)$, focus at $(1, 4)$
29. Directrix $x = 0$, focus at $(2, -1)$
30. Vertex $(2, 3)$, directrix $x = -2$

Sketch and find the equation of the ellipses satisfying the given conditions.

31. Lower vertex at $(-2, -3)$, left vertex at $(-7, 1)$
32. Upper vertex at $(6, 2)$, right vertex at $(12, -1)$
33. Left vertex at $(1, 1)$, upper vertex at $(3, 6)$
34. Right vertex at $(5, -2)$, lower vertex at $(3, -7)$
35. Upper vertex at $(2, 4)$, right vertex at $(4, 3)$
36. Lower vertex at $(5, -1)$, left vertex at $(3, 5)$

Sketch and find the equation of each hyperbola satisfying the given conditions. Do the same for its conjugate.

	Opens	Centered at	a	b
37.	Vertically	$(-1, -4)$	1	5
38.	Horizontally	$(6, -9)$	4	2
39.	Vertically	$(8, 1)$	10	6
40.	Horizontally	$(2, -7)$	4	6
41.	Horizontally	$(-10, 8)$	9	10
42.	Vertically	$(9, -3)$	7	5

Sketch the graphs of the given equations. Label the vertices and foci. If a parabola, indicate the directrix; if a hyperbola, indicate the asymptotes.

43. $2x^2 - 4x - y + 1 = 0$

44. $25x^2 + 9y^2 - 100x + 54y = 44$

45. $x^2 + 16y^2 + 4x - 32y + 16 = 0$

46. $9y^2 - 16x^2 + 64x + 54y + 161 = 0$

47. $2y^2 - 16y + x + 4 = 0$

48. $4y^2 - 9x^2 - 36x + 24y = 36$

49. $x^2 - 2x + y + 2 = 0$

50. $16y^2 - 4x^2 + 48y - 28x = 171$

51. $y^2 - x^2 + 2y - 4x - 7 = 0$

52. $x - 4y + 2y^2 + 2 = 0$

53. $4x^2 - 25y^2 + 16x + 150y = 109$

54. $16x^2 - 9y^2 + 64x - 18y - 89 = 0$

55. $4x^2 + 9y^2 + 16x - 36y + 16 = 0$

56. $5x^2 - 20x + y - 1 = 0$

57. $9x^2 + 4y^2 - 18x + 16y = 11$

58. Sketch and write the equation of the locus of points equidistant from the line $y = -3$ and the point $(2, -1)$.

59. Sketch and write the equation of the locus of points, the sum of whose distances from $(2, 1)$ and $(8, 1)$ is 10.

60. Sketch and write the equation of the locus of points, the difference of whose distances from $(2, -6)$ and $(2, 4)$ is 6.

3.7 CHAPTER REVIEW

TERMS AND CONCEPTS

- Graph

  A pictorial representation of a relationship between two or more quantities. The graph of an equation or relationship between two variables x and y consists of those points (x, y) that satisfy the relationship.

- Cartesian Coordinates

  Enable us to plot points in an x-y plane.

- Intercepts

  Points in which a graph meets the x- and y-axes

RULES AND FORMULAS

- Midpoint Formula

  The midpoint of the line segment joining $P = (x_1, y_1)$ and $Q = (x_2, y_2)$ has coordinates
  $$\left(\frac{x_1 + x_2}{2}, \frac{y_1 + y_2}{2} \right)$$

- Distance Formula

  The distance between two points $P = (x_1, y_1)$ and $Q = (x_2, y_2)$ is given by
  $$d(P, Q) = \sqrt{(x_2 - x_1)^2 + (y_2 - y_1)^2}$$

- Circles

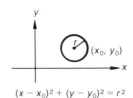

  $(x - x_0)^2 + (y - y_0)^2 = r^2$

  Center at (x_0, y_0), radius $= r$.

- Lines

  1. Slope $\quad m = \dfrac{y_2 - y_1}{x_2 - x_1}$ $\qquad\qquad$ $(x_1, y_1), (x_2, y_2)$ on the line
  2. Point-slope formula $\qquad\qquad$ slope $= m$, (x_0, y_0) on the line
     $(y - y_0) = m(x - x_0)$
  3. Slope-intercept formula $\qquad\qquad$ slope $= m$, y-intercept $= b$
     $y = mx + b$
  4. Vertical line formula $\qquad\qquad$ $x = a;\ a = x$-intercept
  5. General formula $\qquad\qquad$ $Ax + By + C = 0;\ A,\ B$ not both 0
  6. Parallel lines $\qquad\qquad$ $m_1 = m_2$; slopes $m_1,\ m_2$
  7. Perpendicular lines $\qquad\qquad$ $m_1 m_2 = -1$; slopes $m_1,\ m_2$

- Parabolas

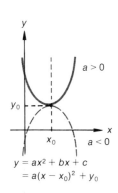

$$y = ax^2 + bx + c$$
$$= a(x - x_0)^2 + y_0$$

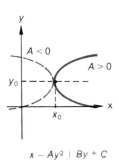

$$x = Ay^2 + By + C$$
$$= A(y - y_0)^2 + x_0$$

- Ellipses
$$\frac{x^2}{a^2} + \frac{y^2}{b^2} = 1$$

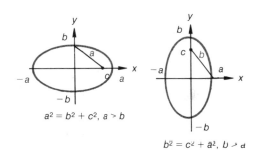

$a^2 = b^2 + c^2,\ a > b$

$b^2 = c^2 + a^2,\ b > a$

- Hyperbolas

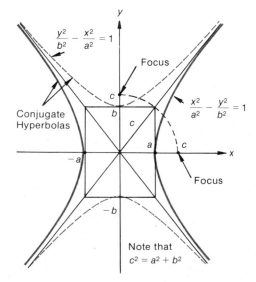

$$\frac{y^2}{b^2} - \frac{x^2}{a^2} = 1$$

$$\frac{x^2}{a^2} - \frac{y^2}{b^2} = 1$$

Conjugate Hyperbolas

Focus

Focus

Note that
$c^2 = a^2 + b^2$

TECHNIQUES

- Solving quadratic Inequalities by Graphing
  1. Find the roots of the equation $ax^2 + bx + c = 0$.
  2. The corresponding parabola $y = ax^2 + bx + c$ crosses the x-axis at these points.
  3. Determine from this parabola the region or regions in which the inequality is satisfied.
- Graphing $Ax^2 + By^2 + Cx + Dy + E = 0$
  1. Complete the squares in x and y.
  2. Substitute $x' = x - h$, $y' = y - k$ and rewrite the equation in terms of x' and y'.
  3. Sketch an x'-y' coordinate system centered at (h, k) on the x-y coordinate system.
  4. Graph the equation in the x'-y' coordinate system.

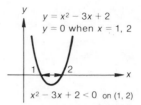

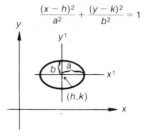

3.8 SUPPLEMENTARY EXERCISES

For each of the following pairs of points
  a. *Plot them*
  b. *Find the distance between them*
  c. *Find the midpoint of the segment joining them*
  d. *Write the equation of the line passing through them*
  e. *Write the equation of the perpendicular bisector of the segment joining them*

  1. $(-3, 0)$, $(-7, 0)$
  2. $(0, 2)$, $(0, 5)$
  3. $(6, 5)$, $(4, 2)$ 4. $(5, 2)$, $(-1, -2)$
  5. $(1, 0)$, $(0, 2)$

Sketch and write the equation of the indicated circle.

  6. center $(1, 3)$, radius 2
  7. center $(-3, 4)$, radius 5
  8. center $(-4, -1)$, radius 3
  9. center $(4, 1)$, passing through $(6, 2)$
  10. center $(-10, 7)$, passing through $(-7, 10)$
  11. center $(-1, -2)$, passing through $(-3, -6)$
  12. $(8, 26)$ and $(-6, -22)$ are at opposite ends of a diameter

In each of the following, determine whether the given points fall on a line.

  13. $(-7, -6)$, $(18, 4)$, $(3, -2)$
  14. $(-2, 6)$, $(9, -2)$, $(3, 2)$
  15. $(-8, 3)$, $(2, -1)$, $(12, -5)$

In each of the following, find an equation of the line satisfying the stated conditions; m denotes slope, b = y-intercept, and a = x-intercept.

  16. $m = 2$, $b = 6$ 17. $m = -6$, $b = 3$
  18. $m = -4$, $b = -2$
  19. $m = 0$, passing through $(4, 2)$
  20. $m = -1/4$, passing through $(-4, 3)$
  21. $m = -1/2$, passing through $(-4, -5)$
  22. $a = 6$, $b = 10$ 23. $a = -3$, $b = 4$
  24. $a = 0$, $b = 5$
  25. tangent to the circle $x^2 + y^2 = 25$ at the point $(3, 4)$
  26. tangent to the circle $x^2 + y^2 = 169$ at the point $(-12, 5)$

Determine whether the following pairs of lines are parallel, perpendicular, or neither.

  27. $3x + 2y - 7 = 0$
      $9x + 6y + 3 = 0$
  28. $3x - y + 2 = 0$
      $x + 3y - 2 = 0$
  29. $11x - 3y + 5 = 0$
      $-22x + 6y - 3 = 0$
  30. $7x - 3y - 4 = 0$
      $3x - 7y + 4 = 0$
  31. $-4x + 2y - 6 = 0$
      $-3x - 6y = 0$

Write the equation of the line through the given point that is parallel to the given line and also for the line that is perpendicular to the given line.

  32. $(2, 3)$, $2x + y - 5 = 0$
  33. $(-1, 5)$, $y = -2$
  34. $(2, 1)$, $3x - 9y - 6 = 0$

35. $(0, 7)$, $x = 3$

36. $(1, -2)$, $x - 2y - 3 = 0$

Graph the following lines. Find their slopes and intercepts.

37. $x - 5y - 10 = 0$

38. $2x + y + 8 = 0$

39. $4x = 6$ **40.** $10y = 0$

41. $3x - y = 9$

42. $-5x + 10y - 10 = 0$

Graph each of the following parabolas. In each, find the vertex, the line of symmetry, and the intercepts.

43. $y = -2x^2 - 1$ **44.** $x = 2y^2 - 4y$

45. $y = -2x^2 + 2x - 1$

46. $x = 2y^2 - 12y + 8$

47. $y = -2x^2 + 4x + 10$

48. $y = -x^2 + 2x - 2$

49. $x = y^2 - 2y + 6$

Sketch and write the equation of the following parabolas.

50.   Vertex at $(3, 4)$, directrix $y = 2$

51.   Vertex at $(-2, -1)$, focus at $(2, -1)$

52.   Focus at $(1, 3)$, directrix $y = 5$

Solve the following inequalities.

53. $x^2 - x - 6 \leq 0$

54. $2x^2 + 4 < 6x$

55. $2x^2 - 6x - 10 > -x^2 + x + 10$

56. $4x^2 - 8x - 20 \geq 2x^2 - 10x + 4$

57. $x^4 - x^2 - 6 < 0$

58. $4x - 3\sqrt{x} - 1 \geq 0$

Sketch the graphs of the following relations.

59. $x^2 + y^2 = 4$ **60.** $x^2 + y^2 \leq 4$

61. $x = |y|$ **62.** $|x| = y^2$

63. $x < |y|$ **64.** $|x| < y^2$

65. $v^2 = 2u^2$ **66.** $y > 2x^2$

67. $x^2 - 4x + y^2 - 6y = 12$

68. $x^2 + 6x + y^2 + 10y + 30 = 0$

69. $x^2 + y^2 < 4x + 8y + 29$

70. $y^2 + 6y \geq 4x - x^2 + 12$

71. $x^2 - 10x + y^2 + 6y + 18 > 0$

72. $x^2 + 10y < 10x - y^2 - 25$

Sketch and write the equation of the locus of points equidistant from

73.   Points $(-1, 3)$ and $(3, 3)$

74.   Points $(2, 6)$ and $(6, 2)$

75.   Points $(3, 4)$ and $(7, 2)$

76.   Points $(-4, -6)$ and $(2, 4)$

77.   Point $(2, -2)$ and the line $y = 4$

78.   Point $(5, 2)$ and the line $x = 1$

Sketch the graphs of the following equations. Indicate vertices and foci. For hyperbolas, indicate the asymptotes and sketch and write the equation of the conjugate hyperbola. If a parabola, also indicate the directrix.

79. $x^2/36 + y^2/49 = 1$

80. $x^2 + 4y^2 = 4$

81. $25x^2 + 16y^2 = 400$

82. $x^2/169 + y^2/144 = 1$

83. $49x^2 + 625y^2 = 30{,}625$

84. $x^2/4 - y^2/25 = 1$ **85.** $x^2 - 9y^2 = 9$

86. $y^2/576 - x^2/49 = 1$

87. $36y^2 - 9x^2 = 324$

88. $4y^2 - 3x^2 = 24$

89. $16x^2 - 25y^2 = 400$

90. $8x + 4y - y^2 = 12$

91. $9x^2 + 4y^2 + 36x - 24y + 36 = 0$

92. $4x^2 - 9y^2 + 24x + 36y = 144$

93. $4x^2 + 9y^2 + 16x - 18y = 11$

94. $x^2 + 6x = 4y - 1$

95. $y^2 - 4x^2 - 2y + 8x = 7$

96. $9x^2 - 4y^2 - 54x - 16y + 61 = 0$

97. $25x^2 + 9y^2 - 150x - 90y + 225 = 0$

98. $x^2 - 2x + y = 3$

Find the equation and sketch the graph of the central ellipse satisfying the following conditions. Label the vertices and foci.

99.   Focus at $(0, 3)$ and horizontal axis of length 8

100.   Focus at $(0, 4)$ and minor axis of length 6

101.   Vertex at $(-2, 0)$ and focus at $(\sqrt{3}, 0)$

102.   Passing through $(0, 5)$ and $(12/5, 4)$

103.   Vertex at $(5, 0)$ and passing through $(-4, 39/5)$

Find the equation and sketch the graph of the central hyperbola satisfying the following conditions. Label the vertices and foci.

104.   Vertex at $(0, 4)$, asymptote $y = 2x$

105.   Vertex at $(-5, 0)$, asymptote $y = -4x/5$

106.   Vertex at $(-2, 0)$, asymptote $x = -2y$

107.   Vertex at $(0, 2)$, passing through $(2, 3)$

108.   Vertex at $(0, 3)$, passing through $(-4, -6)$

109.   Vertex at $(4, 0)$, passing through $(-8, 10)$

110.   Vertex at $(2, 0)$, focus at $(-4, 0)$

111.   Focus at $(0, \sqrt{10})$, asymptote $x = 3y$

Sketch and write the equation of the ellipse satisfying the given conditions.

112.   Lower vertex at $(7, -7)$, left vertex at $(4, -2)$

113.   Upper vertex at $(-3, 6)$, left vertex at $(-8, 2)$

114.   Upper vertex at $(3, 3)$, right vertex at $(8, 0)$

115.   Lower vertex at $(-1, -5)$, right vertex at $(4, 8)$

Sketch and write the equation of the hyperbola satisfying the given conditions. Do the same for its conjugate.

	Opens	Centered at	a	b
116.	Horizontally	$(-7, 2)$	8	6
117.	Vertically	$(3, 5)$	3	9
118.	Horizontally	$(-2, -1)$	2	3
119.	Vertically	$(4, -6)$	7	3

In the following problems, let h be the altitude in feet of a missile t seconds after it is fired. Determine its maximum altitude, the time at which it attains its maximum altitude, and the time at which it hits the ground.

120. $h = 480t - 16t^2$

121. $h = 32t - 16t^2 + 128$

122. $h = 800t - 16t^2 + 4400$

4 *Functions and Their Graphs*

In this chapter, we consider those relationships in which only *one* value of a second quantity corresponds to any given value of a first quantity. The second quantity is then said to be a **function** of the first.

In this age of scarce energy sources, advertising claims that certain cars can travel 50 miles on a gallon of gas. In such a car, we should be able to travel 100 miles on 2 gallons of gas, 250 miles on 5 gallons, and so on. The distance an automobile can travel without a fuel stop is *determined by,* or *is a function of,* the number of gallons its gas tank holds.

On the other hand, the relationship "*y* is the square root of *x*" does not describe a function; each positive real number has *two* square roots, not just one. If two quantities are related in some way that may or may not describe a function, we say that we have a **relation** between the two quantities.

In this chapter, we will also develop additional graphing techniques, the central idea being to modify and move simpler graphs in order to obtain the graphs of more complicated relations.

4.1 FUNCTIONS

Mail order sales companies generally impose a shipping charge in addition to the cost of the item purchased. For instance, the cost of shipping an item from Chicago to Phoenix might be $1.10 for the first 4 pounds and 11¢ for each additional pound. The shipping charge on a 3-pound item is $1.10, and the charge on a 6-pound item is $1.32. The shipping cost is expressed in terms of the weight of the item.

If we travel at a rate of 50 miles per hour, the relationship $D = 50t$ expresses distance in terms of travel time. The distance traveled depends on the amount of time available.

In each of these examples, a quantity—cost or distance—is *uniquely* determined by another quantity—the weight of an item or the travel time, respectively. These examples illustrate functional relationships or **functions.** Informally, a function is a rule that enables us to compute exactly one "new number" from each "given number." However, the concept of function is not limited to numbers. For instance, each house in a city has a unique street address; each person has a unique name at any given time (legally at least). Some teachers assign students certain seats for class. Each of these also represents a functional relationship: "street address" is a function of "house," "name" is a function of "person," and "seat location" is a function of "student in the class."

To facilitate our definition of function, we introduce the terminology of sets. Briefly, a **set** is any collection of objects. These objects may be numbers, students, houses, and so on. The objects that are in a given set are said to be **elements** of that set.

DEFINITION

A **function** is a *correspondence* or *rule* that assigns to each element x of some set D a unique element y of a set I. We say that y *is a function of x.* If y is a function of x, then y is called a **dependent variable** and x an **independent variable.**

If y is a function of x, we sometimes write $y = f(x)$. The symbol $f(x)$ is read "f of x" or "function of x." If an equation or formula is used to express the relationship between x and y, then we find the value of a function for some particular value of x by substituting that value of x into the functional equation. For instance, if

$$y = f(x) = x^2 + 2x - 3$$

then the value of the function for $x = -1$ is

$$f(-1) = (-1)^2 + 2(-1) - 3 = -4$$

and for $x = 20$ is

$$f(20) = (20)^2 + 2(20) - 3 = 437$$

In actual practice, various letters and symbols such as F, G, H, f, g, h, and θ are also used to denote functions. Letters other than x and y are used to denote independent and dependent variables, too. The letters used are not important. What matters is the functional relationship. Thus,

$$s = f(t) = t^2 - 1$$
$$y = g(x) = x^2 - 1$$
$$c = H(\$) = \$^2 - 1$$
$$\Delta = G(\square) = \square^2 - 1$$

all describe the same function; even though different symbols are used, exactly the same relationship or computational process is described. Each value of the independent variable is squared and then 1 is subtracted from the result. For this reason, the symbols used to describe the variables are sometimes called **dummy variables.** The *relationship* described by these symbols is what matters.

Some students find it helpful to visualize a function as a machine in which

1. A given object is fed into the machine.

2. The machine performs some operations on the object

3. The result then comes out of the machine

See Figure 4–1.

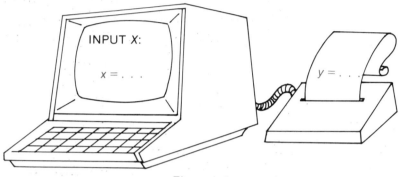

Figure 4–1

Most of the functions and relations we use will be described by an equation or formula. Sometimes more than one formula is needed to describe a function; different formulas may be necessary for various elements of the domain. For instance, in the shipping cost example, the cost of shipping a mail order item depended on the weight of the item. Letting

$$c = \text{shipping cost}$$
$$w = \text{weight of the item in pounds}$$

we have

$$c = f(w) = \begin{cases} \$1.10 & 0 < w \le 4 \\ \$1.10 + 0.11(w - 4) & w \ge 4 \end{cases}$$

EXAMPLE 1 The altitude of a projectile t seconds after firing is

$$h = f(t) = -16t^2 + v_u t + h_0$$

where v_u is the initial upward component of its velocity and h_0 is the altitude from which it is fired. Here, h is a function of t. Is t also a function of h?

Solution Solving for t in terms of h, we use the quadratic formula to get

$$t = \frac{-v_u \pm \sqrt{v_u^2 - 4(-16)(h_0 - h)}}{2(-16)}$$
$$= \frac{-v_u \pm \sqrt{v_u + 64(h_0 - h)}}{-32}$$

The two values of t indicate that the missile is at the same height on the way up and later on the way down. Thus, t is not a function of h; knowing the altitude of the projectile does not enable us to determine the time that has elapsed since it was fired. ◼

Domain and Range

> **DEFINITION**
>
> If $y = f(x)$ describes y as a function of x, the set of x values for which the function is defined is the **domain** of the function. The **range** of f is the set of all y values that can be found by considering all x values in the domain.

Consider again an automobile getting 50 miles per gallon of gasoline and let

$$x = \text{number of gallons of gasoline the tank holds}$$
$$y = \text{distance that can be traveled}$$

The relationship between x and y is expressed by the equation

$$y = f(x) = 50x$$

The domain of this function is the set of all $x \geq 0$; we do not consider a negative amount of gasoline. The range consists of all $y \geq 0$ since presumably we could travel any distance if we had a large enough gas tank.

Functions are frequently described by giving only the functional equation, such as $f(x) = \sqrt{4 - x^2}$, without indicating its domain or range. The domain is then implicitly taken to be as large as possible.

> If the domain of a function $y = f(x)$ is not specified, it is understood that the domain consists of all x's for which the expression or equation makes sense; the range is then determined from these x values.

EXAMPLE 2 Find the implied domain and corresponding range for the function defined by $y = f(x) = \sqrt{4 - x^2}$ if x and y are to be real numbers.

Solution In order for the square root to be a real number, we must have

$$0 \leq 4 - x^2$$
$$x^2 \leq 4$$
$$|x| \leq 2$$
$$-2 \leq x \leq 2$$

The domain of this function is the interval $-2 \leq x \leq 2$. The corresponding functional values range between 0 (when $x = \pm 2$) and 2 (when $x = 0$). The range for this function is the interval $0 \leq y \leq 2$. ◼

Graphs and Functions

We have already studied two important classes of functions in detail:

linear functions $\quad y = ax + b$

and

quadratic functions $\quad y = ax^2 + bx + c$

When we purchase several identical items at a specified price per item, the total cost is a linear function of the number of items purchased. So also, the distance traveled at a constant speed is a linear function of the time of travel. Quadratic functions help to describe phenomena such as the effect of gravity on moving objects like baseballs.

Studying the graphs of various functions increases our understanding of the functions themselves. We have seen that the graph of a linear function is a straight line; the graph of a quadratic function is a parabola.

In Section 3.2, we sketched the graph of a relation by plotting those points that satisfied the relation. A function is a special kind of relation—one in which each value of a given variable determines precisely one new value. If y is a function of x, then each x value determines precisely one y value. This means that each vertical line meets the graph in at most one point.

VERTICAL LINE TEST

A graph represents a function $y = f(x)$ if and only if each vertical line meets the graph in at most one point. In this case, the graph consists of those points (x, y) whose altitude above or below the x-axis at each x is given by $y = f(x)$.

Similarly, a given y value determines precisely one x value when no two points on the graph lie on the same horizontal line.

HORIZONTAL LINE TEST

A graph determines x as a function of y if and only if each horizontal line meets the graph in at most one point.

Figure 4–2 shows the graphs of some functions and relations.

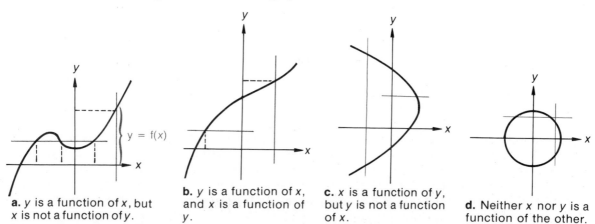

a. y is a function of x, but x is not a function of y.

b. y is a function of x, and x is a function of y.

c. x is a function of y, but y is not a function of x.

d. Neither x nor y is a function of the other.

Figure 4–2

EXAMPLE 3 Determine whether the following equation describes y as a function of x. Is x a function of y? If either is a function, sketch its graph and indicate its domain and range.

$$y - x^2 = 4$$

Solution Since $y = x^2 + 4$, we see that for each x, one and only one value of y can be computed. Thus, y is indeed a function of x. However,

$$x^2 = y - 4$$

so that

$$x = \pm\sqrt{y - 4}$$

For $y = 8$, note that $x = \pm 2$; thus, x is not a function of y. From Section 3.4, we know that the graph of $y = f(x) = x^2 + 4$ is a parabola opening upward with vertex at $(0, 4)$ (see Fig. 4–3). The graph also indicates that x is not a function of y. Note that $f(x) = x^2 + 4$ is defined for all x. Its domain is the entire real number line. On the other hand, $y = f(x) = x^2 + 4 \geq 4$; its range is the interval $4 \leq y < \infty$. ■

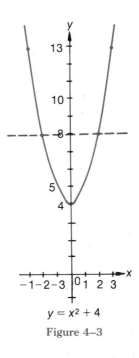

$$y = x^2 + 4$$

Figure 4–3

EXAMPLE 4 The Environmental Equipment Company has developed a new generator. The prototype model has cost the company $10 million in research, development, and manufacturing expenses. It will cost another $40 million to modify the manufacturing facility to make these generators. Then labor and materials for manufacturing and installing one of these generators will cost the company another

$2 million. If each generator can be sold for $3 million, how many must the company sell before it begins to realize a profit?

Solution　The total amount (in millions of dollars) that must be spent to produce x generators is

$$f(x) = 10 + 40 + 2x$$
$$= 50 + 2x$$

From the sale of these generators, the company will receive

$$g(x) = 3x$$

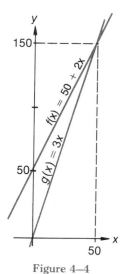

Each of these functions is graphed as a straight line in Figure 4–4. Income exceeds costs when the graph of g rises above the graph of f. This happens when $g(x) = f(x)$:

$$3x = 50 + 2x$$
$$x = 50$$

Figure 4–4

Thus, the company must sell at least 50 generators before it begins to make a profit.　■

We noted earlier that not all functions are described by a single equation. In fact, a function is defined by *any* rule associating a value $f(x)$ to each x in the domain of the function. Of particular importance in calculus and more advanced courses are functions defined by several formulas, each applying in different intervals as illustrated in the next example.

EXAMPLE 5　Sketch the graph of the function defined by

$$f(x) = \begin{cases} x^2, & x < 0 \\ 1, & x = 0 \\ 2 - x, & x > 0 \end{cases}$$

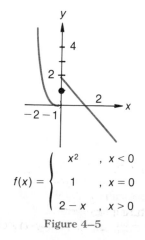

$$f(x) = \begin{cases} x^2 & , \ x < 0 \\ 1 & , \ x = 0 \\ 2 - x & , \ x > 0 \end{cases}$$

Figure 4–5

Solution   The graph of $y = x^2$ was sketched in Example 3, Section 3.2. We shall use the portion of this graph corresponding to $x < 0$. It is easy to plot the value $y = 1$ when $x = 0$. The graph of $y = 2 - x$ is a line of slope -1 with $y = 0$ when $x = 2$. Thus, we sketch the portion of this line corresponding to $x > 0$. The complete graph appears in Figure 4–5. A rounded bracket "(" or ")" is used to indicate that a given endpoint is *not* on the graph.

EXERCISES   4.1

In each of the following, determine whether y is a function of x and whether x is a function of y.

1. $x^2 = y - 1$
2. $2x - 3y = 4$
3. $x < 4y$
4. $y \le 3x^2 + 2$
5. $x = y^2 + 4$
6. $y^2 = x^2 - 1$
7. $y = x^3$
8. $x^2 + 2xy + y^2 = 4$
9. $y^2 + 2y - x = 4$
10. $y^5 = x^2$

Evaluate the following functions at the indicated points.

11. $g(y) = 5y - 6$;   find $g(2)$, $g(3)$, $g(-3)$
12. $H(m) = 8 - 3m$;   find $H(0)$, $H(4)$, $H(1/3)$
13. $f(t) = t^2 - 5t + 1$;   find $f(2)$, $f(3)$, $f(-3)$
14. $N(z) = 5 + 3z - z^2$;   find $N(3)$, $N(4)$, $N(-4)$
15. $\theta(b) = (b^2 + 2b - 3)$;   find $\theta(1)$, $\theta(2^2)$, $[\theta(2)]^2$
16. $h(n) = n^3 - 2n^2 + 5n + 7$;   find $h(0)$, $h(1)$, $h(-1)$
17. $f(x) = \dfrac{1 + x}{1 - x}$;   find $f(0)$, $f(-1)$, $f(2)$
18. $T(z) = \dfrac{2z^2 - z - 1}{z + 2}$;   find $T(1)$, $T(-3)$, $T(1/2)$
19. $r(u) = \sqrt{u^2 - 2u + 1}$;   find $r(4)$, $r(-3)$, $r(2)$
20. $U(t) = \sqrt{t + 2} + \sqrt{t - 3}$;   find $U(7)$, $U(3)$, $U(5)$

In each of the following, find the understood domain of the given function; if indicated, also find the range. Both domain and range are to be restricted to real numbers.

21. $f(x) = \sqrt{x - 1}$; range also
22. $H(s) = \sqrt{s^2 + 2s + 1} + 2$; range also
23. $F(v) = \dfrac{1}{v^2 - 4v + 4}$; range also
24. $p(x) = \dfrac{\sqrt{16 - x^2}}{|x - 3|}$
25. $g(r) = \dfrac{\sqrt{25 - r^2}}{r + 2}$
26. $p(t) = \dfrac{1}{t^2 - 4}$
27. $G(x) = \dfrac{\sqrt{x^2 - 4}}{\sqrt{9 - x^2}}$

28. $K(y) = \dfrac{2y - 3}{\sqrt{y + 1}}$
29. $R(x) = \dfrac{\sqrt{36 - x^2}}{x^2 - x - 2}$
30. $H(b) = \dfrac{3b - 2}{6b^2 + b - 1}$

Sketch the graphs of the following functions.

31. $f(t) = 2t + 1$
32. $g(x) = x^2 - 3$
33. $F(u) = 1/u$
34. $H(v) = v^3$
35. $R(t) = |t^3|$
36. $T(u) = 1 - u^2$
37. $G(s) = s^2 + 1$
38. $H(t) = t^2 - 2t + 1$
39. $g(x) = x^2 + 2x + 2$

40. $f(x) = \begin{cases} 1, & x \le -1 \\ 2, & x > -1 \end{cases}$

41. $g(x) = \begin{cases} 0, & x \le 0 \\ 1, & 0 < x < 1 \\ 2, & x \ge 1 \end{cases}$

42. $F(u) = \begin{cases} 2u + 1, & u \le 1 \\ 2, & 1 < u \le 2 \\ 1 - u, & u > 2 \end{cases}$

43. $G(t) = \begin{cases} t, & t < 0 \\ 1, & t = 0 \\ t, & 0 < t \le 2 \\ 4 - t, & t > 2 \end{cases}$

44. $H(v) = \begin{cases} 1 - v, & v \le 0 \\ v^2 + 1, & v > 0 \end{cases}$

45. Which of the following graphs define y as a function of x? Which define x as a function of y?

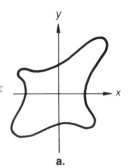

a.

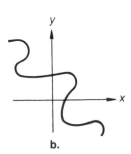

b.

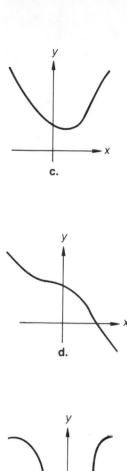

c.

d.

e.

f.

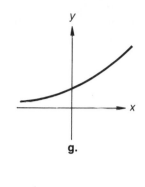

g.

h.

i.

j.

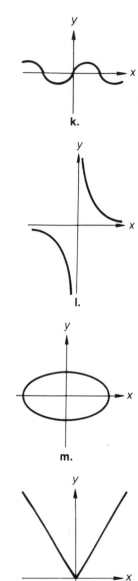

k.

l.

m.

n.

o.

46. For stability in a certain wildlife population, there must be at least 25 rabbits for each fox.
  a. Express this statement as a relation between the number of rabbits and the number of foxes.
  b. Does this relationship express either the number of rabbits or the number of foxes as a function of the other?

47. Ann drives her automobile at 55 miles per hour.
  a. Express the distance traveled as a function of time.
  b. How far does she travel in $2\frac{1}{2}$ hours?
  c. Describe the domain and range of this function.

48. On a recent vacation trip, Bill traveled at a rate of 20 miles per hour for 3 hours while he enjoyed the beautiful sights. He then continued at a rate of 60 miles per hour.
  a. Express the distance he traveled as a function of time.
  b. Determine the time when he averages 40 miles per hour.
  c. How far did he travel in 8 hours?
  d. Describe the domain and range of this function.

49. A grocer must place an order for canned goods. The price is $2 per case for 9 or fewer cases, $1.75 per case for 10 to 99 cases, and $1.50 per case for 100 or more cases.
  a. Express his cost as a function of the number of cases ordered.
  b. Does it make sense for him to buy 9 cases.
  c. At what point should he order 100 cases even though he does not need that many?
  d. Describe the domain and range of this function.

50. A snowball of radius 10 inches is left in the sun. It melts at a rate that decreases its radius by 1 inch each hour.
  a. Express its radius, volume, and surface area as a function of time. (See Exer. 57, Sect. 2.1.)
  b. Find the domain and the range of each of these functions.

51. The Gas and Electric Company estimates Ms. Walsh's monthly utility bill to be $3.75 times the difference between 60°F and the average temperature during the month.
  a. Express her estimated utility bill as a function of the average monthly temperature.

  b. What is her estimated bill when the average temperature is 80°F?
  c. 20°F?
  d. Describe the domain and range of this function.

52. The U-No-Pedal Moped Sales Company must pay $200 for each moped plus $200 per month to the manufacturer for advertising expenses. They sell each moped for $400, but this is offset by $4000 per month of overhead.
  a. Express expenses and income as functions of the number of mopeds sold in a given month.
  b. Determine the level of sales activity that generates a profit for the dealer.
  c. Describe the domain and range of the profit function.

53. A department store charges interest on credit card purchases as follows. The finance charge on a balance of $600 or less is $1\frac{1}{2}$ percent per month; but there is a minimum service charge of 75¢ on balances less than $50. Of course, no charge is levied when the balance is $0. On the balance that exceeds $600, the service charge is 1 percent per month.
  a. Express the finance charge as a function of the outstanding balance.
  b. What is the finance charge on a balance of $750?
  c. On a balance of 93¢?
  d. Describe the domain and range of this function.

54. An advertising agency finds that each minute of commercial television time for advertising beer generates $2.6 million in sales for its client.
  a. If each commercial minute costs $100,000, express the net revenue from beer sales generated by television commercials as a function of the number of commercial minutes purchased.
  b. How much revenue is generated by a commerical blitz of 20 minutes in a given week?
  c. Describe the domain and range of this function.

55. A heart patient enters the hospital with a resting heart rate of 120 beats per minute. Administration of 2 milligrams of medication per hour reduces the rate by 5 beats per minute.
  a. Express the heart rate of this patient as a function of the amount of medication prescribed.

b. What should the prescription be in order to reduce the rate to 70 beats per minute?

c. Assuming that a patient dies when the pulse rate reaches 10 beats per minute or 250 beats per minute, what is the domain of this function if the physician wants to keep the patient alive?

d. What is the range?

56. The altitude of a falling object dropped from an altitude of h feet is given by $f(t) = h - 16t^2$.

a. Two girls decide to measure the height of a tall building by dropping a pebble from its top. It takes the pebble 10 seconds to reach the ground. How high is the skyscraper?

b. Find the domain and range of this function.

4.2 GRAPHING TECHNIQUES

Often the graph of a seemingly complicated relation can be found by "stretching," "compressing," "reflecting," and/or "shifting" a well-known graph. These techniques and others will be developed and illustrated in this section.

Consider, for example, the function whose graph is sketched in Figure 4–6. Various modifications of this function are discussed in the following examples.

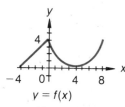

$y = f(x)$

Figure 4–6

Translation

EXAMPLE 1 With $y = f(x)$ as graphed in Figure 4–6, sketch the graph of

a. $y = f(x) + 2$

b. $y = f(x - 2)$

Solution **a.** The y-coordinate of each point in Figure 4–6 is simply increased two units; that is, the graph of $y = f(x)$ is moved two units upward. Similarly, the graph of $y = f(x) - 2$ would be obtained by moving the graph of $y = f(x)$ two units downward (see Fig. 4–7).

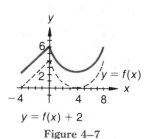

$y = f(x) + 2$

Figure 4–7

b. This graph is simply the translation of the original graph two units to the right. For we plot

$$y = f(-4) \quad \text{when} \quad x = -2$$
$$y = f(0) \quad \text{when} \quad x = 2$$
$$y = f(4) \quad \text{when} \quad x = 6$$
$$y = f(8) \quad \text{when} \quad x = 10$$

In the same way we see that the graph of $y = f(x + 2)$ would be obtained by shifting the original graph two units to the left (see Fig. 4–8).

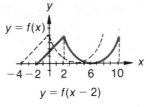

$$y = f(x - 2)$$

Figure 4–8

Modifications of the type discussed in Example 1 are called **translations.** These ideas are summarized in the table.

Translation

Algebraic Change		Change in Graph
1. $y = f(x) + c$	1.	The graph of $y = f(x)$ is moved vertically by c units. The graph moves up if $c > 0$ and down if $c < 0$ (see Fig. 4–7).
2. $y = f(x - a)$	2.	Moves the graph of $y = f(x)$ horizontally. The graph moves right if $a > 0$ and left if $a < 0$. Begin by plotting $y = f(0)$ when $x = a$ (see Fig. 4–8).

Reflection

EXAMPLE 2   With $y = f(x)$ as graphed in Figure 4–6, sketch the graph of

a. $y = -f(x)$ **b.** $y = f(-x)$

Solution   **a.**   The y-coordiante of each point on the graph of $y = f(x)$ changes sign; that is, the graph of $y = f(x)$ is reflected through the x-axis (see Fig. 4–9).

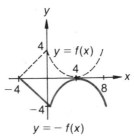

$$y = -f(x)$$

Figure 4–9

b. Obviously, we still plot $y = f(0)$ when $x = 0$. But we plot

$$y = f(-4) \quad \text{when} \quad x = 4$$
$$y = f(4) \quad \text{when} \quad x = -4$$
$$y = f(8) \quad \text{when} \quad x = -8$$

The graph of $y = f(x)$ is reflected through the y-axis (see Fig. 4–10).

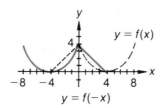

Figure 4–10

Reflection

Algebraic Change		Change in Graph
1. $y = -f(x)$	**1.**	The graph of $y = f(x)$ is reflected through the x-axis (see Fig. 4–9).
2. $y = f(-x)$	**2.**	The graph of $y = f(x)$ is reflected through the y-axis (see Fig. 4–10).

Scaling

EXAMPLE 3 Again, with $y = f(x)$ as graphed in Figure 4–6, sketch the graph of

a. $y = 2f(x)$

b. $y = f(2x)$

Solution **a.** Clearly, the y coordinate of each point on the graph of $y = f(x)$ is doubled. The graph in Figure 4–6 is stretched vertically. Similarly, the graph of $y = \frac{1}{2}f(x)$ would be obtained from the graph of $y = f(x)$ by vertical compression (see Fig. 4–11).

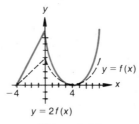

Figure 4–11

b. Observe that we must plot

$$y = f(-4) \quad \text{when} \quad x = -2$$
$$y = f(0) \quad \text{when} \quad x = 0$$
$$y = f(4) \quad \text{when} \quad x = 2$$
$$y = f(8) \quad \text{when} \quad x = 4$$

The graph of $y = f(x)$ is compressed in the horizontal direction. Similarly, the graph of $y = f(\frac{1}{2}x)$ would be obtained by stretching the graph of $y = f(x)$ in the horizontal direction (see Fig. 4–12). ∎

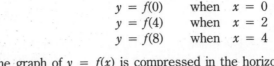

Figure 4–12

This example illustrated changes in the horizontal and vertical scales of a graph. The techniques are summarized in the table.

Scaling

Algebraic Change	Change in Graph
1. $y = Af(x)$	**1.** A vertical stretch or compression is effected on the graph of $y = f(x)$.
2. $y = f(ax)$	**2.** Exerts a horizontal stretch or compression on the graph of $y = f(x)$.

Note the differing stretch-compression behavior.

$$\text{Stretch:} \quad |A| > 1 \quad \text{or} \quad |a| < 1$$
$$\text{Compression:} \quad |A| < 1 \quad \text{or} \quad |a| > 1$$

If $A < 0$ or $a < 0$, a reflection is also involved.

Note that in all the changes discussed so far, changes in the independent variable cause horizontal modifications of the graph, whereas changes in the dependent variable cause vertical modifications.

The principles behind the techniques developed here can also be adapted to sketching graphs of relations giving x in terms of y. We shall do this in part (b) of the following example. In Example 4, we show essentially that the graph of any quadratic expression

$$y = ax^2 + bx + c, \quad a \neq 0$$

or

$$x = Ay^2 + By + C, \quad A \neq 0$$

is a parabola. We shall begin by completing the square as in Section 3.4 but now a basic graph will be modified and moved until we find the final graph.

EXAMPLE 4 Use the graphing techniques developed in this section to sketch the graphs of

a. $y = 2x^2 - 4x + 3$

b. $x = y^2 + 2y - 1$

Solution   **a.**   Completing the square, we obtain

$$
\begin{aligned}
y &= 2x^2 - 4x + 3 \\
&= 2(x^2 - 2x\qquad) + 3 \\
&= 2(x^2 - 2x + 1) + 3 - 2 \\
&= 2(x - 1)^2 + 1.
\end{aligned}
$$

Then we sketch the following sequence of graphs (see Fig. 4–13). Note that the vertex is $(1, 1)$ in agreement with our observations in Section 3.4.

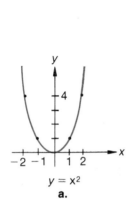
$y = x^2$
a.

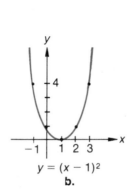
$y = (x - 1)^2$
b.

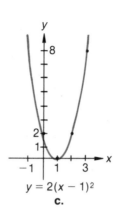
$y = 2(x - 1)^2$
c.

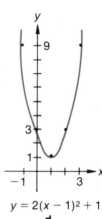
$y = 2(x - 1)^2 + 1$
d.

Figure 4–13

b.

$$
\begin{aligned}
x &= y^2 + 2y - 1 \\
&= (y^2 + 2y\qquad) - 1 \\
&= (y^2 + 2y + 1) - 1 - 1 \\
&= (y + 1)^2 - 2
\end{aligned}
$$

In plotting the sequence of graphs that follow, we obtain Figure 4–14 (b) from (a) by plotting $x = 0$ at $y = -1$ rather than at $y = 0$. Then (c) is obtained from (b) by subtracting 2 from each x-coordinate in (b); that is, the graph of (b) is moved two units to the left. Again, the vertex is $(-2, -1)$ in agreement with our observations in Section 3.4.

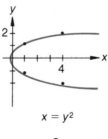
$x = y^2$

a.

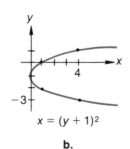
$x = (y + 1)^2$

b.

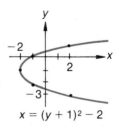
$x = (y + 1)^2 - 2$

c.

Figure 4–14

EXAMPLE 5   With $y = f(x)$ as in Figure 4–6, sketch the graph of the following.

$$
y = -2f(2x + 4) + 8
$$

Solution In order to effect a horizontal translation, we must write $2x + 4$ as $2(x + 2)$. Thus we write

$$y = -2f[2(x + 2)] + 8$$

We then sketch the sequence of graphs shown in Figure 4–15. Each graph is obtained from its predecessor by using the techniques illustrated earlier.

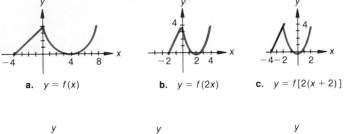

a. $y = f(x)$ **b.** $y = f(2x)$ **c.** $y = f[2(x + 2)]$

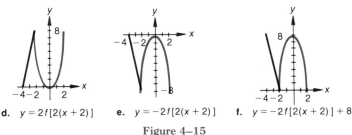

d. $y = 2f[2(x + 2)]$ **e.** $y = -2f[2(x + 2)]$ **f.** $y = -2f[2(x + 2)] + 8$

Figure 4–15

It is important that the steps illustrated in parts (b) and (c) of Figure 4–15 not be reversed. (Try it!) That is, account for any horizontal stretch or compression before shifting or translating.

> Horizontal and vertical changes are independent of one another, but in either case, stretch, compression, and/or reflection must be taken care of before translation.

After finishing, it is a good idea to check the results at the endpoints. For instance, in Example 5, we have

$$x = -4: \quad -2f(2x + 4) + 8 = -2f(-8 + 4) + 8$$
$$= -2f(-4) + 8 = -2 \cdot 0 + 8 = 8$$
$$x = 2: \quad -2f(2x + 4) + 8 = -2f(4 + 4) + 8 = -2f(8) - 8$$
$$= -2 \cdot 4 + 8 = 0$$

These are in agreement with the values indicated on our graph (see Fig. 4–15*f*).

Symmetry

The graphing techniques as illustrated up to this point enable us to obtain new graphs from known graphs. Symmetry properties, on the other hand, enable us to obtain a complete graph from a portion of itself.

Symmetry

	Geometric Description	Algebraic Description	Illustration
1.	Symmetry about the y-axis	1. $(-x, y)$ is on the graph when (x, y) is.	1.
2.	Symmetry about the x-axis	2. $(x, -y)$ is on the graph when (x, y) is.	2.
3.	Symmetry about the origin	3. $(-x, -y)$ is on the graph when (x, y) is.	3.

If the relationship between x and y is expressed as a function $y = f(x)$, then symmetry about the y-axis can be expressed as $f(-x) = f(x)$ and symmetry about the origin can be expressed as $f(-x) = -f(x)$.

DEFINITION

1. A function f for which

$$f(-x) = f(x)$$

is called an **even function.** Its graph is symmetric about the y-axis.

2. A function f for which

$$f(-x) = -f(x)$$

is called an **odd function.** Its graph is symmetric about the origin.

Alternatively, a function $y = f(x)$ is even if replacing x by $-x$ is equivalent to the original function; it is odd if replacing x by $-x$ is equivalent to replacing y by $-y$.

If f is an even function or an odd function, we need only sketch the branch of its graph corresponding to $x > 0$; this branch is then reflected through the y-axis or the origin as the case may be. For example, if $y = f(x)$ is graphed as in Figure 4–16(a) for $x > 0$, then it can be completed as in Figure 4–16(b) or (c) if it is even or odd, respectively. Of course a function could be neither even nor odd as in Figure 4–16 (d).

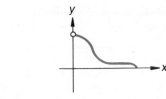

a. $y = f(x)$

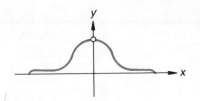

b. $y = f(x)$, f is even: $f(-x) = f(x)$

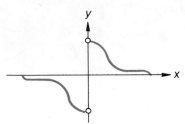

c. $y = f(x)$, f is odd: $f(-x) = -f(x)$

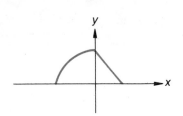

d. $y = f(x)$: f is neither even nor odd.

Figure 4–16

Graphing Absolute Values

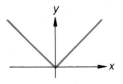

Figure 4–17
$f(x) = |x|$

The relation $f(x) = |x|$ defines an even function of x since $|-x| = |x|$. Its graph appears in Figure 4–17. To graph a relation involving absolute values, it helps to graph the part inside the absolute value symbols first. The absolute value operation then reflects any portion below the y-axis into the corresponding positive altitudes. This technique is illustrated in the next two examples.

EXAMPLE 6   Sketch the graph of the following equation.

$$y = |12 - 3x| - 3$$

Solution   The graph is sketched in the steps illustrated in Figure 4–18.

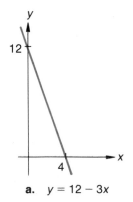

a. $y = 12 - 3x$

b. $y = |12 - 3x|$

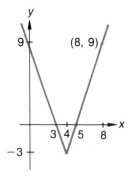

c. $y = |12 - 3x| - 3$

Figure 4–18

EXAMPLE 7 Mary is programming a computer to sketch a hungry fish chasing minnows. After much deliberation, she decides that the outline of the fish should satisfy the equation

$$|y^2 - 4y + 3| = x - 2$$

Sketch the fish.

Solution The equation can be rewritten as

$$x = |y^2 - 4y + 3| + 2$$

The part inside the absolute value can be written as

$$y^2 - 4y + 3 = (y^2 - 4y + 4) + 3 - 4$$
$$= (y - 2)^2 - 1$$

The graph of $x = y^2 - 4y + 3$ is a parabola opening to the right with vertex at $(2, -1)$ and line of symmetry $y = 2$. The fish is sketched by using the graphing techniques as discussed earlier and is shown in Figure 4–19.

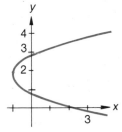

a. $x = y^2 - 4y + 3 = [y - 2]^2 - 1$

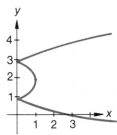

b. $x = |y^2 - 4y + 3|$

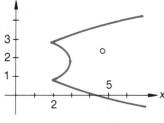

c. $x = |y^2 - 4y + 3| + 2$

Figure 4–19

EXERCISES 4.2

1. Use the graph of $y = f(x)$ as given by the figure to sketch the graph of each of the following functions.

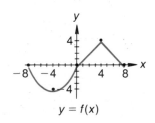

$y = f(x)$

a. $y = f(x - 4)$
b. $y = f(x + 4)$

c. $y = f(x) + 4$
d. $y = f(x) - 4$
e. $y = -3f(x)$
f. $y = \frac{1}{3}f(x)$
g. $y = f(2x)$
h. $y = f(-2x)$
i. $y = f(2x - 2)$
j. $y = f(-2x - 2)$
k. $y = f(-2x + 8)$
l. $y = f(\frac{1}{2}x - 4)$
m. $y = 3f(2x + 2)$
n. $y = -3f(-2x + 2) + 4$
o. $y = -\frac{1}{2}f(\frac{1}{2}x + 4) - 2$

p. $y = |f(x)|$

q. $y = f(|x|)$

2. Repeat Exercise 1 with the following figure.

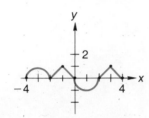

Use the techniques discussed in this section to sketch the graphs of the indicated functions.

3. **a.** $y = x$ **b.** $y = 2x$

    c. $y = -\frac{1}{2}x$ **d.** $y = x - 3$

    e. $y = 2x - 4$ **f.** $y = -3x + 1$

    g. $y = |x - 3|$ **h.** $y = |x + 2|$

    i. $y = |2x + 2|$ **j.** $y = |2x - 4|$

    k. $y = |2 - 4x|$ **l.** $y = |-3x + 1|$

    m. $y = |3x - 6|$

4. **a.** $y = x^2$ **b.** $y = 4x^2$

    c. $y = (x - 1)^2$ **d.** $y = (x + 2)^2$

    e. $y = -2(x + 1)^2$

    f. $y = 3(x - 2)^2 - 3$

    g. $y = |3(x - 2)^2 - 3|$

    h. $y = |-2(x + 1)^2 + 8|$

    i. $x = y^2$ **j.** $x = (y - 1)^2$

    k. $x = (y - 1)^2 - 1$

    l. $x = |(y - 1)^2 - 1|$

    m. $x = 2|(y - 2)^2 - 4|$

5. **a.** $y = x^3$ **b.** $y = x^3/8$

    c. $y = (x + 1)^3$ **d.** $y = (x - 2)^3$

    e. $y = -\frac{1}{2}(x - 1)^3$

    f. $y = \frac{1}{4}(x + 2)^3 + 2$

    g. $y = |x^3|$ **h.** $y = |(x + 1)^3|$

    i. $y = |x - 2|^3$

    j. $y = |\frac{1}{4}(x + 2)^3 + 2|$

    k. $y = x^4$ **l.** $y = (x - 1)^4$

    m. $y = |(x - 1)^4 - 1|$

Sketch the graphs of the following relations.

6. $y = 3 - |1 - 2x|$ **7.** $y = |2x - 4| - 4$

8. $x = |y + 2| - 3$ **9.** $x = -|3y - 6| + 6$

10. $|2y - x| = 1$ **11.** $|x + y| = 4$

12. $y = |x^2 - 2x|$

13. $y = 2 - |x^2 - 2x - 3|$

14. $x = -|y^2 - 4y + 1| + 2$

15. $x = |y^2 + 2y - 3| - 2$

16. $y = |-2x^2 + 8x - 4| - 4$

17. $x = |3y^2 + 6y - 2| - 2$

*Extend each of the following graphs under the condition that it represents an **a.** even function; **b.** odd function.*

18.

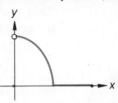

19.

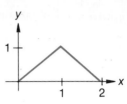

20.

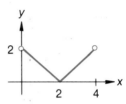

21.

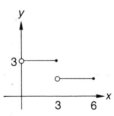

22.

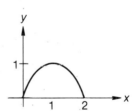

Indicate whether each of the following is even $[f(-x) = f(x)]$ *or odd* $[f(-x) = -f(x)]$ *or neither.*

23. $y = 2x^2 + 3$ **24.** $y = |-2x^3|$

25. $y = \sqrt[3]{x}$ **26.** $y = \sqrt[5]{x^2}$

27. $y = \dfrac{1}{x + 1}, \quad x \neq -1$

28. $y = (x^4 + 1)\sqrt[3]{x}$

29. $y = |x^2 - 2x - 3| + 2$

30. $y = \left| 2x^3 + 3x^5 - \dfrac{1}{x^7} \right|$

31. $y = 2x^3 + 3x - \dfrac{1}{x}$

32. $y = \dfrac{-2}{x^3}$

33. $y = \dfrac{x^2}{|x| + 1}$

34. Show that if f and g are even (or odd) functions, then the following are also even (or odd) functions.
    a. $f + g$
    b. $f - g$
    c. cf for a constant c

35. Show that the product of two even functions or of two odd functions is an even function.

36. Show that the product of an even function and an odd function is an odd function.

37. The French Zipper Company's sales graph for 1980 is given in the figure. The company expects to double this sales volume by 1985. If the same monthly variations are experienced in 1985 as in 1980, sketch its projected sales graph for 1985.

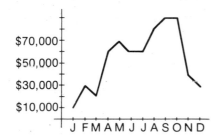

38. Suppose that the concentration of pollution in Indian Lake varies seasonally as sketched in the following graph. The Water Resources Division takes steps to combat these pollution levels and finds that its efforts are only 50 percent effec-

tive. Unforeseen reactions to their additives also delay the seasonal variations by 6 weeks. Sketch the resulting pollution-level curve.

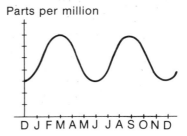

Parts per million

D J F M A M J J A S O N D

39. Injection of a certain drug into a laboratory animal cuts its heart rate in half as indicated in the graph. The antidote increases its heart beat to its normal rate just as rapidly as the drug decreases it. Graph the heart rate of an animal from noon to 6 P.M. if the drug is administered at 1 P.M. and the antidote is administered at 4 P.M.

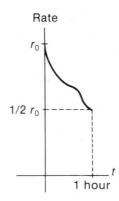

Rate

4.3 OPERATIONS ON FUNCTIONS

We are often asked to combine several functions algebraically. For example, several different kinds of expenses contribute to the cost of operating an automobile. If the car consumes 11¢ worth of gas and oil for each mile driven, then the total fuel cost for driving x miles is

$$G(x) = 0.11x$$

Maintenance and tire costs might be estimated at 4¢ per mile. For x miles, these costs total

$$M(x) = 0.04x$$

Another major expense is insurance, which is a fixed fee, say $500 per year, regardless of the number of miles driven. Annual insurance expense is then given by the constant function

$$I(x) = 500$$

We shall simplify the problem by ignoring other expenses such as depreciation and interest; the total cost of driving the vehicle x miles per year is the sum of the preceding functions:

$$C(x) = G(x) + M(x) + I(x)$$
$$= 0.11x + 0.04x + 500$$
$$= 500 + 0.15x$$

If f and g are both real-valued functions, we can clearly add, subtract, multiply, or divide the functional values. Thus, we make the following definitions.

DEFINITION

Let f and g both be real-valued functions and let $\mathcal{D}$ consist of those points common to the domains of f and g. Then the following functions are defined on $\mathcal{D}$.

1. For a constant a, af is the function whose value at x is

    $$(af)(x) = af(x)$$

    (This function is defined for all x in the domain of f.)

2. $f + g$ is the function whose value at x is

    $$(f + g)(x) = f(x) + g(x).$$

3. $f - g$ is the function whose value at x is

    $$(f - g)(x) = f(x) - g(x)$$

4. $f \cdot g$ is the function whose value at x is

    $$(f \cdot g)(x) = f(x) \cdot g(x)$$

5. f/g is the function whose value at x is

    $$\left(\frac{f}{g}\right)(x) = \frac{f(x)}{g(x)}$$

    (The domain of f/g must be restricted to preclude division by 0. Thus, the domain of f/g consists of those x's in $\mathcal{D}$ for which $g(x) \neq 0$.)

EXAMPLE 1   Let

$$f(x) = 4x^3 - 2x$$
$$g(x) = 2x^2 - 1$$

then,

$$
\begin{aligned}
(f + g)(x) = f(x) + g(x) &= (4x^3 - 2x) + (2x^2 - 1) \\
&= 4x^3 + 2x^2 - 2x - 1 \\
(f - g)(x) = f(x) - g(x) &= (4x^3 - 2x) - (2x^2 - 1) \\
&= 4x^3 - 2x^2 - 2x + 1 \\
(f \cdot g)(x) = f(x) \cdot g(x) &= (4x^3 - 2x) \cdot (2x^2 - 1) \\
&= 8x^5 - 8x^3 + 2x \\
\left(\frac{f}{g}\right)(x) = \frac{f(x)}{g(x)} &= \frac{4x^3 - 2x}{2x^2 - 1} = 2x \quad \text{when } x \neq \pm 1/\sqrt{2} \\
(2f)(x) = 2f(x) &= 2(4x^3 - 2x) = 8x^3 - 4x \quad\blacksquare
\end{aligned}
$$

Composition of Functions

As we indicated earlier, the value of a function $y = f(x)$ at any particular value of x, say, at $x = a$, is found by substituting a for x in the expression for $f(x)$. For the function $f(x) = 3x^2 - 2x + 1$, we have

$$
f(1) = 3 \cdot 1^2 - 2 \cdot 1 + 1 = 2
$$

and

$$
f(-\sqrt{2}) = 3(-\sqrt{2})^2 - 2(-\sqrt{2}) + 1 = 7 + 2\sqrt{2}
$$

There is nothing to prevent us from substituting other unknown expressions for x also. Thus,

$$
f(t) = 3t^2 - 2t + 1
$$

and

$$
\begin{aligned}
f(t + 2) &= 3(t + 2)^2 - 2(t + 2) + 1 \\
&= 3t^2 + 10t + 9
\end{aligned}
$$

In fact, there is nothing to prevent us from substituting into the expression for $f(x)$ a number that is itself computed as a value for some other function. Thus, if $g(x) = 2x + 1$ and $f(x) = 3x^2 - 2x + 1$, we have

$$
\begin{aligned}
f[g(x)] &= 3 \cdot [g(x)]^2 - 2[g(x)] + 1 \\
&= 3[2x + 1]^2 - 2[2x + 1] + 1 \\
&= 12x^2 + 8x + 2
\end{aligned}
$$

On the other hand,

$$
\begin{aligned}
g[f(x)] &= 2[f(x)] + 1 \\
&= 2[3x^2 - 2x + 1] + 1 \\
&= 6x^2 - 4x + 3
\end{aligned}
$$

Note that $f[g(x)] \neq g[f(x)]$.

This procedure is important enough to warrant the following definition.

> **DEFINITION**
>
> For two functions f and g, the **composition** $g \circ f$ is defined as the function whose value at any x is given by
>
> $$g \circ f(x) = g[f(x)]$$
>
> In order for the composition $g \circ f(x)$ to be defined, x must lie in the domain for f; in turn, $f(x)$ must be permissible for substitution into the expression for g; that is, $f(x)$ must be in the domain of g.

The previous illustration showed that sometimes $g \circ f(x) \neq f \circ g(x)$; that is, $g \circ f \neq f \circ g$. Composition of functions is not a commutative operation. Considering function machines we should not even expect the composition of functions to be commutative. For if f paints an object red and g paints it green, then

$$f \circ g(x) = f[g(x)] = \text{Red object}$$
$$g \circ f(x) = g[f(x)] = \text{Green object}$$

The final color of the object depends only on the last paint job it received. The order in which the operations are performed can make a difference (Fig. 4–20).

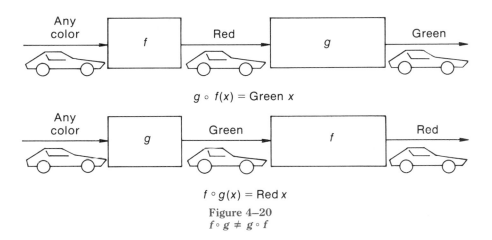

$$g \circ f(x) = \text{Green } x$$

$$f \circ g(x) = \text{Red } x$$

Figure 4–20
$f \circ g \neq g \circ f$

EXAMPLE 2 Let $R(u) = u^2 + 1$ and $S(v) = v - 1$; find $R \circ S(t)$ and $S \circ R(t)$. When are they equal?

Solution

$$R \circ S(t) = R(S(t)) = R(t - 1)$$
$$= (t - 1)^2 + 1$$
$$= t^2 - 2t + 1 + 1$$
$$= t^2 - 2t + 2$$
$$S \circ R(t) = S(R(t)) = S(t^2 + 1)$$
$$= (t^2 + 1) - 1$$
$$= t^2$$

$R \circ S(t) = S \circ R(t)$ only when

$$t^2 = t^2 - 2t + 2$$
$$2t = 2$$
$$t = 1$$

$R \circ S(t) \neq S \circ R(t)$ except when $t = 1$. ∎

EXAMPLE 3   Find functions f and g to describe the following functions as $g \circ f(t)$.

$$h(t) = \sqrt[3]{(t + 1)^2}$$

Solution   Since $h(t) = \sqrt[3]{(t + 1)^2} = (t + 1)^{2/3}$, we could let $f(t) = (t + 1)$ and $g(u) = u^{2/3}$. Then,

$$g \circ f(t) = g[f(t)] = g(t + 1)$$
$$= (t + 1)^{2/3}$$

On the other hand, we could also let $f(t) = (t + 1)^2$ and $g(u) = \sqrt[3]{u}$. Then

$$g \circ f(t) = g[f(t)] = g[(t + 1)^2]$$
$$= \sqrt[3]{(t + 1)^2}$$ ∎

EXAMPLE 4   The Environmental Protection Agency has determined that in a certain section of the country the average level of air pollution is $0.5\sqrt{P + 10{,}000}$ parts per million (ppm), where P is the population. The 1980 census predicts that the population t years after 1980 will be $7000 + 40t^2$.

a.   Express the pollution level t years after 1980 as a composite function and reduce the composite function to a function of t.

b.   What pollution level can be expected in 1990?

c.   In 2000?

Solution   **a.**   The pollution level is expressed as a function of population P as

$$f(P) = 0.5\sqrt{P + 10{,}000}$$

The population is expressed as a function of time t as

$$P(t) = 7000 + 40t^2$$

The pollution level t years after 1980 is given by the composite function

$$f \circ P(t) = f[P(t)]$$
$$= 0.5\sqrt{P(t) + 10{,}000}$$
$$= 0.5\sqrt{7000 + 40t^2 + 10{,}000}$$
$$= 0.5\sqrt{17{,}000 + 40t^2}$$

b. In 1990, $t = 10$ and the pollution level is estimated as

$$f \circ P(10) = 0.5\sqrt{17,000 + 40 \cdot 10^2}$$
$$= 0.5\sqrt{21,000}$$
$$\approx 72 \text{ parts per million} \quad \text{(Use a calculator.)}$$

c. In the year 2000, the pollution level is predicted to be

$$f \circ P(20) = 0.5\sqrt{17,000 + 40 \cdot 20^2}$$
$$= 0.5\sqrt{33,000}$$
$$\approx 91 \text{ parts per million}$$

Difference Quotient

For any function $f(x)$, the *difference quotient* for f is used to estimate the slope of its graph at a given point. The difference quotient is defined as

$$D(x) = \frac{f(x + h) - f(x)}{h}, \quad \text{when } h \neq 0$$

Example 5 Find the difference quotient for $f(x) = x^2 + 2$.

Solution
$$D(x) = \frac{f(x + h) - f(x)}{h} = \frac{[(x + h)^2 + 2] - [x^2 + 2]}{h}$$
$$= \frac{[(x^2 + 2xh + h^2) + 2] - [x^2 + 2]}{h}$$
$$= \frac{2xh + h^2}{h} = \frac{h(2x + h)}{h}$$
$$= 2x + h$$

EXERCISES 4.3

In each of the following, find expressions for $f(x) + g(x)$, $f(x) - g(x)$, $f(x) \cdot g(x)$, $f(x)/g(x)$, and $af(x)$ for the indicated value of a.

1. $f(x) = x^2 + 5$, $g(x) = x^2 + x - 6$; $a = 2$

2. $f(s) = s^2 - 2s - 3$, $g(t) = 6 - t - t^2$, $a = -1$

3. $f(u) = u^3 - 3u$, $g(v) = 2v^2 + v - 6$; $a = 3$

4. $f(x) = x^3 + 3$, $g(y) = 2y^2 + y - 10$; $a = -2$

5. $f(a) = a^3 - a^2 - a + 1$, $g(r) = r^2 - 1$; $a = 4$

In each of the following, find the requested expressions.

6. $A(w) = 2w + 3$; find $A(x - 1)$, $A(2y - 1)$, $A(w - 2)$.

7. $f(x) = x + \dfrac{1}{x}$; find $f(1/x)$, $f(x^2)$, $f(2x)$.

8. $F(z) = z^2 + 6z + 20$; find $F(x)$, $F(2t)$, $F(z - 1)$.

9. $g(t) = 3t^2 - t + 2$; find $g(1/t)$, $g(-x)$, $g(t + 2)$.

10. $s(h) = 2h^2 + h - 1$; find $s(x - 2)$, $s(2y - 3)$, $s(h - 1)$.

11. $G(x - 2) = x^2 - 5x + 6$; find $G(y - 2)$, $G(x)$, $G(0)$, $G(2)$. {Hint: $G(x) = G[(x + 2) - 2]$.}

12. $Y(t + 1) = t^2 + 5$; find $Y(t)$, $Y(x)$, $Y(2)$, $Y(1)$.

13. $P(x + 1) = x^2 + 2x + 1$; find $P(1)$, $P(2)$, $P(a + 1)$, $P(a)$.

14. $I(w + 1) = 3w^2 + 2w + 1$; find $I(0)$, $I(2)$, $I(b)$, $I(y - 2)$.

15. $h(x + 2) = x^2 + 6x + 5$; find $h(4)$, $h(t)$, $h(2x + 1)$, $h(y - 3)$.

In each of the following, find an expression for $f \circ g$, $g \circ f$, and evaluate $f \circ g(a)$, $g \circ f(a)$ for the given value of a.

16. $f(x) = 2x$, $g(x) = 3x^2$, $a = -2$
17. $f(t) = 3t^2 + 2t + 1$, $g(u) = 2u$, $a = 0$
18. $f(s) = s^2 + s$, $g(s) = 3s - 2$, $a = 1$
19. $f(r) = r^2 - 4r + 1$, $g(r) = r + 2$, $a = -1$
20. $f(y) = 2y - y^2$, $g(x) = \dfrac{1}{x}$, $a = 4$

Find functions f, g and write the given function as $g \circ f$.

21. $h(t) = (t - 1)^2$ **22.** $F(x) = \sqrt{x + 1}$

23. $u(n) = \dfrac{1}{(n + 1)^2}$

24. $H(w) = (w^2 + 1)^{3/2}$

25. $G(u) = 3(u^3 + 2)^2 + 2(u^3 + 2) + 4$

The operations combining two different functions or dependent variables should not be confused with operations on independent variables. In general, there is no relationship between operations on independent variables and the corresponding operations on dependent variables. Comparing the values requested in the following exercises should illustrate the situation sufficiently.

26. Let $H(z) = z^2 - 10z + 11$;   find and compare
   a. $H(1) + H(2)$ and $H(1 + 2)$
   b. $\dfrac{H(1)}{H(2)}$ and $H\left(\dfrac{1}{2}\right)$
   c. $H(2 - 1)$ and $H(2) - H(1)$

27. Let $L(s) = 3s - 4$;   find and compare
   a. $L(2^2)$ and $[L(2)]^2$
   b. $L(\sqrt{4})$ and $\sqrt{L(4)}$
   c. $L(-1)$ and $-L(1)$

28. Let $G(x) = x^2 + 2x - 10$;   find and compare
   a. $G(1 \cdot 2)$ and $G(1) \cdot G(2)$
   b. $G(1/2)$, $G(1)/G(2)$, and $1/G(2)$
   c. $G(2x)$ and $2G(x)$

29. Let $F(x) = x^2 + 1$;   find and compare
   a. $F(1 + 1)$ and $F(1) + F(1)$
   b. $F(1 - 1)$ and $F(1) - F(1)$
   c. $F(1 \cdot 1)$ and $F(1) \cdot F(1)$

30. Let $P(t) = 8 + 2t$;   find and compare
   a. $P(\sqrt{9})$ and $\sqrt{P(9)}$
   b. $P(3^2)$ and $[P(3)]^2$
   c. $P(3t)$ and $3P(t)$

Find the difference quotient for the following functions.

31. $f(x) = 3$
32. $g(x) = 4x - 1$
33. $h(x) = 2 - 3x$
34. $F(x) = 5x + 100$

35. $G(x) = 2x^2 - 1$
36. $H(x) = 3 - 4x^2$
37. $f(x) = x^2 + 2x - 5$
38. $g(x) = 3x^2 - 5x + 7$
39. $h(x) = 1 + 4x - 5x^2$
40. $F(x) = x^3$
41. $G(x) = x^4$
42. $H(x) = 1/x$

43. An employee deposits a certain sum of money at 5 percent interest in the company credit union. In order to encourage thrift, the company announces a bonus equal to 7 percent of the value of each employee's savings account. Express both the interest and the bonus as functions of the value of the savings account. Express the net return as a combination of these two functions.

44. Viki paddles her canoe at a rate of 5 miles per hour. The stream is flowing at a rate of 2 miles per hour. Express the distance Viki rows *in the water* as a function of time. Express the distance the water flows as a function of time. If she paddles upstream, express her net distance traveled as a combination of these two functions. What if she rows downstream?

45. The efficiency of a jet engine is affected by the quality of fuel used; the plane can travel faster and longer on better grades of fuel. With a tankful of fuel of quality q, the plane can travel $3q^2 + 2q$ miles per hour and can travel for $2q + 1$ hours. Express the distance the plane can travel on a tankful as a combination of these two functions of fuel quality.

46. The kinetic energy (KE) of an object is given in terms of its mass m and velocity v as $KE = mv^2/2$. Its momentum (M) is given by $M = mv$. If the kinetic energy at time t is $50t^2 + 50t - 300$ and the momentum at time t is $25t + 75$, express the velocity as a function of t by writing it as the quotient of two functions of t. (*Hint*: Show that $v = 2KE/M$.)

47. A vacuum cleaner sales representative earns a commission consisting of 50 percent of the negotiated sales price. At the end of the year, the company gives to each representative a bonus of $1000 plus 10 percent of the commissions earned during the year. Express the representative's bonus as a function of the total amount of sales he concluded by writing the bonus as a composition of two functions. Express his total income for the year in terms of his total sales.

48. Bertha deposits money in the bank at 6 percent interest but must then pay 20 percent of her interest income to the Internal Revenue Service. Express the amount of interest income she pays to the government in terms of her bank balance by writing it as a composition of two functions. What is her net gain from her savings account?

49. The rabbit population in the Midwest depends on the vegetation available as food. The available vegetation in turn depends on the amount of rainfall. The fox population depends on the number of rabbits. There are 20 percent as many foxes as rabbits and each rabbit requires 200 pounds of vegetation per year. Each inch of rainfall in Ogden County supplies 20 tons of ex-

cess vegetation, which is available to the rabbits. Express the number of foxes in Ogden County as a function of rainfall by writing it as a composition of three functions.

50. In the following, determine whether $f \circ g = g \circ f$.

a. Let f be the operation of combing your hair and g be the operation of washing your hair.

b. Let f be the operation of studying for an exam and g be the operation of taking the exam.

c. Let f be the operation of eating a steak and g the operation of grilling the steak.

d. Let f denote the operation of mowing the lawn and g the operation of sweeping the floor.

e. Let f denote buying a piece of property and g denote inspection of the property.

4.4 INVERSE FUNCTIONS

When a dependent variable is given as a function of an independent variable, it is sometimes necessary to determine which value or values of the independent variable generate a certain value of the dependent variable. For example, if the altitude in feet of a projectile t seconds after firing is given by

$$s(t) = 240t - 16t^2$$

we might want to know how long it takes the projectile to reach an altitude of 800 feet. Thus, for a known value of $s = 800$, we are to determine which values of t yield $s(t) = 800$. Solving for t,

$$800 = 240t - 16t^2$$
$$16t^2 - 240t + 800 = 0$$
$$t^2 - 15t + 50 = 0$$
$$(t - 10)(t - 5) = 0$$
$$t = 5, 10$$

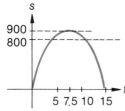

Figure 4–21
$s = 240t - 16t^2$

These answers tell us that 5 seconds after firing, the projectile is 800 feet above the ground on its way up; 10 seconds after firing, it is again 800 feet above the ground, this time on its way down. In this example, t is not a function of s since, for some values of s there are two values of t related to s. In fact, for any nonnegative s value less than 900 there are exactly two values of t that yield that s value (see Fig. 4–21).

In the preceding illustration, we saw that even though s is a function of t, it does not necessarily follow that t is a function of s. For t to be a function of s, each value of s much be related to only one t value; no two points of the graph can lie on the same horizontal line. Recall from Section 4.1 the horizontal line test.

HORIZONTAL LINE TEST

A function $y = f(x)$ determines x as a function of y if and only if each horizontal line crosses its graph in at most one point.

DEFINITION

If v is a function of u given by $v = f(u)$ and u is, in turn, a function of v, the function that determines u in terms of v is called the **inverse** of the function f and denoted as f^{-1}. Thus, if $v = f(u)$ and u is also a function of v, we write $u = f^{-1}(v)$. The domain of f^{-1} is the range of f.

$$f^{-1}(v) \text{ is that } u \text{ such that } f(u) = v$$

The horizontal line test says that in order for f^{-1} to exist, we cannot have $f(x_1) = f(x_2)$ for distinct values $x_1 \neq x_2$. In fact, f^{-1} exists if and only if $f(x_1) = f(x_2)$ implies $x_1 = x_2$. In the preceding illustration $s(5) = s(10)$; thus, this particular altitude function $s(t)$ has no inverse.

The inverse of a function "undoes" the operation of the original function. From u we compute $v = f(u)$ and then $f^{-1}(v)$ takes us back to u. In the same way, f undoes the operation of f^{-1}: applying f to $u = f^{-1}(v)$ takes us back to v. This process is illustrated in Figure 4–22.

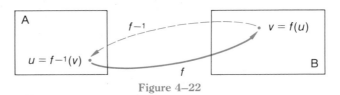

Figure 4–22

If f and g are functions with

$$\text{Domain } f = \text{range } g = A$$
$$\text{Domain } g = \text{range } f = B$$

and

$$g[f(x)] = x \qquad \text{for every } x \text{ in } A$$
$$f[g(x)] = x \qquad \text{for every } x \text{ in } B$$

(i.e., each function "undoes" the other), then g and f are inverses of one another:

$$g = f^{-1} \qquad \text{and} \qquad f = g^{-1}$$

> Whenever f^{-1} exists,
>
> **1.** $f^{-1}(f(t)) = t$ for all t in the domain of f
>
> **2.** $f(f^{-1}(t)) = t$ for all t in the domain of f^{-1} ($=$ range of f)

For example, the cubing and cube root functions are inverses of one another:

$$t \xrightarrow{\;(\;)^3\;} t^3 \xrightarrow{\;\sqrt[3]{\;}\;} t$$

and

$$t \xrightarrow{\;\sqrt[3]{\;}\;} \sqrt[3]{t} \xrightarrow{\;(\;)^3\;} t$$

EXAMPLE 1   If $v = f(u) = 2u + 1$, determine whether u is a function of v. If so, find f^{-1} and verify that f and f^{-1} each "undoes" the effect of the other.

Solution   We can solve for u in $f(u) = 2u + 1$.

$$2u + 1 = v$$
$$2u = v - 1$$
$$u = \frac{v - 1}{2}$$

Thus, u is indeed a function of v:

$$u = f^{-1}(v) = \frac{v - 1}{2}$$

Then,

$$f(f^{-1}(t)) = 2f^{-1}(t) + 1$$
$$= 2\left(\frac{t - 1}{2}\right) + 1$$
$$= t$$

and

$$f^{-1}(f(t)) = \frac{f(t) - 1}{2}$$
$$= \frac{(2t + 1) - 1}{2}$$
$$= t$$

EXAMPLE 2 Let $y = f(x) = 2x^2 + 1$ be defined only for $x \geq 0$.

    a. Show that x is a function of y.

    b. Find $f^{-1}(t)$; $f^{-1}(x)$.

    c. Show that f and f^{-1} each "undoes" the effect of the other.

Solution **a.** Writing $y = 2(x - 0)^2 + 1$, we see that its graph is a parabola with vertex at $(0, 1)$. But since the given function is defined only for $x \geq 0$, its graph is only the right-hand branch of this parabola as shown in Figure 4–23.

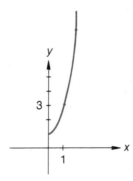

Figure 4–23 $y = 2x^2 + 1$, $x \geq 0$

Since each horizontal line meets the graph in at most one point, x can be computed in terms of y; f^{-1} exists.

    b. Solving for x:

$$2x^2 + 1 = y$$

$$x^2 = \frac{y - 1}{2}$$

$$x = \sqrt{\frac{y - 1}{2}}$$

Note that we take only the positive square root since the function was defined only for $x \geq 0$. Hence,

$$x = f^{-1}(y) = \sqrt{\frac{y - 1}{2}}, \qquad y \geq 1$$

We have found the functional relationship describing f^{-1}. Since the variable used is immaterial (a dummy variable), we have

$$f^{-1}(t) = \sqrt{\frac{t - 1}{2}}, \qquad t \geq 1$$

or, for that matter,

$$f^{-1}(x) = \sqrt{\frac{x - 1}{2}}, \qquad x \geq 1$$

c.
$$f^{-1}[f(t)] = \sqrt{\frac{f(t) - 1}{2}} = \sqrt{\frac{(2t^2 + 1) - 1}{2}}$$
$$= \sqrt{t^2}$$
$$= t \qquad \text{(since } f(t) \text{ is defined only for } t \geq 0)$$
$$f[f^{-1}(t)] = 2[f^{-1}(t)]^2 + 1$$
$$= 2\left(\sqrt{\frac{t - 1}{2}}\right)^2 + 1 = 2 \cdot \frac{t - 1}{2} + 1$$
$$= t \qquad \blacksquare$$

Had some $x < 0$ been allowed in the domain of the function in Example 2, we would not have obtained an inverse function. For instance, if the domain of f is expanded to $x \geq -1$, the graph becomes that of Figure 4–24. Since some horizontal line meets the graph in two points, x is then not a function of y.

In Figure 4–23 we see that $y = f(x)$ increases as x moves to the right and thus f has an inverse.

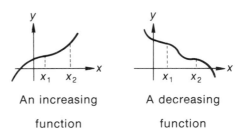

Figure 4–24
$y = 2x^2 + 1$,
$x \geq -1$

DEFINITION

A function f is said to be

1. **Increasing** if $f(x_2) > f(x_1)$ when $x_2 > x_1$

2. **Decreasing** if $f(x_2) < f(x_1)$ when $x_2 > x_1$

(f is said to be **nondecreasing** if $f(x_2) \geq f(x_1)$ when $x_2 > x_1$; **nonincreasing** is defined similarly.)

An increasing
function

A decreasing
function

An increasing or decreasing function clearly has an inverse since $f(x_2) \neq f(x_1)$ when $x_2 \neq x_1$. The graph of an increasing (decreasing) function rises (respectively falls) as x moves to the right. The function graphed in Figure 4–24 has no inverse; it is neither an increasing nor a decreasing function since its graph falls on the interval $[-1, 0]$ and rises on the interval $[0, \infty)$. However, if we restrict our attention to the interval $[0, \infty)$ on which f is increasing, the restricted function has an inverse; see Figure 4–23 again.

While increasing functions and decreasing functions do have inverses, a function need not be strictly increasing or decreasing in order to have an inverse. For example, the horizontal line test indicates that the function graphed in Figure 4–25 has an inverse.

**Graphing Inverse
Functions**

Let f be a function that has an inverse f^{-1}. If $v = f(u)$, then $u = f^{-1}(v)$. Specifically, if $f(2) = 4$, then $f^{-1}(4) = 2$. This means that if the point $(2, 4)$ is on the graph of f, then $(4, 2)$ is on the graph of f^{-1}. More generally, if (u, v) is on the graph of f, then (v, u) is on the graph of f^{-1}.

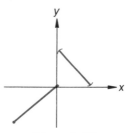

Figure 4–25

The graph of f^{-1} can be obtained by reversing the order of the coordinates of each point on the graph of f.

But reversing the order of the coordinates simply reflects the point through the line $y = x$; see Figure 4–26(a). Alternatively, we could rotate the graph of f about the line $y = x$ in order to sketch the graph of f^{-1} as illustrated in Figure 4–26(b).

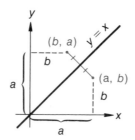

a. Reversing coordinates is equivalent to reflection through the line $y = x$.

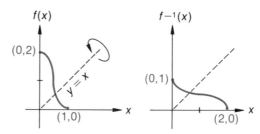

b. Rotating the graph of f about the line $y = x$ obtains the graph of f^{-1}.

Figure 4–26

EXAMPLE 3 If $y = f(x) = \sqrt{2x - 2}$ for $x \geq 1$, sketch the graphs of both the function and its inverse.

Solution Setting

$$y = \sqrt{2x - 2} \geq 0,$$

we find

$$2x - 2 = y^2$$

Thus,

$$x = f^{-1}(y) = \frac{y^2 + 2}{2}, \qquad y \geq 0$$

To graph this functional relationship, we write

$$f^{-1}(x) = \frac{x^2 + 2}{2}, \qquad x \geq 0.$$

(Remember, the variable used to describe the function is immaterial.)

In this case, it is more convenient to graph $y = f^{-1}(x)$ first and then "reflect through the diagonal" to obtain the graph of $y = f(x)$. Figure 4–27(b) shows the graph of $y = f^{-1}(x) = \frac{(x^2 + 2)}{2}$. Because of the restriction $x \geq 0$, we use only the right half of this graph. The graph of $y = f(x) = \sqrt{2x - 2}$ is then sketched in Figure 4–27(a) by reflecting (b) through the line $y = x$.

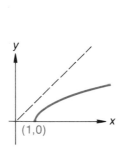

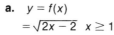

a. $y = f(x)$
$= \sqrt{2x - 2} \; x \geq 1$

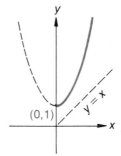

b. $y = f^{-1}(x) = \frac{x^2 + 2}{2} \; x \geq 0$

Figure 4–27

EXERCISES 4.4

In each of the following, determine whether an inverse function exists. If so, find it.

1. $f(x) = -4x + 1$
2. $H(u) = 2u - 8$
3. $g(v) = 32v^5$
4. $F(r) = 1/4r^4, \quad r \neq 0$
5. $P(s) = 16s^4, \; s \geq 0$
6. $F(x) = -\sqrt[3]{2x}$
7. $g(y) = \sqrt[6]{y^2}$

8. $Q(u) = \dfrac{2}{\sqrt[3]{3u}}, \quad u \neq 0$

9. $h(t) = 2t^2 - 1$

10. $G(x) = x^2 - 4x + 3$

11. $s(u) = u^2 - 4u + 2, \quad u \geq 2$

12. $s(t) = 2t^2 + 4t - 5, \quad t \geq -2$

13. $s(t) = \sqrt{2t - 3}, \quad t \geq 3/2$

14. $g(u) = \sqrt[3]{u^2 - 1}$

15. $F(x) = \dfrac{2x + 1}{x - 2}, \quad x \neq 2$

16. $G(s) = \dfrac{s + 1}{s - 1}, \quad s \neq 1$

17. $p(t) = \dfrac{3 - 2t}{2 - 3t}, \quad t \neq 2/3$

18. $Q(r) = \dfrac{1 - r}{1 + r}$

Show that the following functions are inverses of one another by showing that $g[f(x)] = x$ and $f[g(x)] = x$ for all appropriate values of x.

19. $f(x) = -x; \quad g(x) = -x$

20. $f(x) = 1/x, \quad x \neq 0; \quad g(x) = 1/x, \quad x \neq 0$

21. $f(x) = x + 2; \quad g(x) = x - 2$

22. $f(x) = 2x + 2; \quad g(x) = (x - 2)/2$

23. $f(x) = \sqrt{x - 1}, \quad x \geq 1; \quad g(x) = x^2 + 1, \quad x \geq 0$

24. $f(x) = x^2 - 2x + 1, \quad x \geq 1; \quad g(x) = \sqrt{x} + 1, \quad x \geq 0$

25. $f(x) = 1/(x - 1), \quad x \neq 1; \quad g(x) = (1 + x)/x, \quad x \neq 0$

26. $f(x) = 2/(x^2 - 1), \quad x > 1; \quad g(x) = \sqrt{(2 + x)/x}, \quad x > 0$

27. $f(x) = (x - 2)/(x + 2), \quad x \neq -2; \quad g(x) = 2(x + 1)/(x - 1), \quad x \neq 1$

28. $f(x) = (2x - 1)/(2x + 1), \quad x \neq -1/2; \quad g(x) = (x + 1)/[2(1 - x)], \quad x \neq 1$

In each of the following
   a. *Find the inverse function*
   b. *Show that $f[f^{-1}(t)] = t$*
   c. *Show that $f^{-1}[f(t)] = t$*
   d. *Sketch the graphs of both f and f^{-1}*

29. $f(x) = 3x - 2$ 30. $f(x) = 6 - 2x$

31. $f(x) = x^2 - 4x + 3, \quad x \geq 2$

32. $f(x) = -x^2 + 6x - 10, \quad x \geq 3$

33. $f(x) = \sqrt{6 - 2x}, \quad x \leq 3$

34. $f(x) = \sqrt{x + 4}, \quad x \geq -4$

35. $f(x) = 1/2x, \quad x \neq 0$

36. $f(x) = -1/8x^3, \quad x \neq 0$

37. $f(x) = \sqrt[3]{8x}$

38. $f(x) = -2/\sqrt[3]{8x}, \quad x \neq 0$

39. The following are graphs of functions. Determine where each is increasing and where it is decreasing.

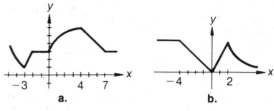

a.

b.

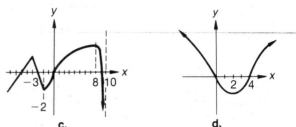

c.

d.

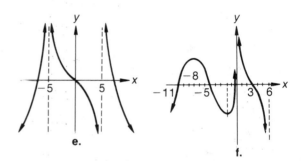

e.

f.

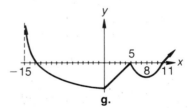

g.

Determine the regions where the following functions are increasing and where they are decreasing.

40. $f(x) = 4 - 3x$

41. $g(u) = 2u + 6$

42. $f(t) = -\dfrac{2}{t}$

43. $h(v) = \dfrac{2}{(v^4 + v^2)}$

44. $F(w) = -w^2 - 2w + 5$

45. $G(t) = 3t^2 - 12t + 7$

46. $f(x) = \sqrt{x}$

47. $g(r) = |r| + r$

48. Show that the sum of two increasing (decreasing) functions is increasing (respectively decreasing).

49. The function that expresses Fahrenheit temperature in terms of Celsius temperature is given by $F(C) = (9/5)C + 32$. Find the inverse function expressing Celsius temperature in terms of Fahrenheit temperature.

50. The legal speed limit is now 88 kilometers per hour. Express the distance that can be traveled at this speed as a function of time. Find the inverse function expressing the time required for a given trip as a function of the distance involved.

51. The interest earned on an investment equals the principal multiplied by the interest rate and the elapsed time. At 5 percent interest, express the amount earned each year as a function of the amount invested. Find the inverse function expressing the investment required to earn a certain sum each year.

52. An object falling freely toward the earth will have fallen $16t^2$ feet t seconds after it is dropped. Express the distance fallen as a function of time. Find the inverse function expressing the time required to fall a certain distance. How long will it take to fall 40,000 feet?

53. The sales graph for the Energetic Battery Company is given below. Can the Internal Revenue Service subpoena the company's records for the month in which it sold $300,000 worth of batteries? That is to say, is time a function of sales receipts? Are the sales receipts a function of time?

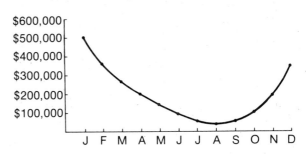

54. The average net yearly income per farm in Ohio is graphed here. During what periods did farm income increase? Decrease?

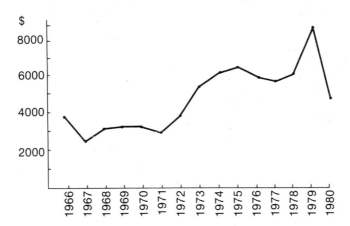

4.5 CHAPTER REVIEW

TERMS AND CONCEPTS

• Function

$y = f(x)$ A correspondence or rule that assigns to each element x of some set D a unique element y of a set I.

• Domain

The set D of x-values for which a function is defined.

- Range
  The set of all y-values that can be found by considering all x-values in the domain.
- Independent Variable $\left. \begin{matrix} x \\ y \end{matrix} \right\}$ in $y = f(x)$
- Dependent Variable
- Inverse of a Function $\qquad u = f^{-1}(v) \qquad$ means $\qquad v = f(u)$

- Composition of Functions
  $g \circ f(x) = g[\, f(x)]$

RULES AND FORMULAS

- Vertical Line Test $\qquad$ A graph represents y as a function of x if and only if each vertical line meets the graph in at most one point.
- Horizontal Line Test $\qquad$ A graph determines x as a function of y if and only if each horizontal line meets the graph in at most one point.
- Inverse Functions $\qquad f^{-1}[\, f(t)] = t \qquad$ and $\qquad f[\, f^{-1}(t)] = t$

GRAPHING TECHNIQUES

- Translation

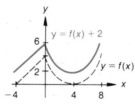

Vertical translation
$y = f(x) + c$

Horizontal translation
$y = f(x - a)$

- Reflection

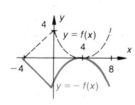

Vertical reflection
$y = -f(x)$

Horizontal reflection
$y = f(-x)$

- Scaling

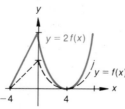

Vertical scaling
$y = Af(x)$

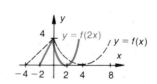

Horizontal scaling
$y = f(ax)$

- Symmetry

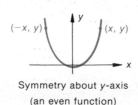 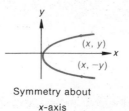 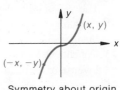

- Absolute Values   Any point with a negative y-value is reflected through the x-axis to the point with the corresponding positive y-value.

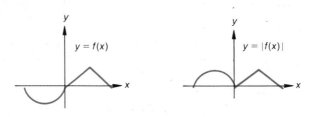

- Inverse Functions   The graph of f^{-1} can be obtained by reversing the order of the coordinates of each point on the graph of f; that is, by reflecting the graph of f through the line $y = x$.

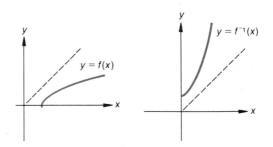

4.6 SUPPLEMENTARY EXERCISES

Which of the following equations define y as a function of x? Which define x as a function of y?

1. $x - y^2 = 3x$
2. $2x^2 - 3y^2 = 4$
3. $x^3 = y^3$
4. $x^2 = y^2$
5. Which of the following graphs define y as a function of x? Which define x as a function of y?

In each of the following, find the requested expressions.

6. $y(x) = 3x - 1$;   find $y(1)$, $y(4)$, $y(-4)$.
7. $g(y) = y^2 + 2y + 1$;   find $g(5)$, $g(2)$, $g(-2)$.
8. $M(s) = 6s^3 + 3s - 5$;   find $M(0)$, $M(1)$, $M(-1)$.

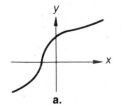

a.

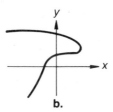

b.

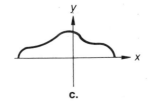

c.

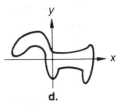

d.

9. $f(v) = \dfrac{2v + 1}{4v - 1}$; find $f(0)$, $f(2)$, $f(1/2)$.

10. $J(c) = 3 + 2c$; find $J(a - 2)$, $J(2a + 1)$, $J(2c - 3)$.

11. $w(x) = x^2 + 2x + 3$; find $w(2x)$, $w(x^2)$, $w(1/x)$.

12. $p(r + 2) = r^2 - 1$; find $p(0)$, $p(2)$, $p(r)$, $p(x + 1)$.

13. $g(t + 5) = t^2 - 3t + 10$; find $g(0)$, $g(1)$, $g(t)$, $g(t - 1)$.

14. $H(s) = 2s - 5$; find and compare
    a. $H(1 + 1)$, and $H(1) + H(1)$.
    b. $H(1 - 1)$ and $H(1) - H(1)$.
    c. $H(1 \cdot 2)$ and $H(1) \cdot H(2)$.
    d. $H(1/2)$, $H(1)/H(2)$, and $1/H(2)$.

15. $w(k) = 3k + 4$; find and compare
    a. $w(2^2)$ and $[w(2)]^2$
    b. $w(\sqrt{4})$ and $\sqrt{w(4)}$
    c. $w(2x)$ and $2w(x)$
    d. $w(-2)$ and $-w(2)$

In the following exercises, find the understood domain of the given function. Both domain and range are to be restricted to real numbers.

16. $g(u) = \dfrac{u^2 - 2u + 1}{u + 2}$

17. $f(w) = \dfrac{\sqrt{w + 1}}{2w - 2}$

18. $f(x) = \dfrac{\sqrt{16 - x^2}}{\sqrt{x^2 - 1}}$

19. $G(v) = \dfrac{\sqrt{4 - v^2}}{\sqrt{v^2 - 16}}$

Sketch the graphs of the following functions.

20. $f(x) = \begin{cases} x, & x < 2 \\ 0, & x = 2 \\ 1, & x > 2 \end{cases}$

21. $g(x) = \begin{cases} x^2, & x < 1 \\ 0, & x = 1 \\ 2 - x^2, & x > 1 \end{cases}$

Use the graph of $y = f(x)$ as given here to sketch the graph of each of the following.

22. $y = f(x - 3)$ 23. $y = -2f(x)$
24. $y = f(x) - 2$ 25. $y = f(x/3)$
26. $y = 2f(2x + 6) - 4$

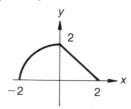

Use the techniques developed in this chapter to sketch the graphs of each of the following.

27. $y = 2x + 3$ 28. $y = -(x - 2)^2$
29. $y = (2 - x)^3$
30. $y = 2(x + 1)^2 - 2$

31. $y = -\dfrac{1}{2}(x - 2)^3 + 3$

32. $y = |x + 2| - 2$
33. $y = 3 - |x - 1|$
34. $|x - y| = 2$
35. $y = 2 - |2 - x|$
36. $x = |y^2 + 2y - 3|$
37. $y = -|x^2 - 4x + 1| + 3$
38. $y = |x^2 - 2x - 3| - 4$
39. Extend the given graph under the condition that it be
    a. even   b. odd

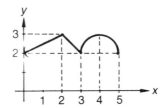

Indicate whether each of the following functions is even $[f(-x) = f(x)]$ or odd $[f(-x) = -f(x)]$ or neither.

40. $y = |x^2 + 2x + 1|$

41. $y = (x^2 + 1)/\sqrt[3]{x}$

42. $y = \sqrt[3]{x}\,\sqrt[5]{x}$

Find the difference quotient for the following functions.

43. $f(x) = -2$
44. $g(x) = 2x + 5$
45. $h(x) = 4x^2 - 2x + 1$
46. $F(x) = 2x^3 - x + 1$
47. $G(x) = 1/x^2$

In each of the following, find expressions for $f + g$, $f - g$, $f \cdot g$, f/g, $f \circ g$, $g \circ f$, and af for the indicated value of a. Also evaluate $f \circ g\,(a)$ and $g \circ f\,(a)$ for this a.

48. $f(x) = 2x - 3$, $g(t) = 5t - 3$, $a = -1$
49. $f(u) = 3u$, $g(v) = 3 - 2v$, $a = 2$
50. $f(r) = r^2 + r + 1$, $g(w) = 2 - 4w$, $a = 3$
51. $f(z) = z^2 - z - 1$, $g(a) = 2a + 1$, $a = -3$
52. $f(t) = 16t^2$, $g(s) = (s - 1)/(s + 1)$, $a = -2$
Find functions, f, g, and write the given function as $g \circ f$.

53. $F(t) = \left(\dfrac{t + 1}{t - 1}\right)^2$

54. $G(x) = (x^2 - 1)^5 + 3(x^2 - 1)^3 + (x^2 - 1) + 2$

55. $V(u) = (u^2 + 2u + 5)^{2/3}$

In each of the following, determine whether an inverse function exists. If so, find it.

56. $G(w) = 3/w^7, \quad w \neq 0$

57. $H(r) = r^2$

58. $F(v) = -3\sqrt[4]{v^2}$

59. $p(t) = 2\sqrt[7]{t}$

60. $f(x) = \dfrac{(2 - 3x)}{(x + 2)} \qquad x \neq -2$

61. $g(u) = \dfrac{(1 + u)}{(1 - u)} \qquad u \neq 1$

Show that f and g are inverses of one another by showing that $g[f(x)] = x$ and $f[g(x)] = x$ for all appropriate x.

62. $f(x) = x + 3; \quad g(x) = x - 3$

63. $f(x) = \sqrt{x} - 1, \quad x \geq 0; \quad g(x) = x^2 + 2x + 1, \quad x \geq -1$

64. $f(x) = \dfrac{2}{(x - 2)}, \quad x \neq 2; \quad g(x) = \dfrac{2(x + 1)}{x}, \quad x \neq 0$

65. $f(x) = \dfrac{(x + 3)}{(x - 1)}, \quad x \neq 1; \quad g(x) = \dfrac{(x + 3)}{(x - 1)}, \quad x \neq 1$

In each of the following, find the inverse function f^{-1}, show that $f^{-1}[f(x)] = x$, show that $f[f^{-1}(x)] = x$, and graph both f and f^{-1}.

66. $f(x) = 1 - 4x$

67. $f(x) = x^2 - 4x + 2, \quad x \geq 2$

68. $f(x) = 2\sqrt{x - 2}, \quad x \geq 2$

69. $f(x) = 3/x^5, \quad x \neq 0$

70. $f(x) = -3\sqrt[4]{x}, \quad x \geq 0$

71. Determine where the function defined by the following graph is increasing and where it is decreasing.

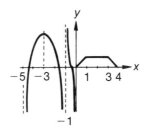

Determine the regions where the following functions are increasing and where they are decreasing.

72. $f(t) = 1 - 2t$

73. $g(x) = \dfrac{1}{x + 1}$

74. $h(r) = \dfrac{-2}{r^2 + |r| + 1}$

75. $f(x) = |x^2 - 2x - 3| + 2$

5 Polynomial and Rational Functions

A function that can be expressed as a polynomial is called a **polynomial function.** The general form of a polynomial function is $P(x) = a_n x^n + \ldots + a_1 x + a_0$. Examples of polynomial functions include the expressions for the volume and surface area of a sphere in terms of its radius x:

$$V(x) = \frac{4}{3}\pi x^3 \quad \text{and} \quad S(x) = 4\pi x^2$$

respectively.

A function that can be expressed as a quotient of two polynomials is called a **rational function:** $R(x) = P(x)/Q(x)$ where P and Q are polynomials. The intensity of illumination of a lamp varies inversely as the square of the distance from the lamp: $I(x) = k/x^2$ and the current carried by an electrical wire is inversely proportional to the resistance of the wire: $C(x) = k/x$. Each of these relationships is an example of a rational function.

We have already studied the polynomial functions of degree ≤ 2 in detail, the linear and quadratic functions, $P(x) = ax + b$ and $P(x) = ax^2 + bx + c$. We now examine polynomial functions of degree > 2 and rational functions. In the final sections of this chapter, we shall graph these functions. Locating the x-intercepts will be a crucial step in our graphing efforts. Techniques for finding these intercepts will be described in the earlier sections of the chapter.

5.1 ROOTS, FACTORS, AND SYNTHETIC DIVISION

If $P(r) = 0$, then r is called a **zero** or **root** of the polynomial $P(x)$. The x-intercepts of the graph of a polynomial function $P(x)$ occur at its real roots.

Most of the techniques for finding roots of polynomials will depend in some way on the division of one polynomial by another.

Remainder Theorem

An analysis of the long-division procedure (Section 1.6) shows an important result of division by a polynomial of the form $(x - a)$, where a is a given constant. In this case the remainder must be a constant; no x terms will remain. If we divide a polynomial $P(x)$ by $(x - a)$ and obtain a quotient $Q(x)$ with a remainder R, we have

$$\frac{P(x)}{x - a} = Q(x) + \frac{R}{x - a}$$

which can be written in the form

$$P(x) = Q(x)(x - a) + R$$

Evaluating the polynomial at $x = a$ yields

$$P(a) = Q(a)\underbrace{(a - a)}_{= 0} + R$$

$$= R$$

The remainder is simply the value of the polynomial at $x = a$. This result is embodied in the

REMAINDER THEOREM

> If a polynomial is divided by $(x - a)$ where a is a constant, the remainder is the value of the polynomial at $x = a$.

EXAMPLE 1 Check the validity of the remainder theorem by dividing $(x^3 + 2x^2 - 3x + 4)$ by $(x - 2)$.

Solution

$$
\begin{array}{r}
x^2 + 4x + 5 \\
x - 2 \overline{)x^3 + 2x^2 - 3x + 4} \\
\underline{x^3 - 2x^2} \\
4x^2 - 3x \\
\underline{4x^2 - 8x} \\
5x + 4 \\
\underline{5x - 10} \\
14
\end{array}
$$

The remainder is 14. Evaluating the polynomial at $x = 2$ also yields 14 as we see:

$$2^3 + 2 \cdot 2^2 - 3 \cdot 2 + 4 = 8 + 8 - 6 + 4$$
$$= 14$$

■

Factor Theorem

Note that the remainder in a long-division problem $\dfrac{P(x)}{D(x)}$ is 0 if and only if the denominator $D(x)$ is a factor of the numerator $P(x)$. For if $\dfrac{P(x)}{D(x)} = Q(x)$, then $P(x) = D(x)Q(x)$. Combining this observation with the remainder theorem, we obtain the

FACTOR THEOREM

> $(x - a)$ is a factor of a polynomial $P(x)$ if and only if $P(a) = 0$.

EXAMPLE 2 Without carrying out the long division, show that $(x + 2)$ is a factor of

$$P(x) = (x^4 + 2x^3 - 13x^2 - 14x + 24).$$

Solution Since $(x + 2) = [x - (-2)]$, it follows that $(x + 2)$ is a factor of $P(x)$ if and only if the value of the polynomial at $x = -2$ is 0. Evaluating at $x = -2$ we obtain

$$P(-2) = (-2)^4 + 2(-2)^3 - 13(-2)^2 - 14(-2) + 24$$
$$= 16 - 16 - 52 + 28 + 24$$
$$= 0$$

$(x + 2)$ is indeed a factor of the given polynomial. This fact can also be verified by the long-division procedure but this task is left to the reader. ■

Synthetic Division

To use the remainder and factor theorems, we first "guess" at a root $x = a$ of a polynomial $P(x)$ and then check to see whether $P(a) = 0$. If so, $x = a$ is a root and $x - a$ is a factor of $P(x)$. We then find the corresponding factor by using the long-division process to divide $P(x)$ by $(x - a)$.

The division of a polynomial by an expression of the form $(x - a)$ can be accomplished without writing out all the details of the long-division process. We can find and record the coefficients by position only by using a process called **synthetic division.**

Consider, for example, the long division of $2x^4 + 22x - 9x^3 + 20$ by $(x - 3)$. As usual, we arrange the terms in decreasing order and include any missing terms. This division is displayed on the left in the following illustration. This division is then rewritten in the middle display but with the variable suppressed; the circled entries are also suppressed since each one is simply a repetition of the entry directly above it. Finally, this format is condensed into the form on the right.

$$
\begin{array}{r}
2x^3 - 3x^2 - 9x - 5 \\
x - 3\overline{)\,2x^4 - 9x^3 + 0x^2 + 22x + 20} \\
\underline{2x^4 - 6x^3} \\
-3x^3 + 0x^2 \\
\underline{-3x^3 + 9x^2} \\
-9x^2 + 22x \\
\underline{-9x^2 + 27x} \\
-5x + 20 \\
\underline{-5x + 15} \\
5 = R
\end{array}
$$

$$
\begin{array}{r}
-3\overline{)\,2 \quad -3 \quad -9 \quad -5} \\
2 \quad -9 \quad 0 \quad 22 \quad 20 \\
\underline{-6} \\
-3 \\
\underline{9} \\
-9 \\
\underline{27} \\
-5 \\
\underline{15} \\
5 = R
\end{array}
$$

$$
\begin{array}{r}
-3\overline{)\,2 \quad -3 \quad -9 \quad -5} \\
2 \quad -9 \quad 0 \quad 22 \quad 20 \\
\underline{-6 \quad 9 \quad 27 \quad 15} \\
-3 \quad -9 \quad -5 \quad 5 = R
\end{array}
$$

In the preceding display on the right, we can eliminate the "answer" row provided we include its leading 2 in the bottom row. The bottom row then includes both the quotient and the remainder. This is illustrated on the left in the following display.

$$
\begin{array}{r}
-3\overline{)\,2 \quad -9 \quad 0 \quad 22 \quad 20} \\
\underline{-6 \quad 9 \quad 27 \quad 15} \\
2 \quad -3 \quad -9 \quad -5 \quad 5 = R
\end{array}
\qquad
\begin{array}{r}
3\overline{)\,2 \quad -9 \quad 0 \quad 22 \quad 20} \\
\underline{6 \quad -9 \quad -27 \quad -15} \\
2 \quad -3 \quad -9 \quad -5 \quad 5 = R
\end{array}
$$

Finally, we can change the subtraction to addition by changing the signs of the -3 and of each entry in the second row as illustrated on the right. Thus, to divide by $(x - 3)$, we place 3 to the left of the indicated division process and bring down the leading 2. Then $3 \cdot 2 = 6$ and 6 is placed below the -9. Add $-9 + 6 = -3$ and then $3 \cdot (-3) = -9$ is placed below the 0. We continue this process until we run out of terms. The "answer" is read from the last row as $2x^3 - 3x^2 - 9x - 5$ with a remainder of 5. This condensed version of the long-division process is called **synthetic division.**

EXAMPLE 3   Use synthetic division to show that $x = 2$ is a root of $P(x) = 3x^5 - 8x^3 - 6x^2 + x - 10$. Factor $P(x)$ into $(x - 2)Q(x)$.

Solution   Using synthetic division to divide $P(x)$ by $(x - 2)$, we have the following display:

$$
\begin{array}{r}
2\overline{)\,3 \quad 0 \quad -8 \quad -6 \quad 1 \quad -10} \\
\underline{6 \quad 12 \quad 8 \quad 4 \quad 10} \\
3 \quad 6 \quad 4 \quad 2 \quad 5 \quad 0 = R
\end{array}
$$

Since the remainder is 0, $(x - 2)$ is indeed a factor of $P(x)$ and $P(2) = 0$ as was to be shown. Furthermore, the last row represents the quotient:

$$
\frac{3x^5 - 8x^3 - 6x^2 + x - 10}{x - 2} = 3x^4 + 6x^3 + 4x^2 + 2x + 5
$$

Thus, we have the factorization

$$
\begin{aligned}
& 3x^5 - 8x^3 - 6x^2 + x - 10 \\
& = (x - 2)(3x^4 + 6x^3 + 4x^2 + 2x + 5)
\end{aligned}
$$

∎

By using synthetic division in Example 3, we were able to factor $P(x)$ almost as fast as we could have evaluated $P(2)$ to determine *if* $(x - 2)$ were a factor. Thus, synthetic division is a shortcut to finding roots and simultaneously factoring out linear factors $(x - a)$.

EXERCISES 5.1

Use synthetic division and verify the remainder theorem in the following.

1. $(4a^2 + 10a - 8) \div (a + 3)$
2. $(x^2 + 2x + 5) \div (x - 2)$
3. $(u^2 + 5u + 7) \div (u + 2)$
4. $(3x^3 + 2x^2 - x) \div (x - 2)$
5. $(t^3 - 2t^2 + 3t - 4) \div (t - 1)$
6. $(3u^3 + 2u^2 + u - 1) \div (u - 1)$
7. $(3y^3 - 4y^2 + 2y - 10) \div (y - 3)$
8. $(5x^4 + 3x^2 - 2) \div (x + 1)$
9. $(2t^4 - t^3 - 3t^2 - t - 10) \div (t + 2)$
10. $(4x^4 + 3x^3 + 2x^2 + x - 2) \div (x + 1)$
11. $(3w^4 + 6w^3 + 2w^2 - w - 3) \div (w + 2)$
12. $(x^4 - x^3 - x^2 + x) \div (x - 3)$
13. $(v^4 + 6v^3 + v^2 - 8) \div (v - 1)$
14. $(t^5 - 10t^3 + 7t + 6) \div (t - 3)$
15. $(x^6 + x^5 + x^2 + x + 1) \div (x + 1)$

Use the factor theorem to determine whether the second

expression is a factor of the first; if so, use synthetic division to factor out the second expression.

16. $(r^3 + r^2 + r + 6),\quad (r + 2)$
17. $(3s^3 + 2s^2 - s - 4),\quad (s + 1)$
18. $(3t^3 + 2t^2 + t + 2),\quad (t + 1)$
19. $(2a^3 - a^2 - 6a - 3),\quad (a - 4)$
20. $(r^4 - 10r^2 + 10r - 1),\quad (r - 1)$
21. $(b^4 - 6b^3 + 6b^2 + 2b + 4),\quad (b - 2)$
22. $(2v^4 - v + 6),\quad (v - 5)$
23. $(x^5 - 32),\quad (x - 2)$
24. $(u^5 - 17u + 2),\quad (u - 2)$
25. $(w^5 - 10w^3 + 6w - 36),\quad (w + 2)$
26. $(z^5 + 3z^4 + 2z^2 - 4),\quad (z + 1)$
27. $(y^6 - 2y^4 + 10y - 12),\quad (y + 2)$
28. $(5u^6 - 7u^3 - 6u + 8),\quad (u - 1)$
29. $(p^6 + 2p^4 + 2p^3 + 30p - 20),\quad (p + 2)$
30. $(s^{100} + s^{67} + s^{13} + s^7 + 1),\quad (s + 1)$

5.2 REAL ROOTS OF POLYNOMIALS

We found synthetic division in conjunction with the factor and remainder theorems to be useful in testing for roots of polynomials. In this section, we address the problem of determining possible roots to be tested.

Rational Roots

We are especially interested in factors of the form $(ax + b)$, where a and b are integers. But $(ax + b) = a[x + (b/a)]$ so that then $x + (b/a)$ is a factor and $x = -b/a$ is a root of the polynomial. Thus, factors of the form $(ax + b)$ with a and b integers correspond to rational roots $x = -b/a$ of the polynomial.

$(ax + b)$ is a factor of a polynomial $P(x)$

if and only if

$-b/a$ is a root of the polynomial: $P\left(-\dfrac{b}{a}\right) = 0$

For a polynomial with integer coefficients, consideration of its first and last coefficients will limit appreciably the rational numbers, which need be considered as potential roots. For if $x = k/m$ is a rational root in lowest terms of $P(x) = a_n x^n + a_{n-1} x^{n-1} + \cdots + a_1 x + a_0$, then

$$a_n \left(\frac{k}{m}\right)^n + a_{n-1}\left(\frac{k}{m}\right)^{n-1} + \cdots + a_1\left(\frac{k}{m}\right) + a_0 = 0$$

Transposing the a_0 term and multiplying by m^n yields

$$a_n k^n + a_{n-1} k^{n-1} m + \cdots + a_1 k m^{n-1} = -a_0 m^n$$
$$k(a_n k^{n-1} + a_{n-1} k^{n-2} m + \cdots + a_1 m^{n-1}) = -a_0 m^n$$

This says that k is a factor of $-a_0 m^n$. But k/m was written in lowest terms. Thus, k cannot be a factor of m; k must be a factor of a_0. In the same way, it can be shown that m is a factor of a_n.

RATIONAL ROOT THEOREM

> If a rational number k/m in lowest terms is a root of a polynomial $a_n x^n + \cdots + a_1 x + a_0$ with integer coefficients, then k is a factor of a_0 and m is a factor of a_n.

Note that the rational root theorem does not indicate that any particular rational number is actually a root. It merely restricts the list of candidates from which we need search for rational roots.

EXAMPLE 1 Find all the rational roots of the following expression and factor out the corresponding linear factors.

$$P(x) = 3x^4 - 7x^3 - 3x^2 - 7x - 6$$

Solution According to the rational root theorem, for k/m to be a rational root in lowest terms

$$k \text{ must be a factor of } -6: \quad k = \pm 1, \ \pm 2, \ \pm 3, \ \pm 6$$
$$m \text{ must be a factor of } 3: \quad m = \pm 1, \ \pm 3$$

Thus, we obtain potential roots

$$\frac{k}{m} = \pm 1, \ \pm 2, \ \pm 3, \ \pm 6, \ \pm \frac{1}{3}, \ \pm \frac{2}{3}$$

Each of these must be tested by direct evaluation, long division, or synthetic division. For example, $P(\pm 1) = 3 \mp 7 - 3 \mp 7 - 6 = -20$ or 8; thus, $x = \pm 1$ are not roots of $P(x)$. It is left to the reader to show that $x = \pm 2$ likewise are not roots. On the other hand, synthetic division by $(x - 3)$ yields

$$3\overline{)3\ \ -7\ \ -3\ \ -7\ \ -6}$$
$$\underline{\ \ \ \ \ 9\ \ \ \ 6\ \ \ \ 9\ \ \ \ 6}$$
$$3\ \ \ \ 2\ \ \ \ 3\ \ \ \ 2\ \ \ \ 0 = R$$

Thus, $x = 3$ is a root and

$$P(x) = (x - 3)(3x^3 + 2x^2 + 3x + 2)$$

Any further roots of $P(x)$ must also be roots of

$$Q(x) = 3x^3 + 2x^2 + 3x + 2$$

Applying the rational root theorem to $Q(x)$ still leaves $k/m = \pm 1/3, \pm 2/3$ as potential roots. We have now eliminated -3 and ± 6 from consideration in addition to the previously eliminated ± 1 and ± 2. Testing $x = -2/3$ as a root for $Q(x)$, we have

$$-\tfrac{2}{3}\overline{)3\ \ \ \ 2\ \ \ \ 3\ \ \ \ 2}$$
$$\underline{\phantom{-\tfrac{2}{3}}\ \ -2\ \ \ \ 0\ \ -2}$$
$$3\ \ \ \ 0\ \ \ \ 3\ \ \ \ 0 = R$$

Thus, $(x + 2/3)$ is a factor of $Q(x)$ and

$$Q(x) = \left(x + \frac{2}{3}\right)(3x^2 + 3)$$
$$= \left(x + \frac{2}{3}\right)3(x^2 + 1)$$

Now $x^2 + 1$ has no real roots. Thus, $x = 3$ and $x = -2/3$ are the only rational (even real) roots of $P(x)$ and

$$P(x) = (x - 3)\left(x + \frac{2}{3}\right)3(x^2 + 1)$$
$$= (x - 3)(3x + 2)(x^2 + 1)$$ ■

Descartes' Rule of Signs

René Descartes, who developed the Cartesian coordinate system, has also given us a rule to help in locating real roots of polynomials. We shall simply state his rule here without proof.

DESCARTES' RULE OF SIGNS

1. The number of positive real roots of a polynomial $P(x)$ is less than or equal to the number of sign changes in $P(x)$.

2. The number of negative real roots is limited in the same way by the number of sign changes in $P(-x)$.

In each of these computations, the terms of the polynomial must be arranged in descending order and missing terms are not inserted. In either case, if the number of roots is not equal to the number of sign changes, it is less than this by an even number.

If $P(x) = (x - a)^2 Q(x)$, then $x = a$ is counted twice as a root in Descartes' rule of signs. A similar comment holds for other multiple roots.

EXAMPLE 2 Use Descartes' rule of signs to discuss the number of positive and negative roots of

$$P(x) = 4x^5 - 3x^3 - 2x^2 + x + 1$$

Solution There are two changes of signs in $P(x)$; thus, $P(x)$ has two or zero positive roots. On the other hand,

$$\begin{aligned} P(-x) &= 4(-x)^5 - 3(-x)^3 - 2(-x)^2 + (-x) + 1 \\ &= -4x^5 + 3x^3 - 2x^2 - x + 1 \end{aligned}$$

has three sign changes. Thus, $P(x)$ has three or one negative roots. In particular, this polynomial has at least one real root. ■

Bounds on Real Roots

The synthetic division process can also be used to find ranges in which all real roots of a polynomial must lie.

RANGE LIMITATION TEST

1. If $b > 0$ and the third row of the synthetic division of $P(x)$ by $(x - b)$ contains no negative entry, then all real roots of $P(x)$ are less than or equal to b.

2. If $a < 0$ and the third row of the synthetic division of $P(-x)$ by $(x - |a|)$ contains no negative entry, then all real roots of $P(x)$ are greater than or equal to a.

To establish the limitation on positive roots, let

$$P(x) = (x - b)Q(x) + R$$

The third row of the synthetic division of $P(x)$ by $x - b$ indicates the coefficients of $Q(x)$ and the remainder R. If all coefficients of $Q(x)$ are ≥ 0, then $Q(x) > 0$ for all $x > 0$ unless $P(x) = R$, a constant. (Why?) If also $R \geq 0$, then for any $x^* > b$ we have

$$P(x^*) = \underbrace{(x^* - b)}_{> 0} \underbrace{Q(x^*)}_{> 0} + \underbrace{R}_{\geq 0} \neq 0$$

that is, no $x^* > b$ can be a root of $P(x)$.

The limitation on negative roots follows since r is a root of $P(x)$ if and only if $-r$ is a root of $P(-x)$; that is, $P[-(-r)] = 0$ if and only if $P(r) = 0$.

EXAMPLE 3 Establish a range that includes all real roots of

$$P(x) = 3x^4 - 4x^3 - x^2 + 4x - 1$$

Solution To make the range as small as possible, we start by trying $b = 1, 2$, and so on. Synthetic division by $(x - 1)$ yields negative entries in the last row. On the other hand, for $(x - 2)$ we have

$$
\begin{array}{r|rrrrr}
2) & 3 & -4 & -1 & 4 & -1 \\
   &   & 6 & 4 & 6 & 20 \\
\hline
   & 3 & 2 & 3 & 10 & 19 = R
\end{array}
$$

Thus, all roots of $P(x)$ are < 2 (since $x = 2$ is obviously not a root).

To limit roots on the negative side, we consider

$$
\begin{aligned}
P(-x) &= 3(-x)^4 - 4(-x)^3 - (-x)^2 + 4(-x) - 1 \\
      &= 3x^4 + 4x^3 - x^2 - 4x - 1
\end{aligned}
$$

Synthetic division by $(x - 1)$ yields

$$
\begin{array}{r|rrrrr}
1) & 3 & 4 & -1 & -4 & -1 \\
   &   & 3 & 7 & 6 & 2 \\
\hline
   & 3 & 7 & 6 & 2 & 1 = R
\end{array}
$$

Thus, all roots of $P(x)$ are > -1 [again, 1 is obviously not a root of $P(-x)$].

We have established the range $-1 < r < 2$ for the roots r of $P(x)$. ■

In regard to the range limitation test, we should mention that if either $P(x)$ or $P(-x)$ has a negative leading coefficient, the bottom row of the synthetic division process will contain this negative entry. In this case, we consider $-P(x)$ or $-P(-x)$, respectively, in order to have a positive leading coefficient. This is permissible since $P(x) = 0$ if and only if $-P(x) = 0$.

Irrational Roots

The range limitation test and Descartes' rule of signs apply to *all* real roots, rational and irrational. To locate irrational roots, we resort to the method of successive approximations. It is based on the simple property that if $P(a)$ and $P(b)$ differ in sign, then $P(x) = 0$ for some x between a and b as illustrated in Figure 5–1.

SUCCESSIVE APPROXIMATION OF ROOTS

1. Find two values $a, b,$ for which $P(a)$ and $P(b)$ differ in sign; there must be a root between a and b.

2. Break down the interval already obtained to isolate a root in a narrower interval.

3. Continue the process of refinement until a sufficient degree of accuracy is achieved.

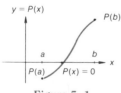

Figure 5–1

Roots will sometimes be missed if only integer values of x are tested. In Figure 5–2, for example, we have $P(1) > 0$ and $P(2) > 0$, yet there are two roots of $P(x)$ between 1 and 2. Thus, if not all roots are found, smaller ranges should be tested.

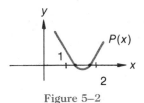

Figure 5–2

Needless to say, a calculator is essential for the method of successive approximations. In fact, a small computer would be even better as the number of computations required can be burdensome. In calculus, faster methods of locating roots are developed.

EXAMPLE 4   Find all real roots of the following correct to the nearest tenth.

$$P(x) = 4x^5 - 2x^4 + 10x^3 - 5x^2 - 6x + 3$$

Solution   In this case

$$P(-x) = -4x^5 - 2x^4 - 10x^3 - 5x^2 + 6x + 3$$

Thus, Descartes' rule of signs indicates zero, two, or four positive roots and exactly one negative root.

Next, we shall limit the range in which the roots must lie. Dividing $P(x)$ by $(x - 1)$ yields

$$
\begin{array}{r|rrrrrr}
1) & 4 & -2 & 10 & -5 & -6 & 3 \\
   &   & 4 & 2 & 12 & 7 & 1 \\
\hline
   & 4 & 2 & 12 & 7 & 1 & 4 = R \\
\end{array}
$$

Thus, all real roots are < 1. Since the leading coefficient of $P(-x)$ is negative, we consider $-P(-x)$. Dividing $-P(-x)$ by $(x - 1)$ yields

$$
\begin{array}{r|rrrrrr}
1) & 4 & 2 & 10 & 5 & -6 & -3 \\
   &   & 4 & 6 & 16 & 21 & 15 \\
\hline
   & 4 & 6 & 16 & 21 & 15 & 12 = R \\
\end{array}
$$

Hence all real roots of $P(x)$ are > -1; the roots of $P(x)$ lie in the interval $(-1, 1)$. Testing the integers in this range, we have

$$
\begin{aligned}
P(-1) &= -12 \\
P(0) &= 3 \\
P(1) &= 4
\end{aligned}
$$

Again, a root is indicated between -1 and 0, but we get no further information regarding the existence of roots between 0 and 1.

Possible rational roots are $k/m = \pm 3/4, \pm 1/2, \pm 1/4$ since $\pm 1, \pm 3$, and $\pm 3/2$ lie outside the range $(-1, 1)$. Synthetic division can be used to eliminate all possibilities except $x = 1/2$. For $x = 1/2$ we have

$$
\begin{array}{r|rrrrrr}
\tfrac{1}{2})4 & -2 & 10 & -5 & -6 & -3 \\
 & 2 & 0 & 5 & 0 & -3 \\
\hline
4 & 0 & 10 & 0 & -6 & 0 = R
\end{array}
$$

Thus, $x = 1/2$ is a root and

$$
\begin{aligned}
P(x) &= \left(x - \frac{1}{2}\right)(4x^4 + 10x^2 - 6) \\
&= (2x - 1)(2x^4 + 5x^2 - 3)
\end{aligned}
$$

There must be at least one more positive root (the number of positive roots is zero, two, or four). Any further roots of $P(x)$ must be roots of

$$
Q(x) = 2x^4 + 5x^2 - 3
$$

But if r is a root of $Q(x)$, then $-r$ is also, since only even powers of x appear in $Q(x)$. Thus, there can be only one more positive root since there is only one negative root.

Using the method of successive approximations to find this positive root, we shall essentially halve the interval each time to minimize the computations necessary to close in on the root.

$$
\begin{aligned}
Q(0) &= -3 \\
Q(1) &= 4 && [\text{a root in } (0,1)] \\
Q(0.5) &= -1.625 && [\text{a root in } (0.5,1)] \\
Q(0.8) &\approx 1.02 && [\text{a root in } (0.5,0.8)] \\
Q(0.7) &= -0.07 && [\text{a root in } (0.7,0.8)] \\
Q(0.75) &\approx 0.45
\end{aligned}
$$

Thus, there is a root between 0.7 and 0.75. To the nearest tenth, the roots of $P(x)$ are $x = 0.5, 0.7$, and -0.7. There are no other real roots. Were we to use the techniques of Section 2.3, we would find the roots of $Q(x)$ to be $x = \pm\sqrt{2}/2 \approx \pm 0.707$. ■

EXERCISES 5.2

Find all rational roots of the following.
1. $v^3 - 3v^2 - 6v + 8$
2. $2w^3 + 3w^2 - 32w + 15$
3. $12s^3 + 16s^2 - 5s - 3$
4. $2t^3 + 5t^2 - 8t - 6$
5. $3u^3 + 7u^2 + 5u + 1$
6. $12y^4 - 4y^3 - 3y^2 + y$
7. $t^4 + 2t^3 - 2t^2 - 8t - 8$

8. $18z^3 + 21z^2 - 10z - 8$

Use Descartes' rule of signs to indicate the possible numbers of positive and negative roots of the following.
9. $x^3 + 3x^2 - 2x - 1$
10. $4s^4 - 2s^3 + s - 2$
11. $2u^5 + u^4 - u^3 - u^2 + 2u + 5$
12. $3y^6 - 10y^4 + 2y^2 - 3$
13. $x^5 + x^3 - x + 1$

14. $2t^4 + 3t^3 - 2t^2 + t - 1$
15. $v^7 - 10v^4 + 4v^3 - v + 3$ 0-2-4) -1 neg
16. $-w^5 + w^4 - 4w^3 - 2w^2 + w + 1$

Use the range limitation test to find the smallest interval with integer endpoints containing all real roots of the following.

17. $u^4 + 4u^3 - 4u - 4$
18. $4s^4 + 2s^3 - s - 2$
19. $3x^6 + 10x^4 - 2x^2 - 3$
20. $t^4 - 4t^3 - 2t^2 + 12t - 3$
21. $2x^4 - 3x^3 + 2x^2 + x - 1$
22. $v^5 - v^4 - 4v^3 + 2v^2 - v + 1$
23. $2w^5 + 5w^4 - 2w - 5$
24. $-12y^4 + 4y^3 - 3y^2 + 2y - 5$

Estimate a real root of the following to within the nearest tenth and within the indicated interval.

25. $u^3 - u^2 - 6u + 2$, $(-3, -2)$
26. $v^3 + 5v^2 - 3$, $(-5, -4)$
27. $y^3 - 5y - 3$, $(-1, 0)$
28. $x^3 + 2x + 7$, $(-2, -1)$
29. $2x^4 - x^3 + 4x^2 + 6x - 4$, $(-2, -1)$
30. $t^3 + t + 1$, $(-1, 0)$

31. $s^3 + 5s^2 - 3$, $(-1, 0)$
32. $y^3 + 5y^2 + 4y + 5$, $(-5, -2)$

Find all real roots of the following. Estimate irrational roots to the nearest tenth.

33. $x^3 - 5x^2 + 2x + 12$
34. $s^3 - s^2 - 5s + 2$
35. $8x^3 + 12x^2 - 66x - 35$
36. $2x^4 - 3x^3 + 6x^2 + x - 15$
37. $t^3 - 2t + 7$
38. $u^3 + 3u^2 + 4u + 5$
39. $2y^5 + 5y^4 - 7y^3 - 2y^2 + 20y - 9$
40. $x^4 - 4x^3 + 3x^2 + 4x - 4$
41. $x^5 - x^4 - 5x^3 + 5x^2 + 6x - 6$
42. $8x^3 - 8x + 1$
43. $s^3 - 3s^2 - 4s + 13$
44. The graph of $P(x) = x^4 + ax^3 + bx^2 + cx + d$ intersects the x-axis at $x = 1, -2, 3, -4$. Find $P(0)$.
   Hint: Use the factor theorem
45. The graph of $P(x) = x^4 + ax^3 + bx^2 + cx + d$ intersects the x-axis at $x = -1, 2, -3$. If $P(4) = 70$, find the y-intercept.

5.3 COMPLEX ROOTS OF POLYNOMIALS

The roots of a quadratic polynomial $P(x) = ax^2 + bx + c$ are given by the quadratic formula (Section 2.2) as

$$x = \frac{-b \pm \sqrt{b^2 - 4ac}}{2a}.$$

There are two real roots if $b^2 - 4ac > 0$ and two complex roots if $b^2 - 4ac < 0$. On the other hand, if $b^2 - 4ac = 0$, we obtain just one root $x_0 = -b/2a$. The factor theorem then tells us that

$$P(x) = (x - x_0)Q(x)$$

with $Q(x)$ of degree 1; that is, $Q(x) = ax - D = a(x - d)$ for some number d. But then $(x - d)$ is a factor, and d is a root of $P(x)$. As there is only one root x_0, this means that $d = x_0$ and

$$P(x) = ax^2 + bx + c = a(x - x_0)^2$$

Since the factor $(x - x_0)$ appears twice, x_0 is called a **double root** of $P(x)$. Higher order roots of a polynomial are defined similarly.

Fundamental Theorem of Algebra

Every second-degree polynomial having real coefficients has at least one real or complex root; in fact, it will have exactly two if we count double roots twice. A similar statement holds for every polynomial having real coefficients. Although we are primarily interested in polynomials having real coefficients, the results of this section also apply to polynomials having complex coefficients. The following theorem is stated without proof.

FUNDAMENTAL THEOREM OF ALGEBRA

> Every polynomial with real coefficients has at least one real or complex root.

The remainder and factor theorems hold for polynomials with complex roots as well as for those having real roots. Thus, if r_1 is a root of $P(x)$, then $P(r_1) = 0$, and by the factor theorem we can factor

$$P(x) = (x - r_1)Q(x)$$

If $P(x)$ has degree n, it should be clear that $Q(x)$ has degree $n - 1$. The fundamental theorem of algebra can then be applied to $Q(x)$ to slice off another factor $(x - r_2)$. Continuing in this way we eventually write

$$P(x) = (x - r_1)(x - r_2) \cdots (x - r_n)A$$

for precisely n factors. We stop only when $Q(x)$ has degree 0; that is $Q(x)$ is a constant. We thus make two observations.

> Let $P(x)$ be a polynomial with real coefficients.
>
> **1.** $P(x)$ can be factored completely into linear factors:
> $$P(x) = (x - r_1)(x - r_2) \cdots (x - r_n)A$$
> where A is real and $r_1, \cdots, r_n$ are real or complex.
>
> **2.** $P(x)$ has precisely n roots; some of these may be repeated roots.

Conjugate Roots

We should observe that if z is a root of $P(x)$, then its conjugate $\bar{z}$ is also a root. For if

$$a_n z^n + \cdots + a_0 = 0$$

then

$$\overline{a_n z^n + \cdots + a_0} = \bar{0} = 0$$

$$\overline{a_n z^n} + \cdots + \overline{a_0} = 0 \qquad \text{(The conjugate of a sum or product is the sum or product of the conjugates. See Exercises 58 and 59 in Section 1.7.)}$$

$$a_n \bar{z}^n + \cdots + a_0 = 0 \qquad (a_k = \bar{a}_k \text{ since } a_k \text{ is real.})$$

Hence, $\bar{z}$ is indeed a root. In this case, writing $z = a + bi$ and $\bar{z} = a - bi$, we have

$$
\begin{aligned}
P(x) &= (x - z)(x - \bar{z})Q(x) \\
&= (x - [a + bi])(x - [a - bi])Q(x) \\
&= (x^2 - [a + bi]x - [a - bi]x + [a + bi][a - bi])Q(x) \\
&= (x^2 - 2ax + [a^2 + b^2])Q(x)
\end{aligned}
$$

But $2a$ and $[a^2 + b^2]$ are real numbers. Thus, $P(x)$ has a real quadratic factor corresponding to the complex conjugate roots. We have obtained the following.

> The complex roots of a polynomial with real coefficients occur in conjugate pairs.
>
> Every polynomial with real coefficients can be factored into real linear and quadratic factors.

EXAMPLE 1　Find a polynomial of lowest degree with real coefficients and having roots 2, -1, $1 + i$, $2 - i$.

Solution　From the preceding observations, we know that there are at least two additional roots: $1 - i$ and $2 + i$. The factor theorem says that r is a root of $P(x)$ if and only if $(x - r)$ is a factor of $P(x)$. Thus, $(x - 2)$, $(x - [-1])$, $(x - [1 + i])$, $(x - [1 - i])$, $(x - [2 - i])$, $(x - [2 + i])$ are all factors of $P(x)$. Writing

$$
\begin{aligned}
P(x) &= (x - 2)(x + 1)(x - [1 + i])(x - [1 - i])(x - [2 - i])(x - [2 + i]) \\
&= (x - 2)(x + 1)(x^2 - 2x + 2)(x^2 - 4x + 5) \\
&= x^6 - 7x^5 + 19x^4 - 21x^3 - 2x^2 + 26x - 20
\end{aligned}
$$

This is the polynomial of lowest degree having the required roots since any polynomial having these given roots must have the indicated factors.　∎

　　The task of finding real roots of polynomials was addressed in the preceding section. In Section 2.2, we saw that the quadratic formula can be used to find the real or complex roots of quadratic polynomials. Since any polynomial can be factored into real linear and quadratic factors, we can theoretically find all roots of any real polynomial. However, it sometimes requires a certain amount of ingenuity to find the quadratic factors. These correspond to the complex roots of the polynomial.

EXAMPLE 2　Find all the roots of the following polynomial and factor $P(x)$ into real linear and quadratic factors.

$$
P(x) = 3x^6 + 5x^5 + x^4 + 5x^3 + x^2 + 5x - 2
$$

Solution　The only possible rational roots are ± 1, ± 2, $\pm 1/3$, and $\pm 2/3$. Clearly, $P(1) \neq 0$ and $P(-1) \neq 0$. Testing $x = -2$ by synthetic division, we find

$$\begin{array}{r} -2\overline{)3 \quad 5 \quad 1 \quad 5 \quad 1 \quad 5 \quad -2} \\ \underline{-6 \quad 2 \quad -6 \quad 2 \quad -6 \quad 2} \\ 3 \quad -1 \quad 3 \quad -1 \quad 3 \quad -1 \quad 0 = R \end{array}$$

Thus, -2 is a root and

$$P(x) = (x + 2)Q(x)$$

where $Q(x) = 3x^5 - x^4 + 3x^3 - x^2 + 3x - 1$. Any remaining roots of $P(x)$ must be roots of $Q(x)$. The only possible rational roots of $Q(x)$ are $\pm 1/3$ since ± 1 were previously eliminated. Testing $x = 1/3$ we find

$$\begin{array}{r} \tfrac{1}{3}\overline{)3 \quad -1 \quad 3 \quad -1 \quad 3 \quad -1} \\ \underline{1 \quad 0 \quad 1 \quad 0 \quad 1} \\ 3 \quad 0 \quad 3 \quad 0 \quad 3 \quad 0 = R \end{array}$$

Hence, $x = 1/3$ is a root and

$$\begin{aligned} P(x) &= (x + 2)\left(x - \frac{1}{3}\right)(3x^4 + 3x^2 + 3) \\ &= (x + 2)(3x - 1)(x^4 + x^2 + 1) \end{aligned}$$

Clearly, $x^4 + x^2 + 1$ has no real roots since it is a sum of even powers of x. To determine its quadratic factors, we complete the square and write

$$\begin{aligned} x^4 + x^2 + 1 &= (x^4 + 2x^2 + 1) - x^2 \\ &= (x^2 + 1)^2 - x^2 \\ &= (x^2 + 1 + x)(x^2 + 1 - x) \end{aligned}$$

Thus,

$$P(x) = (x + 2)(3x - 1)(x^2 + x + 1)(x^2 - x + 1)$$

The complex roots can now be obtained from the quadratic formula. They are

$$x = \frac{-1 \pm \sqrt{1 - 4}}{2} = \frac{-1 \pm \sqrt{3}i}{2}$$

and

$$x = \frac{1 \pm \sqrt{1 - 4}}{2} = \frac{1 \pm \sqrt{3}i}{2}$$

The complete list of six roots is

$$\begin{array}{ccc} -2 & \dfrac{1 + \sqrt{3}i}{2} & \dfrac{-1 + \sqrt{3}i}{2} \\[2ex] \dfrac{1}{3} & \dfrac{1 - \sqrt{3}i}{2} & \dfrac{-1 - \sqrt{3}i}{2} \end{array}$$

EXERCISES 5.3

Find the polynomial of lowest degree with real coeffi-
cients having the indicated roots.

1. $1, 2i$
2. $-1, -i$
3. $i - 1, i + 1$
4. $i, i - 2$
5. $1, 2, i$
6. $-2, 3, 1 + i$
7. $1, 2 - i, i$
8. $-1, i - 1, -i$
9. $2 + i, 1 + i, i$
10. $1 - i, 2, 2 + i$

Find all the roots of the following polynomials and fac-
tor them into real linear and quadratic factors.

11. $t^3 + 3t^2 + 25t + 75$
12. $u^3 - u^2 + 2$
13. $x^4 + 4$
14. $v^3 + 5v^2 + 4v - 10$
15. $2w^3 - w^2 + 2w - 1$

16. $x^4 + 16$
17. $x^3 - 1$
18. $x^3 - x^2 - 4x - 6$
19. $4x^3 - 2x + 2$
20. $w^4 - 3w^3 + 6w^2 - 12w + 8$, $2i$ is a root
21. $t^3 - t^2 - 4t - 6$, 3 is a root
22. $3u^4 + 16u^3 + 24u^2 - 44u - 39$, $2i - 3$ is a root
23. $y^5 - 2y^4 + 6y^3 + 24y^2 + 5y + 28$; $(2 - 3i)$ is a root.
24. $z^4 - 10z^3 + 35z^2 - 50z + 34$; $(4 + i)$ is a root.
25. $s^4 - 2s^3 + 4s^2 + 4s - 12$; $(1 + \sqrt{5}i)$ is a root.

5.4 GRAPHING POLYNOMIAL FUNCTIONS

A polynomial function
$$y = P(x) = a_n x^n + a_{n-1} x^{n-1} + \cdots + a_1 x + a_0$$
can be graphed by making a table of values and plotting lots of points. But it soon becomes evident that a small computer or at least a programmable calculator would be an immense help for generating the values in the table. Even without resorting to such laborious methods, we can sometimes ascertain the *qualitative* nature of the function and its graph. That is to say, by plotting only a few selected points on the graph, we can sometimes sketch a rough graph that will indicate the behavior of the function.

We have already seen the graphs of the simple linear and quadratic polynomials $y = x$ and $y = x^2$ as illustrated in Figures 5–3(a) and (b). It is a simple matter to graph $y = x^3$ and $y = x^4$ as in Figure 5–3(c) and (d). In fact, the graph of $y = x^n$ resembles that of Figure 5–3(c) for n odd and (d) for n even.

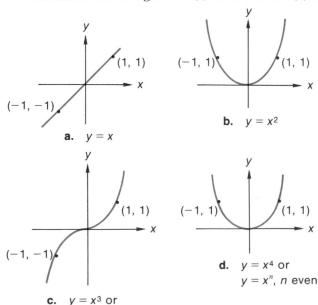

a. $y = x$

b. $y = x^2$

c. $y = x^3$ or $y = x^n$, n odd

d. $y = x^4$ or $y = x^n$, n even

Figure 5–3

Note that each of the graphs $y = x^n$ passes through $(0, 0)$ and $(1, 1)$. These graphs are symmetric about the y-axis or the origin depending on whether n is even or odd, respectively. Increasing n decreases the values of x^n for x between 0 and 1 and increases the values of x^n for $x > 1$ (Figure 5–4).

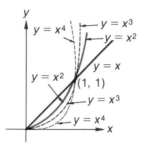

Figure 5–4

The graph of the polynomial

$$P(x) = x^2 - 2x + 2$$

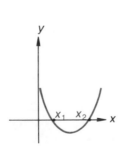

y = x² – 2x + 2

Figure 5–5

is given in Figure 5–5. Note that it does not meet the x-axis; $P(x) > 0$ for every real x. Finding the roots by the quadratic formula, we have

$$x = \frac{-(-2) \pm \sqrt{(-2)^2 - 4 \cdot 2}}{2}$$

$$= \frac{2 \pm \sqrt{-4}}{2} = \frac{2 \pm 2i}{2}$$

$$= 1 \pm i$$

The roots are complex. There could be no real roots since the graph does not cross the x-axis: $P(x) \neq 0$ for all real x. Any quadratic polynomial $P(x) = ax^2 + bx + c$ with real coefficients whose graph does not meet the x-axis will have only complex roots (see Fig. 5–6).

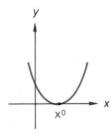

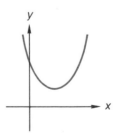

a. Two real roots: x_1 and x_2

b. One real root: x_0 (a double root) no complex roots

c. Two complex roots; no real roots

Figure 5–6 $y = ax^2 + bx + c, a \neq 0$

An effective tool in graphing a polynomial is the determination or estimation of its roots since the real roots are the x-intercepts of its graph. The Fundamental Theorem of Algebra tells us that a polynomial of degree n has exactly n roots, some of which may be complex roots and/or repeated roots. Thus, the graph of a polynomial of degree n can have at most n x-intercepts; these intercepts correspond to its distinct real roots. Plotting these intercepts and points between and beyond the intercepts can already give us enough information to sketch a rough graph of the polynomial.

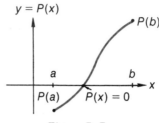

Figure 5–7

**Behavior Between
Roots**

We note in Figure 5–7 that if a polynomial $P(x)$ changes sign from $x = a$ to $x = b$, then there is a root or intercept between a and b. Thus, if all real roots are located, we know that $P(x)$ cannot change sign between two successive roots; for to do so would imply the existence of still another root. Hence, between successive real roots and beyond the largest and smallest real roots, the graph of the polynomial lies entirely above or below the x-axis. Evaluating the polynomial at some point in each region indicates whether the graph in that region lies above or below the x-axis. Synthetic division and the remainder theorem can be used to facilitate these evaluations.

EXAMPLE 1 Sketch the following graphs.

a. $y = P(x) = x^3 - 2x^2 - x + 2$

b. $y = -P(x)$

c. $y = P(x) - 2$

Solution a. The possible rational roots are ± 1 and ± 2. It is easy to show that $P(1) = P(-1) = P(2) = 0$. There can be only three roots for this cubic polynomial; thus $x = -1, 1, 2$ represents the complete set of roots. In the interval $(-1, 1)$, it is easy to evaluate $y = P(0) = 2$. Then $y > 0$ for all x between -1 and 1. Using synthetic division, we find $P(3/2) = -5/8$:

$$\frac{3}{2} \overline{)\begin{array}{rrr} 1 & -2 & -1 & 2 \\ & \frac{3}{2} & \frac{3}{4} & -\frac{21}{8} \\ \hline 1 & -\frac{1}{2} & -\frac{7}{4} & -\frac{5}{8} \end{array}} = R = P\left(\frac{3}{2}\right)$$

Thus $y < 0$ for x between 1 and 2. Similarly, we find $P(3) = 8$, $P(-2) = -12$ and observe that $P(x) > 0$ for $x > 2$ and $P(x) < 0$ for $x < -1$.

For relatively large values of x, we write

$$P(x) = x^3 - 2x^2 - x + 2$$
$$= x^3\left(1 - \frac{2}{x} - \frac{1}{x^2} + \frac{2}{x^3}\right)$$

Note that all terms in the parentheses except the first "1" become smaller and smaller as x gets larger in magnitude. Thus,

$$P(x) \approx x^3 \qquad \text{for } x \text{ having large magnitude.}$$

In particular, $P(x)$ gets large (positively or negatively) as x does (positively or negatively, respectively). We write

$$P(x) \to \infty \qquad \text{as } x \to \infty$$

and

$$P(x) \to -\infty \qquad \text{as } x \to -\infty.$$

We can sketch a rough graph of $y = P(x)$ as in Figure 5–8(a).

b. The graph of $y = -P(x)$ is just the reflection through the x-axis of the graph of $y = P(x)$ (see Fig. 5–8(b)).

c. The graph of $y = P(x) - 2$ is obtained by lowering the first graph two units; it is sketched in Figure 5–8(c). The x-intercepts can be found by observing that $P(x) - 2 = 0$ when

$$x^3 - 2x^2 - x = 0$$
$$x(x^2 - 2x - 1) = 0$$
$$x = 0, \; 1 \pm \sqrt{2} \qquad \text{(by the quadratic formula)} \qquad \blacksquare$$

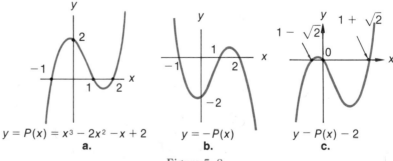

$y = P(x) = x^3 - 2x^2 - x + 2$ $y = -P(x)$ $y - P(x) - 2$
a.   **b.**   **c.**

Figure 5–8

The leading term $a_n x^n$ is important in graphing polynomials. As we saw in Example 1, the term $a_n x^n$ dominates the polynomial for large positive or negative values of x; the values of $a_n x^n$ tend to override any contribution by smaller powers of x. Thus, as $|x|$ gets large, $P(x)$ gets large and the sign of a_n determines whether these values are positive or negative. Specifically, for large positive x's, $P(x) > 0$ for $a_n > 0$ and $P(x) < 0$ if $a_n < 0$. For large negative x values, we simply note that for even n the extreme branches of the curve point in the same direction (up or down) whereas for odd n they point in opposite directions. For instance, if $a_n > 0$ and n is odd, then $a_n x^n < 0$ (and hence $P(x) < 0$) for large negative x.

Figure 5–8 illustrates the general shape of the graph of a third degree or "cubic" polynomial. Note that it has two "turnaround" points. These are called local maxima or minima. Precise location of such maximum and minimum points requires methods of calculus and will not be addressed here. If it is necessary to estimate the maxima and/or minima, we can do so by plotting more points. For instance, in Example 1(a), we note that $P(-0.1) = 2.079 > P(0)$. This indicates a maximum point in the interval $(-1, 0)$ since $P(-0.1) > P(-1)$ and $P(-0.1) > P(0)$.

A useful theorem concerning maximum and minimum points is stated here without proof. It can be proved quite easily by using methods of calculus.

THEOREM

A polynomial of degree n can have at most $n - 1$ local maxima and/or minima.

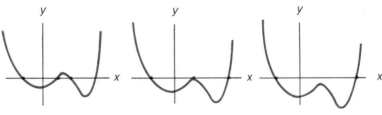

| a. | 4 distinct roots | b. | 3 distinct roots (one is a double root) | c. | 2 distinct real roots (2 complex roots) |

$$y = a_4 x^4 + a_3 x^3 + a_2 x^2 + a_1 x + a_0$$

At most 4 distinct roots;
at most 3 local maxima and/or minima

Figure 5–9

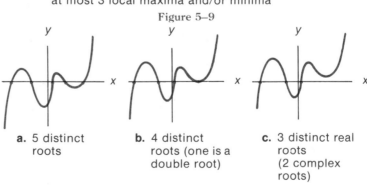

| a. 5 distinct roots | b. 4 distinct roots (one is a double root) | c. 3 distinct real roots (2 complex roots) |

$$y = a_5 x^5 + a_4 x^4 + a_3 x^3 + a_2 x^2 + a_1 x + a_0$$

At most 5 distinct roots;
at most 4 local maxima and/or minima

Figure 5–10

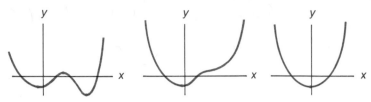

a. 3 maxima/minima   **b.** 1 minimum   **c.** 1 minimum

$$y = a_4x^4 + a_3x^3 + a_2x^2 + a_1x + a_0$$

Figure 5–11

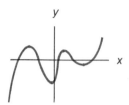 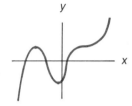

a. 4 maxima/minima   **b.** 2 maxima/minima

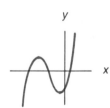 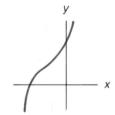

c. 2 maxima/minima   **d.** no maxima/minima

$$y = a_5x^5 + a_4x^4 + a_3x^3 + a_2x^2 + a_1x + a_0$$

Figure 5–12

This theorem gives us an insight into graphs of more general polynomials. For instance, typical fourth- and fifth-degree polynomials (quartics and quintics) are sketched in Figures 5–9 to 5–12.

EXAMPLE 2   Sketch the graph of

$$y = P(x) = 2x^4 + 3x^3 - 12x^2 - 7x + 6$$

Solution   The possible rational roots of $P(x)$ are ± 1, ± 2, ± 3, ± 6, $\pm 1/2$, $\pm 3/2$. It is easy to verify that $r = -1$ is a root. Thus, $(x + 1)$ is a factor of $P(x)$ and we can write

$$P(x) = (x + 1)Q(x)$$

We can find $Q(x)$ by synthetic division:

$$
\begin{array}{r|rrrrr}
-1) & 2 & 3 & -12 & -7 & 6 \\
& 0 & -2 & -1 & 13 & -6 \\
\hline
& 2 & 1 & -13 & 6 & 0
\end{array}
$$

Thus, $Q(x) = 2x^3 + x^2 - 13x + 6$ and

$$P(x) = (x + 1)(2x^3 + x^2 - 13x + 6)$$

The possible rational roots of $Q(x)$ are the same as for $P(x)$. It is easy to verify that $r = \pm 1$ are not roots but that $r = 2$ is a root of $Q(x)$. Thus, $(x - 2)$ is a factor of $Q(x)$; again, we use synthetic division to write

$$Q(x) = (x - 2)(2x^2 + 5x - 3)$$

Finally,

$$\begin{aligned}
P(x) &= (x + 1)Q(x) \\
&= (x + 1)(x - 2)(2x^2 + 5x - 3) \\
&= (x + 1)(x - 2)(2x - 1)(x + 3)
\end{aligned}$$

The roots of $P(x)$ are $x = -1, 2, 1/2, -3$. We evaluate $P(x)$ at points between successive real roots:

$$P(-2) = -20, \qquad P(0) = 6, \qquad P(1) = -8$$

Using a calculator or synthetic division, we can find several other $P(x)$ values if necessary. Since the leading term is $2x^4$, we note that $P(x) \to \infty$ as $x \to \pm\infty$. Finally, we sketch a rough graph as in Figure 5–13.

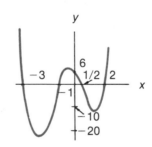

$$y = P(x) = 2x^4 + 3x^3 - 12x^2 - 7x + 6$$

Figure 5–13

EXERCISES 5.4

Sketch rough graphs of the following polynomials.

1. $y = x^4 - 1$
2. $y = x^4 + 1$
3. $y = 4 - x^4$
4. $y = x^3 - 8$
5. $y = 1 - x^3$
6. $y = (x - 1)^4$
7. $y = (x - 2)^4 - 3$
8. $y = 3 - (x + 1)^3$
9. $y = (x - 1)(x - 2)(x - 3)$
10. $y = (x + 1)(x - 1)(2x - 1)$
11. $y = (3x - 2)(1 - 2x)(2x + 4)$
12. $y = (4x + 2)(2 - 4x)(x - 2)$

13. $y = (x + 1)(x - 2)(x + 3)(x - 4)$
14. $y = (x - 1)(x + 2)(x - 3)(x + 1)^2$
15. $y = (x^2 - 1)(x^2 - 4)$
16. $y = (x^2 - 1)(x^2 - 4) - 4$
17. $y = (x - 1)^2(x - 2)^2$
18. $y = 4 - (x - 1)^2(x - 2)^2$
19. $y = x^3 - 4x$
20. $y = 4x^2 - x^3$
21. $y = x^3 - x^5$
22. $y = x^4 - 5x^2 + 4$

23. $y = x^3 - x^2 - 9x + 9$

24. $y = x^3 + 2x^2 - x - 2$

25. $y = 2x^4 + 4x^3 - 2x^2 - 4x$

26. $y = x^5 - x^4 - x^3 + x^2$

27. $y = x^4 + x^3 - 6x^2 - 4x + 8$

28. $y = x^4 - 5x^3 + 5x^2 + 5x - 6$

29. $y = x^4 + 7x^3 + 13x^2 - 3x - 18$

30. $y = x^5 + x^4 - 9x^3 - x^2 + 20x - 12$

5.5 RATIONAL FUNCTIONS

A function such as

$$R(x) = \frac{2x - 1}{x^2 + 3} \quad \text{or} \quad S(x) = \frac{x^{10} + 10x - 3}{2x + 1}$$

which can be expressed as a quotient of two polynomials is called a **rational function.** The roots or zeros of its numerator and denominator are useful in graphing a rational function. The roots of a rational function can occur only at the roots of the numerator. These determine the x-intercepts. The y-intercept is found by setting $x = 0$.

Roots of the denominator that are not also roots of the numerator are called **poles** of the rational function. The graph cannot cross a vertical line at a pole since the pole is not in the domain of the function. We shall see that the graph does get arbitrarily close to such vertical lines, however, and that $y \to \pm\infty$ as this approach takes place.

ASYMPTOTES

1. If $|f(x)| \to \infty$ as $x \to a$, the line $x = a$ is called a **vertical asymptote** of the graph.

2. If the points (x, y) on the graph get arbitrarily close to some line l as $x \to \pm\infty$, the line l is called an **asymptote** of the graph.

   a. If the line is horizontal, it is called a **horizontal asymptote.**

   b. If the line has slope $m \neq 0$, it is called a **slant asymptote.**

Vertical asymptotes occur at the poles of a rational function. Slant asymptotes were obtained with hyperbolas (Section 3.5).

EXAMPLE 1 Sketch the graph of the following rational functions.

a. $y = f(x) = \dfrac{1}{x}$

b. $y = g(x) = \dfrac{1}{x^2}$

c. $y = r(x) = \dfrac{1}{(x - 2)}$

d. $y = s(x) = \dfrac{1}{(x + 1)^2}$

Solution Note that (a) represents an odd function and (b) an even function. It is a simple matter to plot points and generate the graphs in Figures 5–14 (a) and (b); (a) is symmetric about the origin, and (b) is symmetric about the y-axis. Each has a pole at $x = 0$ and $y \to 0$ as $|x| \to \infty$ in each. Thus, we obtain the vertical and horizontal asymptotes indicated. The sketches in Figures 5–14 (c) and (d) are obtained by shifting the first two graphs horizontally.

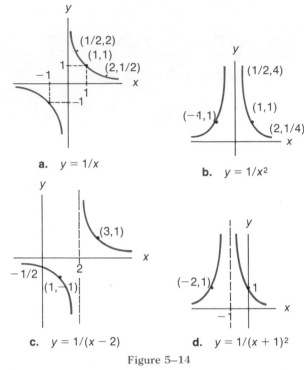

a. $y = 1/x$

b. $y = 1/x^2$

c. $y = 1/(x - 2)$

d. $y = 1/(x + 1)^2$

Figure 5–14

In describing asymptotes such as those that appear in Figure 5–14 (c), we write, for example, $y \to \infty$ as $x \to 2^+$; that is, y gets arbitrarily large as x approaches 2 from the right. Similarly, $y \to -\infty$ as $x \to 2^-$.

EXAMPLE 2 Sketch the graphs of the following rational functions.

a. $y = \dfrac{12}{(x^2 + 4)}$ **b.** $y = \dfrac{12}{(x^2 - 4)}$ **c.** $y = \dfrac{x}{(x^2 + 4)}$

Solution **a.** This is an even function; the graph will be symmetric about the y-axis. Clearly, $y = 3$ at $x = 0$ and $0 < y \le 3$. There is no x-intercept, but note that $y \to 0$ as $x \to \pm\infty$. The graph is sketched in Figure 5–15.

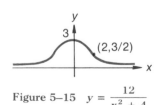

Figure 5–15 $y = \dfrac{12}{x^2 + 4}$

b. This graph too will be symmetric about the y-axis (another even function). Clearly, $y = -3$ when $x = 0$. Again, y is never 0 (no x-intercept) and $y \to 0$ as $|x| \to \infty$. Writing

$$y = \frac{12}{(x^2 - 4)} = \frac{12}{(x - 2)(x + 2)}$$

we see that poles occur at $x = \pm 2$. The behavior of y near these poles is analyzed in the following chart. For instance, if x is slightly larger than 2, then $x - 2$ is a small positive number and $\dfrac{1}{(x - 2)}$ is a large positive number. Then $y = \dfrac{1}{x - 2} \cdot \dfrac{12}{x + 2} \approx \dfrac{1}{x - 2} \cdot 3$; that is, y is large and positive. The graph is sketched in Figure 5–16.

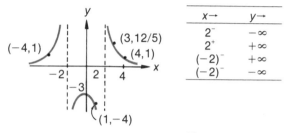

$x \to$	$y \to$
2^-	$-\infty$
2^+	$+\infty$
$(-2)^-$	$+\infty$
$(-2)^-$	$-\infty$

Figure 5–16 $y = \dfrac{12}{x^2 - 4}$

c. This is an odd function; the graph will be symmetric about the origin. Clearly, $y = 0$ when $x = 0$. There are no other intercepts. There are no poles and hence no vertical asymptotes. To analyze the behavior of the graph as x gets large, we write

$$y = \frac{x}{x^2 + 4} = \frac{1}{x + \dfrac{4}{x}} \qquad \text{(Divide numerator and denominator by } x.)$$

For x large, $4/x$ becomes negligible in comparison to x so $y \approx 1/x$; that is, $y \to 0$ as $x \to \infty$.

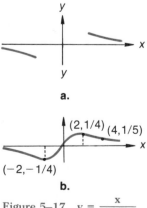

a.

b.

Figure 5–17 $y = \dfrac{x}{x^2 + 4}$

This enables us to sketch a partial graph as indicated in Figure 5–17(a). The graph can be completed by plotting several intermediate points (see Fig. 5–17(b)). Determination of the exact location of the local maxima/minima requires methods of calculus. ■

It is not always the case that a root of the denominator indicates a vertical asymptote, as we see in the following example. In this case the root of the denominator is not a pole.

EXAMPLE 3   Sketch the graph of the rational function

$$r(x) = \frac{x^2 - 5x + 6}{x - 2}$$

Solution   Write

$$r(x) = \frac{x^2 - 5x + 6}{x - 2}$$
$$= \frac{(x - 2)(x - 3)}{(x - 2)}$$
$$= x - 3$$

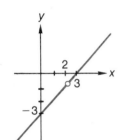

Its graph is the straight line $y = x - 3$ with one point deleted since $x = 2$ is not in the domain of the function (see Fig. 5–18). ■

The following steps should be helpful in graphing rational functions.

Figure 5–18
$$y = \frac{x^2 - 5x + 6}{x - 2}$$

GRAPHING RATIONAL FUNCTIONS

1.   Check for symmetry (even or odd function).

2.   Determine x- and y-intercepts by finding the roots of the numerator and by evaluating the function at $x = 0$.

3.   Find horizontal and slant asymptotes by analyzing the behavior of y as $x \to \pm\infty$.

4.   Find poles by factoring the denominator. Vertical asymptotes can occur only at the roots of the denominator but such roots do not always yield a vertical asymptote.

5.   Determine if (and approximately where) the graph crosses the asymptotes. The techniques for finding zeros of polynomials may be useful here.

6.   Use a calculator to plot additional points if necessary. Then, connect the dots and pieces to sketch a rough graph. Precisely locating "turnaround" points and other important points requires calculus and need not be done at this stage.

To determine the behavior of the graph near the roots of the denominator of a rational function and for $|x| \to \infty$, it is generally best to express the rational function in terms of a **proper fraction,** one in which the degree of the numerator is less than the degree of the denominator. Long division of polynomials or synthetic division can be used if necessary.

EXAMPLE 4   Sketch the graph of the following rational function.

$$y = \frac{x^2 - 5}{x^2 - 2x - 3}$$

Solution   1.   No symmetry is immediately obvious this time.

2.   The x and y intercepts are at $x = \pm\sqrt{5}$ and $y = 5/3$.

3.   Note that as $|x|$ gets large, we have

$$y = \frac{x^2 - 5}{x^2 - 2x - 3} = \frac{1 - 5/x^2}{1 - \dfrac{2}{x} - \dfrac{3}{x^2}} \approx 1$$

(Divide numerator and denominator by x^2.)

that is, $y = 1$ is a horizontal asymptote.

4.   Now write

$$\frac{x^2 - 5}{x^2 - 2x - 3} = \frac{(x^2 - 2x - 3) + 2x - 2}{x^2 - 2x - 3}$$

$$= 1 + \frac{2x - 2}{x^2 - 2x - 3}$$

$$= 1 + \frac{2(x - 1)}{(x - 3)(x + 1)}$$

(This could also have been obtained by long division.)

Vertical asymptotes are located at $x = 3$ and at $x = -1$. We consider the signs of the factors $(x - 1)$, $(x - 3)$, and $(x + 1)$ to develop the chart concerning the behavior near the vertical asymptotes.

5.   To find where the graph crosses the horizontal asymptote, we solve

$$y = \frac{x^2 - 5}{x^2 - 2x - 3} = 1$$

$$x^2 - 5 = x^2 - 2x - 3$$

$$2x = 2$$

$$x = 1$$

we see that $y = 1$ when $x = 1$; the graph crosses the horizontal asymptote at only one point $x = 1$.

6.   A calculator can be used to plot several additional points if desired. We thus obtain the graph sketched in Figure 5–19.

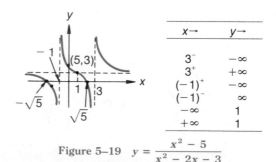

$x \rightarrow$	$y \rightarrow$
3^-	$-\infty$
3^+	$+\infty$
$(-1)^+$	$-\infty$
$(-1)^-$	∞
$-\infty$	1
$+\infty$	1

Figure 5–19 $y = \dfrac{x^2 - 5}{x^2 - 2x - 3}$

EXAMPLE 5 Sketch the graph of the following rational function.

$$y = \frac{x^2 - x - 1}{x - 2}$$

Solution 1. Again, there is no obvious symmetry.

2. To find the y intercept, we set $x = 0$; then $y = 1/2$. The quadratic formula can be used to find the zeros of the numerator of the rational function; these are the x intercepts:

$$x = \frac{1 \pm \sqrt{1 + 4}}{2} = \frac{1 \pm \sqrt{5}}{2}$$

3. We can use long division or synthetic division to write

$$y = \frac{x^2 - x - 1}{x - 2} = (x + 1) + \frac{1}{(x - 2)}.$$

As $|x| \to \infty$ the term $\dfrac{1}{(x - 2)}$ becomes negligible so $y \approx (x + 1)$ for large $|x|$; the line $y = x + 1$ is an asymptote. If $x > 2$, then $y > x + 1$, and if $x < 2$, then $y < x + 1$.

4. Near $x = 2$, the term $(x + 1)$ is negligible in comparison to $\dfrac{1}{(x - 2)}$; $x = 2$ is a vertical asymptote. Clearly, $y \to -\infty$ as $x \to 2^-$ and $y \to \infty$ as $x \to 2^+$.

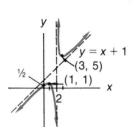

Figure 5–20
$y = \dfrac{x^2 - x - 1}{x - 2}$

5. **and 6.** The graph is sketched in Figure 5–20. It never crosses the asymptotes. ∎

In Example 5 with the degree of the numerator exceeding the degree of the denominator by 1, y is asymptotic to a linear function; $y \approx x + 1$ as $x \to \pm\infty$. Similarly, if the degree of the numerator exceeds the degree of the denominator by 2, y will be asymptotic to a quadratic function: $y \approx ax^2 + bx + c$ as $x \to \pm\infty$, and so on.

EXERCISES 5.5

Sketch the graph of each of the following rational functions.

1. $y = \dfrac{1}{x - 3}$

2. $y = \dfrac{1}{(x - 3)^2}$

3. $y = \dfrac{-1}{x + 4}$

4. $y = \dfrac{2}{1 - x}$

5. $x = \dfrac{2}{1 - y}$

6. $x = \dfrac{2}{(1 - y)^4}$

7. $y = \dfrac{2}{(x - 1)^3}$

8. $y = \dfrac{4}{2x - 1}$

9. $y = \dfrac{3}{x^2 + 1}$

10. $y = \dfrac{5}{x^2 - 4x - 5}$

11. $y = \dfrac{3}{x^2 - 1}$

12. $x = \dfrac{2}{y^4 + 1}$

13. $x = \dfrac{2}{y^3 + 1}$

14. $y = \dfrac{x}{x^2 + 3}$

15. $y = \dfrac{2x}{4 - x^2}$

16. $y = \dfrac{2x + 4}{x^2 + 1}$

17. $y = \dfrac{x - 1}{2x^2 + x + 1}$

18. $x = \dfrac{3y + 6}{y^2 + 3y + 2}$

19. $y = \dfrac{x^2 - x - 2}{x + 1}$

20. $y = \dfrac{x^2 + 2x - 3}{x - 1}$

21. $x = \dfrac{y^2 - 3y + 2}{y - 2}$

22. $y = \dfrac{x - 4}{2x - 2}$

23. $y = \dfrac{4x + 2}{x}$ **24.** $y = \dfrac{2x}{x + 1}$

25. $y = \dfrac{x^2 - 1}{x^2 + 1}$ **26.** $y = \dfrac{x^2 - 3x + 2}{x^2 - 4}$

27. $y = \dfrac{x^2}{x^2 + x - 6}$ **28.** $y = \dfrac{x^3}{x^2 - 1}$

29. $y = \dfrac{x^2 + 2}{x - 1}$ **30.** $y = \dfrac{x^2 - x - 6}{x}$

31. $x = \dfrac{y^2 - 4y + 4}{y - 1}$ **32.** $y = \dfrac{x^3}{x + 1}$

33. $y = \dfrac{x^3 - 2}{x + 1}$ **34.** $y = \dfrac{x^3 - 4x}{x - 1}$

35. $y = \dfrac{x^3 - 2x + 1}{x - 1}$ **36.** $y = \dfrac{x^2 - 3x + 2}{x^2 - x}$

37. $y = \dfrac{x^2 - x - 2}{9x - x^3}$ **38.** $y = \dfrac{x^2 - x}{x^3 + x^2 - 2x}$

39. $y = \dfrac{x^2}{x^2 - 7x + 12}$ **40.** $y = \dfrac{x^2 - 1}{x^2 - 4}$

5.6 CHAPTER REVIEW

TERMS AND CONCEPTS

- Polynomial Function
  $P(x) = a_n x^n + \cdots + a_1 x + a_0$
- Rational Function

  A quotient of polynomials: $R(x) = \dfrac{P(x)}{Q(x)}$
- Root
  r is a root of $P(x)$ if $P(r) = 0$
- Asymptotes
  1. vertical
  2. horizontal
  3. slant

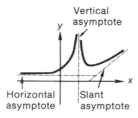

$|f(x)| \to \infty$ as $x \to a$

$f(x) \to A$ as $x \to \pm\infty$

Points on graph approach a line as $x \to \pm\infty$

RULES AND FORMULAS

- Remainder Theorem
  If a polynomial is divided by $(x - a)$ where a is a constant, the remainder is the value of the polynomial at $x = a$.
- Factor Theorem
  $(x - a)$ is a factor of a polynomial $P(x)$ if and only if $P(a) = 0$.
- Rational Root Theorem
  If a rational number k/m in lowest terms is a root of a polynomial $a_n x^n + \cdots + a_1 x + a_0$ with integer coefficients, then k is a factor of a_0 and m is a factor of a_n.
- Fundamental Theorem of Algebra
  Let $P(x)$ be a polynomial of degree n with real coefficients.
  1. $P(x)$ can be factored completely into linear factors:

  $$P(x) = (x - r_1)(x - r_2) \cdots (x - r_n)A$$

  where A is real and $r_1, \ldots, r_n$ are real or complex.
  2. $P(x)$ has precisely n roots; some of these may be repeated roots.
  3. The complex roots occur in conjugate pairs.
  4. $P(x)$ can be factored into real linear and quadratic factors.

- Descartes' Rule of Signs
  1. The number of positive real roots of a polynomial $P(x)$ is less than or equal to the number of sign changes in $P(x)$.
  2. The number of negative real roots is limited in the same way by the number of sign changes in $P(-x)$.
  3. In either case, if the number of roots is not equal to the number of sign changes, it is less than this by an even number.

TECHNIQUES

- Synthetic Division
  A short way of writing the steps in the long division of a polynomial by $x - a$.
- Range Limitation Test
  1. If $b > 0$ and the third row of the synthetic division of $P(x)$ by $(x - b)$ contains no negative entry, then all real roots of $P(x)$ are less than or equal to b.
  2. If $a < 0$ and the third row of the synthetic division of $P(-x)$ by $(x - |a|)$ contains no negative entry, then all real roots of $P(x)$ are greater than or equal to a.
- Successive Approximation of Roots
  1. Find two values a, b for which $P(a)$ and $P(b)$ differ in sign; there must be a root between a and b.
  2. Break down the interval already obtained to isolate a root in a narrower interval.
  3. Continue the process of refinement until a sufficient degree of accuracy is achieved.
- Graphing Polynomial Functions
  1. A polynomial of degree n can have at most
     a. n real roots; these correspond to the x-intercepts on its graph.
     b. $n - 1$ local maxima and/or minima; these represent "turnaround" points on its graph.
  2. Locate the roots of the polynomial by using the factor theorem, the rational root theorem, Descartes' rule of signs, the range limitation test, and successive approximation, if necessary.
  3. Evaluate the polynomial between two successive roots to determine whether the graph lies above or below the x-axis in that interval.
  4. Synthetic division can be used to facilitate this evaluation and the location of roots.
- Graphing Rational Functions
  1. Check for symmetry.
  2. Find x- and y-intercepts.
  3. Find horizontal and slant asymptotes; let $x \to \pm\infty$
  4. Factor the denominator to find possible vertical asymptotes.
  5. Determine if and where the graph crosses the asymptotes.
  6. Connect the pieces and dots to sketch a rough graph.

5.7 SUPPLEMENTARY EXERCISES

Use synthetic division to verify the validity of the remainder theorem in each of the following.

1. $(v^4 - 10v^2 + 25v - 2) \div (v + 4)$
2. $(5w^6 - 7w^3 - 6w + 8) \div (w - 1)$
3. $(2s^5 - s^3 + 3s - 2) \div (s + 1)$
4. $(3u^4 + 10u^3 - 4u + 5) \div (u + 4)$

Use the Factor Theorem to determine whether the second expression is a factor of the first; if so, use synthetic division to factor out the second expression.

5. $r^4 + 2r^2 - 3r + 5$, $(r - 3)$

6. $2t^5 - 5t^2 - t + 6$, $(t + 1)$
7. $s^5 + 4s^4 + s^3 + s - 1$, $(s + 1)$
8. $w^3 - 3w^2 + w + 1$, $(w - 2)$

Find all rational roots of the following.

9. $3x^4 - 5x^3 + 4x^2 - 10x - 4$
10. $2y^4 + y^3 - 3y^2 + 2y - 2$
11. $u^4 - u^3 - u^2 + 3u - 6$
12. $12v^3 + 20v^2 + v - 3$

Use Descartes' Rule of Signs to indicate the possible numbers of positive and negative roots of the following.

13. $r^3 - 6r^2 + 11r - 6$

14. $3s^3 + 2s^2 - 7s + 6$

15. $t^5 - 3t^4 + 4t^3 - 4t^2 + 3t - 1$

16. $u^5 + u^4 - u^3 - u^2 + u + 5$

Use the range limitation test to find the smallest interval with integer endpoints containing all real roots of the following.

17. $2u^4 + 3u^3 + 2u^2 - u + 1$

18. $w^4 - 4w^3 + 4w + 4$

19. $2x^5 - 5x^4 - 2x + 5$

20. $t^7 + 10t^4 + 4t^3 - t - 3$

Estimate a real root of the following to within the nearest tenth and within the indicated interval.

21. $r^3 - 2r^2 - r + 1$, $(-1, 0)$

22. $u^3 - 5u^2 + 4u - 5$, $(2, 5)$

23. $2s^4 + s^3 + 4s^2 - 6s - 4$, $(1, 2)$

24. $t^5 - t^4 + t^3 + 2t^2 - 2t + 2$, $(-1, -2)$

Find all real roots of the following; estimate all irrational roots to the nearest tenth.

25. $x^3 + 2x^2 - x - 1$

26. $t^4 - t^3 - 5t^2 + 3t + 6$

27. $3u^3 + u^2 - 8u + 2$

28. $v^4 + 5v^3 + 1$

Find the polynomial of lowest degree with real coefficients having the indicted roots.

29. $-2, 3 - i$ **30.** $(1 + i)(2 - i)$

31. $1, -1, 1 - i$ **32.** $i, 1 - i, 0, 1$

Find all real and complex roots of the following.

33. $z^3 - 2z^2 + 4z - 8$

34. $2x^4 - 7x^3 + 11x^2 - 8x + 2$

35. $t^4 - t^3 + t^2 - t$

36. $2s^3 + s^2 + 2s + 1$; i is a root

37. $w^4 + 3w^3 + 3w^2 - 2$; $(-1 - i)$ is a root

Factor the following by finding the roots and using the factor theorem.

38. $x^3 - 7x + 6$

39. $2w^3 - 12w^2 + 13w + 12$

40. $u^4 - 2u^3 + u - 2$

41. $v^4 + 2v^3 + v^2 - 2v - 2$

Sketch a rough graph of the following polynomials.

42. $y = x^6 - 1$

43. $y = 16 - (x - 1)^4$

44. $y = x^3 + 3x^2 - x - 3$

45. $y = (x + 1)(x - 2)(x + 4)(x - 5)$

46. $y = x^4 + x^3 - 3x^2 - x + 2$

47. $y = 64 - x^6$

48. $y = 4x^2 - x^4$

49. $y = (x - 2)(x + 5)(x - 3)$

50. $y = 16 - (x^2 - 4)(x - 4)$

51. $y = x^4 - 13x^2 + 36$

Sketch the graph of each of the following functions.

52. $y = \dfrac{-2}{x + 3}$

53. $y = \dfrac{2x}{x^2 + 2x + 1}$

54. $y = \dfrac{2x^2}{x^2 + 2x + 1}$

55. $x = \dfrac{y^2 + 5y + 4}{y + 2}$

56. $y = \dfrac{4}{x^2 - 5x + 6}$

57. $y = \dfrac{x^2 + x - 6}{x + 3}$

58. $y = \dfrac{x^2}{x + 1}$

6

Exponential and Logarithmic Functions

In many natural phenomena, the rate at which something grows or decays depends on the amount of it that is present. The secretion of medications from the bloodstream by the kidneys is an example of this behavior. A typical situation is for half the medication to be removed every 4 hours (approximately). Thus if an initial dose of 400 milligrams (mg) of the medication is given at noon, then there are

$$400 \cdot \left(\frac{1}{2}\right) = 200 \text{ mg remaining at 4 P.M.}$$

$$400 \cdot \left(\frac{1}{2}\right) \cdot \left(\frac{1}{2}\right) = 100 \text{ mg remaining at 8 P.M.}$$

$$400 \cdot \left(\frac{1}{2}\right) \cdot \left(\frac{1}{2}\right) \cdot \left(\frac{1}{2}\right) = 50 \text{ mg remaining at midnight}$$

and so on. We see that after n of these 4-hour periods, the amount of medication remaining in the bloodstream is given by $A(n) = 400 \cdot (1/2)^n$. But this secretion does not occur suddenly every 4 hours on the hour. Rather, it is a *continuous* process. Thus, after 3 1/2 of these four-hour periods (2 A.M.), the amount of medication remaining should be

$$A\left(3\frac{1}{2}\right) = 400 \cdot \left(\frac{1}{2}\right)^{3\frac{1}{2}} \approx 35.4 \text{ mg}$$

In fact, the amount of medication remaining in the bloodstream after t 4-hour periods have elapsed is expressed as

$$A(t) = 400 \cdot \left(\frac{1}{2}\right)^t$$

for *any* value of *t*, not just for integer or rational values.

This kind of function is called an **exponential function** and has many practical applications. Exponential functions are used in biology (population growth), physics (radioactive decay), and in modern banking for computations regarding continuously compounded interest. In this chapter we develop and study these exponential functions and their inverses, the **logarithmic functions**.

Logarithms were developed as a computational tool. They can be used to transform a multiplication or division problem into a simpler exercise in addition or subtraction. Although these computations are now performed by inexpensive, sophisticated hand calculators, logarithmic functions are still important and useful by virtue of their relationship to the exponential functions.

6.1 EXPONENTIAL FUNCTIONS

In Chapter 1 we introduced integer exponents by defining $b^n = b \cdot b \cdots b$ (n times), $b^{-n} = 1/b^n$, and $b^0 = 1$. Rational exponents were introduced by defining $b^{1/n} = \sqrt[n]{b}$ and $b^{m/n} = (\sqrt[n]{b})^m$. As long as $\sqrt[n]{b}$ is a real number, the usual properties of exponents are then satisfied. Recall that $\sqrt[n]{b}$ is not a real number if n is even and $b < 0$; for example, $\sqrt{-4} = 2i$.

In this section we introduce numbers of the form b^x, where b is a positive real number and x is an arbitrary real number. Only positive b's are considered since, for example, $b^{1/2} = \sqrt{b}$ is not a real number when $b < 0$. We shall give meaning to numbers of the form $2^{\sqrt{3}}$, 3^π, $2^{-\sqrt{3}}$, and so on but shall not consider numbers of the form $(-2)^{\sqrt{3}}$.

A completely satisfactory introduction of these ideas requires a certain amount of calculus. The concepts of "limit" and "continuity," which are quite simply the idea of "getting better and better approximations to something," play a central role.

To estimate $10^{\sqrt{2}}$, we can write $\sqrt{2}$ in terms of its decimal expansion to as many places as desired: $\sqrt{2} = 1.414213562 \ldots$ Whatever $10^{\sqrt{2}}$ is, it should be between $10^{1.4}$ and $10^{1.5}$, between $10^{1.41}$ and $10^{1.42}$, between $10^{1.414}$ and $10^{1.415}$, $\ldots$, between $10^{1.414213}$ and $10^{1.414214}$, $\ldots$ A hand calculator yields the following estimates.

$25.1 \ldots$	$\approx 10^{1.4}$	$< 10^{\sqrt{2}} < 10^{1.5}$	$\approx 31.6 \ldots$
$25.7 \ldots$	$\approx 10^{1.41}$	$< 10^{\sqrt{2}} < 10^{1.42}$	$\approx 26.3 \ldots$
$25.94 \ldots$	$\approx 10^{1.414}$	$< 10^{\sqrt{2}} < 10^{1.415}$	$\approx 26.001 \ldots$

$$25.95452\ldots = 10^{1.414213} < 10^{\sqrt{2}} < 10^{1.414214} = 25.95458\ldots$$

It should be evident that $10^{\sqrt{2}}$ can be obtained to within any desired degree of accuracy by using sufficiently accurate rational approximations of $\sqrt{2}$. We can write $10^{\sqrt{2}} \approx 25.955$ and be assured of three-place accuracy.

For any $b > 0$ and irrational x, b^x is approximated by numbers of the form b^r, where r is a rational number approximating x. This means that if we plot the points b^r for rational r, then b^x is "where it should be" on the graph; it is the "limit" of numbers b^r.

With this interpretation of b^x, the earlier rules of exponents can be extended to include arbitrary real exponents.

LAWS OF EXPONENTS

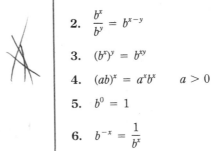

For $b > 0$, x and y real numbers

1. $b^x b^y = b^{x+y}$

2. $\dfrac{b^x}{b^y} = b^{x-y}$

3. $(b^x)^y = b^{xy}$

4. $(ab)^x = a^x b^x \qquad a > 0$

5. $b^0 = 1$

6. $b^{-x} = \dfrac{1}{b^x}$

For $b > 0$ and r an integer or rational number, $b^r > 0$. Thus $b^x > 0$ for all x. If $b > 1$, then b, b^2, b^3, . . . are increasing in magnitude; in fact, $b^x \to \infty$ as $x \to \infty$. But then $b^{-u} = 1/b^u \to 0$ as $u \to \infty$; that is, $b^x \to 0$ as $x \to -\infty$. If $0 < b < 1$ we can write $a = 1/b > 1$. Then

$$b^{-t} = \frac{1}{b^t} = \left(\frac{1}{b}\right)^t = a^t \to \begin{cases} \infty \text{ as } t \to \infty \\ 0 \text{ as } t \to -\infty \end{cases}$$

Thus $b^x \to \infty$ as $x \to -\infty$ and $b^x \to 0$ as $x \to \infty$ when $0 < b < 1$.

For $b > 0$

1. $b^x > 0$ for all real x

2. If $b > 1$, then $b^x \to \begin{cases} \infty \text{ as } x \to \infty \\ 0 \text{ as } x \to -\infty \end{cases}$

3. If $0 < b < 1$, then $b^x \to \begin{cases} 0 \text{ as } x \to \infty \\ \infty \text{ as } x \to -\infty \end{cases}$

DEFINITION

If $b > 0$, $b \neq 1$, the function f defined by

$$f(x) = b^x \qquad x \text{ a real number}$$

is called an **exponential function** with **base** b.

The base $b = 1$ was excluded from the definition of an exponential function because it would simply yield the constant function $1^x \equiv 1$.

EXAMPLE 1 Graph the following exponential functions

a. $f(x) = 2^x$ b. $g(x) = \left(\dfrac{1}{2}\right)^x$

Solution Make a table of values, plot several points, and then "connect the dots" to sketch the graph as illustrated in Figure 6–1. Note that the graph in (a) "flattens" as x moves to the left and rises ever more rapidly as x moves to the right. The reverse holds true for (b).

x	2^x	$(1/2)^x$
0	1	1
1	2	1/2
2	4	1/4
3	8	1/8
4	16	1/16
-1	1/2	2
-2	1/4	4
-3	1/8	8
-4	1/16	16

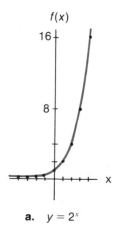

a. $y = 2^x$

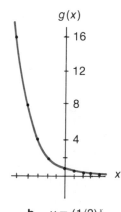

b. $y = (1/2)^x$

Figure 6–1

In the preceding example,

$$g(x) = \left(\frac{1}{2}\right)^x = \frac{1}{2^x} = 2^{-x} = f(-x)$$

Thus the graph of $y = (1/2)^x$ could be obtained by reflecting the graph of $y = 2^x$ through the x-axis.

The graphs of the exponential functions $y = b^x$ all pass through $(0, 1)$ and have the general shape that is indicated in Figure 6–2.

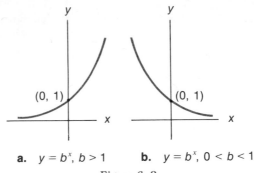

a. $y = b^x, b > 1$ **b.** $y = b^x, 0 < b < 1$

Figure 6–2

EXAMPLE 2 Sketch the following graphs.

     a. $y = 2^{|x|}$

     b. $y = 2^{1-x}$

Solution **a.** Note that $y = 2^{|x|} = \begin{cases} 2^x, & x \geq 0 \\ 2^{-x} = \left(\dfrac{1}{2}\right)^x, & x \leq 0 \end{cases}$

Thus this graph is obtained by using the right half ($x \geq 0$) of Figure 6–1(a) and the left half ($x \leq 0$) of Figure 6–1(b). Or we could observe that this is an even function so that the graph must be symmetric about the y-axis. In any case, it is sketched as Figure 6–3(a).

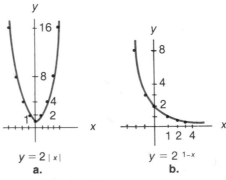

$y = 2^{|x|}$
   a.

$y = 2^{1-x}$
   b.

Figure 6–3

b. Write

$$y = 2^{1-x}$$
$$= \frac{1}{2^{x-1}}$$
$$= \left(\frac{1}{2}\right)^{x-1}$$
$$= g(x - 1)$$

where g is the function of Example 1(b). We obtain this graph by shifting the graph of Example 1(b) one unit to the right (see Fig. 6–3(b)).

The graphs of the exponential functions as illustrated in Figure 6–2 indicate that an exponential function with base b is increasing if $b > 1$ and decreasing if $b < 1$. (Recall a function is *increasing* if $f(x_2) > f(x_1)$ when $x_2 > x_1$; *decreasing* is defined in a similar fashion.) In particular, this means that if $0 < b \neq 1$ and $r \neq s$, then $b^r \neq b^s$. Alternatively, for $0 < b \neq 1$.

$$\text{if } b^r = b^s, \qquad \text{then} \qquad r = s$$

This property enables us to solve for x in equations that include x in the exponent. To solve any such equation, however, requires expressing both sides of the equation in terms of the same base. Later in the chapter we shall learn additional techniques to be used when no such common base is apparent.

EXAMPLE 3 Solve each of the following.

    a. $3^x = 81$

    b. $2^{3(1+x)} = 64$

    c. $4^{2-v} = \dfrac{1}{8^{4/3}}$

Solution **a.** Since $81 = 3^4$, we write

$$3^x = 3^4$$

Thus

$$x = 4$$

    b. Since $64 = 2^6$, we have

$$2^{3(1+x)} = 2^6$$

Thus

$$3(1 + x) = 6$$
$$1 + x = 2$$
$$x = 1$$

c. Both 4 and 8 are powers of 2; thus we rewrite the equation as follows:

$$4^{2-v} = \frac{1}{8^{4/3}}$$

$$(2^2)^{(2-v)} = \frac{1}{(2^3)^{4/3}}$$

$$2^{4-2v} = \frac{1}{2^4} = 2^{-4}$$

$$4 - 2v = -4$$

$$-2v = -8$$

$$v = 4$$

Functions of the form $g(x) = b^{-x}$ can be used to describe situations in which there is a rapid initial decrease that tends to level off. This situation is illustrated in the following example.

EXAMPLE 4 When an object is immersed in a temperature-controlled environment, it tends to take on the temperature of that environment. Specifically, Newton's Law of Cooling states that the temperature $T(t)$ of the object at time t after immersion is given by

$$T(t) = T_0 + (T_1 - T_0)b^{-at}$$

where T_0 is the temperature of the surrounding medium and T_1 is the initial temperature of the object; b and a are determined by properties of the object and its environment. Graph the temperature as a function of time for a piece of 340°F steel that is placed in 40°F water; the constants are $b = 2$, $a = 3$. Use the graph to estimate the time required to reduce the temperature of the steel to 150°F.

Solution The temperature function is given by

$$T(t) = 40° + (340° - 40°) \cdot 2^{-3t}$$
$$= 40 + 300 \cdot (2^{-3t})$$

The graph is best obtained in stages. First, we graph $y = 2^{-3t}$. This graph is similar to Figure 6–1(b) except that the horizontal dimension is "compressed"; at $t = 1$ we plot $2^{-3} = g(3)$, at $t = 2$ we plot $2^{-6} = g(6)$, and so on as in Figure 6–4(a). Next, we plot $y = 300 \cdot (2^{-3t})$ by multiplying each y-coordinate of Figure 6–4(a) by 300; see Figure 6–4(b). Finally, we raise the graph 40 units to obtain $y = 40 + 300 \cdot (2^{-3t})$ as in Figure 6–4(c). This graph is sketched only for $t \geq 0$ since the object is presumed to be immersed at $t = 0$. The graph should indicate that even before $t = 1/2$, the temperature is reduced to 150°F. In fact, for $t = 1/2$,

$$T\left(\frac{1}{2}\right) = 40 + 300(2^{-3/2})$$
$$= 40 + 300/2\sqrt{2}$$
$$\approx 146°F \quad \text{(Use a calculator.)}$$

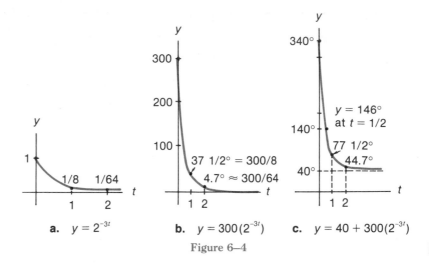

a. $y = 2^{-3t}$ **b.** $y = 300(2^{-3t})$ **c.** $y = 40 + 300(2^{-3t})$

Figure 6–4

In calculus and in many of the applications of mathematics, it is convenient to use a certain irrational number as a base. This number is

$$e = 2.71828 \ldots$$

The number e occurs naturally in the study of continuous growth processes such as bacteria growth (or growth of other populations), radioactive decay, and even continuously compounded interest. A discussion of compound interest and its relationship to the number e will be given in Section 6.5. The function $y = e^x$ is called the **natural exponential function** and e is called the **natural expo-**

HISTORICAL PERSPECTIVE

The symbol e was introduced by Leonhard Euler (1707–1783). Euler was one of the greatest mathematicians of his time and the most prolific mathematician of *all time*. His name occurs in every branch of mathematics. His life's work in mathematics comprises over 100 large volumes. In addition to the symbol e, he introduced the functional notation $f(x)$ that we use today. At an early age, he lost the sight of one eye, and 17 years before his death he became totally blind. Even so, he remained lively and cheerful and devoted to his family of 13 children. His powers of concentration were so great that even after going blind he produced volumes of important mathematics and could still perform mental calculations accurately to 15 figures.

CALCULATOR COMMENTS

Most calculators with a "power" key also have an $\boxed{e^x}$ key or an $\boxed{\exp}$ key. To find e^3 on such a calculator

1. Punch $\boxed{3}$.

2. Punch $\boxed{e^x}$ or $\boxed{\exp}$.

3. Read the result $e^3 \approx \boxed{20.08553692}$ . . .

Some calculators also have a $\boxed{10^x}$ key, which is used in the same way for base 10. For all other bases, the $\boxed{y^x}$ key must be used; note that it operates differently. You must

1. Enter the base y.

2. Punch $\boxed{y^x}$.

3. Enter the exponent x.

4. Punch $\boxed{=}$ and read the result.

---

nential base. The graph of $y = e^x$ has the same shape as the graph in Figure 6–2(a) since $e > 1$. The details are left to the exercises.

EXERCISES 6.1

Use a hand calculator with a power key to find an approximation, correct to the third decimal place, of each of the following.

1. $2^{\sqrt{3}}$, use $\sqrt{3} = 1.732050808$ . . .

2. $3^{\sqrt[3]{4}}$, use $\sqrt[3]{4} = 1.587401052$ . . .

3. $14^{\sqrt{5}}$, use $\sqrt{5} = 2.236067977$ . . .

4. $5^{\sqrt[3]{5}}$, use $\sqrt[3]{5} = 1.709975947$ . . .

5. $7^{\sqrt[3]{3}}$, use $\sqrt[3]{3} = 1.44224957$ . . . 16.55168875

Sketch the graph of the exponential functions $f(x) =$

6. 7^x

7. 3^x

8. $\left(\dfrac{1}{7}\right)^x$

9. $\left(\dfrac{1}{3}\right)^x$

10. 7^{-x}

11. 3^{-x}

12. $3 \cdot 7^{-x}$

13. $2 \cdot 3^{-x}$

14. $7^{|x|}$

15. $3^{|x|}$

16. $7^x - 1$

17. $3^x - 2$

18. 7^{x-1}

19. 3^{x-2}

20. $7^{|x-1|}$

21. $1 - 3^x$

22. $2 - 7^x$

23. 3^{1-x}

24. 7^{2-x}

25. $3^{|1-x|}$

26. 7^{x+1}

27. 3^{x+1}

28. $7^x + 1$

29. $3^x + 1$

30. $7 - 7^{x+1}$

31. $3 - 3^{x+1}$

32. $-3 \cdot 2^x$

33. 4^x

34. $\left(\dfrac{1}{4}\right)^x$

35. 4^{-x}

36. 2^{-2x}

37. $-4 \cdot 2^{-2x}$

38. 2^{2x}

39. 2^{2x-1}

40. 2^{x^2}

41. 2^{-x^2}

42. $2^{x/2}$

43. $2^{-|x|}$

44. $\dfrac{2}{(\sqrt{2})^x}$

45. $2 - 2^{-|x|}$

46. $3 - 2^{-x}$

Use a calculator to compile a table of values for e^x and e^{x^2} (or you may use Table 1 at the end of this book). Then graph the following functions $f(x) =$

47. e^x

48. e^{-x}

49. e^{x^2}

50. e^{-x^2}

51. e^{2x}

52. e^{x-2}

53. e^{2-x}

54. $e^{|x-2|}$

55. $e^{|x|}$

56. $e^{|x-1|}$

57. $e^{|x|} - 1$

58. $|e^x - 1|$

Solve the following equations.

59. $2^{2x} = 256$

60. $5^{8x} = \dfrac{1}{625}$

61. $5^{9v} = 125^6$

62. $27^r = 9$

63. $3^{-2w} = 81$

64. $7^{10} = 49^y$

65. $7^{4x} = 49^x$

66. $5^{y+1} = 125$

67. $7^{3z} = 49^3$

68. $3^{2x+1} = 243$

69. $25^{x+1} = 5^{3x}$

70. $10^{z-2} = 100^{2z-1}$

71. $2^{x-3} = \dfrac{1}{16}$

72. $3^{u^2+1} = 9^{2.5}$

73. $5^{2-3r} = \dfrac{1}{25^r}$

74. $3^{w^2+w} = (81)9^{w/2}$

75. $7^{t^2-1} = 343$

76. $2^{3u+2} = 4^{u+2}$

77. $2^{1-v^2} = \dfrac{2}{16^4}$

78. $2^{7+3y} = 8^y \cdot 512$

79. $81^{5-s} = 9^{s/2}$

80. $16^{2t-1} = 4^{2(t+1)}$

81. Bodies in cemeteries are usually buried 6 feet deep because at that depth the temperature remains fairly stable year round at 55°F. From the surface down to this depth, the temperature varies exponentially. This can be expressed by writing the temperature x feet below the surface as

$$T(x) = 55 + (T_s - 55)2^{-x}$$

where T_s denotes the average daily high temperature.

a. What is the temperature 6 feet down when the temperature generally reaches 105°F? -50°F?

b. Answer (a) for a depth of 1 foot and for a depth of 3 feet.

c. Graph temperature as a function of depth for $T_s = 105$°F and for $T_s = -50$°F.

82. An injection of 30 milligrams of anesthesia per kilogram of body weight anesthetizes a patient for surgery. The body metabolizes this anesthesia so that it decreases exponentially after the injection. The amount remaining t minutes after the injection is given by

$$A(t) = A_0 \cdot 2^{-t/180}$$

where A_0 is the amount of the injection.

a. How much anesthesia must be injected to prepare an 80-kilogram patient for surgery?

b. After 1 hour, the effects of anesthesia have noticeably diminished. What should the amount of a second injection be to anesthetize the patient at the original level?

c. If the surgery lasts 5 hours and hourly injections are given as needed, sketch a curve indicating the amount of anesthesia in the patient at any given time t.

83. After a patient quits smoking, her body begins to eliminate nicotine. The nicotine level in the blood stream of a person who smokes n packs of cigarettes per day for t years is

$$N(t) = 500(1 - 2^{-nt})$$

and the nicotine level in the bloodstream q years after quitting is

$$N^*(q) = N_0 2^{-q}$$

where N_0 is the nicotine level at the time she quits.

a. Sketch the graph of the nicotine level in the bloodstream of a person who smokes two packs per day for 10 years and then quits.

b. From the graph, estimate how long it takes for her to get down to the level she achieved after 5 years of smoking.

c. Repeat (b) for her level after 1 year of smoking.

6.2 LOGARITHMIC FUNCTIONS AND THEIR GRAPHS

In Example 4 of Section 6.1 we sketched a graph in order to estimate the time needed to cool an object to 150°F if its temperature at time t is given by

$$T(t) = 40 + 300(2^{-3t})$$

An exact determination of this time would require that we solve the following equation for t:

$$40 + 300(2^{-3t}) = 150$$
$$300(2^{-3t}) = 110$$
$$2^{-3t} = \frac{110}{300} = \frac{11}{30}$$
$$\left(\frac{1}{8}\right)^{t} = \frac{11}{30}$$

We must find an exponent t such that b^t takes on a given value. This is precisely the task to be addressed in this section. Find an exponent u that will yield a given value for b^u.

Considering the graphs of the exponential functions $v = b^u$, we see that *for any $v > 0$ there is precisely one u for which $v = b^u$; this number u is called the* **logarithm** *of v to the* **base** b *and is denoted* $\log_b v$. (See Fig. 6–5 for the case $b > 1$.) Note that $\log_b v$ is defined only for $v > 0$ and $0 < b \neq 1$.

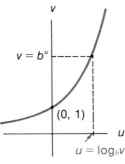

Figure 6–5

DEFINITION

For $v > 0$ and $0 < b \neq 1$,

$$u = \log_b v \qquad \text{if and only if} \qquad v = b^u.$$

The function

$$f(x) = \log_b x, \qquad x > 0$$

is called the **logarithmic function with base b.**

The following box shows some exponent-logarithm relationships.

1. $16 = 4^2$ $\log_4 16 = 2$
2. $16 = 2^4$ $\log_2 16 = 4$
3. $\dfrac{1}{25} = 5^{-2}$ $\log_5\left(\dfrac{1}{25}\right) = -2$
4. $1 = 29^0$ $\log_{29} 1 = 0$
5. $\sqrt[4]{3} = 3^{1/4}$ $\log_3 \sqrt[4]{3} = \dfrac{1}{4}$

Usually the base is not specifically indicated when $b = 10$ or $b = e$. For base 10 we simply write log x and for the natural base e we write ln x:

$$\log x = \log_{10} x$$
$$\ln x = \log_e x$$

Log x is called the **common logarithm** of x and ln x is called the **natural logarithm** of x. Logarithm tables are available or calculators can be used to find common logarithms and natural logarithms. See Tables 2 and 3 at the end of this book.

The logarithmic and exponential functions with the same base are inverses of one another; each "undoes" the effect of the other and returns the original argument. Specifically

$$u = \log_b v \qquad \text{and} \qquad v = b^u$$

are equivalent statements. Substituting u and v from each of these into the other yields

$$u = \log_b b^u \qquad \text{and} \qquad v = b^{\log_b v}$$

respectively. In particular, the functions $f(x) = e^x$ and $g(x) = \ln x$ are inverses of one another as are $f(x) = 10^x$ and $g(x) = \log x$.

Since the logarithmic and exponential functions are inverses of one another, the graph of a logarithmic function is simply the reflection of the corresponding exponential graph through the line $y = x$.

EXAMPLE 1   Sketch the graph of $f(x) = \log_2 x$.

Solution   We shall list and plot several points and then connect the dots. Figure 6–6 shows the graph. The corresponding exponential function $y = 2^x$ is sketched as the broken curve.

x	1	2	1/2	4	1/4	8	1/8
$x = 2^?$	2^0	2^1	2^{-1}	2^2	2^{-2}	2^3	2^{-3}
$\log_2 x$	0	1	-1	2	-2	3	-3

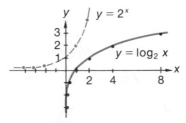

Figure 6–6 $y = \log_2 x$

EXAMPLE 2   Sketch the graph of $f(x) = \log_2(-x)$.

Solution   This graph can be obtained from the graph in Example 1 by simply reflecting it through the y-axis. For example, $f(-2) = \log 2$, $f(-4) = \log 4$, and so on. For $x \geq 0$, $f(x)$ does not exist. The graph appears in Figure 6–7.

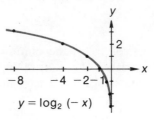

Figure 6–7 $y = \log_2(-x)$

EXAMPLE 3   Sketch the graph of $g(x) = \log_{1/2} x$.

Solution   We could plot points as in Example 1, but observe that if

$$v = g(x) = \log_{1/2} x$$

then

$$x = \left(\frac{1}{2}\right)^v = \frac{1}{2^v}$$
$$= 2^{-v}$$
$$-v = \log_2 x$$
$$v = -\log_2 x.$$

Thus

$$g(x) = -f(x)$$

where f is the function of Example 1. Consequently, this graph is obtained by reflecting the graph of Example 1 through the x-axis (see Figs. 6–6 and 6–8). Positive y-coordinates in Figure 6–6 become negative for the graph of g and negative altitudes in Figure 6–6 become positive for Figure 6–8.

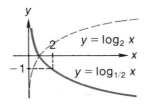

Figure 6–8 $y = \log_{1/2} x = -\log_2 x$

Figures 6–6 and 6–8 illustrate the typical shapes taken by graphs of logarithmic functions as indicated in Figure 6–9.

Note that $\log_b x$ is an increasing function if $b > 1$; it is a decreasing function if $0 < b < 1$.

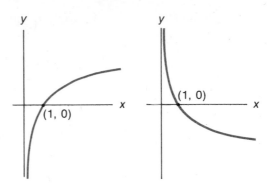

a. $y = \log_b x,\ b > 1$ **b.** $y = \log_b x,\ 0 < b < 1$

Figure 6–9

$$\text{For } b > 1, \quad \log_b x \to \begin{cases} \infty \text{ as } x \to \infty \\ -\infty \text{ as } x \to 0^+ \end{cases}$$

$$\text{For } b < 1, \quad \log_b x \to \begin{cases} -\infty \text{ as } x \to \infty \\ \infty \text{ as } x \to 0^+ \end{cases}$$

EXAMPLE 4   Sketch the following graphs.

a. $y = \log_2 |x|$

b. $y = \log_2 |x - 1|$

Solution   **a.** This function is defined for all $x \neq 0$ since $|x| > 0$ for all x. Note that here we have an even function so that the graph must be symmetric about the y-axis. Now $y = \log_2 |x| = \log_2 x$ for $x > 0$, and this graph was sketched in Figure 6–6. Reflecting this portion of the graph through the y-axis, we obtain the complete graph as in Figure 6–10(a).

b. The graph of $y = \log_2 |x - 1|$ will have the same shape as the graph of $y = \log_2 |x|$ with "1" playing the role of "0"; that is, we simply translate the graph of $y = \log_2 |x|$ one unit to the right to obtain the graph of $y = \log_2 |x - 1|$ as in Figure 6–10(b).

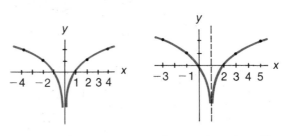

a. $y = \log_2 x$ **b.** $y = \log_2 |x - 1|$

Figure 6–10

EXAMPLE 5   As each day passes, students will remember less and less of today's lecture. With no further reinforcement, the percentage P of students who could recall the important features of today's lecture x days later decreases logarithmically. P might be given by

$$P(x) = 95 - 30 \log_2 x$$

a. After how many days do fewer than half the students recall the important features of the lecture?

b. After how many days will everyone have forgotten?

c. Graph this function.

Solution **a.** If only 50 percent of the students recall, then

$$50 = 95 - 30 \log_2 x$$
$$30 \log_2 x = 45$$
$$\log_2 x = \frac{45}{30} = \frac{3}{2}$$
$$x = 2^{3/2} = 2\sqrt{2} \approx 2(1.414)$$
$$\approx 2.828$$

After 3 days, fewer than half the students recall the important points of the lecture.

b. Everyone will have forgotten when

$$0 = 95 - 30 \log_2 x$$
$$\log_2 x = \frac{95}{30} = \frac{19}{6}$$
$$x = 2^{19/6}$$
$$\approx 8.98 \qquad \text{(Use a calculator.)}$$

After 9 days, everyone has forgotten.

c.

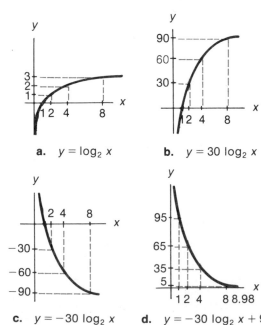

a. $y = \log_2 x$ b. $y = 30 \log_2 x$

c. $y = -30 \log_2 x$ d. $y = -30 \log_2 x + 95$

Figure 6–11

The graph is obtained from the graph of $\log_2 x$ (Fig. 6–6) in three steps as indicated in Figure 6–11. Note that the graph does not extend beyond $x = 8.98$ since we cannot have $P(x) < 0$. ∎

EXERCISES 6.2

Write the equivalent logarithmic form of each of the following.

1. $10^5 = 100,000$

2. $3^4 = 81$

3. $\dfrac{1}{125} = 5^{-3}$

4. $\sqrt[3]{125} = 5$

5. $(0.001)^0 = 1$

Write the equivalent exponential form of each of the following.

6. $\log_7 49 = 2$

7. $\log_4 8 = \dfrac{3}{2}$

8. $\log_{1/2} \dfrac{1}{8} = 3$

9. $\log_{1/6} 36 = -2$

10. $\log_{81} \dfrac{1}{9} = -\dfrac{1}{2}$

Evaluate each of the following without using a calculator.

11. $5^{\log_5 16}$

12. $\log_3 (3^{163})$

13. $1000^{\log_{10} 3}$

14. $3^{\log_9 5}$

15. $\log_3 (81^{-2})$

16. $\log_4 (2^8)$

17. $\dfrac{\ln e^6}{3}$

18. $4e^{\ln \sqrt{e}}$

Find the indicated logarithms.

19. $\log_{10} 100,000,000$

20. $\log_2 256$

21. $\log_3 81 = 4$

22. $\log_{10} (0.000001) = -6$

23. $\log_7 49$

24. $\log_{49} 7$

25. $\log_{25} 125$

26. $\log_{27} 243$

27. $\log_2 1/4$

28. $\log_{10} \sqrt{10,000}$

29. $\log_2 \sqrt{8}$

30. $\log_{10} \sqrt[5]{0.001}$

31. $\log_2 \sqrt[3]{256}$

32. $\log_2 (\sqrt[3]{2})^{1/5}$

33. $\log_3 [(1/3) \cdot (\sqrt[3]{3})]$

34. $\ln (\sqrt{e^3})$

Sketch the graphs of the following functions $f(x) =$

35. $2 \log_2 x$

36. $\log_3 x$

37. $\log_2 (1 - x)$

38. $\log_3 3x$

39. $|\log_2 x|$

40. $\log_3 x/3$

41. $|\log_2 |x||$

42. $\log_3 (x - 1)$

43. $\log_2 (x - 2)$

44. $\log_3 (1 - x)$

45. $\log_2 (2 - x)$

46. $\log_3 |x - 1|$

47. $\log_2 |x - 2|$

48. $|\log_3 (x - 1)|$

49. $|\log_2 (x - 2)|$

50. $|\log_3 |x - 1||$

51. $|\log_2 |x - 2||$

52. $\log_3 (x - 2)$

53. $\log_2 (x + 2)$

54. $|\log_3 (x - 2)|$

55. $\log_2 |x + 2|$

56. $\log_3 |x - 2|$

57. $|\log_2 (x + 2)|$

58. $|\log_3 |x - 2||$

59. $|\log_2 |x + 2||$

Use a calculator or Table 2 at the end of this book to compile a table of values for $\log x$ and $\ln x$ and then sketch the graphs of

60. $f(x) = \log x$

61. $f(x) = \ln x$

62. $g(x) = \log (-x)$

63. $h(x) = \ln |x|$

64. $F(x) = |\log x|$

65. $G(x) = \ln 2x$

66. $H(x) = \log (x - 1)$

67. Certain tasks involve the consecutive performance of several different steps. The time needed to learn these steps is a logarithmic function of the number of steps required. In a certain situation, this might be given as

$$T(t) = 302 - 50 \log_2 (65 - t)$$

where $T(t)$ is the number of hours required to learn t steps.

a. How many hours does it take to learn 64 steps? 1 step?

b. Sketch the graph of this function.

c. Express the number of steps that can be learned as a function of the time available.

68. Our earth depends on a layer of ozone in the atmosphere for protection from radiation. The strength of this ozone layer depends on many factors, one of which is the amount of forestland. For the sake of argument, let

$$S(x) = \log_{10} (x + 1)^2$$

where x represents the amount of forestland in billions of acres.

a. Sketch the graph of this function.

b. How does the strength of the ozone layer change if forest acreage is halved from 4 billion acres to 2 billion acres? (Use a calculator or Table 3 at the back of this book.)

c. If a 10 percent decrease in the strength of the ozone layer can cause serious problems, how much reduction in forest acreage from 4 billion acres can be permitted?

6.3 PROPERTIES OF LOGARITHMS; LOGARITHMIC EQUATIONS

The properties of exponents and the relationship between exponents and logarithms can sometimes be used to obtain information regarding logarithms. This occurs when the base b and the number x in $\log_b x$ are powers of the same number.

EXAMPLE 1   Find $\log_{27} 9$.

Solution   We are to find the exponent x for which $27^x = 9$. This x can be found by writing $27 = 3^3$, $9 = 3^2$ and proceeding as follows.

$$\log_{27} 9 = x$$
$$27^x = 9$$
$$(3^3)^x = 3^2$$
$$3^{3x} = 3^2$$
$$3x = 2 \qquad \text{(Property of exponential functions.)}$$
$$x = \frac{2}{3}$$

Thus, $\log_{27} 9 = 2/3$; that is,

$$27^{2/3} = 9$$

This can easily be checked since $27^{2/3} = (\sqrt[3]{27})^2 = 3^2 = 9$. ∎

In the preceding example, we converted the logarithmic equation to an exponential equation. As with our earlier work on exponential equations, we expressed both sides of the equation in terms of a common base ($b = 3$ in this case). But in solving for t in $(1/8)^t = 11/30$ (see Section 6.2), we are as yet unable to progress beyond $t = \log_{1/8} (11/30)$ since no common base for $1/8$ and $11/30$ is readily apparent. Although logarithm tables or calculators can be used to find logarithms to the base e and to the base 10, the base $b = 1/8$ is another matter entirely. To find a numerical value or estimate for t in this case requires other properties of logarithms.

Since logarithms are defined in terms of exponents, the properties of exponents translate into analogous properties of logarithms as indicated in the box on the next page.

To establish L1, for instance, let $x = \log_b u$ and $y = \log_b v$. Then, $u = b^x$ and $v = b^y$;

$$uv = b^x b^y = b^{x+y}$$

Consequently, $\log_b uv = x + y = \log_b u + \log_b v$.

Similarly, Properties L2 and L3 are established by considering u/v and u^r and applying the properties of exponents E2 and E3, respectively. Properties L4 to L6 should be obvious from E4 to E6. The proofs of all these properties are left as exercises.

Exponents	Logarithms
E1. $b^x b^y = b^{x+y}$	L1. $\log_b(uv) = \log_b u + \log_b v$
E2. $\dfrac{b^x}{b^y} = b^{x-y}$	L2. $\log_b\left(\dfrac{u}{v}\right) = \log_b u - \log_b v$
E3. $(b^x)^r = b^{xr}$	L3. $\log_b(u^r) = r \log_b u$
E4. $b^0 = 1$	L4. $\log_b 1 = 0$
E5. $b^1 = b$	L5. $\log_b b = 1$
E6. $\dfrac{1}{b^x} = b^{-x}$	L6. $\log_b\left(\dfrac{1}{u}\right) = -\log_b u$
E7. $(ab)^x = a^x b^x$	L7. $\log_b u = \dfrac{\log_a u}{\log_a b}$
E8. $x = \log_b(b^x)$	L8. $u = b^{\log_b u}$
for $0 < a \neq 1,\quad 0 < b \neq 1,\quad u > 0,\quad v > 0$	

Properties E7 and L7 relate exponential and logarithmic functions having different bases a and b; L7 is called the "change-of-base formula" for logarithms. To establish L7, let $v = \log_b u$. Then $u = b^v$ and $\log_a u = v \log_a b$ (by L3). Finally, $v = (\log_a u)/(\log_a b)$.

Properties E8 and L8 were established in Section 6.2.

For common and natural logarithms, Properties E8 and L8 take the form

$$x = \log 10^x \qquad\qquad x = \ln e^x$$
$$u = 10^{\log u} \qquad \text{and} \qquad u = e^{\ln u}$$

The preceding properties are used in the following examples. Note that the table in Example 2 gives logarithms only for prime numbers. The logarithm of any positive integer can be found in terms of the logarithms of its factors.

EXAMPLE 2 Consider the following table of logarithms to some base b.

n	$\log_b n$
2	0.301
3	0.477
5	0.699
7	0.845
11	1.041

Use this table to estimate $\log_b$ of each of the following.

a. 6

b. $7\sqrt{5}$

c. $350/\sqrt[10]{63}$

Solution **a.** $\log_b 6 = \log_b(2 \cdot 3)$

$\qquad\qquad = \log_b 2 + \log_b 3 \qquad$ (By Property L1)

$\qquad\qquad \approx 0.301 + 0.477$

$\qquad\qquad = 0.778$

b. $\log_b 7\sqrt{5} = \log_b(7 \cdot 5^{1/2})$

$\qquad\qquad = \log_b 7 + \log_b 5^{1/2} \qquad$ (By Property L1)

$\qquad\qquad = \log_b 7 + \dfrac{1}{2}\log_b 5 \qquad$ (By Property L3)

$\qquad\qquad \approx 0.845 + \dfrac{1}{2}(0.699)$

$\qquad\qquad = 0.845 + 0.3495$

$\qquad\qquad = 1.1945$

c. $\log_b(350/\sqrt[10]{63})$

$\qquad = \log_b \dfrac{5 \cdot 7 \cdot 5 \cdot 2}{(7 \cdot 3^2)^{1/10}}$

$\qquad = \log_b(5^2 \cdot 7 \cdot 2) - \log_b(7^{1/10}3^{2/10}) \qquad$ (By Property L2)

$\qquad = 2\log_b 5 + \log_b 7 + \log_b 2 - \dfrac{1}{10}\log_b 7 - \dfrac{2}{10}\log_b 3$

$\qquad = 2\log_b 5 + \dfrac{9}{10}\log_b 7 + \log_b 2 - \dfrac{1}{5}\log_b 3 \qquad$ (By L2 and L3)

$\qquad \approx 2(0.699) + 0.9(0.845) + 0.301 - 0.2(0.477)$

$\qquad = 2.3641$ ∎

EXAMPLE 3 Evaluate

a. $125^{\log_5 2}$

b. $\log_3(9\sqrt{3})^5$

Solution **a.** $125^{\log_5 2} = (5^3)^{\log_5 2}$

$\qquad\qquad = 5^{3\log_5 2} \qquad$ (By Property E3)

$\qquad\qquad = 5^{\log_5(2^3)} \qquad$ (By Property L3)

$\qquad\qquad = 2^3 \qquad$ (By Property L8)

$\qquad\qquad = 8$

b. $\log_3(9\sqrt{3})^5 = \log_3(3^2 \cdot 3^{1/2})^5$

$\qquad\qquad = \log_3(3^{5/2})^5$

$\qquad\qquad = \log_3 3^{25/2}$

$\qquad\qquad = \dfrac{25}{2} \qquad$ (By Property E8) ∎

CALCULATOR COMMENTS

Most scientific calculators are programmed to provide common and natural logarithms. These operations are generally provided through a $\boxed{\log}$ key and an $\boxed{\ln}$ key. On some calculators, these are one and the same key; if so, one of the operations must then be preceded by punching a $\boxed{\text{2nd}}$ key. See your calculator manual. To obtain logarithms on such a calculator

1. Enter the number x, for example, 29.063, into the calculator:

$$\boxed{29.063}$$

2. Punch the $\boxed{\log}$ or $\boxed{\ln}$ key and read $\log x$ or $\ln x$ on the display:

$$\log 29.063 \approx \boxed{1.463340442} \ldots$$

or

$$\ln 29.063 \approx \boxed{3.369465888} \ldots$$

If you calculator does not have logarithm capabilities, it can still be of assistance in finding logarithms. The logarithm of a complicated expression can be broken into components by using the logarithm properties L1 to L7. The logarithms of the various components can sometimes be found by using Tables 2 or 3 at the back of this book. A simple four-function $(+, -, \cdot, \div)$ calculator can then be used to combine these components into a single logarithm value. See Example 2, part (c) for instance.

EXAMPLE 4 Use the natural logarithm to find $\log_{1/8}(11/30)$.

Solution

$$\log_{1/8}\left(\frac{11}{30}\right) = \frac{\log_e\left(\dfrac{11}{30}\right)}{\log_e\left(\dfrac{1}{8}\right)} \qquad \text{(By Property L7)}$$

$$= \frac{\ln 11 - \ln 30}{\ln 1 - \ln 8} \qquad \text{(By Property L2)}$$

$$\approx \frac{2.398 - 3.401}{0 - 2.079} \qquad \text{(Use a calculator or Table 2.)}$$

$$\approx 0.482$$

Thus, in Example 4 of Section 6.1, the steel cools to 150°F when $t \approx 0.482$. ■

Logarithmic Equations

Science and engineering often require solving equations with an unknown quantity that appears in either an exponent or a logarithm. Several techniques can be used here. The simplest is to convert a logarithmic equation to an exponential equation.

EXAMPLE 5 If $\log_{10} 4x = 3$, find x.

Solution

$$\log_{10} 4x = 3$$
$$4x = 10^3 = 1000$$
$$x = 250$$

EXAMPLE 6   Solve for v if

$$\log_6 4 + 2 \log_6 v = 2$$

Solution

$$\log_6 4 + 2 \log_6 v = \log_6 4 + \log_6 v^2 \qquad \text{(By L3)}$$
$$= \log_6 4v^2 \qquad \text{(By L1)}$$

Thus

$$\log_6 4v^2 = 2$$
$$4v^2 = 6^2$$
$$= 36$$
$$v^2 = 9$$
$$v = \pm 3$$

But logarithms of negative numbers do not exist; hence they only solution is

$$v = 3$$

We observed earlier that the logarithmic and exponential functions are inverses of one another. When an inverse function f^{-1} exists, $f(x_1) \neq f(x_2)$ when $x_1 \neq x_2$. Consequently,

$$\boxed{\text{if} \quad \log_b u = \log_b v; \quad \text{then} \quad u = v}$$

EXAMPLE 7   Find w if

$$\log_b(3w + 2) + \log_b 4 = \log_b 64 + \log_b 2 - \log_b(3w - 2)$$

Solution   We can rewrite the equation as

$$\log_b(3w + 2) + \log_b(3w - 2) = \log_b 64 + \log_b 2 - \log_b 4$$
$$\log_b(3w + 2)(3w - 2) = \log_b \frac{64 \cdot 2}{4} \qquad \text{(By L1 and L2)}$$
$$\log_b(9w^2 - 4) = \log_b 32$$

Then by the preceding property,

$$9w^2 - 4 = 32$$
$$9w^2 = 36$$
$$w^2 = 4$$
$$w = \pm 2$$

But if $w = -2$, then $3w + 2 = -4 < 0$ and there is no $\log_b(-4)$. Thus, the only solution is

$$w = 2 \qquad \blacksquare$$

EXAMPLE 8 Find t if $3^{t-1} = 2^{2t+4}$.

Solution Begin by taking the $\log_{10}$ of both sides. Then solve for t as indicated.

$$
\begin{aligned}
3^{t-1} &= 2^{2t+4} \\
\log 3^{t-1} &= \log 2^{2t+4} \\
(t - 1) \log 3 &= (2t + 4) \log 2 \qquad \text{(By Property L3)} \\
t \log 3 - 2t \log 2 &= 4 \log 2 + \log 3 \\
t(\log 3 - 2 \log 2) &= \log 2^4 + \log 3 \\
t(\log 3 - \log 2^2) &= \log 16 + \log 3 \\
t \log\left(\frac{3}{2^2}\right) &= \log 16 \cdot 3 \\
t &= \frac{\log 48}{\log(3/4)} \\
&\approx \frac{1.6812}{-0.1249} \qquad \text{(Use a calculator or Table 3.)} \\
&\approx -13.460 \qquad \blacksquare
\end{aligned}
$$

EXAMPLE 9 The basic unit of sound measurement is called a bel (in honor of Alexander Graham Bell). A decibel (one tenth of a bel) is the smallest increase in intensity detectable by the human ear. The ear does not respond directly to the intensity of the sound but rather to the logarithm of the intensity. The number of decibels difference in level between sounds I_1 and I_2 is given by

$$D = 10 \log_{10} \frac{I_1}{I_2}$$

where I denotes the intensity or power of the sound in watts/per cubic centimeter. A sound rated at a certain number of decibels is compared to the threshold of human hearing $I_0 = 10^{-16}$. Sounds are rated in decibels from 0 to 120. Various levels are given in the table. How many times louder is an air hammer than normal conversation? How much more intense?

Sound	Number of Decibels
Threshold of hearing	0
Normal conversation	50
Street noise	70
Air hammer	90
Rock concert	110

Solution Let I_1 and I_2 be the intensities of the air hammer and conversation, respectively. Then for the air hammer we have

$$D_1 = 10 \log_{10} \frac{I_1}{I_0} = 10 (\log I_1 - \log I_0) = 90$$

and for conversation

$$D_2 = 10 \log_{10} \frac{I_2}{I_0} = 10 (\log I_2 - \log I_0) = 50$$

Thus

$$\log I_1 = \log I_0 + 9$$
$$\log I_2 = \log I_0 + 5$$
$$\log_{10} \frac{I_1}{I_2} = \log I_1 - \log I_2$$
$$= (\log I_0 + 9) - (\log I_0 + 5)$$
$$= 4$$

The difference in sound levels is

$$D = 10 \log_{10} \frac{I_1}{I_2} = 40$$

the air hammer is 40 times as loud as normal conversation. On the other hand,

$$\log_{10} \frac{I_1}{I_2} = 4$$
$$\frac{I_1}{I_2} = 10^4 = 10,000$$
$$I_1 = 10,000 I_2$$

the air hammer is 10,000 times as intense as normal conversation. ■

The properties of logarithms can also be used as an aid in graphing logarithmic functions as illustrated here.

EXAMPLE 10 Sketch the graph of $f(x) = \log_2(x - 1)^2$.

Solution We note that $(x - 1)^2 \geq 0$ for all x; hence $\log_2(x - 1)^2$ exists for all $x \neq 1$. Now

$$\log_2(x - 1)^2 = \log_2 |x - 1|^2$$
$$= 2 \log_2 |x - 1|$$

We begin by graphing

$$y = \log_2|x| \qquad \text{[Figure 6–10(a), page 251]}$$

and then translate this graph one unit to the right to obtain the graph of

$$y = \log_2|x - 1| \qquad \text{[Figure 6–10(b), page 251]}$$

HISTORICAL PERSPECTIVE

The incorporation of new technological developments into the daily operations of business, commerce, and science requires that computations be performed ever more rapidly and accurately. It is these demands that led to the development of the modern electronic computer and to the more recent development of sophisticated and inexpensive hand calculators and desk-top computers. Earlier demands led to the development of the slide rule. Even earlier, all calculations had to be done by hand. The development of our Hindu-Arabic system of numbers and decimal notation was a great improvement over hieroglyphics and Roman numerals; hand calculations were simplified immensely by the organization inherent in the decimal system.

In the early seventeenth century, John Napier developed a system of logarithms. The details of Napier's development need not concern us here, but his logarithm was

$$\log_{\text{nap}} y = 10^7 \log_{1/e}(y/10^7)$$

We do not use these logarithms today. Toward the end of his life, Napier collaborated with Henry Briggs. Together they decided that these logarithms were too cumbersome and developed the system of logarithms to the base 10.

To make these logarithms useful, extensive tables were needed. Developing these tables by hand was a very laborious task. Once the computations were completed and tabulated in a table of logarithms, the logarithms themselves became a very useful tool in simplifying calculations. In fact, they were so useful that they came to be known as common logarithms. Multiplying or dividing two numbers could then be accomplished by adding or subtracting their logarithms and then finding the

number that had this value as its logarithm. Raising a number to a power or extracting a root was achieved by multiplying its logarithm by the power. By taking logarithms, we lower the level of difficulty of an operation by one notch: from multiplication to addition or from exponentiation to multiplication as indicated by Properties L1 and L3 and illustrated in Example 2.

In present times, however, even the simplest calculators can dispatch multiplication with ease. Calculators with a power key and/or root extraction capabilities are also very inexpensive. Thus tables of logarithms have almost become obsolete as a computational tool. The irony is that computing machines can easily generate logarithms but at the same time they render logarithms obsolete as a computational device.

Although we know logarithms as exponents, they were actually discovered before any use was made of exponential functions. Leonhard Euler (see page 245) first expressed logarithms in terms of exponents.

Finally, we double the y-coordinates of this second graph to obtain the graph of

$$y = 2 \log_2|x - 1| \qquad \text{(Figure 6–12)}$$

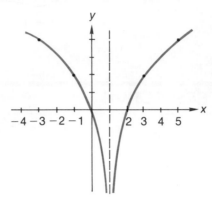

$$y = \log_2 (x - 1)^2 = 2 \log_2 |x - 1|$$

Figure 6–12

EXERCISES 6.3

Use the logarithm properties and the table given in Example 2 to find $\log_b$ of the following.

1. 40

2. $\dfrac{3}{25}$

3. $\dfrac{700}{49}$

4. $\dfrac{(30 \cdot 21)}{27}$

5. $\dfrac{(10^{-6} \cdot 4^2)}{7^3}$

6. $\sqrt{10}$

7. $\sqrt[3]{36}$

8. $\dfrac{1}{\sqrt[3]{25}}$

9. $\dfrac{\sqrt{6}}{14}$

10. $\sqrt[4]{15} \cdot \sqrt{12}$

Use common logarithms or natural logarithms and the change-of-base formula L7 to find the following. Use a calculator to do the computations and approximate the requested value.

11. $\log_3 7$

12. $\log_5 45$

13. $\log_3 72$

14. $\log_7 \dfrac{127 \cdot 243}{24}$

15. $\log_5 64 \sqrt[3]{9}$

16. $\log_4 15$

17. $\log_{1/5} \dfrac{2}{3}$

18. $\log_{\sqrt{2}} 24$

19. $\log_{36} 100$

20. $\log_{3/2} \dfrac{1}{4}$

Use the properties of exponents and logarithms to express each of the following as a single logarithm.

21. $\log_b 100 - \log_b 25 + \log_b \dfrac{1}{4}$

22. $\dfrac{2}{3} \log_b 3 + \dfrac{1}{2} \log_b 6 - 2 \log_b 9$

23. $\log_b \dfrac{1}{4} - 3 \log_b 2 + \log_b \sqrt{16}$

24. $\log_2(2x + 1) + \log_2(2x - 1)$

25. $\log_3(y + 1) + \log_3(y - 1) - \log_3(y^2 - 1)$

26. $3 \log_3(3v - 9) - \log_3(v - 3)^3$

27. $4 \log_e \sqrt{u + 1} - \log_e(u + 1)$

Solve the following equations.

28. $\log_3 2v + \log_3 v = \log_3 100 - \log_3 2$

29. $\log_2(2w + 1) + \log_2(2w - 1)$
$$= \log_2 11 + 2 \log_2 3$$

30. $\ln 6 + \ln 4 = 5 \ln 2 + \ln 3 - \ln x$

31. $\log_3 2 - \log_3(y - 1) = \log_3(y + 1)$

32. $4 \log_8 \sqrt{r + 1} - \log_8(r + 1)$
$$= \log_8 3 + \log_8 4$$

33. $3 \log_5 s^2 - 2 \log_5 s^3 + \log_4(s + 1) = \log_4 3$

34. $\log_b(z + 1) - \log_b(z - 1) = \log_b 2$

35. $\log_5 x = -3$

36. $\log_8 y = \dfrac{2}{3}$

37. $\log_{1/2}(-2v) = 4$

38. $\log_2(3r + 5) = 5$

39. $\log_3(u^2 + 2) = 3$

40. $\log_7 16\sqrt{t} = 0$

41. $\log_z 3 = \dfrac{1}{4}$

42. $\log_r 16 = 4$

43. $\log_t 1000 = -3$

44. $\log_t 9 = \dfrac{2}{3}$

45. $\log_k \dfrac{1}{32} = -\dfrac{5}{3}$

46. $\log_x 64 = \dfrac{3}{2}$

47. $\log_3(z + 1) = 2$ **48.** $\frac{1}{3}\log_3(2v + 1) = 1$

49. $2\log_3 9w^2 = 8$ HW

50. $\log_2(u - 1) + \log_2(u + 1) = 3$ HW

51. $\log_{10}(2z - 4) - \log_{10}(z - 2) = 1$

52. $\log_2(2y - 4) + \log_2(y - 2) = 1$

53. $\log_2(r - 1) + \log_2(r + 2) = 2$

Solve the following equations explicitly and then use a calculator to approximate the solution.

54. $2^{3-x} = 3^{2-x}$ **55.** $10^{3-y} = 3^{2-y}$

56. $4^{2r+1} = 5^{r+2}$ HW   **57.** $3^{2z-1} = 4^{z+5}$

58. $5^{2s+1} = 4^{s+5}$ **59.** $3^{2-t} = 2^{1+t}$

60. $5^{u+4} = 3^{1-u}$ HW   **61.** $7^{1-v} = 2^{1+v}$

62. $9^{2w} = 5^{4w}$ **63.** $11^{x-1} = 13^{x+2}$

64. Prove Property L2.

65. Prove Property L3.

66. Prove Property L4.

67. Prove Property L5.

68. Prove Property L6.

Sketch the following graphs $f(x) =$

69. $\frac{1}{2}\log_2 4x$ **70.** $\log_2 x^2$

71. $\log_2(x + 2)^2$ **72.** $\log_3(x - 2)^2$

73. $\log_2\sqrt{x + 2}$ **74.** $\log_3\sqrt{x - 2}$

75. $\log_2\dfrac{1}{(x + 2)}$ **76.** $\log_3\dfrac{1}{(x - 2)}$

77. $\log_2\dfrac{1}{|x + 2|}$ **78.** $\log_3\dfrac{1}{|x - 2|}$

79. $\log_2\dfrac{1}{x^2}$ **80.** $\log_{1/2}\dfrac{1}{x^2}$

81. $\log_{1/2}\dfrac{1}{x}$ **82.** $\log_2\dfrac{1}{x}$

83. In a chain-letter scheme a letter contains five names. A new participant purchases a letter by mailing a $10 money order to the name at the top of the list, in the presence of the person selling the letter. The buyer is then instructed to remove the top name, place his name at the bottom, and make and sell five copies of the letter.

  a. What is the payoff if the chain is unbroken?

  b. How many steps can the chain take until the entire U.S. population (225 million) is involved?

  c. How many steps can the chain take until the population of the entire world ($\approx$ 4 billion) is involved?

84. Betty wants to replace the antifreeze in her car. Her cooling system has a 16-liter capacity. When she drains the radiator, 4 liters remain in the engine block. How many times must she fill the radiator with fresh water, run the engine to mix the old coolant with the fresh water, and then redrain in order to have less than 5 percent polluted coolant in the final mixture?

85. How much louder is a rock concert than an air hammer? How much more intense? (See Example 9.)

86. The pH of a chemical substance is defined as

$$pH = \log_{10}\frac{1}{H}$$

where H is the concentration (moles/liter) of hydrogen ions in a water solution of the substance. Distilled water has a pH of 7. The substance is *acidic* if its pH is less than 7 and *basic* if its pH is greater than 7. Determine the pH of each of the substances listed below and state which are acidic.

Substance	H
Beer	$3.16(10^{-5})$
Crackers	$3.16(10^{-8})$
Eggs	$1.58(10^{-8})$
Dill pickles	$3.98(10^{-4})$
Rhubarb	$7.08(10^{-4})$
Soft drinks	$1.12(10^{-3})$
Tuna	$9.55(10^{-7})$
Blood plasma	$3.98(10^{-8})$
Gastric contents	$9.77(10^{-3})$
Bile	$1.26(10^{-7})$

87. The Richter scale relates the magnitude of the shock waves of an earthquake to the energy released by the quake. If an earthquake releases an energy E (in watts per cubic centimeter), then the Richter scale measures a magnitude

$$M = \frac{\log_{10} E - 11.4}{1.5}$$

(The relationship is of the form $\log_{10} E = A + B \cdot M$, where A and B are still being refined. The values given here are those that were accepted in 1978.)

One full unit difference on the Richter scale represents ten times the amount of motion.

  a. How many times more energy is represented by one unit on the Richter scale?

b. How many times more motion and how many times more energy is represented by two units on the Richter scale?

c. Compare the 1964 Alaska earthquake ($M = 8.5$), the 1906 San Francisco quake ($M = 8.3$), and a 1975 Yellowstone Park quake ($M = 6.7$).

d. Express the energy released in terms of the magnitude M measured by the Richter scale.

M	Effect
2	Can be felt
4.5	Slight damage
6	Moderate damage
8.5	Largest recorded

88. The exponential law of cooling can be used to determine the time of death of accident or murder victims. The temperature of a body t hours after death is

$$T(t) = T_0 + (T_1 - T_0)(0.97)^t$$

where T_0 is the air temperature and T_1 is the body temperature at the time of death. Assuming normal body temperature (98.6°F) and room temperature of 70°F, this formula becomes

$$T(t) = 70 + (98.6 - 70)(0.97)^t$$
$$= 70 + 28.6(0.97)^t$$

a. Express t as a logarithmic function of temperature.

b. At 10 P.M., John Doe was found murdered. His body temperature at that time was 82°F. When was he killed?

6.4 USING TABLES (OPTIONAL)

The tables at the end of this book list values for e^x and for the natural and common logarithms. Since each of these tables is used in the same manner, we shall concentrate on only one (Table 3: Common Logarithms) in the following discussion.

These tables list values correct to four decimal places. The column headings are used for further refinement of the argument. For instance, in the portion of Table 3 that is reproduced in Figure 6–13 we see that $\log 2.37 \approx 0.3747$.

x	0	1	2	3	4	5	6	7	8	9
.	.	.	.	.	.	.	.	.	.	.
.	.	.	.	.	.	.	.	.	.	.
.	.	.	.	.	.	.	.	.	.	.
2.3	.3617	.3636	.3655	.3674	.3692	.3711	.3729	.3747	.3766	.3784
.	.	.	.	.	.	.	.	.	.	.
.	.	.	.	.	.	.	.	.	.	.
.	.	.	.	.	.	.	.	.	.	.

Figure 6–13

The process of **linear interpolation** is used to estimate values of a function at an argument between two of those listed in the table.

LINEAR INTERPOLATION

For x between x_1 and x_2, linear interpolation estimates $f(x)$ by adding to (or subtracting from) $f(x_1)$ that portion of the increase (or decrease) $[f(x_2) - f(x_1)]$ determined by the percentage of the distance that x has traveled from x_1 to x_2.

EXAMPLE 1   Find log 2.346

We use Table 3 to find

$$\left.\begin{array}{l} \log 2.34 \approx 0.3692 \\ \log 2.35 \approx 0.3711 \end{array}\right\} \quad \text{Increase} = 0.0019$$

As x moves from 2.34 to 2.35, log x increases from 0.3692 to 0.3711; the increase of log x is 0.0019. Now 2.346 is 60 percent of the way from 2.34 to 2.35. Thus

$$\begin{aligned} \log 2.346 &\approx \log 2.34 + (0.6)(0.0019) \\ &\approx 0.3692 + 0.00114 \\ &\approx 0.3692 + 0.0011 \\ &\approx 0.3703 \end{aligned}$$

In this example, we rounded 0.00114 off to 0.0011 and obtained 0.3703 rather than 0.37034 as the value of log 2.346. The reason for this is that the values in the table are already rounded off to four places and the linear interpolation procedure is also just an estimate of the intermediate value; hence we cannot expect to obtain logarithms correct to five places from a four-place table. For the reasons we interpolate to only one more decimal place in the argument; for example, we interpolate between 2.34 and 2.35 to 2.346 but not to 2.3463.

These tables can be used in reverse, too. Again, interpolation may be necessary.

EXAMPLE 2   Find x for which log $x = 0.3741$.

Solution   We bracket log x between two consecutive values in Table 3 as follows:

$$\left.\begin{array}{l} \log 2.37 \approx 0.3747 \\ \log \ \ x \approx 0.3741 \\ \log 2.36 \approx 0.3729 \end{array}\right\} \ \left.0.0012 \ \right\} 0.0018$$

log x is approximately $0.0012/0.0018 = 2/3$ of the way from log 2.36 to log 2.37. Hence, x must be approximately 2/3 of the way from 2.36 to 2.37; that is,

$$x \approx 2.36 + \frac{2}{3} \cdot (0.01) = 2.36 + 0.00666 \ . \ . \ .$$
$$\approx 2.36 + 0.007$$
$$= 2.367$$

Again, because of the roundoff errors inherent in Table 3 and because of errors introduced in the interpolation process, we write $x \approx 2.367$ rather than 2.36666 . . .

Any real number x is the $\log_b$ of precisely one real number u. The number

u that has x as its $\log_b$ is called the **antilogarithm to the base b of x;** it is denoted $\text{antilog}_b \, x$. In view of the equivalence of the statements

$$x = \log_b u \qquad \text{and} \qquad u = b^x$$

we see immediately that

$$\text{antilog}_b \, x = b^x$$

that is, the antilogarithm function is just the exponential function.

As with common logarithms, whenever the base $b = 10$, it will not be mentioned. Thus

$$\text{antilog} \, x = \text{antilog}_{10} \, x = 10^x$$

For instance, we see in Figure 6–13 (Table 3) that

$$10^{0.3747} = \text{antilog}(0.3747) \approx 2.37$$

In Chapter 1 we saw that any positive number x could be written in scientific notation as

$$x = b \cdot 10^n \qquad 1 \le b < 10$$

Then we have

$$\log x = \log b + \log 10^n$$
$$= \log b + n$$

that is, the logarithm of any number can be expressed as an integer plus the log of some number between 1 and 10. For this reason, Table 3 lists values of log x only for x between 1 and 10.

The integer part of log x is called its **characteristic;** log b is called the **mantissa** of log x. Note that the mantissa is a **positive decimal fraction:**

$$\log 1 \le \log b < \log 10$$
$$0 \le \log b < 1;$$
$$\log b = 0.a_1 a_2 a_3 \, . \, . \, .$$

The mantissa of log x depends only on the order of the digits in the decimal representation for x rather than on the location of the decimal point. Moving the decimal point only changes the characteristic of the logarithm. For example, if log $2.346 = 0.3703$, then

$$\log 2346 = \log 2.346(10^3)$$
$$= \log 2.346 + \log 10^3$$
$$= 0.3703 + 3$$
$$= 3.3703$$

$$\text{and} \qquad \log 234600 = \log 2.346(10^5)$$
$$= 5.3703$$

EXAMPLE 3 Find log $(0.00365)^4$

Solution

$$\log(0.00365)^4 = 4 \cdot \log(0.00365) \qquad \text{(By Property L3)}$$
$$= 4 \cdot \log[(3.65) \cdot 10^{-3}]$$
$$= 4(-3 + \log 3.65)$$
$$\approx 4(-3 + 0.5623) \qquad \text{(Table 3)}$$
$$= -12 + 2.2492$$
$$= -9.7508 \qquad\qquad\blacksquare$$

Before using Table 3 in reverse, we must write the logarithm in a form with a positive decimal part; that is, we must find its characteristic and mantissa.

EXAMPLE 4 Find x if log $x = -3.6259$

Solution We write

$$\log x = -3.6259$$
$$= (4 - 3.6259) - 4$$
$$= 0.3741 - 4$$

Thus

$$x = 10^{0.3741 - 4}$$
$$= (10^{0.3741})(10^{-4})$$

Now $10^{0.3741}$ is the number whose log is 0.3741. In Example 2, we found this to be 2.367. Thus

$$x \approx (2.367)(10^{-4})$$
$$= 0.0002367 \qquad\qquad\blacksquare$$

Tables 1 and 2 (at the end of this book) need not be used in reverse, since each is, in effect, the reverse of the other.

As we mentioned earlier, logarithms are a useful computational tool in the absence of a sophisticated hand calculator. They can also be used to extend the capabilities of a simple four-function ($+$, $-$, $\times$, $\div$) calculator to permit computation of roots and powers.

EXAMPLE 5 Use logarithms to calculate

$$\frac{(2.367)^{26} \cdot \sqrt[17]{0.0003471}}{\sqrt[5]{(6027)^{13}}}$$

Solution Let x denote the value of the given expression. Then

$$\log x = 26 \cdot \log (2.367) + \frac{1}{17} \cdot \log (0.0003471) - \frac{13}{5} \cdot \log 6027$$

$$\approx 26(0.3742) + \frac{1}{17}(0.5405 - 4) - \frac{13}{5}(3.7801)$$

$$\approx 9.7292 - (-0.2035) - 9.8283$$

$$\approx -0.3026 = 0.6974 - 1$$

$$x \approx \text{antilog} \ (0.6974 - 1)$$

$$\approx 4.982 \ (10^{-1}) = 0.4982$$

EXERCISES 6.4

Use Table 3 (at the end of this book) to find the logarithms of the following numbers.

1. 123, 12300, 0.00123
2. $(421)^4$, $(42100)^{-3}$, $(0.000421)^3$
3. $\sqrt[4]{0.0888}$, $(88.8)^3$, $(888,000)^{-6}$
4. $(105)^{63}$, $\sqrt{10,500}$, $(0.0105)^{-3}$
5. $\sqrt[3]{4720}$, $\dfrac{1}{\sqrt[3]{4720}}$, $\dfrac{1}{\sqrt[3]{47.2}}$
6. $(7270)^2$, $\dfrac{1}{\sqrt[3]{7270}}$, $(0.00727)^{1/3}$
7. $(1540)^{-2}$, $(0.00154)^2$, $\sqrt{154}$
8. $(0.0436)^{20}$, $(43600)^{20}$, $\sqrt[3]{436}$
9. $(71300)^3$, $(0.0713)^{1/3}$, $\dfrac{1}{\sqrt[3]{713}}$
10. $(306)^{18}$, $(3060)^{1.8}$, $(3.06)^{180}$

Use Table 3 to find x for $\log x$ as given in each of the following exercises.

11. 3.7168, 1.7168, 6.7168
12. 1.7642, 0.7642 − 5, 0.7642 − 3
13. −2.2197, −3.2197, −5.2197
14. 3.8681, 0.8681 − 3, −3.1319
15. 2.9058, 6.9058, 3.9058
16. −0.0640, −2.0640, −1.0640
17. −1.0405, −3.0405, 3.9595
18. 7.2148, 1.2148, 0.2148 − 3
19. 2.4281, 1.4281, 0.4281
20. 2.5403, 6.5403, −2.4597

Approximate the logarithms of the following numbers by linear interpolation from Table 3.

21. 2.631
22. 7.642
23. 676.7
24. 0.04586
25. 31430
26. 642,700
27. 0.0006475
28. 0.000006175

Interpolate from Table 3 to find x if $\log x$ is given as follows.

29. 0.2230
30. 0.0921
31. −0.3430
32. 2.6704
33. −3.7568
34. 4.2516
35. −4.6824

Use logarithms from Table 3 to calculate the following expressions.

36. $(1237)(0.4945)(0.007654)$
37. $\dfrac{(839,200)(0.3044)}{0.06237}$
38. $\dfrac{\sqrt[3]{74.99} \ \sqrt{5042}}{\sqrt[3]{(6.718)^5}}$
39. $\dfrac{(3.01)^{15}(2.34)^{-5}}{(16.89)^6}$
40. $\dfrac{(4.579)^2\sqrt{82.71}}{\sqrt[3]{0.1224}}$

6.5 APPLICATIONS OF EXPONENTIAL AND LOGARITHMIC FUNCTIONS

Exponential functions are useful in studying, among other things, population growth, radioactive decay, and compound interest. Because of their inverse relationship to exponential functions, logarithmic functions are often helpful in these situations as well. For example, determination of the time required to raise a certain sum of money in a compound interest situation requires logarithms.

The applications in this section illustrate bona fide uses of logarithms in the sense that one or more logarithms must be found in order to answer a certain question. Logarithms are not used here merely as a computational device; calculators serve this purpose better anyway.

EXAMPLE 1 A certain bacteria colony is known to double its population every hour. If 10,000 bacteria are present at noon, at what time will the population be 15,000?

Solution Under these conditions, the number $f(x)$ of bacteria present x hours after noon is

$$f(x) = 10,000(2^x)$$

We must solve $10,000(2^x) = 15,000$:

$$2^x = \frac{15,000}{10,000}$$
$$= \frac{3}{2}$$
$$x \log 2 = \log 3 - \log 2$$
$$(x + 1)\log 2 = \log 3$$
$$(x + 1) = \frac{\log 3}{\log 2}$$
$$\approx \frac{0.4771}{0.3010}$$
$$\approx 1.5850$$
$$x \approx 0.5850$$

Now, 0.585 hours is approximately 35 minutes. Hence, the population is 15,000 at approximately 12:35 P.M. ◼

Most banks now compound the interest on savings accounts. This means simply that the interest is paid into the account after a certain period of time; it then begins to earn interest, too. Assume that a principle of P_0 dollars is invested at I percent interest per year and that the interest is to be compounded N times per year. Write $r = I/100$, (e.g., if $I = 5\%$, then $r = 0.05$). After one compounding period or $1/N$th of a year, the interest payment is

$$P_0 \cdot r \cdot \left(\frac{1}{N}\right)$$

The new principal is then

$$P_0 + P_0 \cdot r \cdot \left(\frac{1}{N}\right) = P_0\left(1 + \frac{r}{N}\right)$$

After another compounding period, this principal is multiplied by $[1 + (r/N)]$ again. After t years ($= Nt$ compounding periods), the principal increases to

$$P(t) = P_0\left(1 + \frac{r}{N}\right)^{Nt} = P_0\left[\left(1 + \frac{r}{N}\right)^N\right]^t$$

Continuous compounding can be approximated by compounding over very small time intervals such as every minute or every second. This yields a very large value of N. For instance, there are $N = 365 \cdot 24 \cdot 60 \cdot 60 = 31{,}536{,}000$ seconds in 1 year. It can be shown that as N gets larger and larger, the number $[1 + (r/N)]^N$ becomes a better and better estimate of e^r, where $e \approx 2.71828 \ldots$ (see Exercise 33). Thus after t years of continuous compounding, we accumulate a principal of

$$P(t) \approx P_0\left[\left(1 + \frac{r}{N}\right)^N\right]^t \qquad \text{for } N \text{ very large}$$
$$\approx P_0[e^r]^t = P_0 e^{rt}$$

Let an initial principal of P_0 be invested at I percent and let $r = I/100$. The principal resulting after compounding the interest for t years is given by

$$P(t) = \begin{cases} P_0\left(1 + \dfrac{r}{N}\right)^{Nt} & \text{compounding } N \text{ times per year} \\ P_0 e^{rt} & \text{compounding continuously} \end{cases}$$

In the equation $P(t) = P_0 e^{rt}$, t may be any real number; it may even represent an irrational number of years. Most scientific calculators are programmed to compute the powers e^{rt} needed to make these calculations. These values can be estimated from Table 1 if you do not have such a calculator.

EXAMPLE 2 If $500 is invested at 6 percent interest, how much capital is accumulated after 5 years with

   a. No compounding

   b. Yearly compounding

   c. Quarterly compounding

   d. Daily compounding

   e. Continuous compounding

Solution **a.** With no compounding, the principal after 5 years is

$$\$500 + \$500 \cdot (0.06) \cdot 5 = \$650$$

b. Compounding yearly, we have

$$\$500 \cdot (1 + 0.06)^5 \approx \$669$$

c. With quarterly compounding, the interest is compounded 20 times in 5 years. The capital appreciates to

$$(\$500)\left(1 + \frac{0.06}{4}\right)^{20} \approx \$673$$

d. Under daily compounding, the interest is compounded $5 \cdot 365 = 1825$ times. The investment is worth

$$(\$500)\left(1 + \frac{0.06}{365}\right)^{1825} \approx \$674.91$$

e. Under continuous compounding, the investment is worth

$$\$500e^{(0.06) \cdot 5} \approx \$500 \cdot (1.34986) = \$674.93 \qquad \blacksquare$$

Lending institutions frequently advertise passbook savings accounts or certificates of deposit (CDs) paying a certain rate of interest, which is compounded regularly. Then they state an *effective* interest rate, which is slightly higher. This is called an APR for "annual percentage rate." This effective rate is the interest rate that generates the same annual return with no compounding.

EXAMPLE 3 A passbook savings account pays 5 percent interest. Find the effective interest rates for quarterly and continuous compounding.

Solution With quarterly compounding for a full year, an investment of P_0 dollars will grow to

$$P_0\left(1 + \frac{0.05}{4}\right)^4 \approx 1.0509 \, P_0$$

Compounding continuously for 1 year yields a principal of

$$P_0(e^{0.05})^1 \approx 1.05127 \, P_0$$

The effective interest rates are 5.09 and 5.127 percent for quarterly and continuous compounding, respectively. $\qquad \blacksquare$

EXAMPLE 4 If we wish to double our investment in 5 years, what interest rate must we receive if the interest is compounded continuously?

Solution In 5 years, the value of an investment at I percent compounded continuously is

$$P(5) = P_0 e^{5I/100}$$

To double our investment, we must have

$$P_0 e^{5I/100} = 2P_0$$
$$e^{I/20} = 2$$
$$\frac{I}{20} = \ln 2$$
$$I = 20 \ln 2$$
$$\approx 20 \cdot (0.693)$$
$$\approx 13.86 \text{ percent}$$

EXAMPLE 5 Compare the lengths of time required to double an investment at 7 percent interest under the following compounding situations:

a. Quarterly

b. Daily

c. Continuously

Solution Given an initial principal of P_0, we are to determine the time that yields $2P_0$

a. Compounding quarterly for n quarters is to double the investment:

$$P_0\left(1 + \frac{0.07}{4}\right)^n = 2P_0$$
$$(1.0175)^n = 2$$
$$n \log (1.0175) = \log 2$$
$$n = \frac{\log 2}{\log 1.0175}$$
$$\approx \frac{0.30103}{0.00753}$$
$$\approx 39.97$$

It takes 40 quarters or 10 years to double an investment at 7 percent compounded quarterly.

b. Compounding daily for n days should yield

$$P_0\left(1 + \frac{0.07}{365}\right)^n = 2P_0$$
$$n \log\left(1 + \frac{0.07}{365}\right) = \log 2$$
$$n = \frac{\log 2}{\log\left(1 + \frac{0.07}{365}\right)}$$
$$\approx 3615$$

It takes 3615 days to double the investment at 7 percent compounded daily.

c. Compounding continuously for t years should yield

$$P_0(e^{0.07})^t = 2P_0$$
$$0.07t = \ln 2$$
$$t = \frac{\ln 2}{0.07}$$
$$\approx \frac{0.69315}{0.07}$$
$$\approx 9.902 \text{ years}$$

Continuous compounding doubles a 7 percent investment in 9.9 years or 3614 days. Thus continuous compounding is almost indiscernible from daily compounding. Daily compounding, in turn, doubles a 7 percent investment only 35 days earlier than quarterly compounding. ■

The **half-life** of a radioactive substance is the time in which a given amount of the substance will decay to half the original amount. If the half-life is very long, it can be used for dating prehistoric artifacts.

EXAMPLE 6 When archaeologists discover a new site, they determine the approximate time during which its inhabitants lived by a process known as carbon-14 dating. Carbon-14 and carbon-12 exist in equilibrium in the atmosphere and in living organisms; the ratio of carbon-14 to carbon-12 being $(1.3)10^{-12}$. Carbon-14 decays with a half-life of 5730 years. As long as a being lives, it continues to replace its lost carbon-14 through interaction with the atmosphere by such activities as eating and breathing. Once it dies, however, its carbon-14 content continues to decay but can no longer be replaced. Its carbon-12 content remains the same. Determine the age of a bone in which the ratio of carbon-14 to carbon-12 is $(3.25)10^{-13}$.

Solution Let

A_0 = amount of carbon-14 present in the organism while it lives
B_0 = amount of carbon-12 present in the organism while it lives
$A(t)$ = amount of carbon-14 present in the organism t half-lives
(for carbon-14) after it dies

Then

$$A(t) = A_0\left(\frac{1}{2}\right)^t$$

$$\frac{A_0}{B_0} = (1.3)10^{-12}$$

$$\frac{A(t)}{B(t)} = (3.25)10^{-13} \quad \text{(presently)}$$

Substituting $A(t) = A_0(1/2)^t$ and $B(t) = B_0$ into this last equation gives

$$\frac{A_0\left(\frac{1}{2}\right)^t}{B_0} = (3.25)10^{-13}$$

But $A_0/B_0 = (1.3)10^{-12}$:

$$(1.3)(10^{-12})\left(\frac{1}{2}\right)^t = 3.25(10^{-13})$$

$$\left(\frac{1}{2}\right)^t = \frac{3.25(10^{-13})}{1.3(10^{-12})}$$

$$\approx (2.5)10^{-1}$$

$$= 0.25$$

Taking the logarithm of both sides, we get

$$t\left(\log\frac{1}{2}\right) \approx \log(0.25)$$

$$t \approx \frac{\log(0.25)}{\log\frac{1}{2}}$$

$$= \frac{\log(0.25)}{-\log 2}$$

$$\approx \frac{-0.602}{-0.301}$$

$$\approx 2$$

The bone is from an animal that has been dead for approximately two half-life periods or $2 \cdot 5730 = 11,460$ years. ∎

EXERCISES 6.5

1. A manufacturer of heavy duty trucks claims that if it produces T_0 trucks in a given year, then $T_0 \cdot 2^{(-0.012)t}$ of these trucks will be in service t years later.
   a. If 50,000 trucks were produced 10 years ago, how many should still be in service?
   b. After how many years are 90 percent of its trucks still serviceable?
   c. 75 percent?

2. The intensity of illumination (brightness) of a camping lantern is $I_0 \cdot 3^{(-0.05)t}$ lumens, where I_0 represents the initial brightness and t is the number of hours that fuel has been used from its tank.
   a. After 5 hours of operation, how will its brightness have been affected?
   b. After how many hours has the brightness been halved?

3. A department store can make a yearly profit of $\$100,000 + (500,000)4^{(-0.2)n}$ on a popular gadget n years after the introduction of the gadget. How much profit does it make on these gadgets after they have been on the market 10 years?

4. If crops are not rotated and fertilized, the yield declines from year to year. If each succeeding year's crop will be only 80 percent that of the preceding year, how long does it take for the yield to be halved?

5. A municipal government finds that each time it raises taxes by 10 percent, 5 percent of its residents will move to the suburbs. If it raises taxes 10 percent each year for 10 years, how much will its population decrease in this time?

6. When a cold front moved in, Mr. Waters observed that the temperature of the water in his pool decreased by 5 percent in 15 minutes. Assuming that the temperature continues to drop at this rate and that the temperature is now 50°F, when will the water freeze (32°F)?

7. A flu virus triples the number of its victims every 10 days. If it has infected 25,000 people when $t = 5$ days, how many will it have affected when $t = 20$ days?

8. The half-life of carbon-14 is approximately 5730 years. If 3 milligrams of carbon-14 are present in a fossil that is 3000 years old, how much carbon-14 was present in this fossil when it lived?

9. The half-life of radium is approximately 1600 years. If 2000 units are present now,
   a. When were there 5000 units present?
   b. When will there be only 200 units present?
   c. How much will be present 50 years from now?

10. A given sample of radium consists of 100 units.
   a. When will it contain only 75 units?
   b. How much will there be 1000 years from now?
   c. When were 150 units present?
   d. When will one third of the original amount remain?

11. The doubling time of a bacteria culture is 2 hours. If the culture now consists of 100 bacteria,
   a. How long will it take to reach 750 members?
   b. When were there 10 bacteria present?
   c. What is the population $5\frac{1}{2}$ hours from now?

12. If the doubling time of a bacteria culture is known to be 3 hours and there are 1000 bacteria present now,
   a. When will the population reach 2500 members?
   b. When were only 100 members present?

   c. How many will there be 5 hours from now?

13. The charge on a capacitor dissipates at a rate proportional to the charge present. The half-life of the charge is 10 minutes. The charge now is 50 coulombs.
   a. When will the charge be 40 coulombs?
   b. When was the charge 60 coulombs?
   c. What will the charge be 2 minutes from now?
   d. What was the charge 5 minutes ago?

14. Water flows into a tank in such a way that its volume triples every hour. The tank now contains 10 gallons.
   a. When will it contain 15 gallons?
   b. When did the tank contain only 5 gallons?
   c. How much will it contain 15 minutes from now?
   d. How long does it take for the volume to double?

15. The Goodyear blimp has a leak and 5 percent of its gas is escaping each minute. Presently, the blimp holds 500,000 cubic feet of gas.
   a. When will 250,000 cubic feet of gas remain in the blimp?
   b. When will 50,000 cubic feet of gas remain?
   c. When were 700,000 cubic feet present?
   d. How much was present $2\frac{1}{4}$ minutes ago?

16. How fast must your salary double if your buying power is not to suffer and the inflation rate is
   a. 3 percent compounded semiannually?
   b. 5 percent compounded quarterly?
   c. 6 percent compounded monthly?
   d. 7 percent compounded continuously?
   e. 9 percent compounded semiannually?
   f. 12 percent compounded annually?

17. In each of the following, find the effective annual interest rate.
   a. A passbook pays $5\frac{1}{2}$ percent compounded quarterly.
   b. A certificiate of deposit (CD) pays 8 percent compounded semiannually.
   c. A CD pays 7 percent compounded daily.
   d. A CD pays 7 percent compounded continuously.
   e. A CD pays $7\frac{1}{2}$ percent compounded daily.

18. Five years ago Dan invested a certain sum of money at 7 percent interest compounded continuously. How much did he invest if his principal is now worth $30,000?

19. Paula invested a certain sum of money at 6 per-

cent compounded continuously. Two years later, her investment was worth $10,000.
a. How much is it worth 5 years after this?
b. How much did she initially invest?

20. If Verna invests a given sum at 5 percent interest, how long does it take to triple her investment if the interest is compounded
a. Not at all b. Yearly c. Daily d. Continuously

21. Dale invests $5000 and the interest is compounded quarterly. How long does it take to double his investment if the interest rate is
a. 5 percent b. 6 percent c. 7 percent

22. If Bernice needs $7500 5 years from now and has $5000 to invest now, what interest rate compounded daily does she need?

23. If we need $10,000 10 years from now and have $5000 to invest now, what interest rate do we need if it is compounded continuously?

24. If Joan needs $5000 12 years from now and has $2000 to invest now, what interest rate compounded quarterly does she need?

25. What principal must be invested now at $6\frac{1}{2}$ percent interest compounded quarterly to achieve a value of $15,000 in 10 years? What sum is needed with continuous compounding? The **present value** of a sum of money s to be received at a certain time in the future is the principal P, which if invested now would grow to s in that time.

26. What principal must be invested now at 7 percent interest compounded daily to achieve a value of $6000 in 5 years? What sum is needed with continuous compounding?

27. What principal must be invested now at 5 percent interest compounded semiannually to achieve a value of $2000 in 6 years? What sum is needed with continuous compounding?

28. What principal must be invested now at 6 percent interest compounded annually to achieve a

value of $15,000 in 8 years. What sum is needed with continuous compounding?

29. Scientists can estimate the age of the earth by comparing the relative amounts of two kinds of uranium: U^{235} and U^{238}. U^{235} has a half-life of $7.1(10^8)$ years, while the half-life of U^{238} is $4.5(10^9)$ years. There is now 140 times as much U^{238} in the earth as there is U^{235}. If it is determined that U^{238} was 3.2 times as plentiful as U^{235} when the earth was formed, what is the age of the earth?

30. If the world population sustains its doubling time of 35 years and the population is now 4 billion,
a. When will the population reach 5 billion?
b. When will the population reach 6 billion?
c. What will the population be 10 years from now?

31. If the U.S. population sustains its doubling time of 60 years and the current U.S. population is 225 million, when will the U.S. population reach 250 million? 300 million? 400 million?

32. In 1626 Peter Minuit purchased Manhattan Island from the Manhattan Indians for about $24. Had they invested this money at 6 percent interest compounded annually, what would their investment now be worth? How about 4 percent (Use a calculator.)

33. To see that $(1 + r/N)^N \approx e^r$ for large N, write

$$\left(1 + \frac{r}{N}\right)^N = \left(1 + \frac{1}{N/r}\right)^N$$
$$= \left[\left(1 + \frac{1}{N/r}\right)^{N/r}\right]^r$$
$$= \left[\left(1 + \frac{1}{x}\right)^x\right]^r$$

for $x = N/r$. As N gets large, x gets large. Now use a calculator to evaluate $(1 + 1/x)^x$ for $x = 10^n$, $n = 1, 2, 3, 4, 5, 6$.

6.6 CHAPTER REVIEW

TERMS AND CONCEPTS

- Logarithms

    $u = \log_b v$ if and only if $v = b^u$ $(v > 0$ and $0 < b \neq 1)$

RULES AND FORMULAS $\quad (0 < a \neq 1,\ 0 < b \neq 1,\ u > 0,\ v > 0)$

- Exponents

  1. $b^x b^y = b^{x+y}$ $\qquad\qquad\qquad$ $2^2 2^3 = 2^5$

  2. $\dfrac{b^x}{b^y} = b^{x-y}$ $\qquad\qquad\qquad$ $\dfrac{2^5}{2^2} = 2^3$

  3. $(b^x)^y = b^{xy}$ $\qquad\qquad\qquad$ $(2^2)^3 = 2^6$

  4. $b^0 = 1$ $\qquad\qquad\qquad$ $2^0 = 1$

  5. $b^1 = b$ $\qquad\qquad\qquad$ $2^1 = 2$

  6. $\dfrac{1}{b^x} = b^{-x}$ $\qquad\qquad\qquad$ $\dfrac{1}{2^2} = 2^{-2}$

  7. $(ab)^x = a^x b^x$ $\qquad\qquad\qquad$ $(2 \cdot 3)^2 = 2^2 \cdot 3^2$

  8. $x = \log_b (b^x)$ $\qquad\qquad\qquad$ $3 = \log_2 2^3$

- Logarithms

  1. $\log_b(uv) = \log_b u + \log_b v$ $\qquad$ $\log_2 (2 \cdot 4) = \log_2 2 + \log_2 4 = 1 + 2 = 3$

  2. $\log_b\left(\dfrac{u}{v}\right) = \log_b u - \log_b v$ $\qquad$ $\log_2\left(\dfrac{4}{2}\right) = \log_2 4 - \log_2 2 = 2 - 1 = 1$

  3. $\log_b (u^r) = r \log_b u$ $\qquad$ $\log_2 (4^2) = 2 \log_2 4 = 2 \cdot 2 = 4$

  4. $\log_b 1 = 0$ $\qquad$ $\log_2 1 = 0$

  5. $\log_b b = 1$ $\qquad$ $\log_2 2 = 1$

  6. $\log_b\left(\dfrac{1}{u}\right) = -\log_b u$ $\qquad$ $\log_2\left(\dfrac{1}{4}\right) = -\log_2 4 = -2$

  7. $\log_b u = \dfrac{\log_a u}{\log_a b}$ $\qquad$ $\log_4 16 = \dfrac{\log_2 16}{\log_2 4} = \dfrac{4}{2} = 2$

  8. $u = b^{\log_b u}$ $\qquad$ $3 = 2^{\log_2 3}$

GRAPHING TECHNIQUES

- Exponential Functions
- Logarithmic Functions

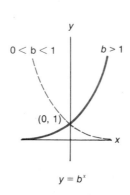

$0 < b < 1 \qquad b > 1$

$(0, 1)$

$y = b^x$

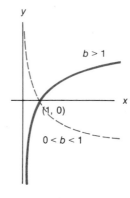

$b > 1$

$(1, 0)$

$0 < b < 1$

$y = \log_b x$

6.7 SUPPLEMENTARY EXERCISES

Write the equivalent logarithmic form of each of the following.

1. $5^3 = 125$ $\qquad$ **2.** $4^{-3} = \dfrac{1}{64}$

3. $16^0 = 1$

Write the equivalent exponential form of each of the following.

4. $\log_{10} 100 = 2$ $\qquad$ **5.** $\log_{64} 4 = \dfrac{1}{3}$

6. $\log_{1/2} 8 = -3$

Evaluate each of the following without using a calculator.

7. $7^{\log_7 132}$

8. $\ln(e^{\sqrt{4}})$

9. $49^{\log_7 5}$

10. $2^{\log_4 3}$

11. $\log_2 4^{-3}$

12. $\log_9 3^3$

13. $\log_9 81$

14. $\log_{16} 256$

15. $\log_{27} 3$

16. $\log_{16} \dfrac{1}{2}$

17. $\log_3(\sqrt{3})^7$

18. $\log_7 \dfrac{1}{(\sqrt[4]{7})^5}$

19. $3e^{\ln 6}$

Use the properties of exponents and logarithms to express each of the following as a single logarithm.

20. $2 \log_6 9 - 3 \log_6 \left(\dfrac{1}{3}\right) + 2 \log_6 \sqrt{12}$

21. $\log_2(2t - 4) - \log_2(t - 2)$

22. $\left(\log_b 27 + 3 \log_b \dfrac{1}{3}\right)(\log_{126} 4729 - \log_{372} 628)$

Use the logarithm properties and the table given in Example 2 of Section 6.3 to find $\log_b$ of each of the following.

23. 10

24. $\dfrac{49}{8}$

25. $(10^6)(14^5)$

26. $\sqrt[3]{4}$

27. $\sqrt{\dfrac{5}{48}}$

Use common logarithms or natural logarithms and the change-of-base formula L7 to find the following logarithms. You will also need a calculator or Tables 2 or 3.

28. $\log_5 \dfrac{2}{3}$

29. $\log_2 \dfrac{63}{\sqrt{250}}$

30. $\log_9 25$

31. $\log_{\sqrt{3}} 6$

32. $\log_6 21$

Sketch the graphs of the following functions $f(x) =$

33. 5^x

34. $\left(\dfrac{1}{5}\right)^x$

35. 5^{-x}

36. 5^{3+x}

37. 5^{2-x}

38. $1 - 5^x$

39. $5^{|x|}$

40. $5^{|x-2|}$

41. $1 - e^x$

42. $(-4) \cdot 5^{2-x}$

43. $\log_2(x + 3)$

44. $\log_2|x + 3|$

45. $\log_2(x + 3)^2$

46. $\log_2\sqrt{x + 3}$

47. $\log_2\left(\dfrac{1}{x + 3}\right)$

48. $|\log x|$

49. $|\log_2(-x)|$

50. $|\log_2|x||$

Solve the following equations.

51. $7^{s-1} = 343$

52. $3^{2s-4} = (81)^2$

53. $8^z = 16$

54. $100^{2y+4} = 10^{3y}$

55. $9^{1-2w} = 27^{-2/3}$

56. $5^{4-r} = \dfrac{1}{25}$

57. $(125)^{3u} = 5^{(3u)^2}$

58. $\log_4 z = 3$

59. $\log_{27} r = \dfrac{4}{3}$

60. $\log_v 25 = \dfrac{1}{2}$

61. $\log_u\left(\dfrac{1}{64}\right) = -3$

62. $\log_y 343 = \dfrac{3}{2}$

63. $\log_2 4t^3 = 5$

64. $\log_3(2 - 5w) = 3$

65. $\log_6(2x + 6) = 2$

66. $3 \ln 2 - \ln 4 + 2 \ln x = 3 \ln 3 + \ln 2 - \ln 3$

67. $\log_2(2x - 4) + \log_2(x - 2) = 1$

68. $\log_{\sqrt{3}}(z + 1) - \log_{\sqrt{3}}(z - 1) = 2$

69. $\log_3(2z) - \log_3(z + 4)$
$$= \log_3(z - 4) - \log_3(z - 5)$$

70. $3^{3-t} = 2^{2-t}$

71. $5^{2x+1} = 4^{x+2}$

72. $2^{1-y} = 7^{1+y}$

73. $15^{u-1} = 12^{u+2}$

74. The charge on a capacitor dissipates at a rate that is proportional to the charge present. The half-life of the charge is 10 minutes. The charge is now 40 coulombs.
  a. What will the charge be 7 minutes from now?
  b. What was the charge 3 minutes ago?
  c. If the charge is 60 coulombs when $t = 4$, what will it be when $t = 8$?
  d. When will the charge be 50 coulombs?

75. Water flows into a tank in such a way that its volume triples every hour. The tank now contains 10 gallons.
  a. How much will it contain 4 hours, 15 minutes from now?
  b. How much did it contain 2 hours, 10 minutes ago?
  c. If the tank contains 20 gallons when $t = 2$, how much will it contain when $t = 2.5$?
  d. How long does it take for the volume to double?

76. Five years ago Mary invested some money at 7 percent interest compounded continuously. If her investment is now worth $32,000, how much did she invest initially? How much will her investment be worth 10 years from now?

77. In each of the following problems, let the principal P be invested at the annual interest rate I for the number n years. For each, find the resulting principal assuming no compounding, yearly compounding, semiannual compounding, quarterly compounding, daily compounding, and continuous compounding. If you do not have a calculator, leave your answers in the form $P \cdot x^n$.

a. $P = \$100$, $I = 5$ percent, $n = 100$ years

b. $P = \$1000$, $I = 6$ percent, $n = 10$ years

c. $P = \$2000$, $I = 8$ percent, $n = 5$ years

d. $P = \$4500$, $I = 6\frac{1}{4}$ percent, $n = 5$ years

e. $P = \$5000$, $I = 7\frac{1}{2}$ percent, $n = 5$ years

Optional Table Exercises.

78. *Use Table 3 (at the end of this book) to find the logarithms of the following numbers*

a. 26.4,   2640,   .0264

b. $\sqrt{798}$, $\sqrt[3]{0.00798}$, $\sqrt[3]{0.0798}$

c. $(0.0631)^{-2}$, $(63.1)^{1/2}$, 6310^2

d. $(2060)^3$, $\dfrac{1}{(2060^{30})}$, $(2060)^{1/3}$

e. $(59600)^{10}$, $(0.00596)^{10}$, 596^{10}

79. *Use Table 3 (at the end of this book) to find x for log x as given in each of the exercises that follow.*

a. 7.6656,   3.6656,   2.6656

b. 6.3345, -3.6655, -2.6655

c. -1.1831, -0.1831, -3.1831

d. 6.9827,   3.9827, -2.0173

e. 2.9926, $0.9926 - 2$, -2.0074

80. *Approximate the logarithms of the following by linear interpolation in Table 3 (at the end of this book).*

a.   5.064   **b.**   964.2   **c.**   0.01588

d.   1,001,000   **e.**   0.00003998

81. *Interpolate in Table III to find x if log x is given here.*

a.   0.9610   **b.**   7.9515   **c.** -3.1541

d.   3.7555   **e.** -2.5227

82. *Use logs to calculate the following expressions.*

a. $(0.0675)^{4.5}(423)^{3.7}$

b. $\dfrac{(31.08)^2\sqrt[3]{0.7926}}{(68.37)^3}$

7 Systems of Equations and Inequalities

Scientists, engineers, economists, and politicians use equations to describe various phenomena. Separate equations must be used to describe several constraints, which may be in effect *simultaneously*.

For instance, in selecting a variety of foods to satisfy certain dietary requirements, a dietitian will find that from a given food group, each serving of a particular food contains certain amounts of vitamins A and B and that each serving of another food contains different amounts of these vitamins. The total number of units of vitamins A and B contained in a monthly diet depends on the number of times the various foods supplying these vitamins are served. For example, if carrots are served x times per month and each serving provides 8,000 International Units (IU) of vitamin A, while spinach is served y times per month with each serving providing 4,000 IU of vitamin A, the total number of units of vitamin A provided by these two foods is $8000x + 4000y$. To satisfy the monthly requirement for vitamin A (say 150,000 IU) with only these two foods requires

$$8000x + 4000y = 150,000$$

Of course, consideration of more foods introduces additional variables. Each dietary requirement results in a similar equation. To provide a balanced diet, each of these equations must then be satisfied simultaneously.

A collection of equations, each of which is to be satisfied simultaneously, is called a **system of equations.** A **system of inequalities** is a collection of inequalities, each of which must be satisfied simultaneously.

$$(1) \quad \begin{cases} 2x + 3y - 4z = 0 \\ 2x^2 - 4y^2 + 2 = 0 \\ 9z - 3w + 4y = 0 \end{cases} \qquad (2) \quad \begin{cases} x + 3y < 10 \\ 2xy + 4 \geq -2 \end{cases}$$

A System of Equations A System of Inequalities

> The **solution** to a system of equations or inequalities consists of all values of the variables that satisfy each of the equations or inequalities in the system.

7.1 TWO EQUATIONS IN TWO UNKNOWNS

In this section we discuss systems of linear equations in two variables: systems involving equations of the form

$$ax + by = c \qquad a, b, c \text{ constants}, \qquad a, b \text{ not both } 0$$

Graphing

The graph of a linear equation in two variables is a straight line (see Section 3.3). Thus the solution to a system of two linear equations in two variables consists of the points lying on each of the lines represented by these equations. Three situations could occur.

1. If the lines intersect in precisely one point, there is precisely one solution; the system is said to be **consistent.**

2. If the lines are parallel, there are no solutions; the system is said to be **inconsistent.**

3. If the lines coincide, every point on this common line is a solution; the system is said to be **dependent.**

These situations are illustrated in Figure 7–1.

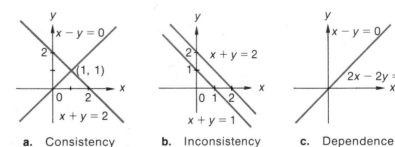

a. Consistency   **b.** Inconsistency   **c.** Dependence

Figure 7–1

Recall that two lines have the same slope if and only if they are parallel or coincide. A system of two linear equations in two unknowns is consistent if and only if the associated lines have different slopes.

EXAMPLE 1   Determine whether the following systems are consistent, inconsistent, or dependent. If the system is consistent, estimate the solution from the graphs.

a. $\begin{cases} 3x - 2y = 1 \\ -6x + 4y = -1 \end{cases}$ b. $\begin{cases} x - y = 4 \\ 2y - 2x = -8 \end{cases}$ c. $\begin{cases} x + 2y = 3 \\ x + 4y = 6 \end{cases}$

Solution   **a.**   The equations may be rewritten as

$$\begin{cases} 2y = 3x - 1 \\ 4y = 6x - 1 \end{cases} \quad \text{or} \quad \begin{cases} y = \dfrac{3}{2}x - \dfrac{1}{2} \\ y = \dfrac{3}{2}x - \dfrac{1}{4} \end{cases}$$

The latter version clearly indicates parallel lines each having slope $m = 3/2$; the system is inconsistent.

b.   This system is equivalent to

$$\begin{cases} y = x - 4 \\ 2y = 2x - 8 \end{cases} \quad \text{or} \quad \begin{cases} y = x - 4 \\ y = x - 4 \end{cases}$$

The lines coincide; the system is dependent.

c.   Again, rewrite the system in a form that indicates the slope of each line:

$$\begin{cases} 2y = -x + 3 \\ 4y = -x + 6 \end{cases} \quad \text{or} \quad \begin{cases} y = -\dfrac{1}{2}x + \dfrac{3}{2} \\ y = -\dfrac{1}{4}x + \dfrac{3}{2} \end{cases}$$

The slopes are $m_1 = -1/2$ and $m_2 = -1/4$, respectively. The lines are not parallel. The system is consistent; it has precisely one solution. The lines are sketched in Figure 7–2. The intersection of these lines gives us the solution

$$x = 0, \; y = 3/2.$$

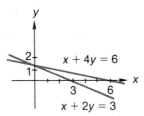

Figure 7–2

Substituting these values into the original equations verifies the solution:

$$0 + 2 \cdot \frac{3}{2} \overset{\checkmark}{=} 3$$

$$0 + 4 \cdot \frac{3}{2} \overset{\checkmark}{=} 6$$

The Substitution Method

It is sometimes difficult to estimate the solution with any great degree of precision by using a graph. On the other hand, one of the equations in a linear system can be used to express some variable in terms of the others. Substituting this expression into the other equation(s) simplifies the problem by reducing the number of variables. This technique is called the **substitution method of solution.** It is especially useful in solving systems of two equations in two unknowns.

EXAMPLE 2   Solve the following systems.

a. $\begin{cases} x - y = 1 \\ 2x + 3y = 7 \end{cases}$ b. $\begin{cases} x - y = 1 \\ 2x - 2y = 2 \end{cases}$ c. $\begin{cases} 2x + 2y = 4 \\ x + y = 1 \end{cases}$

Solution   **a.**   In order for the first equation to be satisfied, we must have

$$x = y + 1$$

Substituting this expression for x into the second equation gives

$$2(y + 1) + 3y = 7$$
$$5y + 2 = 7$$
$$5y = 5$$
$$y = 1$$

Hence $y = 1$ is the only possible y value that can satisfy this system of equations. But then

$$x = y + 1 = 1 + 1 = 2$$

The solution to the system is $x = 2$, $y = 1$, or

$$(x, y) = (2, 1)$$

We check the solution by substituting $x = 2$, $y = 1$ into the original equations:

$$2 - 1 \overset{\le}{=} 1$$
$$2 \cdot 2 + 3 \cdot 1 \overset{\le}{=} 7$$

b.   Again, we have $x = y + 1$ from the first equation. Substituting, the second equation becomes

$$2(y + 1) - 2y = 2$$
$$2 = 2$$

No restrictions are placed on y. The only restriction on the variables is that $x = y + 1$. The system is dependent.

c.   Use the second equation to write $y = 1 - x$ and substitute into the first equation:

$$2x + 2(1 - x) = 4$$
$$2 = 4$$

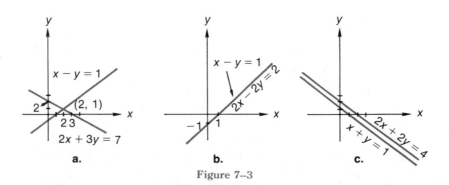

a.   b.   c.

Figure 7–3

If these equations are both satisfied, then $2 = 4$. This cannot be; thus both equations cannot be satisfied simultaneously. There are no solutions.

The graphs are given in Figure 7–3. ∎

EXAMPLE 3 Recall that for complex numbers, $a + bi = c + di$ if and only if $a = c$ and $b = d$. With this in mind, find x and y if

$$x(1 - 2i) + y(1 + i) = 7 + i$$

Solution

$$x(1 - 2i) + y(1 + i) = 7 + i$$
$$x - 2xi + y + yi = 7 + i$$
$$x + y + (-2x + y)i = 7 + i$$

This leads to the system

$$\begin{cases} x + y = 7 \\ -2x + y = 1 \end{cases}$$

which we can easily solve to obtain $x = 2$, $y = 5$. ∎

EXERCISES 7.1

For each of the following systems **a.** *Sketch the graphs.* **b.** *Indicate whether the system is consistent, inconsistent, or dependent.* **c.** *If consistent, estimate the solution from the graphs.* **d.** *Use the substitution method to solve the system exactly.*

1. $\begin{cases} x - y = 2 \\ x + 2y = -1 \end{cases}$

2. $\begin{cases} 14r + 6s = 8 \\ 7r + 3s = 5 \end{cases}$

3. $\begin{cases} p - 5q = 7 \\ 3p + 4q = 2 \end{cases}$

4. $\begin{cases} u - 3v = 2 \\ 7u - 21v = 14 \end{cases}$

5. $\begin{cases} t - 4u = 24 \\ 2t + 3u = -7 \end{cases}$

6. $\begin{cases} 2x + y = 4 \\ x - 3y = 2 \end{cases}$

7. $\begin{cases} 2t - 6u = 8 \\ 3t - 9u = 9 \end{cases}$

8. $\begin{cases} 4p - 3q = 5 \\ 4p - 9q = -1 \end{cases}$

9. $\begin{cases} 2y + z = 1 \\ 4y + 2z = 2 \end{cases}$

10. $\begin{cases} 4v + 2w = 3 \\ 10v + 5w = 8 \end{cases}$

11. $\begin{cases} x + w = 2 \\ 3x + 2w = 1 \end{cases}$

12. $\begin{cases} 4t + 3x = 3 \\ 8t + 6x = 6 \end{cases}$

13. $\begin{cases} 2u + v = 6 \\ u + 2v = 0 \end{cases}$

14. $\begin{cases} 3y - 4u = 5 \\ y - 3u = -5 \end{cases}$

15. $\begin{cases} 5v + 7w = 3 \\ 10v + 14w = 6 \end{cases}$

16. $\begin{cases} 13t + 8y = 6 \\ 7t + 6y = 10 \end{cases}$

17. $\begin{cases} u + v = 0 \\ 2u - 3v = 5 \end{cases}$

18. $\begin{cases} 7x + 6y = 3 \\ 35x + 30y = 15 \end{cases}$

19. $\begin{cases} w - 2z = 5 \\ 2w - 3z = 8 \end{cases}$

20. $\begin{cases} v + 5w = -1 \\ v - 3w = 7 \end{cases}$

21. $\begin{cases} 2t - 3u = 15 \\ 6t - 9u = 5 \end{cases}$

22. $\begin{cases} 3y - 2z = 3 \\ 5y - z = -2 \end{cases}$

23. $\begin{cases} 3x - 2y = 4 \\ 4x - 6y = 2 \end{cases}$

24. $\begin{cases} 4r - s = 5 \\ 4r - 3s = -17 \end{cases}$

25. $\begin{cases} 3v - 2r = 5 \\ 4v - 5r = 2 \end{cases}$

Solve for x and y in the following.

26. $(x - y) + (x + y)i = 1 + 7i$
27. $(x - yi) + (y + xi) = 3 - i$
28. $y + xi = 3i(y + 1) + x + 1$
29. $x(1 + 2i) + y(3i - 1) = 1 + 7i$
30. $x(1 + 2i) + y(3i - 4) = 20 + 7i$
31. $4x(1 + i) - y(1 + 3i) = 5 - 17i$
32. $(x + yi)(3i - 2) = 1 + 5i$
33. The equation for converting Celsius temperatures to Farenheit temperatures is $F = (9/5)C + 32$. One method for finding if and when the two temperature scales agree is to solve the system of equations

$$\begin{cases} F = \dfrac{9}{5}C + 32 \\ F = C \end{cases}$$

a. Graph the two equations on the same coordinate system.

b. Solve the system of equations by the method of substitution.

c. Check your results on the graph

34. When corn sells for $1.50 per bushel, there is a demand for 100,000 metric tons and a supply of only 50,000 metric tons. Each penny increase in the price per bushel stimulates an additional production of 8000 metric tons and decreases the demand by 2000 metric tons.

   a. If equilibrium occurs when supply equals demand, write a system of equations whose solution indicates the equilibrium price for corn.

   b. Sketch the lines represented by these equations.

   c. Solve the system by the method of substitution.

   d. Check your results on the graph.

35. A tortoise travels at a rate of 1 mile in 5 hours; the hare can travel 5 miles in 1 hour.

   a. If the tortoise has an 8-hour head start, write a system of equations whose solution indicates when the hare passes the tortoise and how far each has traveled when this happens.

b. Sketch the lines represented by these equations.

c. Solve the system by the method of substitution.

d. Check your results on the graph.

36. How old are Karen and Bea if 4 years ago Karen was seven times as old as Bea and 4 years from now Karen will be only three times as old as Bea?

37. Four years ago Viki was 22 years younger than her mother. If Viki's mother is now 8 years more than twice as old as Viki, how old are Viki and her mother now?

38. Find a two-digit number such that the sum of the digits is 10 and the square of the tens digit is two more than the units digit.

39. The tens digit of a two-digit number exceeds the units digit by 3. If the digits are reversed, the result is one less than half the original. Find the number. *Hint:* If the tens and units digits are t and u respectively, the number is $10t + u$. For example $27 = 10 \cdot 2 + 7$.

7.2 THE ELIMINATION METHOD OF SOLUTION

The arithmetic involved in the substitution method frequently leads to messy fractions and becomes tedious. On the other hand, sketching the graphs represented by a system of equations and "guesstimating" the coordinates of their intersections can be very inaccurate. In this section, we introduce another method for solving systems of equations. Known as the **elimination method,** it yields precise results and the manipulations are sometimes easier than those of the substitution method.

The key to the elimination method for solving a system of equations lies in the replacement of the given system with another system that has precisely the same solutions. Two systems having the same solutions are said to be **equivalent.**

PRINCIPLES OF THE ELIMINATION METHOD

The following operations on a system of equations yield an equivalent system:

1. Interchanging equations

2. Multiplying an equation by a nonzero constant

3. Adding a nonzero multiple of one equation to another equation

These operations are performed on a system of equations to eliminate one or more of the variables from one or several equations.

Clearly, if two equations in a system are interchanged, the list of equations that must be satisfied is not changed; only the order of the listing has been changed. Regarding the second operation, it should be clear, for example, that

$$x + y = 2$$
$$2x + 2y = 4$$

represent the same line and hence provide the same solutions in a system. Concerning the third, consider

$$\begin{cases} x - y = 2 \\ 2x + 3y = -1 \end{cases} \quad \text{and} \tag{1}$$

$$\begin{cases} x - y = 2 \\ \underbrace{3(x - y)} + 2x + 3y = -1 + \underbrace{3 \cdot 2} \end{cases} \tag{2}$$

System (2) is obtained by adding three times the first equation to the second equation of system (1). Clearly, the two systems have the same solutions since $3(x - y) = 3 \cdot 2$ in either case.

EXAMPLE 1   Solve the following system by the elimination method.

$$\begin{cases} x - y = 1 \\ 2x + 3y = 7 \end{cases}$$

Solution   We can eliminate y by adding three times the first equation to the second equation. This yields the equivalent system:

$$\begin{cases} x - y = 1 \\ 3(x - y) + 2x + 3y = 7 + 3 \cdot 1 \end{cases}$$

which simplifies to

$$\begin{cases} x - y = 1 \\ 5x = 10 \end{cases}$$

From the new second equation we obtain

$$x = 2$$

as the only possible x value that can satisfy the system. Substituting $x = 2$ into the first equation, we obtain

$$2 - y = 1$$
$$1 = y$$

as the only possible y solution. Since the equations in question represent non-parallel lines, the solution must consist of a single point. Thus

$$(x, y) = (2, 1)$$

is the solution to the system. This agrees with the solution we found by substitution in Example 2a of the preceding section. ■

EXAMPLE 2 Solve the following system.

$$\begin{cases} x + 2y = 4 \\ 5x + 10y = 10 \end{cases}$$

Solution Subtracting five times the first equation from the second, we obtain

$$\begin{cases} x + 2y = 4 \\ 0 = -10 \end{cases}$$

Since $0 \neq -10$, there can be no solutions. Of course, this is reasonable since we recognize that the two equations represent distinct parallel lines. ■

EXAMPLE 3 Mixing a given quantity of 30 percent silver alloy with a quantity of 90 percent silver alloy yields a 54 percent silver alloy. However, the mixture would have been only 50 percent silver had 20 units less of the 90 percent alloy been used. How many units of each alloy were used?

Solution In solving any word problem, we begin by labeling the quantities we are expected to find. Thus we let

$$x = \text{number of units of 30 percent alloy}$$
$$y = \text{number of units of 90 percent alloy}$$

The situation is illustrated in Figure 7–4. The quantity of silver in each "container" is listed below each.

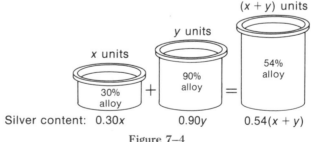

Silver content: $0.30x$ $0.90y$ $0.54(x + y)$

Figure 7–4

Thus we obtain the equation

$$0.30x + 0.90y = 0.54(x + y) \tag{1}$$

Had $(y - 20)$ units of the second alloy been used, computation of the corresponding amounts yields the equation

$$0.30x + 0.90(y - 20) = 0.50[x + (y - 20)] \tag{2}$$

We thus have a system of equations

$$\begin{cases} 0.30x + 0.90y = 0.54(x + y) & (1) \\ 0.30x + 0.90(y - 20) = 0.50(x + y - 20) & (2) \end{cases}$$

Multiplying each equation by 100, collecting terms, and simplifying, we can obtain the equivalent system

$$\begin{cases} -2x + 3y = 0 \\ -x + 2y = 40 \end{cases}$$

Solving this system by the elimination method yields

$$x = 120 \quad \text{and} \quad y = 80$$

Thus 120 units of 30 percent alloy and 80 units of 90 percent alloy were used.

The reader should verify that these values satisfy the conditions of the original problem. ■

EXAMPLE 4 Ron and Ken working together can paint a house in 45 hours. But after 30 hours Ron is stricken with "painter's elbow" and quits. Ken keeps working and finally finishes the job 40 hours later. How long would it have taken each of them to paint the house alone?

Solution Let

x = number of hours needed for Ron to paint the house alone
y = number of hours needed for Ken to paint the house alone

Then Ron paints $1/x$ of the house in 1 hour, and Ken paints $1/y$ of the house in 1 hour. But working together they paint 1/45 of the house in 1 hour. This translates into the equation.

$$\frac{1}{x} + \frac{1}{y} = \frac{1}{45} \qquad (1)$$

After 30 hours, Ron has painted $30 \cdot 1/x$ of the house. Ken works a total of 70 hours, thereby painting $70 \cdot 1/y$ of the house. After this time the entire house is painted. We thus write the equation

$$30 \cdot \frac{1}{x} + 70 \cdot \frac{1}{y} = 1 \qquad (2)$$

where the 1 indicates that the entire house is painted. These two equations give us the system

$$\begin{cases} \dfrac{1}{x} + \dfrac{1}{y} = \dfrac{1}{45} & (1) \\ 30 \cdot \dfrac{1}{x} + 70 \cdot \dfrac{1}{y} = 1 & (2) \end{cases}$$

which can be treated as a system in the variables $1/x$ and $1/y$. The elimination method yields

$$\frac{1}{x} = \frac{1}{72} \quad \text{and} \quad \frac{1}{y} = \frac{1}{120}$$

Thus $x = 72$ and $y = 120$. Ron can paint the house in 72 hours and Ken can paint the house in 120 hours without any help.

Again, verify this result by checking the original statement. ■

EXERCISES 7.2

Solve the following systems by the elimination method.

ODD

1. $\begin{cases} t - u = -2 \\ t + 2u = 1 \end{cases}$

2. $\begin{cases} v + w = 1 \\ v - 2w = 8 \end{cases}$

3. $\begin{cases} 2w + z = 1 \\ 2w - z = 3 \end{cases}$

4. $\begin{cases} 8r + 2s = 3 \\ 32r + 8s = 12 \end{cases}$

5. $\begin{cases} 3x + 5y = 1 \\ x - 7y = 9 \end{cases}$

6. $\begin{cases} 6u - 5v = 10 \\ 9u - 4v = -13 \end{cases}$

7. $\begin{cases} 4x - z = 5 \\ 4x + 3z = 17 \end{cases}$

8. $\begin{cases} x - y = 1 \\ 6x + 5y = 28 \end{cases}$

9. $\begin{cases} 3y + 2z = 6 \\ 6y + 4z = 12 \end{cases}$

10. $\begin{cases} z - 2u = 10 \\ -5z + 10u = 3 \end{cases}$

11. $\begin{cases} 5t - 6v = 7 \\ 6t - 7v = 9 \end{cases}$

12. $\begin{cases} 9w + 4x = -1 \\ 3w + 2x = 1 \end{cases}$

13. $\begin{cases} 2r - 3s = 7 \\ 3r - 4s = 8 \end{cases}$

14. $\begin{cases} 7r - 6y = -10 \\ 3r + 2y = 14 \end{cases}$

15. $\begin{cases} z - 4t = 7 \\ z + 2t = 1 \end{cases}$

16. $\begin{cases} 3w - v = 7 \\ 2w + 4v = 0 \end{cases}$

17. $\begin{cases} 2v - 2r = 4 \\ 5v - 5r = 5 \end{cases}$

18. $\begin{cases} 7r + 6t = 10 \\ 13r + 8t = 6 \end{cases}$

19. $\begin{cases} 6w - 2y = -5 \\ w + 7y = 1 \end{cases}$

20. $\begin{cases} 2s + 3t = 7 \\ s - 4t = -24 \end{cases}$

21. $\begin{cases} 6x - 4y = 7 \\ 5x - 3y = 6 \end{cases}$

22. $\begin{cases} 3u - 2v = 4 \\ 2u - v = 3 \end{cases}$

23. $\begin{cases} r - s = 2 \\ 2r - 3s = -1 \end{cases}$

24. $\begin{cases} a - 5b = 8 \\ 2a + 3b = -10 \end{cases}$

25. $\begin{cases} 3p + 2q = 8 \\ 2p + q = 5 \end{cases}$

26. One number is 9 more than twice another number, whereas twice the other number is 5 more than three times the first. What are the two numbers?

27. A two-digit number is 2 more than six times the sum of the digits. If the digits are reversed the number is decreased by 18. What is the number? (See Exer. 39, Sect. 7.1.)

28. Jean has $2.35 in dimes and quarters. The num-

ber of quarters is five less than twice the number of dimes. How many dimes and how many quarters does she have?

29. If 5 pounds of fudge and 7 pounds of mints cost $5.62, whereas 8 pounds of fudge and 8 pounds of mints cost $8.48, find the prices of each per pound.

30. An 18 percent acid solution is mixed with a 27 percent acid solution to yield a 22 percent acid solution. Had 15 units more of the 18 percent solution been used the resulting mixture would have been 21 percent acid. How many units of each solution were used?

31. Mr. Rammel invests a certain sum of money at 7 percent interest and another sum at 6 percent to achieve an annual yield of $628. Were he to interchange his investments, his annual interest would decrease by $8. What sums did he invest at each rate of interest?

32. Eileen flew 300 miles into the city in 90 minutes and back in 3 hours. What was her airspeed and what was the wind velocity?

33. Two years ago Laara's mother was eight times as old as Laara. Eight years from now her mother will be only three times as old as Laara. What are their ages now?

34. Joseph and Adolph can empty a truck in 12 hours working together. But their schedules do not coincide. Joseph works for 4 hours alone, after which Adolph empties the truck in 18 hours. How many hours would it have taken each of them to empty the truck alone?

35. Two pipes with different capacities are used to fill a pool. If both are working, the pool can be filled in 20 hours. After 8 hours the first well runs dry; the second continues pumping for another 27 hours in order to fill the pool. How

many hours would it take each of them individually to fill the pool?

36. Mr. Frilling rents three farms paying $40, $30, and $25 per acre, respectively. The total rent for the $25-acre farm is one half the total combined rent for the other two farms. If his total rental fee is $15,000 and the total acreage is 480, how much is his rental fee for each farm?

37. Decreasing the length of a rectangle by 3 inches and increasing its width by 2 inches decreases its area by 16 square inches. On the other hand, increasing its length by 4 inches and decreasing its width by 2 inches increases its area by 32 square inches. Find the area of the rectangle.

7.3 SYSTEMS WITH THREE VARIABLES

Both the substitution method and the elimination method can be used on systems involving more than two unknowns or equations. But the substitution method gets to be very messy for large systems. The elimination method is much more advantageous when more than two equations are present. A compact notation or "record-keeping system" for the elimination method will be introduced to handle systems having three or more equations or unknowns.

Just as with two equations in two unknowns, larger systems may be

1. Consistent: have precisely one solution

2. Inconsistent: have no solutions

3. Dependent: have infinitely many solutions.

It can be shown that an equation of the form

$$ax + by + cz = d \qquad a, b, c \text{ not all } 0 \qquad (1)$$

represents a plane in three-dimensional space. Nevertheless, these are called linear equations. In general, an equation involving n variables is said to be linear if it has the form

$$ax_1 + bx_2 + \cdots + cx_n = d, \qquad a, b, \ldots, c \text{ not all } 0$$

The numbers $a, b, \ldots, c$ are called **coefficients** of $x_1, x_2, \ldots, x_n$, respectively.

One linear equation in three unknowns represents a plane and hence has infinitely many solutions. A system of two such equations then represents the intersection of two planes. This intersection could be a line, empty (if the planes are parallel), or a plane (if the planes coincide); see Figure 7–5.

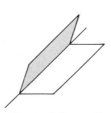

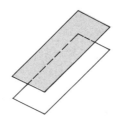

a. Planes intersecting in a line **b.** Parallel planes **c.** Coincident planes

Figure 7–5

In any case, the solution to a system of two linear equations in three unknowns, being the intersection of two planes, cannot be a single point. In fact, any linear system with fewer equations than unknowns never has precisely one solution; it is necessarily dependent or inconsistent.

EXAMPLE 1   Solve the following system.

$$\begin{cases} x + 2y + 2z = 7 \\ 2x - 2y - 5z = -1 \end{cases}$$

Solution   Adding the second equation to the first yields the equivalent system,

$$\begin{cases} 3x \qquad - 3z = 6 \\ 2x - 2y - 5z = -1 \end{cases}$$

Divide the first equation by 3:

$$\begin{cases} x \qquad - z = 2 \\ 2x - 2y - 5z = -1 \end{cases}$$

Subtract twice the first equation from the second:

$$\begin{cases} x \qquad - z = 2 \\ \qquad - 2y - 3z = -5 \end{cases}$$

From the first equation we obtain

$$x = z + 2$$

from the second

$$y = \frac{5 - 3z}{2}$$

For *any* value of z, these corresponding x and y values yield a solution to the system. The solution consists of all points of the form

$$\left(t + 2, \frac{5 - 3t}{2}, t \right) \qquad \text{for any real number } t$$ ■

If we have three equations in three unknowns, we are essentially intersecting a third plane with one of the configurations in Figure 7–5. If two or more of these planes coincide, one of the configurations in Figure 7–5 results. But if no two planes coincide, the possibilities are illustrated in Figure 7–6.

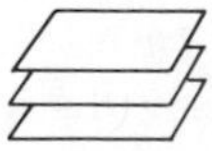

a. Three parallel planes

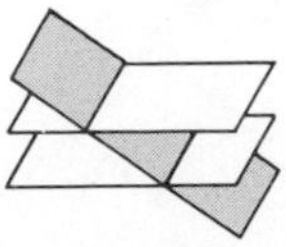

b. Two parallel lines

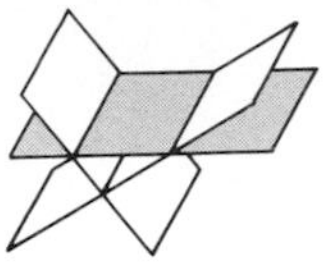

c. Three parallel lines

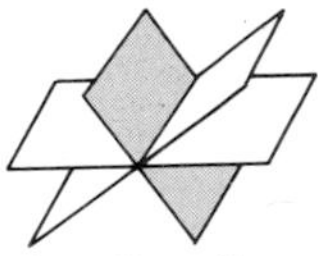

d. One line

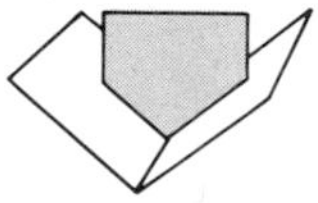

e. A unique point
of intersection

Figure 7–6

Figures 7–5(b) and 7–6(a), (b), and (c) represent inconsistent systems. Dependent systems are illustrated in Figure 7–5(a), (c) and 7–6(d). Only Figure 7–6(e) represents a consistent system.

A linear system with more equations than unknowns could also be consistent, dependent, or inconsistent. To visualize this, consider a system of four linear equations in three variables. Each equation represents a plane. If the first three equations have a unique solution, then the fourth equation may or may not represent a plane through this point; such a system may have exactly one or no solutions. For a dependent system, consider

$$\begin{cases} 2x + 4y + 4z = 14 \\ x + 2y + 2z = 7 \\ 2x - 2y - 5z = -1 \\ 4x - 4y - 10z = -2 \end{cases}$$

The first two equations represent a single plane (hence the same solutions) as do the last two. Figure 7–5(a) results.

EXAMPLE 2   Solve the system

$$\begin{cases} 2x + 4y + 3z = 4 \\ x + 2y + z = 1 \\ 3x - y - z = 2 \end{cases} \qquad \text{(a)}$$

Solution   It will be convenient to do several steps of the elimination method at one time. In order to indicate what is being done we shall refer to the equations as ①, ②, and ③ in that order. The operation of adding twice the third equation to the second will be indicated as $2 \cdot ③ + ②$.

The preceding system (a) is equivalent to the following.

$$\begin{matrix} ① + 3 \cdot ③ \\ ② + ③ \end{matrix} \qquad \begin{cases} 11x + y = 10 \\ 4x + y = 3 \\ 3x - y - z = 2 \end{cases} \qquad \text{(b)}$$

We shall use the same notation in finding another equivalent system. However, ① will now refer to the first equation of system (b). System (b) is equivalent to

$$\begin{matrix} ① - ② \end{matrix} \qquad \begin{cases} 7x = 7 \\ 4x + y = 3 \\ 3x - y - z = 2 \end{cases} \qquad \text{(c)}$$

In system (c), we find

$$x = 1$$

from the first equation. Substituting $x = 1$ into the second equation yields

$$4 \cdot 1 + y = 3$$
$$y = -1$$

Finally, substituting $x = 1$ and $y = -1$ into the third equation yields

$$3 \cdot 1 - (-1) - z = 2$$
$$z = 2$$

The solution is $(x, y, z) = (1, -1, 2)$.

We check by substituting these values into the original equations:

$$2 \cdot 1 + 4(-1) + 3 \cdot 2 \overset{\checkmark}{=} 4$$
$$1 + 2(-1) + 2 \overset{\checkmark}{=} 1$$
$$3 \cdot 1 - (-1) - 2 \overset{\checkmark}{=} 2$$

System (c) of Example 2 is said to be in **echelon form.** Echelon form means that

1. The first equation contains precisely one variable.

2. Each succeeding equation contains only one variable that is not included in its predecessors.

When a system of equations is in echelon form, we can

1. Use the first equation to solve for one of the variables.

2. Substitute this value into the next equation to solve for another variable.

3. Continue substituting known values into successive equations in order to determine more values.

Generally, we do not completely reduce a system of equations to echelon form. We stop when it is evident that simply interchanging equations would result in echelon form. Such a system will be said to be in **modified echelon form.** After this stage, the solution procedure will be the same: solve for one of the variables, use its value to solve for another, and so on.

EXAMPLE 3   Solve the following system of equations.

$$\begin{cases} 2x + y - z = 5 \\ x - 2y + z = 1 \\ -x + 2y + 2z = -4 \end{cases}$$

Solution   Using the notation developed above, we proceed as follows.

$$\begin{array}{c} ① - 2 \cdot ② \\ \\ ③ + ② \end{array} \qquad \begin{cases} 5y - 3z = 3 \\ x - 2y + z = 1 \\ 3z = -3 \end{cases}$$

From the last equation, we find

$$z = -1$$

Substituting $z = -1$ into the first equation gives

$$y = 0$$

Finally, substituting these values into the second equation yields

$$x = 2$$

The solution is

$$(x, y, z) = (2, 0, -1).$$

It is a simple matter to verify these values in the original equations.   ■

EXAMPLE 4   Solve the system

$$\begin{cases} 2x + 4y + 4z = 14 \\ 2x - 2y - 5z = -1 \\ x + 2y + 2z = 7 \end{cases}$$

Solution   The system is equivalent to

$$\begin{array}{c} \text{①} - 2 \cdot \text{③} \end{array} \quad \begin{cases} 0x + 0y + 0z = 0 \\ 2x - 2y - 5z = -1 \\ x + 2y + 2z = 7 \end{cases}$$

Since this first equation is satisfied for all (x, y, z), this system of three equations is equivalent to the following system of two equations.

$$\begin{cases} 2x - 2y - 5z = -1 \\ x + 2y + 2z = 7 \end{cases}$$

This is the system of Example 1; its solution set consists of all points of the form

$$\left(t + 2, \ \frac{5 - 3t}{2}, \ t \right) \qquad \text{for any real number } t \qquad ■$$

EXAMPLE 5   Solve the system

$$\begin{cases} x - 2y - 4z = 7 \\ 2x - 3y + z = 5 \\ 3x - 4y + 6z = 2 \end{cases}$$

Solution   Using the elimination method

$$\begin{array}{c} \text{②} - 2 \cdot \text{①} \\ \text{③} - 3 \cdot \text{①} \end{array} \quad \begin{cases} x - 2y - 4z = 7 \\ y + 9z = -9 \\ 2y + 18z = -19 \end{cases}$$

$$\begin{array}{c} \text{③} - 2 \cdot \text{②} \end{array} \quad \begin{cases} x - 2y - 4z = 7 \\ y + 9z = -9 \\ 0y + 0z = -1 \end{cases}$$

Since the last equation can never be satisfied, there are no solutions to this problem; the solution set is empty.   ■

EXAMPLE 6   A lawn company has three kinds of fertilizer labeled as grades A, B, and C, respectively. Grade A contains 40 percent plant food and 40 percent weed killer;

grade B contains 60 percent plant food and 20 percent weed killer; grade C contains 50 percent plant food and 24 percent weed killer. In mixing a ton (2000 lb) of fertilizer, how many pounds of each must be used to achieve a mixture containing $52\frac{1}{2}$ percent plant food and 26 percent weed killer?

Solution Let a = amount of A used in the mixture; b = amount of B used in the mixture; and c = amount of C used in the mixture. Then

$$(0.4)a + (0.6)b + (0.5)c = \text{total amount of plant}$$
$$\text{food in the mixture}$$
$$(0.4)a + (0.2)b + (0.24)c = \text{total amount of weed}$$
$$\text{killer in the mixture}$$
$$a + b + c = \text{total weight of mixture}$$

In 2000 pounds of mixture containing $52\frac{1}{2}$ percent plant food and 26 percent weed killer, there are $(0.525)2000 = 1050$ pounds of plant food and $(0.26)2000 = 520$ pounds of weed killer. Thus we obtain the equations

$$\begin{cases} (0.4)a + (0.6)b + (0.5)c = 1050 \\ (0.4)a + (0.2)b + (0.24)c = 520 \\ a + b + c = 2000 \end{cases}$$

We solve this sytem as follows

$$10①$$
$$100②$$
$$\begin{cases} 4a + 6b + 5c = 10500 \\ 40a + 20b + 24c = 52000 \\ a + b + c = 2000 \end{cases}$$

$$② - 10①$$
$$\begin{cases} 4a + 6b + 5c = 10500 \\ -40b - 26c = -53000 \\ a + b + c = 2000 \end{cases}$$

$$① - 4③$$
$$-②/2$$
$$\begin{cases} 2b + c = 2500 \\ 20b + 13c = 26500 \\ a + b + c = 2000 \end{cases}$$

$$② - 10①$$
$$③ - ①$$
$$\begin{cases} 2b + c = 2500 \\ 3c = 1500 \\ a - b = -500 \end{cases}$$

$$① - ②/3$$
$$②/3$$
$$\begin{cases} 2b = 2000 \\ c = 500 \\ a - b = -500 \end{cases}$$

We can now read off the solution as $b = 1000$ pounds, $c = 500$ pounds, and $a = 500$ pounds. It is a simple matter to verify that these amounts do satisfy the conditions stated in the problem. ∎

EXERCISES 7.3

Solve the following systems of equations by the elimination method.

1. $\begin{cases} 3x - y + 2z = 3 \\ 2x + 3y + z = -1 \end{cases}$

2. $\begin{cases} t - u + 2v = 7 \\ t - 2u + 3v = 11 \end{cases}$

3. $\begin{cases} 2r + 3s - t = 5 \\ 2r - 6s - 10t = -4 \end{cases}$

4. $\begin{cases} 4v - 5w + 3x = 3 \\ 3v - 4w + 2x = 1 \end{cases}$

5. $\begin{cases} 2x - y + z = -5 \\ x - 2y - 3z = 6 \\ x + y - 2z = 1 \end{cases}$

6. $\begin{cases} p - 2q + r = 8 \\ 5p - q + 3r = 6 \\ 3p - 6q + 3r = 14 \end{cases}$

7. $\begin{cases} t - 2u - 3v = 4 \\ 2t - 3u + v = 7 \\ 2t - 4u - 6v = -1 \end{cases}$

8. $\begin{cases} 2u - v + 2w = 5 \\ -u + v + 3w = 6 \\ 2u - 2v - w = 8 \end{cases}$

9. $\begin{cases} x - 3y + z = -9 \\ -2x + y + 3z = 8 \\ x - 5y + 3z = -13 \end{cases}$

10. $\begin{cases} p + 2q - 3r = 10 \\ p + q - r = 4 \\ 2p - q + 2r = 6 \end{cases}$

11. $\begin{cases} a - b + c = -1 \\ b + 2c = -6 \\ 3a + 2b + c = 3 \end{cases}$

12. $\begin{cases} t - v - w = 0 \\ 2t + v - 2w = 0 \\ t - 2v - w = 0 \end{cases}$

13. $\begin{cases} y - 2z + w = 5 \\ 3y + 2z - 2w = -13 \\ 2y - z + 3w = 11 \end{cases}$

14. $\begin{cases} 3t - 2z + p = 4 \\ 6t - 4z + 2p = 8 \\ 9t - 6z + 3p = 12 \end{cases}$

15. $\begin{cases} 3x + 4y + z = 6 \\ x + 3y - 2z = -7 \\ 2x + y + z = 1 \end{cases}$

16. $\begin{cases} 8a + 2b + 5c = 4 \\ 3a - b + 2c = 7 \\ a - b - 4c = -5 \end{cases}$

17. $\begin{cases} r - 3s + t = 3 \\ r + s + 3t = 3 \\ 2r + s + 2t = -1 \end{cases}$

18. $\begin{cases} x - 2y - 3z = 3 \\ 2x + y - z = 1 \\ x + y + z = -4 \end{cases}$

19. $\begin{cases} 3r - 2s + 5t = 12 \\ 2r - 5s + 3t = 4 \\ r + 3s + 2t = 8 \end{cases}$

20. $\begin{cases} -x + 2y + z = 8 \\ 2x - y + z = -1 \\ x + y + 2z = 7 \end{cases}$

21. $\begin{cases} u - 2v - w = 0 \\ 2u + v - w = 0 \\ 2u + v - 3w = 0 \end{cases}$

22. $\begin{cases} r + s = 2 \\ 3r + 2s = 1 \\ 5r + 4s = 5 \end{cases}$

23. $\begin{cases} 3x - 4y = 5 \\ x - 3y = 10 \\ 2x + 5y = 4 \end{cases}$

24. $\begin{cases} u + v - w = 3 \\ 2u - v + w = 0 \\ 2u + v + 3w = 0 \\ u - 2v - 2w = 0 \end{cases}$

25. $\begin{cases} p - 2q + 3r = 0 \\ 2p + 3q - 2r = 1 \\ 2p + q - r = 2 \\ p + q - r = 1 \end{cases}$

26. Clara has \$20,000 invested, some at 4 percent, some at 5 percent, and some at 8 percent. She collects \$1250 interest yearly. She has twice as much invested at 8 percent as at 5 percent. How much is invested at each rate of interest?

27. A fertilizer company has solutions containing 20 percent, 30 percent, and 70 percent nitrogen, respectively. Because of other characteristics of the liquids, the company wants half of any mixture to consist of the 30 percent solution. How many pounds of each solution should be used to make 1 ton of fertilizer containing 50 percent nitrogen?

28. Henry lost \$56 to the "one-armed-bandits" in Las Vegas. He lost three times as much on the half-dollar machines as on the quarter and nickel machines put together. The total number of coins lost was 224. How many nickels, quarters, and halves did he lose?

29. The sum of the digits of a three-digit number is 12. Reversing the digits increases the number by 99. The tens digit is 2 more than the sum of the units and hundreds digits. Find the number. (*Hint:* A three-digit number like 684 is computed as $684 = 6 \cdot 100 + 8 \cdot 10 + 4$; in general, if h = hundreds digit, t = tens digit, u = units digit, then $100h + 10t + u$ = the number. Reversing the order of the digits yields the number $100u + 10t + h$.)

30. On her paper route, Cindy collected \$5.20 in nickels, dimes, and quarters in a total of 33

coins. The number of quarters is one more than twice the number of nickels. How many of each type of coin does she have?

31. A three-digit number is 12 less than 25 times the sum of its digits. Reversing the digits increases the number by 99. The tens digit is one more than the sum of the hundreds and units digits. Find the number.

32. Find a three-digit number whose digits add up to 11 and whose tens digit is both twice the units digit and three times the hundreds digit.

33. Find a three-digit number whose units digit is one less than its tens digit; the sum of its digits is 10; if the digits are reversed, the new number is 297 more than the original.

34. Tara has $2.00 in nickels, dimes, and quarters. She has 14 coins in all, and there are twice as many dimes as nickels. How many of each type of coin does she have?

35. Kevin has 13 coins in nickels, dimes, and quarters totaling $1.45. If he has one more dime than quarters, find how many coins of each type he has.

36. Doris spent $1.88 for 14 stamps in 8¢, 10¢, and 30¢ denominations. If she purchased twice as many 8¢ stamps as 30¢ stamps, find how many stamps of each denomination she purchased.

7.4 MATRIX METHODS

The solution to a system of equations depends on the coefficients rather than on the specific names given to the variables. For instance, we would solve the two systems

$$\begin{cases} x + y = 2 \\ x - y = 0 \end{cases} \qquad \begin{cases} u + v = 2 \\ u - v = 0 \end{cases}$$

in exactly the same way. The sequence of steps to be used in the elimination method is determined by considering the coefficients of the various variables. In fact, we can solve a system of equations by the elimination method without ever writing the variables, provided we keep track of the coefficients in the proper manner.

For example, in the system of Example 2 of the preceding section, suppress the variables and write only the coefficients in a rectangular array as follows. Such an array is called a **matrix.**

System of equations

$$\begin{cases} 2x + 4y + 3z = 4 \\ x + 2y + z = 1 \\ 3x - y - z = 2 \end{cases}$$

Matrix

$$\begin{bmatrix} 2 & 4 & 3 & | & 4 \\ 1 & 2 & 1 & | & 1 \\ 3 & -1 & -1 & | & 2 \end{bmatrix}$$

The vertical broken line in this matrix is optional. It merely indicates the location of the equals signs ($=$) in the corresponding equations.

The matrix displayed here is said to have 3 (horizontal) **rows** and 4 (vertical) **columns.** It will be called a 3 by 4 (3×4) matrix. In general, a matrix with m rows and n columns is called an m by n ($m \times n$) matrix; if $m = n$ it is called a *square* matrix.

The steps in our solution of this example are repeated here. Included alongside the equivalent systems are their respective matrices.

$$\left.\begin{array}{rl} 11x + y & = 10 \\ 4x + y & = 3 \\ 3x - y - z & = 2 \end{array}\right\} \quad \begin{array}{l} ① + 3 \cdot ③ \\ ② + ③ \end{array} \quad \left[\begin{array}{ccc|c} 11 & 1 & 0 & 10 \\ 4 & 1 & 0 & 3 \\ 3 & -1 & -1 & 2 \end{array}\right]$$

$$\left.\begin{array}{rl} 7x & = 7 \\ 4x + y & = 3 \\ 3x - y - z & = 2 \end{array}\right\} \quad ① - ② \quad \left[\begin{array}{ccc|c} 7 & 0 & 0 & 7 \\ 4 & 1 & 0 & 3 \\ 3 & -1 & -1 & 2 \end{array}\right]$$

It is clear that each succeeding matrix is obtained from its predecessor in exactly the same way that succeeding equivalent systems are. That is to say, a linear system can be solved by manipulating its matrix in the same way that we manipulate the system itself. This method of solution is called the **matrix form** of the elimination method.

If M is the matrix for a system of equations S, then, since a row of M corresponds to an equation in S, the following operations on M yield the matrix for a system of equations equivalent to the original system S:

1. Interchanging two rows

2. Multiplying any row by a nonzero constant

3. Adding a nonzero multiple of one row to another

EXAMPLE 1 Solve the following system of equations by using the matrix form of the elimination method.

$$\begin{cases} 3x + 4y + z = 0 \\ 5x + 3y + z = 1 \\ x - 4y - 3z = 5 \end{cases}$$

Solution Write the matrix for the system and find matrices for equivalent systems as indicated here. The notation of the previous section is adapted so that ① + ③ now means that row ① is added to row ③. To assist our memory, we shall label the columns of the matrix corresponding to the variables x, y, and z.

$$\begin{array}{ccc} x & y & z \end{array}$$
$$\left[\begin{array}{ccc|c} 3 & 4 & 1 & 0 \\ 5 & 3 & 1 & 1 \\ 1 & -4 & -3 & 5 \end{array}\right]$$

$$\begin{array}{l} ② - ① \\ ③ + ① \end{array} \quad \left[\begin{array}{ccc|c} 3 & 4 & 1 & 0 \\ 2 & -1 & 0 & 1 \\ 4 & 0 & -2 & 5 \end{array}\right]$$

$$① + 4 \cdot ② \quad \left[\begin{array}{ccc|c} 11 & 0 & 1 & 4 \\ 2 & -1 & 0 & 1 \\ 4 & 0 & -2 & 5 \end{array}\right]$$

$$[\text{③} + 2 \cdot \text{①}]/13 \qquad \begin{bmatrix} 11 & 0 & 1 & | & 4 \\ 2 & -1 & 0 & | & 1 \\ 2 & 0 & 0 & | & 1 \end{bmatrix}$$

$$\text{②} - \text{③} \qquad \begin{bmatrix} 11 & 0 & 1 & | & 4 \\ 0 & -1 & 0 & | & 0 \\ 2 & 0 & 0 & | & 1 \end{bmatrix}$$

This last matrix corresponds to the system

$$\begin{cases} 11x & + z = 4 \\ \quad - y & = 0 \\ 2x & = 1 \end{cases}$$

Since this system is equivalent to the original, we can read off a partial solution as

$$y = 0 \qquad x = \frac{1}{2}$$

Of course, this partial solution can just as easily be read from the last two rows of the matrix once the appropriate columns are reassociated with their respective variables. In any event, substituting these values into the equation represented by the first row yields

$$11 \cdot \left(\frac{1}{2} \right) + z = 4$$

$$z = \frac{-3}{2}$$

The solution is $(x, y, z) = (1/2, 0, -3/2)$. No matter what solution technique is used, we should verify the solution by substitution into the original equations. That is left to the reader. ∎

EXAMPLE 2 Solve the system

$$\begin{cases} 3w - 4x - 2y + z = 11 \\ 2w - 2x - 3y + 2z = 13 \\ 2w - 2x + 2y + z = 5 \\ 2w + 4x - y + 3z = 14 \end{cases}$$

Solution   Use the matrix method.

$$\begin{array}{cccc} w & x & y & z \\ \begin{bmatrix} 3 & -4 & -2 & 1 & | & 11 \\ 2 & -2 & -3 & 2 & | & 13 \\ 2 & -2 & 2 & 1 & | & 5 \\ 2 & 4 & -1 & 3 & | & 14 \end{bmatrix} \end{array}$$

$$
\begin{array}{c}
② - ③ \\
\\
④ - ③
\end{array}
\qquad
\left[
\begin{array}{cccc|c}
3 & -4 & -2 & 1 & 11 \\
0 & 0 & -5 & 1 & 8 \\
2 & -2 & 2 & 1 & 5 \\
0 & 6 & -3 & 2 & 9
\end{array}
\right]
$$

$$
\begin{array}{c}
① - ② \\
\\
③ - ② \\
④ - 2 \cdot ②
\end{array}
\qquad
\left[
\begin{array}{cccc|c}
3 & -4 & 3 & 0 & 3 \\
0 & 0 & -5 & 1 & 8 \\
2 & -2 & 7 & 0 & -3 \\
0 & 6 & 7 & 0 & -7
\end{array}
\right]
$$

$$
\begin{array}{c}
① - 2 \cdot ③ \\
\\
[③ - ④] \div 2
\end{array}
\qquad
\left[
\begin{array}{cccc|c}
-1 & 0 & -11 & 0 & 9 \\
0 & 0 & -5 & 1 & 8 \\
1 & -4 & 0 & 0 & 2 \\
0 & 6 & 7 & 0 & -7
\end{array}
\right]
$$

$$
\begin{array}{c}
- ① \\
\\
③ + ① \\
2 \cdot ④
\end{array}
\qquad
\left[
\begin{array}{cccc|c}
1 & 0 & 11 & 0 & -9 \\
0 & 0 & -5 & 1 & 8 \\
0 & -4 & -11 & 0 & 11 \\
0 & 12 & 14 & 0 & -14
\end{array}
\right]
$$

$$
\begin{array}{c}
\\
\\
\\
[④ + 3 \cdot ③]/19
\end{array}
\qquad
\left[
\begin{array}{cccc|c}
1 & 0 & 11 & 0 & -9 \\
0 & 0 & -5 & 1 & 8 \\
0 & -4 & -11 & 0 & 11 \\
0 & 0 & -1 & 0 & 1
\end{array}
\right]
$$

$$
\begin{array}{c}
① + 11 \cdot ④ \\
② - 5 \cdot ④ \\
③ - 11 \cdot ④
\end{array}
\qquad
\left[
\begin{array}{cccc|c}
1 & 0 & 0 & 0 & 2 \\
0 & 0 & 0 & 1 & 3 \\
0 & -4 & 0 & 0 & 0 \\
0 & 0 & -1 & 0 & 1
\end{array}
\right]
$$

The solution is

$$
w = 2 \qquad z = 3 \qquad x = 0 \qquad y = -1
$$

Checking the solution, we have

$$
3 \cdot 2 - 4 \cdot 0 - 2(-1) + 3 \overset{\leq}{=} 11
$$

$$
2 \cdot 2 - 2 \cdot 0 - 3(-1) + 2 \cdot 3 \overset{\leq}{=} 13
$$

$$
2 \cdot 2 - 2 \cdot 0 + 2(-1) + 3 \overset{\leq}{=} 5
$$

$$
2 \cdot 2 + 4 \cdot 0 - (-1) + 3 \cdot 3 \overset{\leq}{=} 14
$$

Checking the results is especially important when long, drawn-out computations are involved in finding the solution. Had an error been indicated when checking the solution in Example 2, the records that we kept along the way (① − ②, ④ − 2 · ②, etc.) would be invaluable in tracking down the error or errors.

We have not listed a specific sequence of steps to follow when using the matrix method. There are generally several different sequences of steps that will yield a solution. Keep in mind, however, that before the matrix is obtained, the variables must be written in the same order in each equation. Each column of the matrix lists the coefficients of a single variable. The following is a *general* outline of the matrix method.

THE MATRIX METHOD OF SOLVING A SYSTEM OF EQUATIONS

1. Write the variables in the same order in each equation.

2. Extract the matrix by suppressing the variables and replacing the equal signs with a broken vertical line. If desired, you may label the columns of the matrix corresponding to the respective variables.

3. Focus on a given column and use the steps of the elimination method to reduce this column to, at most, one nonzero entry.

4. Repeat step 3 but focus on another column and take care so as not to lose any zeros in the previously "zeroed" column(s).

5. When no more columns can be "zeroed" without losing a previously zeroed column, it is time to read off the solution. If desired, the matrix can be rewritten as a system of equations at this stage.

Actually, it is not necessary to "zero" the columns to the extent indicated in steps 3 and 4. Reducing the matrix to echelon form or modified echelon form would also lead to a solution as in Example 1.

Occasionally, one of the rows in the matrix will reduce to something like

$$\begin{bmatrix} 0 & 0 & 0 & 0 & | & 7 \end{bmatrix}$$

Such a row would translate into the equation

$$0 = 7$$

that is, *if* the equations are satisfied, then $0 = 7$. Thus this means there are no solutions.

The system of Example 1 in Section 7.3 gives rise to another situation. Had we solved this system by the matrix method, the final matrix would have been

$$\begin{array}{ccc} x & y & z \\ \begin{bmatrix} 1 & 0 & -1 & | & 2 \\ 0 & 1 & \dfrac{3}{2} & | & \dfrac{5}{2} \end{bmatrix} \end{array}$$

In this case, some of the variables (*x* and *y* here) can only be solved in terms of another variable or variables (z in this illustration). There are infinitely many solutions. In fact, here we have

$$x = z + 2$$
$$y = -\frac{3}{2}z + \frac{5}{2}$$

for any value of z.

The matrix method is especially useful for large systems because it cuts down on the writing needed to solve a problem. Another advantage is that the numbers are easier to see when the variables are suppressed. Since only the coefficients are used and the names of the variables are suppressed, this method lends itself to machine computation. It is very easy to program a computer to solve a system of equations by using the matrix method. Generally, the machine is instructed to put the matrix (or system of equations) in echelon form; it is then instructed to compute the respective values.

EXAMPLE 3   Find the equation of the circle passing through the three points $(2, 2)$, $(-3, 7)$, $(1, 5)$; find its radius and center.

Solution   Recall that the general equation of a circle (Sect. 3.1) is

$$x^2 + y^2 + ax + by + c = 0$$

If this circle is to pass through the given points, the following equations must be satisfied.

$$
\begin{array}{ll}
(2, 2): & 4 + 4 + 2a + 2b + c = 0 \\
(-3, 7): & 9 + 49 - 3a + 7b + c = 0 \\
(1, 5): & 1 + 25 + a + 5b + c = 0
\end{array}
$$

Collecting terms results in the following:

$$
\left\{
\begin{array}{r}
2a + 2b + c = -8 \\
-3a + 7b + c = -58 \\
a + 5b + c = -26
\end{array}
\right.
$$

Thus what originally seems to be a system of quadratic equations is in actuality a linear system since we are seeking a, b, and c. This system can be solved in the usual way to find

$$a = 6 \qquad b = -4 \qquad c = -12$$

The equation of this circle is

$$x^2 + y^2 + 6x - 4y - 12 = 0$$

Completing the squares will recast this equation in a form that indicates the radius and center of the circle:

$$(x^2 + 6x \quad\;) + (y^2 - 4y \quad\;) = 12$$
$$(x^2 + 6x + 9) + (y^2 - 4y + 4) = 12 + 9 + 4$$
$$(x + 3)^2 + (y - 2)^2 = 25$$

The circle is centered at $(-3, 2)$ and has radius 5. ▪

EXERCISES 7.4

Solve the following systems by the matrix method.

1. $\begin{cases} 2x - 2y - z = -7 \\ -x + y + 3z = 6 \\ 2x - y + 2z = 5 \end{cases}$

2. $\begin{cases} 3a + 2b + c = 1 \\ a - b + c = 0 \\ b + 2c = -2 \end{cases}$

3. $\begin{cases} 2r + s + t = 0 \\ r - 3s - 2t = 0 \\ r - s - 2t = 0 \end{cases}$

4. $\begin{cases} 2u - v + w = -3 \\ u - 2v - 3w = 4 \\ u + v - 2w = 5 \end{cases}$

5. $\begin{cases} 2r - s - 3t = 0 \\ r - 5s - 6t = -9 \\ r - 2s - 3t = -3 \end{cases}$

6. $\begin{cases} x + y - z = 4 \\ 2x - y + 2z = 6 \\ x + 2y - 3z = 10 \end{cases}$

7. $\begin{cases} p - 2q + r = 8 \\ 2p - q + 3r = 11 \\ 3p + 2q - 2r = 3 \end{cases}$

8. $\begin{cases} 2u - v + w = 8 \\ -u + 2v + w = -1 \\ u + v + 2w = 7 \end{cases}$

9. $\begin{cases} 2a + 5b = -2 \\ a - 3b = 10 \\ 3a + 4b = 4 \end{cases}$

10. $\begin{cases} x + y - z = 1 \\ 2x + 3y - 2z = 1 \\ x - 2y + 3z = 0 \\ 2x + y - z = 2 \end{cases}$

11. $\begin{cases} r - 2s - t - 3u = -9 \\ r + s - t = 0 \\ 3r + 4s + u = 6 \\ 2s - 2t + u = 3 \end{cases}$

12. $\begin{cases} -x + y + z - 2w = 1 \\ x - y + 2z - w = -1 \\ 3x + y + 2z - 2w = -2 \\ -2x - 2y + z + 2w = 3 \end{cases}$

13. $\begin{cases} 2r - s + t + u = 1 \\ r + 3s - 2t + 5u = 0 \\ 3r - 2s + 4t - u = 0 \\ -r + s - 3t + 6u = 1 \end{cases}$

14. $\begin{cases} 3u - 14v + 4w + 2x = -1 \\ -2u + 13v + 2w - 2x = 3 \\ u - 5v - 2w + 2x = 2 \\ u - 11v - 4w + 3x = -2 \end{cases}$

15. $\begin{cases} -a + 3b + 2c - d = -6 \\ 3a - 2b + c - 6d = 2 \\ -2a + b + 2c = -2 \\ a - 3b + c - 4d = 4 \end{cases}$

16. $\begin{cases} 2x + y - 2z = 1 \\ -x - 2y + z - 5t = -4 \\ -x - y - 2z + t = 2 \\ 3x + y + 3z - 10t = -1 \end{cases}$

17. $\begin{cases} 2r + s - t = 3 \\ r - 3s + 2t = -4 \\ 3r + s - 3t + u = 1 \\ r + 2s - 4t - u = -2 \end{cases}$

18. $\begin{cases} -2u + v - x = -6 \\ u + 2w - 3x = 4 \\ -2v + 3w - 6x = -2 \\ -u + 3v + w - 2x = 0 \end{cases}$

19.
$$\begin{cases} 4u - 3v + w + 2x + 2y = -4 \\ u - 2v + 2w + 2x - 3y = -8 \\ -2u + v - w + x = 8 \\ -u - 2v + 2w + 3x - 7y = -8 \\ 2u - v - 2w - x - 5y = -4 \end{cases}$$

20.
$$\begin{cases} r + s + t + u + 2v = 3 \\ 3r + s - t - v = 0 \\ 2r - 2s - t + 2u + v = 4 \\ 4r - s - 2t - u - 2v = 0 \\ 2r - 3s + t - u + 4v = 6 \end{cases}$$

In each of the following, sketch and find the equation of the circle passing through the three points. Indicate its center and radius.

21. $(5, 5)$, $(4, 6)$, $(1, 7)$

22. $(-2, -16)$, $(3, 9)$, $(10, 2)$

23. $(0, 0)$, $(48, 14)$, $(17, -17)$

24. $(1, 7)$, $(-2, 8)$, $(-6, 6)$

25. $(-6, -4)$, $(2, -16)$, $(19, 1)$

26. $(4, -20)$, $(22, 4)$, $(21, -3)$

27. Old MacDonald has three farms, which he rents to his neighbors. His holdings total 960 acres. The farms rent for $80, $60, and $50 per acre, respectively. His combined income from the higher priced rentals exactly doubles his income from the $50-per-acre farm. If his total income is $60,000, how large is each farm?

28. Dan, Paul, and Les can build a garage in 24 hours. However, they cannot always work together. If Les works alone for 54 hours, Dan and Paul can finish the job in 15 hours. If Dan works alone for 17 hours, Paul and Les can finish the job in 30 hours. How long would it take each one alone to build a garage? (*Hint:* See Example 4, Sect. 7.2.)

29. Sam Showplace uses three brands of combina- tion weed killer and plant food on his lawn. Su- per Feeder brand is 80 percent plant food, 10 percent weed killer, and 10 percent filler ma- terial. Weed-Go brand contains 30 percent plant food, 33 percent weed killer, and 37 percent filler. In bulk (no label), the mixture is 60 per- cent plant food, 12 percent weed killer, and 28 percent filler. How should Sam prepare his mix- ture in order to obtain 65 percent plant food, 16 percent weed killer, and 19 percent filler? (See Example 3, Sect. 7.2.)

30. Three pipes are used to fill a pool. Pipe 1 and pipe 2 both run for 2 hours; then these are turned off and the third is turned on for 1 hour and the pool is filled. Last week someone turned pipe 1 and pipe 2 off after one hour, then turned pipe 3 on. It then took pipe 3 five hours and 30 minutes to finish filling the pool. Yester- day all three were turned on simultaneously. Af- ter 30 minutes, pipe 3 stopped. After another 30 minutes, pipe 2 stopped. Two hours later the pool was filled. How long would it take each pipe alone to fill the pool?

31. Of three numbers, the third is one less than the sum of the other two. The third is also eight more than twice the second. The first is one less than three times the difference of the last two (third minus second). Find the numbers.

32. The units digit of a three-digit number is one more than the sum of the other two digits. If the digits are reversed, the number is in- creased by 792. If the hundreds and tens digits are interchanged, the original number is in- creased by 540. Find the number. (See Exer. 29, Sect. 7.3.)

7.5 DETERMINANTS

In the preceding section, we indicated that the matrix of a system of equations can be used to solve the system; using the matrix eliminates the need to write down the variables repeatedly. There is another way in which the matrix can be used to find the solution. Central to this method of solution is the concept of **determinant** of a matrix. In this section, we shall be working solely with "square" matrices. An $n \times n$ matrix or determinant will be said to have order n.

The determinant of a square matrix will be defined inductively; that is, a determinant of order $(n + 1)$ will be defined in terms of determinants of order *n*. The definitions may seem somewhat arbitrary at first, but we shall show that determinants arise naturally in the solution of linear systems.

DEFINITION

The determinant of a 2×2 matrix is *defined* as

$$\begin{vmatrix} a & b \\ c & d \end{vmatrix} = ad - bc$$

EXAMPLE 1   **a.** $\begin{vmatrix} 1 & 2 \\ 3 & 4 \end{vmatrix} = 1 \cdot 4 - 2 \cdot 3 = -2$

b. $\begin{vmatrix} -2 & 4 \\ 5 & 3 \end{vmatrix} = (-2) \cdot 3 - 4 \cdot 5 = -26$ ∎

To define determinants of higher order, we need the concept of "minor."

DEFINITION

For any square matrix M, the **minor** M_{ij} is the determinant of the matrix that remains when the ith row and the jth column of M are removed.

EXAMPLE 2   Let $M = \begin{bmatrix} 1 & 2 & 0 \\ 3 & -1 & -2 \\ 4 & -3 & 5 \end{bmatrix}$

Then

$$M_{11} = \begin{vmatrix} -1 & -2 \\ -3 & 5 \end{vmatrix} = -5 - 6 = -11$$

$$M_{13} = \begin{vmatrix} 3 & -1 \\ 4 & -3 \end{vmatrix} = -9 + 4 = -5$$

$$M_{32} = \begin{vmatrix} 1 & 0 \\ 3 & -2 \end{vmatrix} = -2$$ ∎

DEFINITION

A 3×3 determinant is defined in terms of 2×2 determinants:

$$\begin{vmatrix} a_{11} & a_{12} & a_{13} \\ a_{21} & a_{22} & a_{23} \\ a_{31} & a_{32} & a_{33} \end{vmatrix} = a_{11}M_{11} - a_{12}M_{12} + a_{13}M_{13}$$

EXAMPLE 3

$$\begin{vmatrix} 1 & 2 & 0 \\ 3 & -1 & -2 \\ 4 & -3 & 5 \end{vmatrix} = 1 \begin{vmatrix} -1 & -2 \\ -3 & 5 \end{vmatrix} - 2 \begin{vmatrix} 3 & -2 \\ 4 & 5 \end{vmatrix} + 0 \begin{vmatrix} 3 & -1 \\ 4 & -3 \end{vmatrix}$$

$$= 1(-5 - 6) - 2(15 + 8) + 0$$

$$= -57$$

■

The determinants of order 4 are defined in terms of determinants of order 3; then determinants of order 5 are defined in terms of determinants of order 4 and so on.

DEFINITION

$$\begin{vmatrix} a_{11} & \cdots & a_{1n} \\ \cdot & & \cdot \\ \cdot & & \cdot \\ \cdot & & \cdot \\ a_{n1} & \cdots & a_{nn} \end{vmatrix} = a_{11}M_{11} - a_{12}M_{12} + \cdots + (-1)^{1+n}a_{1_n}M_{1_n}$$

Since other methods of evaluating determinants will be given later, this particular evaluation will be called "expansion by the first row."

EXAMPLE 4

$$\begin{vmatrix} 1 & 2 & 3 & 4 \\ -1 & -3 & 0 & 2 \\ 0 & 1 & -4 & -1 \\ 2 & 0 & 1 & 5 \end{vmatrix}$$

$$= 1 \begin{vmatrix} -3 & 0 & 2 \\ 1 & -4 & -1 \\ 0 & 1 & 5 \end{vmatrix} - 2 \begin{vmatrix} -1 & 0 & 2 \\ 0 & -4 & -1 \\ 2 & 1 & 5 \end{vmatrix}$$

$$+ 3 \begin{vmatrix} -1 & -3 & 2 \\ 0 & 1 & -1 \\ 2 & 0 & 5 \end{vmatrix} - 4 \begin{vmatrix} -1 & -3 & 0 \\ 0 & 1 & -4 \\ 2 & 0 & 1 \end{vmatrix}$$

$$= \left((-3) \begin{vmatrix} -4 & -1 \\ 1 & 5 \end{vmatrix} - 0 \begin{vmatrix} 1 & -1 \\ 0 & 5 \end{vmatrix} + 2 \begin{vmatrix} 1 & -4 \\ 0 & 1 \end{vmatrix} \right)$$

$$- 2 \left(-1 \begin{vmatrix} -4 & -1 \\ 1 & 5 \end{vmatrix} - 0 \begin{vmatrix} 0 & -1 \\ 2 & 5 \end{vmatrix} + 2 \begin{vmatrix} 0 & -4 \\ 2 & 1 \end{vmatrix} \right)$$

$$+ 3 \left(-1 \begin{vmatrix} 1 & -1 \\ 0 & 5 \end{vmatrix} - (-3) \begin{vmatrix} 0 & -1 \\ 2 & 5 \end{vmatrix} + 2 \begin{vmatrix} 0 & 1 \\ 2 & 0 \end{vmatrix} \right)$$

$$- 4 \left(-1 \begin{vmatrix} 1 & -4 \\ 0 & 1 \end{vmatrix} - (-3) \begin{vmatrix} 0 & -4 \\ 2 & 1 \end{vmatrix} + 0 \begin{vmatrix} 0 & 1 \\ 2 & 0 \end{vmatrix} \right)$$

$$= [(-3)(-20 + 1) - 0 + 2(1)] - 2[(-1)(-20 + 1) - 0 + 2(8)]$$
$$+ 3[(-1)(5) + 3(2) + 2(-2)] - 4[(-1)(1) + 3(8) + 0]$$

$$= (57 + 2) - 2(19 + 16) + 3(-5 + 6 - 4) - 4(23)$$

$$= -112 \qquad \blacksquare$$

Obviously the evaluation of fourth-order or larger determinants can be very laborious and time-consuming. In addition, the number of operations that are necessary for the evaluation is very conducive to arithmetical and/or sign errors. Therefore, we shall discuss several properties of determinants that simplify their evaluation. Formal proofs will not be given for determinants of arbitrary order; we shall only establish the validity of the properties for second- or third-order determinants and leave the verification for higher orders to a course in linear algebra.

The **main diagonal** of a matrix runs from its upper left to lower right corners. Reflecting through this diagonal changes rows to columns and vice versa; this is called **transposing** the matrix. In Exercise 30 at the end of this section, you will be asked to show that transposing a third-order matrix does not affect the value of its determinant. For a second-order determinant, this is clear:

$$\begin{vmatrix} a & b \\ c & d \end{vmatrix} = ad - bc = \begin{vmatrix} a & c \\ b & d \end{vmatrix}$$

In fact, this property holds for determinants of any order.

PROPERTY 1

> Transposing a matrix (reflecting through its main diagonal) does not change the value of its determinant.

As a consequence of this property, a determinant can be expanded by its first column as well as by its first row. For instance, in the 3×3 case, we have

$$\begin{vmatrix} a_1 & b_1 & c_1 \\ a_2 & b_2 & c_2 \\ a_3 & b_3 & c_3 \end{vmatrix} = \begin{vmatrix} a_1 & a_2 & a_3 \\ b_1 & b_2 & b_3 \\ c_1 & c_2 & c_3 \end{vmatrix}$$

$$= a_1 \begin{vmatrix} b_2 & b_3 \\ c_2 & c_3 \end{vmatrix} - a_2 \begin{vmatrix} b_1 & b_3 \\ c_1 & c_3 \end{vmatrix} + a_3 \begin{vmatrix} b_1 & b_2 \\ c_1 & c_2 \end{vmatrix}$$

$$= a_1 M_{11} - a_2 M_{21} + a_3 M_{31}$$

This latter form is just the expansion of the original determinant by its first column.

Not only can a determinant be evaluated by its first column, but in virtue of Property 1, *any* statement we make later regarding rows of a determinant must also hold for columns, because columns become rows when the matrix is transposed. Column operations on the original matrix become row operations on the transposed matrix.

PROPERTY 2

> Interchanging two rows or two columns of a determinant changes its sign.

We shall confirm this property for third-order determinants. Interchange the first and second rows of

$$\begin{vmatrix} a_1 & b_1 & c_1 \\ a_2 & b_2 & c_2 \\ a_3 & b_3 & c_3 \end{vmatrix} \qquad \text{to get} \qquad \begin{vmatrix} a_2 & b_2 & c_2 \\ a_1 & b_1 & c_1 \\ a_3 & b_3 & c_3 \end{vmatrix}$$

Expand this latter determinant by its first column.

$$\begin{vmatrix} a_2 & b_2 & c_2 \\ a_1 & b_1 & c_1 \\ a_3 & b_3 & c_3 \end{vmatrix} = a_2 \begin{vmatrix} b_1 & c_1 \\ b_3 & c_3 \end{vmatrix} - a_1 \begin{vmatrix} b_2 & c_2 \\ b_3 & c_3 \end{vmatrix} + a_3 \begin{vmatrix} b_2 & c_2 \\ b_1 & c_1 \end{vmatrix}$$

$$= -\left[a_1 \begin{vmatrix} b_2 & c_2 \\ b_3 & c_3 \end{vmatrix} - a_2 \begin{vmatrix} b_1 & c_1 \\ b_3 & c_3 \end{vmatrix} + a_3(b_1 c_2 - b_2 c_1) \right]$$

$$= -\left(a_1 \begin{vmatrix} b_2 & c_2 \\ b_3 & c_3 \end{vmatrix} - a_2 \begin{vmatrix} b_1 & c_1 \\ b_3 & c_3 \end{vmatrix} + a_3 \begin{vmatrix} b_1 & c_1 \\ b_2 & c_2 \end{vmatrix} \right)$$

$$= - \begin{vmatrix} a_1 & b_1 & c_1 \\ a_2 & b_2 & c_2 \\ a_3 & b_3 & c_3 \end{vmatrix}$$

from the expansion of this final determinant by its first column.

As a consequence of Property 2, we can expand a determinant by any row or any column; with a simple interchange we can make any row the first row or any column the first column. But we must be careful of the signs we affix to the terms $a_{ij}M_{ij}$. (a_{ij} is the entry in row i, column j; and M_{ij} is the corresponding minor.) An easy way to keep track of the appropriate signs is to impose a checkerboard pattern of $+$ and $-$ signs on the matrix:

$$\begin{bmatrix} + & - & + & - & + \\ - & + & - & + & - \\ + & - & + & - & + \\ - & + & - & + & - \\ + & - & + & - & + \end{bmatrix}$$

Expanding by the fourth row of a fifth-order determinant, we have

$$\begin{vmatrix} a_{11} & \cdots & \cdots & \cdots & a_{15} \\ a_{21} & \cdots & \cdots & \cdots & \cdots \\ a_{31} & \cdots & \cdots & \cdots & \cdots \\ a_{41} & a_{42} & a_{43} & a_{44} & a_{45} \\ a_{51} & \cdots & \cdots & \cdots & a_{55} \end{vmatrix}$$

$$= -a_{41}M_{41} + a_{42}M_{42} - a_{43}M_{43} + a_{44}M_{44} - a_{45}M_{45}$$

Note that the sign in front of $a_{ij}M_{ij}$ is just $(-1)^{i+j}$.

The following example illustrates how these properties simplify the evaluation of certain determinants.

EXAMPLE 5   Evaluate

$$\begin{vmatrix} 1 & 0 & 3 & 1 \\ 2 & 1 & 4 & 0 \\ -2 & 0 & 2 & 0 \\ -1 & 0 & 3 & -1 \end{vmatrix}$$

Solution   Expanding by the second column, we obtain

$$-0\ M_{12} + 1 \begin{vmatrix} 1 & 3 & 1 \\ -2 & 2 & 0 \\ -1 & 3 & -1 \end{vmatrix} - 0\ M_{32} + 0\ M_{42}$$

Expanding the only nonzero term by the last column in the determinant yields

$$\begin{vmatrix} 1 & 0 & 3 & 1 \\ 2 & 1 & 4 & 0 \\ -2 & 0 & 2 & 0 \\ -1 & 0 & 3 & -1 \end{vmatrix} = \begin{vmatrix} 1 & 3 & 1 \\ -2 & 2 & 0 \\ -1 & 3 & -1 \end{vmatrix}$$

$$= 1 \begin{vmatrix} -2 & 2 \\ -1 & 3 \end{vmatrix} - 0 \cdot M_{23} + (-1) \begin{vmatrix} 1 & 3 \\ -2 & 2 \end{vmatrix}$$

$$= (-6 + 2) - (2 + 6)$$

$$= -12$$

Another consequence of Property 2 is that *if two rows or two columns of a matrix are identical, its determinant must be 0.* This is so because interchanging these two rows or two columns must change the sign of the determinant; yet the determinant must be unchanged because the matrix is exactly the same as it was before the interchange. Thus $|M| = -|M|$ for such a matrix. But the only number that equals its own negative is 0. Thus $|M| = 0$ when two columns or two rows of M are identical.

PROPERTY 3

Multiplying each entry of any row or column of a matrix by the same number k multiplies its determinant by k; equivalently, a common factor of each element in a given row or column can be factored out of the determinant.

To verify this property for third-order determinants, note that k is a common factor of each term when

$$\begin{vmatrix} a_1 & kb_1 & c_1 \\ a_2 & kb_2 & c_2 \\ a_3 & kb_3 & c_3 \end{vmatrix}$$

is expanded down its second column. Factoring out k leaves the determinant of the matrix with the k's deleted from its second column. Thus

$$\begin{vmatrix} a_1 & kb_1 & c_1 \\ a_2 & kb_2 & c_2 \\ a_3 & kb_3 & c_3 \end{vmatrix} = k \begin{vmatrix} a_1 & b_1 & c_1 \\ a_2 & b_2 & c_2 \\ a_3 & b_3 & c_3 \end{vmatrix}$$

PROPERTY 4

> Adding a multiple of any row (or column) to another row (column, respectively) does not change the value of the determinant.

This property is illustrated in third-order determinants by adding k times row 2 to row 3 and expanding by the new row 3 to obtain

$$\begin{vmatrix} a_1 & b_1 & c_1 \\ a_2 & b_2 & c_2 \\ a_3 + ka_2 & b_3 + kb_2 & c_3 + kc_2 \end{vmatrix}$$

$$= (a_3 + ka_2) M_{31} - (b_3 + kb_2) M_{32} + (c_3 + kc_2) M_{33}$$

$$= a_3 M_{31} - b_3 M_{32} + c_3 M_{33} + k(a_2 M_{31} - b_2 M_{32} + c_2 M_{33})$$

$$= \begin{vmatrix} a_1 & b_1 & c_1 \\ a_2 & b_2 & c_2 \\ a_3 & b_3 & c_3 \end{vmatrix} + k \begin{vmatrix} a_1 & b_1 & c_1 \\ a_2 & b_2 & c_2 \\ a_2 & b_2 & c_2 \end{vmatrix} \qquad \text{(Note: Two rows are equal; this determinant is 0.)}$$

$$= \begin{vmatrix} a_1 & b_1 & c_1 \\ a_2 & b_2 & c_2 \\ a_3 & b_3 & c_3 \end{vmatrix}$$

We shall now illustrate the use of these properties in evaluating determinants. The procedure is to work on one row or column until all but one of its entries is 0; then expand by this row or column. We shall use the notation ① + 2 · ③ from the elimination and matrix methods to indicate the operations we perform on the determinant.

EXAMPLE 6

$$\begin{vmatrix} 4 & 1 & 3 & 1 \\ 3 & 4 & -2 & -2 \\ -2 & 3 & 1 & 5 \\ 2 & -2 & 1 & 3 \end{vmatrix}$$

└── Use this entry to generate 0's in its column.

$$= \begin{vmatrix} -2 & 7 & 0 & -8 \\ 7 & 0 & 0 & 4 \\ -4 & 5 & 0 & 2 \\ 2 & -2 & 1 & 3 \end{vmatrix} \quad \begin{matrix} ① - 3 \cdot ④ \\ ② + 2 \cdot ④ \\ ③ - ④ \end{matrix}$$

└ Expand by this column.

$$= (-1) \begin{vmatrix} -2 & 7 & -8 \\ 7 & 0 & 4 \\ -4 & 5 & 2 \end{vmatrix}$$

$$= (-1) \begin{vmatrix} -2 & 7 & 2 \cdot (-4) \\ 7 & 0 & 2 \cdot 2 \\ -4 & 5 & 2 \cdot 1 \end{vmatrix}$$

↑ ───── Factor 2 out of this column.

$$= (-1) \cdot 2 \begin{vmatrix} -2 & 7 & -4 \\ 7 & 0 & 2 \\ -4 & 5 & 1 \end{vmatrix}$$

$$= (-2) \begin{vmatrix} -18 & 27 & 0 \\ 15 & -10 & 0 \\ -4 & 5 & 1 \end{vmatrix} \quad \begin{matrix} ① + 4 \cdot ③ \\ ② - 2 \cdot ③ \end{matrix}$$

└ Expand down this column.

$$= (-2) \begin{vmatrix} -18 & 27 \\ 15 & -10 \end{vmatrix} = -2 \begin{vmatrix} 9(-2) & 9 \cdot 3 \\ 5 \cdot 3 & 5 \cdot (-2) \end{vmatrix}$$

Factor 9 and 5 out of rows 1 and 2 respectively.

$$= (-2) \cdot 9 \cdot 5 \begin{vmatrix} -2 & 3 \\ 3 & -2 \end{vmatrix}$$

$$= -90(4 - 9)$$

$$= 450$$

EXERCISES 7.5

Evaluate the following determinants.

1. $\begin{vmatrix} 1 & 3 \\ -1 & 2 \end{vmatrix}$

2. $\begin{vmatrix} 1 & 2 \\ -1 & 1 \end{vmatrix}$

3. $\begin{vmatrix} 1 & 6 \\ 3 & -2 \end{vmatrix}$

4. $\begin{vmatrix} 3 & -5 \\ 8 & -2 \end{vmatrix}$

5. $\begin{vmatrix} 1 & 5 \\ 2 & -3 \end{vmatrix}$

6. $\begin{vmatrix} 1 & -2 & 5 \\ 2 & 3 & 1 \\ 4 & -1 & -6 \end{vmatrix}$

7. $\begin{vmatrix} 1 & -1 & -3 \\ 2 & -1 & 2 \\ -2 & 2 & 1 \end{vmatrix}$

8. $\begin{vmatrix} 6 & 2 & 7 \\ 11 & 4 & 9 \\ 2 & 6 & 4 \end{vmatrix}$

9. $\begin{vmatrix} 2 & 0 & 4 \\ 4 & 2 & 2 \\ 1 & 6 & -1 \end{vmatrix}$

10. $\begin{vmatrix} 4 & -1 & 3 \\ -8 & -3 & -1 \\ 6 & 4 & -1 \end{vmatrix}$

11. $\begin{vmatrix} 2 & 1 & -3 \\ -1 & 1 & -2 \\ 1 & -2 & 1 \end{vmatrix}$

12. $\begin{vmatrix} -2 & 2 & -3 \\ 2 & 1 & -1 \\ -1 & 3 & 2 \end{vmatrix}$

13. $\begin{vmatrix} 2 & 11 & 6 \\ 4 & 9 & 7 \\ 3 & 2 & 1 \end{vmatrix}$

14. $\begin{vmatrix} 4 & 2 & -5 \\ 6 & 3 & 0 \\ 7 & -1 & 1 \end{vmatrix}$

15. $\begin{vmatrix} 2 & 1 & 1 \\ 1 & 1 & -2 \\ -1 & 1 & -1 \end{vmatrix}$ **16.** $\begin{vmatrix} -2 & 1 & 2 \\ 1 & -3 & 2 \\ 2 & -1 & -1 \end{vmatrix}$

17. $\begin{vmatrix} 3 & 2 & 5 \\ 4 & 3 & -1 \\ -1 & -1 & -2 \end{vmatrix}$ **18.** $\begin{vmatrix} 2 & 1 & 3 \\ 1 & 5 & -2 \\ 4 & -6 & -1 \end{vmatrix}$

19. $\begin{vmatrix} 3 & -2 & 1 & 5 \\ 2 & 3 & 4 & -2 \\ 1 & -2 & 3 & 1 \\ -2 & 4 & 1 & 3 \end{vmatrix}$

20. $\begin{vmatrix} 3 & -1 & -1 & 2 \\ 3 & 1 & 2 & 1 \\ -1 & 3 & 2 & -1 \\ 1 & 2 & -1 & -1 \end{vmatrix}$

21. $\begin{vmatrix} -1 & -2 & 1 & 3 \\ 3 & 1 & -3 & -2 \\ -3 & -1 & 2 & 1 \\ 2 & 2 & 1 & 1 \end{vmatrix}$

22. $\begin{vmatrix} 2 & 1 & 2 & 1 \\ 1 & 2 & -1 & 0 \\ -1 & 1 & 2 & 2 \\ 2 & -1 & 1 & 1 \end{vmatrix}$

23. $\begin{vmatrix} 3 & 3 & 2 & -1 \\ 1 & 1 & 2 & -2 \\ -3 & -2 & 1 & 1 \\ -2 & -1 & 1 & 3 \end{vmatrix}$

24. $\begin{vmatrix} 4 & 3 & 2 & 1 \\ 2 & -1 & 0 & 3 \\ 1 & 2 & -2 & -1 \\ 2 & 1 & -1 & 2 \end{vmatrix}$

25. $\begin{vmatrix} 3 & 1 & -1 & -6 \\ 2 & 3 & -1 & 1 \\ 5 & -3 & 3 & 2 \\ -1 & -2 & -4 & 3 \end{vmatrix}$

26. $\begin{vmatrix} 2 & -1 & -2 & 0 \\ 5 & 1 & 4 & 3 \\ 4 & 2 & -2 & 1 \\ 0 & 3 & 3 & 1 \end{vmatrix}$

27. $\begin{vmatrix} 2 & 0 & 2 & 0 & 0 \\ 1 & -1 & 0 & 2 & -1 \\ 0 & -2 & 1 & 1 & 1 \\ 1 & 1 & 0 & -2 & -1 \\ 0 & 2 & 1 & -1 & 1 \end{vmatrix}$

28. $\begin{vmatrix} 0 & -1 & 0 & -1 & 0 \\ 1 & 0 & 0 & 1 & 0 \\ -1 & 0 & 1 & 0 & -1 \\ 0 & 1 & 0 & -1 & 0 \\ -1 & 0 & -1 & 0 & 1 \end{vmatrix}$

29. $\begin{vmatrix} 1 & 0 & 0 & 2 & 0 & 1 \\ 0 & 0 & -1 & 0 & 1 & 0 \\ -1 & 1 & 0 & 1 & 0 & 0 \\ 0 & 1 & 0 & 0 & 0 & 2 \\ 2 & 0 & 1 & 1 & 0 & 0 \\ 0 & 0 & 1 & 0 & 1 & 0 \end{vmatrix}$

30. Show that reflecting a third-order matrix through its main diagonal does not affect its determinant.

7.6 CRAMER'S RULE

In a system of two linear equations in two unknowns

$$a_1x + b_1y = c_1$$
$$a_2x + b_2y = c_2$$

we can eliminate y if we multiply the first equation by b_2, the second by b_1 and subtract:

$$b_2a_1x + b_2b_1y = b_2c_1$$
$$b_1a_2x + b_1b_2y = b_1c_2$$
$$(b_2a_1 - b_1a_2)x = (b_2c_1 - b_1c_2)$$

The quantities in parentheses should be recognized as determinants. Thus

$$\begin{vmatrix} a_1 & b_1 \\ a_2 & b_2 \end{vmatrix} x = \begin{vmatrix} c_1 & b_1 \\ c_2 & b_2 \end{vmatrix}$$

It can be shown similarly that

$$\begin{vmatrix} a_1 & b_1 \\ a_2 & b_2 \end{vmatrix} y = \begin{vmatrix} a_1 & c_1 \\ a_2 & c_2 \end{vmatrix}$$

If $\begin{vmatrix} a_1 & b_1 \\ a_2 & b_2 \end{vmatrix} \neq 0$, the solution (x, y) is then given as

$$x = \frac{\begin{vmatrix} c_1 & b_1 \\ c_2 & b_2 \end{vmatrix}}{\begin{vmatrix} a_1 & b_1 \\ a_2 & b_2 \end{vmatrix}} \qquad y = \frac{\begin{vmatrix} a_1 & c_1 \\ a_2 & c_2 \end{vmatrix}}{\begin{vmatrix} a_1 & b_1 \\ a_2 & b_2 \end{vmatrix}}$$

In linear algebra, this work can be generalized to systems of n equations in n unknowns. Put such a system in the form required for solution by the matrix method; that is, the variables should be aligned in columns on the left side of the equal signs and the constant terms should be on the right side of the equal signs. Let Δ be the determinant of the coefficient matrix and Δ_j be the determinant that results when the jth column of Δ is replaced with the column of constants. If x_j is the variable represented by the jth column, it can be shown that

$$\Delta \cdot x_j = \Delta_j$$

If $\Delta \neq 0$, we can then solve for each x_j as

$$x_j = \frac{\Delta_j}{\Delta}$$

If $\Delta = 0$, this method of solution breaks down. We then use the matrix or elimination methods. In fact, it can be shown that the system is either inconsistent or dependent when $\Delta = 0$. Thus a system of n linear equations in n unknowns has a unique solution if and only if the determinant of the system is not 0. The method of solution that we have described above in terms of determinants is called **Cramer's rule.**

CRAMER'S RULE

In a system of n linear equations in n unknowns

$$x_1, \ldots, x_n$$

a. If $\Delta \neq 0$, the solution is given by

$$x_j = \frac{\Delta_j}{\Delta} \qquad j = 1, \ldots, n$$

b. If $\Delta = 0$, the system is dependent or inconsistent. Use the elimination or matrix methods for further clarification of the situation.

EXAMPLE 1 Use Cramer's rule to solve the system

$$\begin{cases} 2x - 3y = 4 \\ 5x - y = -2 \end{cases}$$

Solution The determinant of the coefficient matrix is

$$\Delta = \begin{vmatrix} 2 & -3 \\ 5 & -1 \end{vmatrix} = -2 + 15 = 13$$

Replacing the first and second columns respectively with the column of constants $\begin{pmatrix} 4 \\ -2 \end{pmatrix}$, we obtain

$$\Delta_1 = \begin{vmatrix} 4 & -3 \\ -2 & -1 \end{vmatrix} = -4 - 6 = -10$$

and

$$\Delta_2 = \begin{vmatrix} 2 & 4 \\ 5 & -2 \end{vmatrix} = -4 - 20 = -24$$

Since x is represented by the first column and y by the second column, the solution is given by

$$x = \frac{\Delta_1}{\Delta} = \frac{-10}{13} \qquad y = \frac{\Delta_2}{\Delta} = \frac{-24}{13}$$

Verify by checking. ∎

EXAMPLE 2 Use Cramer's rule to solve the system

$$\begin{cases} 6x + y - z = 1 \\ -2x + 3y + 4z = 4 \\ 9x + 2y - 2z = -1 \end{cases}$$

Solution The notation ② − ① is adopted to keep track of operations on columns as well as on rows. The determinants in Cramer's rule are then evaluated as indicated.

$$x = \frac{\begin{vmatrix} 1 & 1 & -1 \\ 4 & 3 & 4 \\ -1 & 2 & -2 \end{vmatrix}}{\begin{vmatrix} 6 & 1 & -1 \\ -2 & 3 & 4 \\ 9 & 2 & -2 \end{vmatrix}} = \frac{\begin{vmatrix} 1 & 0 & 0 \\ 4 & -1 & 8 \\ -1 & 3 & -3 \end{vmatrix}}{\begin{vmatrix} 0 & 1 & 0 \\ -20 & 3 & 7 \\ -3 & 2 & 0 \end{vmatrix}}$$

② − ①
③ + ①

Expand each across the first row.

③ + ②
① − 6 · ②

$$= \frac{\begin{vmatrix} -1 & 8 \\ 3 & -3 \end{vmatrix}}{(-1)\begin{vmatrix} -20 & 7 \\ -3 & 0 \end{vmatrix}} = \frac{3 - 24}{(-1)(21)} = \frac{-21}{-21} = 1$$

$$y = \frac{\begin{vmatrix} 6 & 1 & -1 \\ -2 & 4 & 4 \\ 9 & -1 & -2 \end{vmatrix}}{\Delta}$$

Expand down the second column.

$$= \frac{\begin{vmatrix} 6 & 1 & -1 \\ -26 & 0 & 8 \\ 15 & 0 & -3 \end{vmatrix} \begin{array}{l} ②-4① \\ ③+① \end{array}}{-21}$$

Note that $\Delta = -21$ was evaluated in the computation for x.

$$= \frac{-\begin{vmatrix} -26 & 8 \\ 15 & -3 \end{vmatrix}}{-21} = \frac{-(78 - 120)}{-21}$$

$$= \frac{42}{-21} = -2$$

$$z = \frac{\begin{vmatrix} 6 & 1 & 1 \\ -2 & 3 & 4 \\ 9 & 2 & -1 \end{vmatrix}}{\Delta}$$

Expand down the third column.

$$= \frac{\begin{vmatrix} 6 & 1 & 1 \\ -26 & -1 & 0 \\ 15 & 3 & 0 \end{vmatrix} \begin{array}{l} ②-4·① \\ ③+① \end{array}}{-21}$$

$$= \frac{\begin{vmatrix} -26 & -1 \\ 15 & 3 \end{vmatrix}}{-21} = \frac{-78 + 15}{-21} = \frac{-63}{-21} = 3$$

The solution is $x = 1$, $y = -2$, $z = 3$. Verify it.   ■

 Although Cramer's rule is quite useful in solving small systems, its use is impractical for large systems because of the large number of high-order determinants that must be evaluated. For large systems, the matrix method is probably the most efficient method of solution. However, for small systems, especially 2 × 2 systems, Cramer's rule is very fast.

 As we mentioned earlier, it is easy to program a computer to solve a system of equations by using its matrix. These programs are very efficient and easy to write. In contrast, a computer can also be programmed to solve a system of equations by using Cramer's rule, but these programs are very inefficient and difficult to write.

EXERCISES 7.6

Use Cramer's rule to solve the following systems of equations.

1. $\begin{cases} x - y = 3 \\ 3x - 2y = 1 \end{cases}$

2. $\begin{cases} 2p - 5q = 2 \\ 4p + 3q = 1 \end{cases}$

3. $\begin{cases} t + 2u = 0 \\ 3t + 2u = 5 \end{cases}$

4. $\begin{cases} 3r - 5s = -3 \\ 2r + 4s = 2 \end{cases}$

5. $\begin{cases} u - 2v = -3 \\ 2u - 3v = 1 \end{cases}$

6. $\begin{cases} x - y = 2 \\ x + 2y = 1 \end{cases}$

7. $\begin{cases} w + z = 0 \\ 2w - 3z = 5 \end{cases}$

8. $\begin{cases} r - 4u = -1 \\ 3r + 2u = 2 \end{cases}$

9. $\begin{cases} t - 2y = 5 \\ 2t - 3y = 8 \end{cases}$

10. $\begin{cases} x - 4z = 24 \\ 2x + 7z = 8 \end{cases}$

11. $\begin{cases} 2x - y + z = -5 \\ x - 2y - 3z = -8 \\ x + y - 2z = 1 \end{cases}$

12. $\begin{cases} u - 2v + w = 8 \\ 5u - v + 3w = 6 \\ 3u - 6v + 3w = 14 \end{cases}$

13. $\begin{cases} 2r - s + 2t = 5 \\ -r + s + 3t = 6 \\ 2r - 2s - t = -7 \end{cases}$

14. $\begin{cases} 3p + 2q + 5r = 12 \\ 2p - 5q + 3r = 4 \\ p + 3q + 2r = 8 \end{cases}$

15. $\begin{cases} x - y + z = 0 \\ y + 2z = -2 \\ 3x + 2y + z = 1 \end{cases}$

16. $\begin{cases} w - 2x - 3y = 3 \\ 2w + x - y = 1 \\ w + x + y = -4 \end{cases}$

17. $\begin{cases} 3w + 4z + u = 0 \\ 5w + 3z + u = 1 \\ w - 4z - 3u = 5 \end{cases}$

18. $\begin{cases} 2x + 3y - 2z = 1 \\ 2x + y - z = 2 \\ x + y - z = 1 \end{cases}$

19. $\begin{cases} -a + b + c - 2d = 1 \\ a - b + 2c - d = -1 \\ 3a + b + 2c - 2d = -2 \\ -2a - 2b + c + 2d = 3 \end{cases}$

20. $\begin{cases} r - 2s - t - 3u = -9 \\ r + s - t = 0 \\ 3r + 4s + u = 6 \\ 2s - 2t + u = 3 \end{cases}$

21. The larger of two numbers is one more than three times the smaller. The smaller plus twice the larger equals 30. Find the numbers.

22. A two-digit number is one less than five times the sum of its digits. If the digits are reversed the number is increased by 18. Find the number. (See Exer. 39, Sect. 7.1.)

23. Mr. Huber invests money in two institutions at 8 percent and 5 percent interest, respectively, and earns $1130 annually. He has $1000 more than twice the smaller investment invested at 8 percent. How much is deposited in each bank?

24. Cindy has $5.15 in nickels and dimes. The number of nickels is eight more than three times the number of dimes. What is the number of each type of coin she has?

25. Two years from now Gideon's father will be four times as old as Gideon. Sixteen years from now the father will be twice as old as Gideon. How old are they now?

7.7 PARTIAL FRACTIONS (OPTIONAL)

We obtain the sum of rational expressions

$$\frac{2}{x - 1} + \frac{1}{x + 1} - \frac{1}{(x + 1)^2}$$

by writing each fraction in terms of the LCD and adding numerators:

$$\frac{2(x + 1)^2}{(x - 1)(x + 1)^2} + \frac{(x - 1)(x + 1)}{(x - 1)(x + 1)^2} - \frac{(x - 1)}{(x - 1)(x + 1)^2}$$

$$= \frac{2(x^2 + 2x + 1) + (x^2 - 1) - (x - 1)}{(x - 1)(x + 1)^2}$$

$$= \frac{3x^2 + 3x + 2}{(x - 1)(x + 1)^2}$$

It is sometimes useful (especially in calculus courses) to reverse this process: separate the combined form into the separate fractions with which we started. This process is called *partial fractions decomposition*.

PARTIAL FRACTIONS

1. If the degree of the numerator is not less than that of the denominator, carry out the long division to obtain a proper fraction.

2. Factor the denominator as far as possible; linear and quadratic factors should be the result. (See Chap. 5.)

3. For each linear factor $(ax + b)$, consider a fraction of the form $\dfrac{A}{ax + b}$.

4. For each linear factor raised to a power $(ax + b)^k$, consider fractions of the form $\dfrac{A_1}{(ax + b)}$, $\dfrac{A_2}{(ax + b)^2}$, $\cdots$, $\dfrac{A_k}{(ax + b)^k}$.

5. For each quadratic factor $(ax^2 + bx + c)$, consider a fraction of the form $\dfrac{Ax + B}{ax^2 + bx + c}$.

6. For each quadratic factor raised to a power $(ax^2 + bx + c)^k$, consider fractions of the form $\dfrac{A_1 x + B_1}{ax^2 + bx + c}$, $\cdots$, $\dfrac{A_k x + B_k}{(ax^2 + bx + c)^k}$.

EXAMPLE 1 Find the partial fractions decomposition of

$$\frac{x^3 - 3x^2 + 5x - 2}{x^3 - 2x^2 - 4x + 8}$$

Solution First, we carry out the long division to obtain a proper fraction:

$$\frac{x^3 - 3x^2 + 5x - 2}{x^3 - 2x^2 - 4x + 8} = 1 + \frac{-x^2 + 9x - 10}{x^3 - 2x^2 - 4x + 8}$$

Then we factor the denominator:

$$\begin{aligned} x^3 - 2x^2 - 4x + 8 &= x^2(x - 2) - 4(x - 2) \\ &= (x^2 - 4)(x - 2) \\ &= (x + 2)(x - 2)^2 \end{aligned}$$

Thus we consider the following partial fractions decomposition of the proper fraction:

$$\frac{-x^2 + 9x - 10}{(x + 2)(x - 2)^2} = \frac{A}{x + 2} + \frac{B}{x - 2} + \frac{C}{(x - 2)^2}$$

The procedure is to "put these fractions back together" and solve for A, B, and C.

$$\frac{-x^2 + 9x - 10}{(x + 2)(x - 2)^2} = \frac{A(x - 2)^2 + B(x + 2)(x - 2) + C(x + 2)}{(x + 2)(x - 2)^2}$$

Equating numerators, we obtain

$$-x^2 + 9x - 10 = A(x^2 - 4x + 4) + B(x^2 - 4) + C(x + 2)$$
$$= x^2(A + B) + x(-4A + C) + (4A - 4B + 2C)$$

This is satisfied when the respective coefficients agree:

$$
\begin{array}{rrrrrcr}
x^2: & A & + & B & & = & -1 \\
x: & -4A & & & + \; C & = & 9 \\
\text{constant:} & 4A & - & 4B & + \; 2C & = & -10
\end{array}
$$

Solving this system, using the techniques of this chapter, we find

$$A = -2, \qquad B = 1, \qquad C = 1$$

Thus

$$\frac{x^3 - 3x^2 + 5x - 2}{x^3 - 2x^2 - 4x + 8} = 1 + \frac{-x^2 + 9x - 10}{(x + 2)(x - 2)^2}$$
$$= 1 + \frac{-2}{x + 2} + \frac{1}{x - 2} + \frac{1}{(x - 2)^2} \qquad \blacksquare$$

EXAMPLE 2 Find the partial fractions decomposition of

$$\frac{3x^4 - 2x^3 + 7x^2 - x + 1}{(x - 1)(x^2 + 1)^2}.$$

Solution We already have a proper fraction with denominator factored. Thus we proceed with the decomposition.

$$\frac{3x^4 - 2x^3 + 7x^2 - x + 1}{(x - 1)(x^2 + 1)^2}$$
$$= \frac{A}{x - 1} + \frac{Bx + C}{x^2 + 1} + \frac{Dx + E}{(x^2 + 1)^2}$$
$$= \frac{A(x^2 + 1)^2 + (Bx + C)(x - 1)(x^2 + 1) + (Dx + E)(x - 1)}{(x - 1)(x^2 + 1)^2}$$

Equating numerators, we obtain

$$3x^4 - 2x^3 + 7x^2 - x + 1$$
$$= A(x^4 + 2x^2 + 1) + B(x^4 - x^3 + x^2 - x) + C(x^3 - x^2 + x - 1)$$
$$+ D(x^2 - x) + E(x - 1)$$
$$= x^4(A + B) + x^3(-B + C) + x^2(2A + B - C + D)$$
$$+ x(-B + C - D + E) + (A - C - E)$$

Equate corresponding coefficients:

$$
\begin{array}{llll}
x^4: & A + B & = 3 \\
x^3: & -B + C & = -2 \\
x^2: & 2A + B - C + D & = 7 \\
x: & -B + C - D + E & = -1 \\
\text{constant:} & A \quad - C \quad - E & = 1
\end{array}
$$

Solving the system, we obtain

$$A = 2, \quad B = 1, \quad C = -1, \quad D = 1, \quad E = 2.$$

$$\frac{3x^4 - 2x^3 + 7x^2 - x + 1}{(x - 1)(x^2 + 1)^2} = \frac{2}{x - 1} + \frac{x - 1}{x^2 + 1} + \frac{x + 2}{(x^2 + 1)^2} \qquad \blacksquare$$

EXERCISES 7.7

Find the partial fractions decomposition of the following.

1. $\dfrac{8x - 9}{(x - 3)(x + 2)}$

2. $\dfrac{3x + 4}{x(x + 2)}$

3. $\dfrac{x + 2}{(x + 1)(x - 2)}$

4. $\dfrac{4x + 6}{(x + 3)(2x + 1)}$

5. $\dfrac{2x^2 - 4x - 1}{x^2 - x - 6}$

6. $\dfrac{4x^2 - 7x - 12}{x(x + 1)(x - 2)}$

7. $\dfrac{x + 11}{(x + 2)(x^2 + 2x + 2)}$

8. $\dfrac{3x^2 + 2x + 2}{(x + 1)(x^2 + 2)}$

9. $\dfrac{x - 2}{x(x^2 - 2)}$

10. $\dfrac{2x^2 + 4x - 2}{(x - 1)(x^2 + 1)}$

11. $\dfrac{4x^2 + 7x + 4}{(x + 2)(x^2 + 2)}$

12. $\dfrac{2x^4 + x^3 + 6x^2 + 3x + 5}{x^3 + 3x}$

13. $\dfrac{2}{x(x + 2)^2}$

14. $\dfrac{-2x^2 + 5x - 4}{(x - 4)(x - 2)^2}$

15. $\dfrac{3x^2 + 12x}{(x - 2)(x + 1)^2}$

16. $\dfrac{2x^2 - 11x + 11}{(x - 1)(x - 2)^2}$

17. $\dfrac{2x^4 - 5x^3 - 3x^2 + 13x - 12}{x^3 - 3x^2}$

18. $\dfrac{2x^2 + 7x + 7}{(x - 1)(x + 1)^2}$

19. $\dfrac{x^4 + 4x^2 - x}{(x - 1)(x^2 - 1)^2}$

20. $\dfrac{1}{x(x^2 + 1)^2}$

21. $\dfrac{-2x^3 - 4x - 3}{(x - 1)(x^2 + 2)^2}$

22. $\dfrac{x^3 + 2x - 1}{(x + 1)(x^2 + 1)^2}$

23. $\dfrac{x^4 + 3x^3 - 3x^2 - 2x + 12}{(x - 2)(x^2 + 2)^2}$

24. $\dfrac{x^3 + 2x^2 + 4}{x(x^2 + 2)^2}$

25. $\dfrac{2x^6 + 4x^5 + 9x^4 + 16x^3 + 12x^2 + 17x + 6}{(x + 2)(x^2 + 2)^2}$

7.8 NONLINEAR SYSTEMS

Although the theory of linear equations is well developed, an engineer, scientist, or other user of mathematics will often encounter systems in which some or all the equations are nonlinear. An example of such a system might be

$$\begin{cases} x^2 + y^2 = 25 \\ 3x - y = 5 \end{cases}$$

These nonlinear systems arise when one wants to satisfy several constraints, not all of which are linear. For example, a frequent task is to find the points in which a given line meets a given circle; this necessitates the solution of a system consisting of a linear equation and a second-degree equation as previously illustrated. Systems involving the logarithmic and exponential functions are also commonplace in the workaday world.

There is no "cookbook" approach to the solution of nonlinear systems. The key word here is *ingenuity*. The operations introduced in the sections on linear systems may be used to replace any system with an equivalent system; the substitution and elimination methods are valid for nonlinear systems as well. However, these operations are generally not sufficient to yield a solution to a nonlinear system. Generally, one must also use special properties of the particular situation. About the closest thing to a set of rules for solving nonlinear systems is the following.

NONLINEAR SYSTEMS

1. Look for a linear equation from which to make a substitution and decrease the size of the system by eliminating a variable.

2. Use the elimination method to whatever extent possible.

3. Call upon your knowledge and past experience to progress to a solution if possible.

With these admonitions and suggestions we shall proceed to work several examples. Note that with nonlinear systems it is possible to have a solution consisting of several points (more than one, but not infinitely many).

EXAMPLE 1   Determine where the line $x + y = 2$ meets the parabola $y = x^2$.

Solution   We must solve the system

$$\begin{cases} x + y = 2 \\ y = x^2 \end{cases}$$

From the first equation, we find

$$y = 2 - x$$

and substitute this into the second equation

$$2 - x = x^2$$
$$0 = x^2 + x - 2$$
$$= (x + 2)(x - 1)$$
$$x = -2, 1$$

From $y = 2 - x$, we find

$$y = 4 \quad \text{when} \quad x = -2$$
$$y = 1 \quad \text{when} \quad x = 1$$

The points of intersection are $(-2, 4)$ and $(1, 1)$ as illustrated in Figure 7–7.

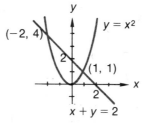

Figure 7–7

EXAMPLE 2   Find the points in which the line $3x - y = 5$ meets the circle $x^2 + y^2 = 25$.

Solution   At the points (x, y) where the line meets the circle, each of the equations

$$\begin{cases} x^2 + y^2 = 25 \\ 3x - y = 5 \end{cases}$$

must be satisfied. We can use the second equation to find

$$y = 3x - 5$$

and substitute this into the first equation:

$$x^2 + (3x - 5)^2 = 25$$
$$x^2 + 9x^2 - 30x + 25 = 25$$
$$10x^2 - 30x = 0$$
$$10x(x - 3) = 0$$
$$x = 0, 3$$

Since $y = 3x - 5$, we find $y = -5$ when $x = 0$ and $y = 4$ when $x = 3$. The solutions are $(0, 5)$ and $(3, 4)$; see Figure 7–8.

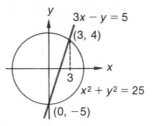

Figure 7–8

EXAMPLE 3 Determine where the parabola $y = x^2$ meets the circle $x^2 + y^2 = 2$.

Solution We must solve the system

$$\begin{cases} y = x^2 \\ x^2 + y^2 = 2 \end{cases}$$

Substituting $y = x^2$ into the second equation, we obtain

$$x^2 + x^4 = 2$$
$$(x^2)^2 + x^2 - 2 = 0$$

Use the quadratic formula to find

$$x^2 = \frac{-1 \pm \sqrt{1 + 8}}{2}$$
$$= \frac{-1 \pm 3}{2}$$

But $x^2 \geq 0$; hence

$$x^2 = \frac{-1 + 3}{2} = 1$$
$$x = \pm 1$$

Since $y = x^2$, the solutions are

$$(1, 1) \quad \text{and} \quad (-1, 1)$$

See Figure 7–9.

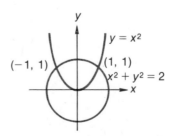

Figure 7–9

Alternate Solution   To illustrate the use of the elimination method, we could rewrite the system as

$$\begin{cases} x^2 - y = 0 \\ x^2 + y^2 = 2 \end{cases}$$

This sytem is equivalent to

$$② - ① \quad \begin{cases} x^2 - y = 0 \\ y^2 + y = 2 \end{cases}$$

Solving for y in the second equation:

$$y^2 + y - 2 = 0$$
$$(y + 2)(y - 1) = 0$$
$$y = -2, 1$$

Since $y = x^2 \geq 0$, the only possible y solution is

$$y = 1$$

Then from the first equation

$$x^2 = y = 1$$
$$x = \pm 1$$

Again, we find the solutions

$$(1, 1) \quad \text{and} \quad (-1, 1) \qquad ■$$

EXAMPLE 4   The Soviet KGB wishes to eliminate an illicit broadcaster. On the basis of the strength of the radio signals transmitted, it is determined that he is located 2 kilometers from KGB station A and approximately $\sqrt{10}$ kilometers from station B. If A is 2 kilometers west and 1 kilometer north of the Kremlin and B is 1 kilometer east and 4 kilometers north of the Kremlin, where should they search for the illicit broadcaster?

Solution   Set up a north-south, east-west coordinate system with the Kremlin at the origin. Then A and B are situated as indicated in Figure 7–10, Let T, the illicit broadcaster, have coordinates (x, y) on this coordinate system. Points A and B have coordinates $(-2, 1)$ and $(1, 4)$, respectively. The distances are

$$d(A, T) = \sqrt{(x + 2)^2 + (y - 1)^2} = 2$$
$$d(B, T) = \sqrt{(x - 1)^2 + (y - 4)^2} = \sqrt{10}$$

Squaring these we obtain

$$\begin{cases} (x + 2)^2 + (y - 1)^2 = 4 \\ (x - 1)^2 + (y - 4)^2 = 10 \end{cases}$$

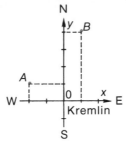

Figure 7–10

Expand the squared terms and rewrite the system as

$$\begin{cases} x^2 + 4x + 4 + y^2 - 2y + 1 = 4 \\ x^2 - 2x + 1 + y^2 - 8y + 16 = 10 \end{cases}$$

Collect terms:

$$\begin{cases} x^2 + y^2 + 4x - 2y = -1 \\ x^2 + y^2 - 2x - 8y = -7 \end{cases}$$

Subtract the second equation from the first:

$$① - ② \begin{cases} 6x + 6y = 6 \\ x^2 + y^2 - 2x - 8y = -7 \end{cases}$$

From the first equation, we find

$$y = 1 - x$$

which we substitute into the second:

$$x^2 + (1 - x)^2 - 2x - 8(1 - x) = -7$$

This is just a quadratic equation in x. It can be simplified as follows:

$$x^2 + (1 - 2x + x^2) - 2x - 8 + 8x = -7$$
$$2x^2 + 4x = 0$$
$$2x(x + 2) = 0$$

Its solutions are found to be $x = 0$ and $x = -2$. Substituting these values into the relationship $y = 1 - x$, we find the solutions to the system:

$$(0, 1) \quad \text{and} \quad (-2, 3)$$

There are only two possible locations for the illicit broadcaster. He is either directly 1 kilometer north of the Kremlin or 2 kilometers west and 3 kilometers north (see Figure 7–11).

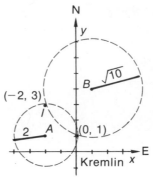

Figure 7–11

Since T is on the circle of radius 2 centered at A and also on the circle of radius $\sqrt{10}$ centered at B, we have in actuality found the intersection of these two circles. ■

EXAMPLE 5 Solve the system

$$\begin{cases} 3x^2 + 2xy + 2x + y & = 0 \\ 2x^2 + 3xy & + y^2 = 0 \end{cases}$$

Solution Remember the admonition to use your ingenuity! Observe that this second equation factors.

$$2x^2 + 3xy + y^2 = (2x + y)(x + y) = 0$$

Thus (x, y) satisfies the given system whenever it satisfies *either* of the following systems.

(1) $\begin{cases} 3x^2 + 2xy + 2x + y = 0 \\ \qquad\qquad\quad 2x + y = 0 \end{cases}$ or   (2) $\begin{cases} 3x^2 + 2xy + 2x + y = 0 \\ \qquad\qquad\qquad x + y = 0 \end{cases}$

In (1) we substitute $y = -2x$ to find

$$3x^2 + 2x(-2x) + 2x + (-2x) = 0$$

This last equation can easily be solved to find $x = 0$; then $y = -2x = 0$, also.
In (2) we substitute $y = -x$ to find

$$3x^2 + 2x(-x) + 2x + (-x) = 0$$

This time the solution yields two values for x: $x = 0$ and $x = -1$. Since $y = -x$, the corresponding y values are 0 and 1.
    The solution to the original system consists of the two points $(0, 0)$ and $(-1, 1)$. ■

In an equation involving an xy term, the preceding method is frequently fruitful. If the constant terms are not both 0, one of them can be made 0 by the elimination method.

EXAMPLE 6   Solve the system

$$\begin{cases} \log_3(x + 1) + y^2 = 5 \\ \log_3(x - 1) - y^2 = -4 \end{cases}$$

Solution   Adding the two equations, we obtain the following

$$\log_3(x + 1) + \log_3(x - 1) = 5 - 4$$
$$\log_3(x + 1)(x - 1) = 1 \qquad \text{(Section 6.3)}$$
$$(x + 1)(x - 1) = 3^1$$
$$x^2 - 1 = 3$$
$$x^2 = 4$$
$$x = \pm 2$$

Since $\log_3(-2 - 1) = \log_3(-3)$ does not exist, $x = 2$ is the only x solution. Substituting this into either of the original equations, we obtain (from the second equation)

$$\log_3(2 - 1) - y^2 = -4$$
$$\log_3(1) - y^2 = -4$$
$$0 - y^2 = -4$$
$$y^2 = 4$$
$$y = \pm 2$$

The solution set consists of two points $(x, y) = (2, 2)$ and $(2, -2)$.   ■

EXERCISES 7.8

Solve the following nonlinear systems.

1. $\begin{cases} x + 2y = 7 \\ x^2 + 4y^2 = 25 \end{cases}$
2. $\begin{cases} x + 3y = 2 \\ x^2 - 2y^2 = -1 \end{cases}$

3. $\begin{cases} 2x^2 - 2xy + y^2 = 8 \\ 2x + y = 4 \end{cases}$

4. $\begin{cases} x - y^2 = 1 \\ x - 3y = 11 \end{cases}$

5. $\begin{cases} x^2 - y^2 = 5 \\ x^2 + 2y^2 = 17 \end{cases}$

(*Hint:* First solve for x^2 and y^2.)

6. $\begin{cases} x^2 - 3y^2 = 1 \\ 3x^2 + 2y^2 = 14 \end{cases}$
7. $\begin{cases} 3x^2 + 2y^2 = 62 \\ 4x^2 - y^2 = 68 \end{cases}$

8. $\begin{cases} x^2 - xy - y^2 = 4 \\ xy - 2y^2 = 0 \end{cases}$

9. $\begin{cases} 3x^2 - xy - y^2 - 4x - 2y = -1 \\ 3x^2 - 5xy - 2y^2 = 0 \end{cases}$

10. $\begin{cases} x^2 + xy + y^2 = 2 \\ x^2 + y^2 = 4 \end{cases}$

11. $\begin{cases} 2x^2 + xy = 4 \\ 3x^2 - 5xy = 6 \end{cases}$

12. $\begin{cases} x^2 + 2xy + y^2 = 1 \\ 2x^2 + xy = 2 \end{cases}$

13. $\begin{cases} \log 2x + y = \log 2 \\ 2 \log x - y = 0 \end{cases}$

14. $\begin{cases} v + \log_2(w + 1) = 3 \\ v - \log_2(w + 1) = -1 \end{cases}$

15. $\begin{cases} \log_2(x + 4) + y^3 = 11 \\ \log_2(x - 3) - y^3 = -8 \end{cases}$

16. $\begin{cases} r - \log_4(3s + 2) = -3 \\ 2r + \log_4(s - 1)^2 = -3 \end{cases}$

17. $\begin{cases} \log_2(7 - p) + q^2 = \log_2 5 \\ \log_2(2 + p) - q^2 = 2 \end{cases}$

18. $\begin{cases} \sqrt{u} + \log_2(3 - v) = 0 \\ \sqrt{u} - \log_2(2 + v) = -2 \end{cases}$

19. $\begin{cases} 3y - 2^{2x} = 2 \\ y - 2^x = 0 \end{cases}$

20. $\begin{cases} e^{2x} + ye^x = 2 \\ 3e^x - y = 0 \end{cases}$

21. $\begin{cases} u - 2^v = 4 \\ u + 2^v = 8 \end{cases}$

22. $\begin{cases} 3^{2v} - r^2 = 1 \\ 3^v + r^2 = 1 \end{cases}$

23. $\begin{cases} 2^{2x} + 3 \cdot 2^x = 28 \\ 3 \cdot 2^x + 2^{x+1} = 20 \end{cases}$

24. $\begin{cases} 3 \cdot 3^{2x} - 2 \cdot 3^x = 225 \\ 2 \cdot 3^{2x} - \quad 3^x = 153 \end{cases}$

25. $\begin{cases} 4^x - (-2)^x = 72 \\ 2^{2x} + 2(-2)^x = 48 \end{cases}$

26. $\begin{cases} 2^{x+1} + 2^x = 6 \\ 2^{x+1} - 2^x = 2 \end{cases}$

Solve for x and y in the following.

27. $x^2 + 3 + (x + 1)i = y + 2x + yi$

28. $x(1 + i) - y(5 + i) = y^2 + 3(2 + i)$

29. $(2x + yi)(3x - yi) = 2xy + 9 + xi$

30. Find the points in which the line $y = 2x$ meets the circle of radius $\sqrt{5}$ centered at $(1, 2)$.

31. Find the points in which the line passing through the origin with slope $m = 7$ meets the circle of radius 5 centered at $(5, 10)$.

32. Determine where the parabola $y = 2x^2$ meets the circle $x^2 + y^2 = 5$.

33. Wildlife experts are treating two problems simultaneously in a river. One chemical is supposed to purify the water; the other is supposed to promote growth of fish foods. Because of adverse reactions between the two chemicals, the relative amounts must be precisely controlled. The product of the amounts must be 168 and the sum of the squares of the amounts must be 340. What treatment levels are permissible?

34. The product of two nonnegative numbers is 65. The sum of their squares is 194. Find the numbers.

35. Find two integers whose sum is 17 and for which the sum of their reciprocals is 17/72.

36. If 3 is added to both numerator and denominator of a fraction, its value is increased by 25 percent. On the other hand, subtracting 1 from both numerator and denominator decreases its value by 25 percent. Find the fraction.

37. If two objects having mass m_1, m_2 and velocities v_1, v_2, respectively, collide head-on, the respective velocities v_1', v_2' after collision must satisfy

$$\begin{cases} m_1(v_1 - v_1') = -m_2(v_2 - v_2') \\ m_1(v_1^2 - v_1'^2) = -m_2(v_2^2 - v_2'^2) \end{cases}$$

(These equations assume total elasticity of the objects.)

a. Simplify this system of equations by factoring the second equation and then dividing the second equation by the first. (The velocities do change unless one of the objects is stationary and unmoved by the collision; thus you will not be dividing by zero.)

b. If an object with a mass $m_1 = 1$ kilogram and velocity $v_1 = 1$ meter per second collides head-on with another object having mass $m_2 = 2$ kilograms and velocity $v_2 = -8$ meters per second, how fast are they traveling after the collision? (Opposite signs were used on v_1 and on v_2 here to indicate that they are approaching each other; that is, they are traveling in opposite directions.)

c. In (b), did either or both objects change direction of travel? Compare the results of (b) with the conclusion in (a).

7.9 THE ALGEBRA OF MATRICES

In Sections 7.4 and 7.6, we saw how matrices were used to simplify solving a system of equations. Matrices are useful in other contexts as well. Thus it is appropriate to develop some rules concerning the algebraic manipulation of matrices. Three algebraic operations are defined for matrices:

1. The sum $A + B$ and difference $A - B$ of two matrices A and B.

2. The product AB of matrices. (Quotients are not defined, however.)

3. The product kA of a real number k times a matrix A.

To add or subtract two matrices, just add or subtract corresponding entries. To multiply a matrix by a real number, multiply each entry of the matrix by that number. For example,

$$\begin{bmatrix} 1 & -2 & -1 \\ 0 & 3 & 1 \end{bmatrix} + \begin{bmatrix} 3 & 1 & -4 \\ -3 & -2 & 1 \end{bmatrix} = \begin{bmatrix} 4 & -1 & -5 \\ -3 & 1 & 2 \end{bmatrix}$$

and

$$2\begin{bmatrix} 1 & 2 & -1 \\ 0 & 3 & 2 \end{bmatrix} = \begin{bmatrix} 2 & 4 & -2 \\ 0 & 6 & 4 \end{bmatrix}$$

We do not add or subtract matrices unless they have the same number of rows and the same number of columns. The preceding matrices each consisted of two rows and three columns.

Since addition of matrices and multiplication of a matrix by a real number are defined in terms of corresponding entries and these entries are real numbers, many of the properties of real numbers carry over to matrices. Specifically, if A, B, and C are matrices and k, l are real numbers, the following properties hold.

---

1. $A + B = B + A$

2. $A + (B + C) = (A + B) + C$

3. $k(A + B) = kA + kB$

4. $(k + l)A = kA + lA$

5. $k(lA) = (kl)A$

EXAMPLE 1   Find $2A - 3B$ if

$$A = \begin{bmatrix} 1 & -1 \\ 3 & 1 \\ 2 & -4 \end{bmatrix} \quad \text{and} \quad B = \begin{bmatrix} -1 & 1 \\ 2 & 4 \\ 1 & -3 \end{bmatrix}$$

Solution

$$2A - 3B = 2\begin{bmatrix} 1 & -1 \\ 3 & 1 \\ 2 & -4 \end{bmatrix} - 3\begin{bmatrix} -1 & 1 \\ 2 & 4 \\ 1 & -3 \end{bmatrix}$$

$$= \begin{bmatrix} 2 & -2 \\ 6 & 2 \\ 4 & -8 \end{bmatrix} - \begin{bmatrix} -3 & 3 \\ 6 & 12 \\ 3 & -9 \end{bmatrix} = \begin{bmatrix} 5 & -5 \\ 0 & -10 \\ 1 & 1 \end{bmatrix}$$

Matrix multiplication is *not* defined in terms of corresponding entries. Rather, a different kind of product of matrices is found to be useful. We let a_{ij} denote the entry in row i, column j of the matrix A. To find the entry in row i, column j of the product $C = AB$ of two matrices, we use row i of A and column j of B as illustrated here.

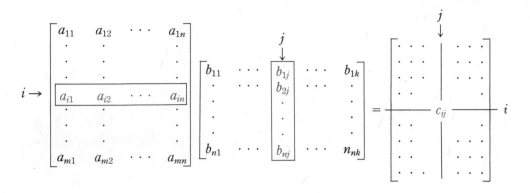

where

$$c_{ij} = a_{i1}b_{1j} + a_{i2}b_{2j} + \cdots + a_{in}b_{nj}$$

Thus, to form the product $C = AB$ of two matrices, the number of columns of A must equal the number of rows of B. If A is an $m \times n$ matrix and B is an $n \times k$ matrix, then $C = AB$ is an $m \times k$ matrix.

EXAMPLE 2   Find AB if

$$A = \begin{bmatrix} -1 & 1 \\ 2 & 4 \\ 3 & -1 \end{bmatrix} \quad \text{and} \quad B = \begin{bmatrix} 2 & 0 & 1 & 3 \\ -1 & 2 & -2 & 4 \end{bmatrix}$$

Solution   Let

$$AB = \begin{bmatrix} -1 & 1 \\ 2 & 4 \\ 3 & -1 \end{bmatrix} \begin{bmatrix} 2 & 0 & 1 & 3 \\ -1 & 2 & -2 & 4 \end{bmatrix} = C$$

The product should have 3 rows and 4 columns. The c_{13} entry is found as illustrated.

$$\begin{bmatrix} \boxed{-1 \quad 1} \\ 2 \quad 4 \\ 3 \quad -1 \end{bmatrix} \begin{bmatrix} 2 & 0 & \boxed{1} & 3 \\ -1 & 2 & \boxed{-2} & 4 \end{bmatrix} = \begin{bmatrix} \cdot & \cdot & c_{13} & \cdot \\ \cdot & \cdot & \cdot & \cdot \\ \cdot & \cdot & \cdot & \cdot \end{bmatrix}$$

where $c_{13} = (-1) \cdot 1 + 1 \cdot (-2) = -3$. Similarly, the c_{22} computation is illustrated.

$$\begin{bmatrix} -1 & 1 \\ \boxed{2 \quad 4} \\ 3 & -1 \end{bmatrix} \begin{bmatrix} 2 & \boxed{0} & 1 & 3 \\ -1 & \boxed{2} & -2 & 4 \end{bmatrix} = \begin{bmatrix} \cdot & \cdot & -3 & \cdot \\ \cdot & c_{22} & \cdot & \cdot \\ \cdot & \cdot & \cdot & \cdot \end{bmatrix}$$

where $c_{22} = 2 \cdot 0 + 4 \cdot 2 = 8$. The other entries are computed in a similar fashion to yield

$$\begin{bmatrix} -1 & 1 \\ 2 & 4 \\ 3 & -1 \end{bmatrix} \begin{bmatrix} 2 & 0 & 1 & 3 \\ -1 & 2 & -2 & 4 \end{bmatrix} = \begin{bmatrix} -3 & 2 & -3 & 1 \\ 0 & 8 & -6 & 22 \\ 7 & -2 & 5 & 5 \end{bmatrix}$$ ∎

Note in Example 2 that we cannot even form a product BA since the sizes do not match up; B has 4 columns and A has only 3 rows. Even if the sizes do match up, it is not necessarily true that $AB = BA$. For instance, let

$$A = \begin{bmatrix} 1 & -1 \\ 0 & 1 \end{bmatrix} \quad \text{and} \quad B = \begin{bmatrix} 2 & 1 \\ 1 & 0 \end{bmatrix}$$

Then

$$AB = \begin{bmatrix} 1 & -1 \\ 0 & 1 \end{bmatrix} \begin{bmatrix} 2 & 1 \\ 1 & 0 \end{bmatrix} = \begin{bmatrix} 1 & 1 \\ 1 & 0 \end{bmatrix}$$

and

$$BA = \begin{bmatrix} 2 & 1 \\ 1 & 0 \end{bmatrix} \begin{bmatrix} 1 & -1 \\ 0 & 1 \end{bmatrix} = \begin{bmatrix} 2 & -1 \\ 1 & -1 \end{bmatrix} \neq AB$$

These are clearly not the same. Nevertheless, matrix multiplication is associative and distributive whenever the sizes match up so that the products can be formed. We mention these properties here without proof.

For matrices A, B, and C

1. $A(BC) = (AB)C$

2. $A(B + C) = AB + AC$

3. $(A + B)C = AC + BC$

4. AB is not necessarily the same as BA; AB and BA may be equal in special cases.

As an example of the use of matrices for storing and using information, consider the following situation. Assume that the Apex Company produces compressors and fans on three machines: a press, a molder, and a welder. Each compressor requires 2 hours on the press, 1 hour on the molder, and 1/2 hour on the welder. Each fan requires 1 hour on the press, $1\frac{1}{2}$ hours on the molder, and 3/4 hour on the welder. This information can be stored in the matrix A as illustrated here.

$$A = \begin{array}{c} \\ \text{compressor} \quad \text{fan} \\ \begin{bmatrix} 2 & 1 \\ 1 & \dfrac{3}{2} \\ \dfrac{1}{2} & \dfrac{3}{4} \end{bmatrix} \begin{array}{l} \text{press} \\ \text{molder} \\ \text{welder} \end{array} \end{array}$$

If a customer orders 12 compressors and 8 fans, the total number of hours each machine must run is given by the matrix product

$$
\begin{bmatrix} 2 & 1 \\ 1 & \dfrac{3}{2} \\ \dfrac{1}{2} & \dfrac{3}{4} \end{bmatrix}
\begin{bmatrix} 12 \\ 8 \end{bmatrix}
=
\begin{bmatrix} 32 \\ 24 \\ 12 \end{bmatrix}
\quad
\begin{matrix} \text{press} \\ \text{molder} \\ \text{welder} \end{matrix}
$$

that is, the press must run for 32 hours to produce 12 compressors and 8 fans, and so on. Furthermore, if 1 hour of operation costs $20 on the press, $40 on the molder, and $30 on the welder, the total production cost to the company for producing these 12 compressors and 8 fans is given by

$$
\begin{bmatrix} 20 & 40 & 30 \end{bmatrix}
\begin{bmatrix} 32 \\ 24 \\ 12 \end{bmatrix}
= \$1960
$$

Inverse of a Matrix

For any size n, the symbol I will denote the $n \times n$ matrix with $a_{ii} = 1$ and $a_{ij} = 0$ for $i \neq j$:

$$
I =
\begin{bmatrix}
1 & 0 & 0 & \cdots & 0 \\
0 & 1 & 0 & \cdots & 0 \\
0 & 0 & 1 & \cdots & 0 \\
\cdot & \cdot & \cdot & \cdots & \cdot \\
\cdot & \cdot & \cdot & \cdots & \cdot \\
\cdot & \cdot & \cdot & \cdots & \cdot \\
0 & 0 & 0 & \cdots & 1
\end{bmatrix}
$$

It can easily be verified that $AI = A$ and $IB = B$ for any matrices A, B for which these products are defined. I is called the $n \times n$ **identity matrix.** For some (but not all) matrices A, there is a matrix B for which $AB = BA = I$. When such a matrix B exists, it is called the **inverse** of A and denoted A^{-1}:

$$
AA^{-1} = A^{-1}A = I
$$

MATRIX INVERSE

To find the inverse of a matrix

1. Write the identity I alongside A.

2. Use the steps of the elimination method on this extended matrix to convert A to the identity matrix.

3. **a.** If A is indeed converted to I, then the original I will have been converted to A^{-1}.

   b. If a row of 0's results in the converted A, then A has no inverse.

EXAMPLE 3 Either find A^{-1} or show that A^{-1} does not exist for the following matrices.

a. $A = \begin{bmatrix} 2 & -1 \\ -5 & 3 \end{bmatrix}$ b. $A = \begin{bmatrix} 1 & -1 & 2 \\ -3 & 2 & 1 \\ 8 & -6 & 2 \end{bmatrix}$

Solution a. We write I alongside A and proceed with the steps of the elimination method as follows:

$$\begin{bmatrix} 2 & -1 & | & 1 & 0 \\ -5 & 3 & | & 0 & 1 \end{bmatrix} \longrightarrow \underset{②+2·①}{} \begin{bmatrix} 2 & -1 & | & 1 & 0 \\ -1 & 1 & | & 2 & 1 \end{bmatrix}$$

$$①+② \begin{bmatrix} 1 & 0 & | & 3 & 1 \\ -1 & 1 & | & 2 & 1 \end{bmatrix} \longrightarrow \underset{②+①}{} \begin{bmatrix} 1 & 0 & | & 3 & 1 \\ 0 & 1 & | & 5 & 2 \end{bmatrix}$$

Thus $A^{-1} = \begin{bmatrix} 3 & 1 \\ 5 & 2 \end{bmatrix}$. Checking, we find

$$AA^{-1} = \begin{bmatrix} 2 & -1 \\ -5 & 3 \end{bmatrix}\begin{bmatrix} 3 & 1 \\ 5 & 2 \end{bmatrix} = \begin{bmatrix} 1 & 0 \\ 0 & 1 \end{bmatrix} = I$$

Similarly, we can check that $A^{-1}A = I$

b.

$$\begin{bmatrix} 1 & -1 & 2 & | & 1 & 0 & 0 \\ -3 & 2 & 1 & | & 0 & 1 & 0 \\ 8 & -6 & 2 & | & 0 & 0 & 1 \end{bmatrix}$$

$$\begin{matrix} ②+3① \\ ③-8① \end{matrix} \begin{bmatrix} 1 & -1 & 2 & | & 1 & 0 & 0 \\ 0 & -1 & 7 & | & 3 & 1 & 0 \\ 0 & 2 & -14 & | & -8 & 0 & 1 \end{bmatrix}$$

$$\begin{matrix} ①-② \\ \\ ③+2② \end{matrix} \begin{bmatrix} 1 & 0 & -5 & | & -2 & -1 & 0 \\ 0 & -1 & 7 & | & 3 & 1 & 0 \\ 0 & 0 & 0 & | & -2 & 2 & 1 \end{bmatrix}$$

The 0 row indicates that A^{-1} does not exist.   ■

The inverse of a matrix can be used to solve a system of equations. To illustrate this, consider the following system:

$$\begin{cases} 2x - y = 1 \\ -5x + 3y = -2 \end{cases}$$

Rewrite this system in matrix form as

$$\begin{bmatrix} 2 & -1 \\ -5 & 3 \end{bmatrix}\begin{bmatrix} x \\ y \end{bmatrix} = \begin{bmatrix} 2x - y \\ -5x + 3y \end{bmatrix} = \begin{bmatrix} 1 \\ -2 \end{bmatrix}$$

Letting

$$A = \begin{bmatrix} 2 & -1 \\ -5 & 3 \end{bmatrix} \quad X = \begin{bmatrix} x \\ y \end{bmatrix} \quad B = \begin{bmatrix} 1 \\ -2 \end{bmatrix}$$

we have

$$AX = B$$

If A^{-1} exists, then

$$A^{-1}AX = A^{-1}B$$
$$IX = A^{-1}B$$
$$X = A^{-1}B$$

Thus if A^{-1} is known, the solution can be computed directly as $A^{-1}B$. In practice, though, finding A^{-1} requires as much work as solving the system by the matrix or elimination methods. However, if basically the same system must be solved repeatedly but with different B columns, this can be a relatively efficient technique.

EXAMPLE 4   Solve the system

$$2x - y = b_1$$
$$-5x + 3y = b_2$$

with

a. $\begin{bmatrix} b_1 \\ b_2 \end{bmatrix} = \begin{bmatrix} 1 \\ -2 \end{bmatrix}$ **b.** $\begin{bmatrix} b_1 \\ b_2 \end{bmatrix} = \begin{bmatrix} 2 \\ 3 \end{bmatrix}$ **c.** $\begin{bmatrix} b_1 \\ b_2 \end{bmatrix} = \begin{bmatrix} -1 \\ 1 \end{bmatrix}$

Solution   Writing this sytem in the form $AX = B$, we note that A is the matrix of Example 3. Thus $A^{-1} = \begin{bmatrix} 3 & 1 \\ 5 & 2 \end{bmatrix}$, $X = A^{-1}B$, and we proceed as follows:

a. $\begin{bmatrix} x \\ y \end{bmatrix} = A^{-1}B = \begin{bmatrix} 3 & 1 \\ 5 & 2 \end{bmatrix} \begin{bmatrix} 1 \\ -2 \end{bmatrix} = \begin{bmatrix} 1 \\ 1 \end{bmatrix}$

$$x = 1, \qquad y = 1$$

b. $\begin{bmatrix} x \\ y \end{bmatrix} = A^{-1}B = \begin{bmatrix} 3 & 1 \\ 5 & 2 \end{bmatrix} \begin{bmatrix} 2 \\ 3 \end{bmatrix} = \begin{bmatrix} 9 \\ 16 \end{bmatrix}$

$$x = 9, \qquad y = 16$$

c. $\begin{bmatrix} x \\ y \end{bmatrix} = A^{-1}B = \begin{bmatrix} 3 & 1 \\ 5 & 2 \end{bmatrix} \begin{bmatrix} -1 \\ 1 \end{bmatrix} = \begin{bmatrix} -2 \\ -3 \end{bmatrix}$

$$x = -2, \qquad y = -3$$

Each of these solutions checks in the original system(s). Verify them!   ■

EXERCISES 7.9

Let $A = \begin{bmatrix} 1 & 2 \\ -1 & 3 \end{bmatrix}$, $B = \begin{bmatrix} 0 & 1 \\ 4 & -1 \end{bmatrix}$, $C = \begin{bmatrix} 2 & 5 \\ -1 & -3 \end{bmatrix}$. Find

OOD

1. $A + B$
2. $A - C$
3. $4A + 2B$
4. $5B - 3C$
5. AB
6. BA
7. $A(B - C)$
8. $AB + AC$
9. $(2A - B)C$
10. $AC - BC$
11. $(AB)C$
12. $A(BC)$

$$\text{Let } A = \begin{bmatrix} 1 & 2 & -1 \\ 3 & 0 & 1 \end{bmatrix}, B = \begin{bmatrix} 2 & 1 \\ 1 & 3 \\ 4 & 0 \end{bmatrix},$$

$$C = \begin{bmatrix} -1 & 0 \\ -3 & 2 \\ 5 & -2 \end{bmatrix}. \text{ Find}$$

13. $B - C$
14. $3C + 4B$
15. AB
16. AC
17. BA
18. CA
19. $A(B + C)$
20. $(B - C)A$
21. $2BA + 3CA$
22. $2AC - 4AB$

$$\text{Let } A = \begin{vmatrix} 1 & 0 & -1 & 2 \\ 0 & 3 & 4 & 1 \\ 1 & -1 & 2 & -2 \\ -3 & -1 & 0 & 5 \end{vmatrix},$$

$$B = \begin{bmatrix} 1 & -1 & 1 & 2 \\ 0 & 3 & 4 & 0 \end{bmatrix},$$

$$\text{and } C = \begin{bmatrix} -2 & 0 & 1 & 4 \\ 3 & 1 & 2 & 5 \end{bmatrix}.$$

23. BA
24. CA
25. $B - 2C$
26. $3BA - 2CA$
27. $B(2A) + C(3A)$
28. $(C + B)A$
29. AA
30. Find a matrix M so that $BA + M = CA$.

For the following matrices either find the inverse or show that no inverse exists.

31. $\begin{bmatrix} 2 & 5 \\ 1 & 3 \end{bmatrix}$
32. $\begin{bmatrix} 2 & 3 \\ -4 & -6 \end{bmatrix}$

33. $\begin{bmatrix} 1 & -1 \\ -6 & 7 \end{bmatrix}$
34. $\begin{bmatrix} 2 & -1 \\ 3 & -1 \end{bmatrix}$

35. $\begin{bmatrix} -2 & 8 \\ 3 & -12 \end{bmatrix}$
36. $\begin{bmatrix} 3 & 7 \\ 1 & 2 \end{bmatrix}$

37. $\begin{bmatrix} 10 & 4 & 1 \\ 5 & 1 & 1 \\ 7 & 2 & 1 \end{bmatrix}$

38. $\begin{bmatrix} -3 & -3 & 8 \\ 5 & -1 & -2 \\ 4 & 1 & -5 \end{bmatrix}$

39. $\begin{bmatrix} 2 & 1 & -1 \\ 1 & 1 & -1 \\ -1 & -2 & 3 \end{bmatrix}$

40. $\begin{bmatrix} 2 & 1 & -1 \\ 3 & 2 & -2 \\ 1 & 2 & -3 \end{bmatrix}$

Write the following systems in the form $AX = B$ and use A^{-1} to solve the system. A^{-1} will have been found in Exercises 31 to 40.

41. $\begin{cases} 2x + 5y = 2 \\ x + 3y = 3 \end{cases}$

42. $\begin{cases} x - y = - \\ -6x + 7y = 2 \end{cases}$

43. $\begin{cases} 2x - y = 3 \\ 3x - y = -2 \end{cases}$

44. $\begin{cases} 3x + 7y = 4 \\ x + 2y = 1 \end{cases}$

45. $\begin{cases} 10x + 4y + z = 1 \\ 5x + y + z = -1 \\ 7x + 2y + z = 1 \end{cases}$

46. $\begin{cases} 2x + y - z = 0 \\ x + y - z = -2 \\ -x - 2y + 3z = 1 \end{cases}$

47. $\begin{cases} 2x + y - z = -2 \\ 3x + 2y - 2z = 1 \\ x + 2y - 3z = -1 \end{cases}$

The following information is used in the remaining exercises.

The Robinson Manufacturing Company produces bicycles, mopeds, and tricycles on four machines: a stamper, a press, a welder, and a painter. Each bicycle requires 2 hours on the stamper, 4 on the press, 1 on the welder, and 1 on the painter. Each moped requires 1, 2, 3, and 2 hours on these machines, respectively, and each tricycle requires 2, 1, 1, and 3 hours respectively. It costs $20 an hour to operate the stamper, $15 an hour for the press, $25 an hour for the welder, and $10 an hour for the painter.

48. Express the total production cost for each product as a product of two matrices. What are these total costs?

49. Express the total number of hours required on each machine to produce 10 bicycles, 5 mopeds, and 20 tricycles as a product of two matrices. How many hours are required on each machine?

50. Express the total cost of producing the items in Exercise 49 as a product of three matrices. What is this total cost?

51. The company constructs another factory for which the hourly costs are $10 for the stamper, $12 for the press, $20 for the welder, and $10 for the painter. Production times are not affected. If 40 percent of its products are manufactured in the new plant, express the average production cost for each item in matrix terms. What are these average costs?

52. If 10 bicycles, 20 mopeds, and 30 tricycles are manufactured at the original plant and 30 bicycles, 15 mopeds, and 20 tricycles are produced at the new plant, express the total production cost in matrix terms. What is the total production cost?

7.10 SYSTEMS OF INEQUALITIES

Sometimes the constraints describing a situation result in a system of inequalities rather than in a system of equations. We have already discussed several procedures for solving a system of equations; among these were Cramer's rule and the elimination method. Unfortunately, there are no correspondingly slick techniques available for solving a system of inequalities.

In a system of inequalities, each inequality must be solved separately, generally by graphing its solution. The points that satisfy the system of inequalities are precisely those that lie on *each* of the individual graphs. The elimination and substitution methods are frequently useful in finding the points in which two boundary curves for the solutions meet.

Although this work can be extended to more than two variables, it can be very difficult to visualize the solutions with three variables and is impossible with four or more variables. For this reason, we shall work with only two variables.

To solve a system of inequalities in two variables:

1. Replace the inequality symbols with = signs.

2. Sketch the resulting graphs. Use a solid line if the original inequality was $\leq$ or $\geq$; use a broken line if the original was $<$ or $>$.

3. The solution to a given inequality consists of one of the regions determined by the graph of the corresponding equality. A broken line would not be included in the solution set; a solid line would be included.

4. The solution set will consist of region(s) common to all the regions determined in Step 3.

5. If possible, find the vertices (corner points) of the solution set.

EXAMPLE 1 Solve the following system of inequalities.

$$\begin{cases} 3x - y \leq 12 \\ 2x - 4y > -12 \end{cases}$$

Solution To illustrate the procedure just described, we shall sketch the graph of each inequality independently.

1. Rewrite

$$3x - y \leq 12$$

as

$$y \geq 3x - 12$$

and sketch the line $y = 3x - 12$. The solution to the first inequality consists of all points on or above this line as indicated in Figure 7–12(a). For instance, (5, 0) lies below the line and does not satisfy the inequality, (5, 5) satisfies the inequality and lies above the line, and (5, 3) is on the edge of the solution region.

2. Then write the second inequality $2x - 4y > -12$ as

$$4y < 2x + 12$$
$$y < \left(\frac{1}{2}\right)x + 3$$

The solution to this inequality consists of all points below the line $y = (1/2)x + 3$ as indicated in Figure 7–12(b). For instance, (4, 4) lies below the line and satisfies the inequality, (4, 7) lies above the line and does not satisfy the inequality, and (4, 5) lies at the edge of the solution region but does not satisfy the inequality. The line in Figure 7–12(b) is broken to indicate that it is not part of the solution; the solid line is used in Figure 7–12(a) to indicate that it is part of the solution.

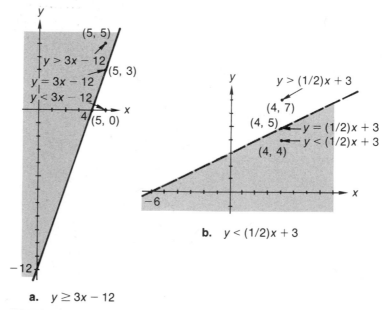

a. $y \geq 3x - 12$

b. $y < (1/2)x + 3$

Figure 7–12

Finally, both inequalities are satisfied in the region that is sketched in Figure 7–13 (next page). The point at which the boundary lines meet can be found by the elimination or substitution methods or even by Cramer's rule. ∎

Systems of inequalities do not always appear originally as "systems." For instance, in the following example, the inequality statement appears on a single line, yet is indeed a system of inequalities. In Example 3, a single $\geq$ is used; yet its solution also requires consideration of a system of inequalities. Any inequality involving an absolute value is also a system in disguise as we shall see in Example 4.

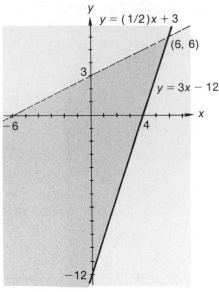

Figure 7–13 $\quad \begin{cases} 3x - y \le 12 \\ 2x - 4y > -12 \end{cases}$

EXAMPLE 2 Graph the following inequality.

$$3x + y < x - y \le 2x + 3y$$

Solution Since both inequalities must be satisfied, we actually have a system of inequalities:

$$\begin{cases} 3x + y < x - y \\ x - y \le 2x + 3y \end{cases}$$

Simplifying each of these expressions yields the equivalent system:

$$\begin{cases} y < -x \\ y \ge -\dfrac{1}{4}x \end{cases}$$

Each of these is then graphed and the solution is found as indicated in Figure 7–14.

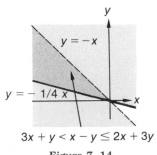

$$3x + y < x - y \le 2x + 3y$$

Figure 7–14

EXAMPLE 3 Graph the following inequality.

$$(x - 2y)(y - 2x) \geq 0$$

Solution In order for the product to be nonnegative, both factors must have the same sign.

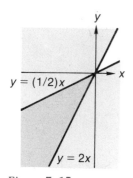

Figure 7–15
$$\begin{cases} x - 2y \geq 0 \\ y - 2x \geq 0 \end{cases}$$

a. If both factors are nonnegative, then

$$\begin{cases} x - 2y \geq 0 \\ y - 2x \geq 0 \end{cases} \quad \text{equivalently} \quad \begin{cases} y \leq \frac{1}{2}x \\ y \geq 2x \end{cases}$$

This solution is sketched in Figure 7–15.

b. If both factors are nonpositive, then

$$\begin{cases} x - 2y \leq 0 \\ y - 2x \leq 0 \end{cases} \quad \text{equivalently} \quad \begin{cases} y \geq \frac{1}{2}x \\ y \leq 2x \end{cases}$$

This solution is sketched in Figure 7–16.

The original inequality is satisfied in both cases. Thus its solution is as illustrated in Figure 7–17.

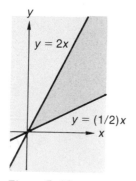

Figure 7–16
$$\begin{cases} x - 2y \leq 0 \\ y - 2x \leq 0 \end{cases}$$

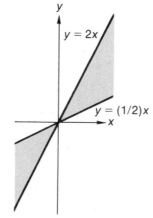

Figure 7–17 $(x - 2y)(y - 2x) \geq 0$

EXAMPLE 4 Graph the following inequality.

$$|y - x| \leq 2$$

Solution This inequality is equivalent to

$$-2 \leq y - x \leq 2$$

Since two inequalities are expressed here, this is actually a system of inequalities:

$$\begin{cases} y - x \geq -2 \\ y - x \leq 2 \end{cases}$$

This is equivalent to the system

$$\begin{cases} y \geq x - 2 \\ y \leq x + 2 \end{cases}$$

Its solution is sketched in Figure 7–18.

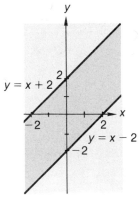

Figure 7–18 $|y - x| \leq 2$

EXAMPLE 5   Solve the following system of inequalities.

$$\begin{cases} y > x^2 + 1 \\ y < x + 3 \end{cases}$$

Solution   Even though one of these inequalities is quadratic, the technique is the same: graph the two inequalities and find the common solution. Without laboring over details, the solution is sketched in Figure 7–19. The techniques developed for systems of equations are used to find the "corner points" of the region.

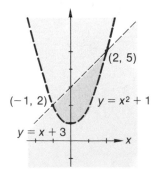

Figure 7–19 $\begin{cases} y > x^2 + 1 \\ y < x + 3 \end{cases}$

EXERCISES 7.10

Graph the solution sets for the following inequalities.

1. $\begin{cases} x - y > 0 \\ 3x - y \leq 0 \end{cases}$ 2. $\begin{cases} x + y > 0 \\ x - 2y < 0 \end{cases}$

3. $\begin{cases} 3x - 2y \leq 16 \\ 2x + 3y > 18 \end{cases}$ 4. $\begin{cases} 5x + 3y \leq 60 \\ 2x + 4y \geq 40 \end{cases}$

5. $\begin{cases} x + 2y < 2 \\ 2y - x \geq 4 \end{cases}$

6. $2x - 3y < 3x + 4y \leq 3y - 2x$

7. $2x + y - 2 \leq x - y + 5 < 8 - 2x - y$

8. $(y - 3x)(2x - y) \geq 0$

9. $(x + y)(x - 2y) \leq 0$

10. $|2x + y| \leq 3$

11. $|x + 2y| < 4$

12. $|x - 2y - 6| < 4$

13. $|x + 6y - 8| \leq 5$

14. $|x + y - 3| \, |2x - y + 4| \leq 0$

15. $\dfrac{2}{x + y} \leq \dfrac{3}{x - y} \leq 0$

16. $\begin{cases} 2x^2 + y < 2 \\ x - y < 1 \end{cases}$

17. $\begin{cases} 4x^2 - 8x + 8 < y \\ y - x > 8 \end{cases}$

18. $\begin{cases} 2x^2 - y \leq 1 \\ 2x^2 + y \leq 1 \end{cases}$

19. $\begin{cases} 3x^2 - 4y \leq 0 \\ 4y - x^2 \leq 8 \end{cases}$

20. $\begin{cases} x^2 + y^2 < 25 \\ x + y \geq 1 \end{cases}$

21. $\begin{cases} x^2 + y^2 < 9 \\ 2x^2 + 5y < 0 \end{cases}$

22. The Universal Recycling Company manufactures a Tin-Can Crusher. The by-products from manufacturing this item are, in turn, recycled and used to make Bottle Crushers. Limitations on the amount of material available require that the number of Bottle Crushers produced daily is no more than the square of the number of Tin-Can Crushers produced during that day. The Fizz Cola Company has succeeded in lobbying the state legislature to forbid the production of more than 90 crushers per day. What legal production schedules are available to this company?

23. Mr. Doerger is treating his meadow with two products. One is designed to decrease the growth of clover; the other is designed to increase the growth of alfalfa. Unfortunately, the products react with one another to kill wildlife in the area. The adverse reactions are held to a minimum if the sum of the squares of the relative amounts per acre does not exceed 625 and the product of these amounts does not exceed 168. What relative treatment levels are permissible?

24. Doris has two part-time jobs; one pays $6 per hour and the other pays $5 per hour. If she wants to earn at least $180 per week, sketch a graph indicating the ways in which she could divide her time between the two jobs.

25. Injecting a patient with x milliliters of a certain medication results in undesirable toxins in the amount of $x^2 + 3x$ micrograms above the patient's existing level of 2 micrograms. All this toxin must be destroyed by administering at least as much antitoxin as there is toxin in the patient. Sketch a graph indicating the permissible combinations of medication and antitoxin that can be given to this patient.

7.11 LINEAR PROGRAMMING

Linear programming is a branch of mathematics that has become increasingly popular in recent years. When one's choices are constrained, linear programming can be very useful in deciding which of various options to choose for maximum benefit. For example, a plant manager might use linear programming to schedule production to maximize profit. A dietitian could use it to purchase foods at the least cost that will satisfy certain dietary requirements. Although some very sophisticated techniques have been developed to solve linear programming problems, the basic ideas are very simple: essentially the problem is solved by graphing the solution to a system of linear inequalities.

    The solution technique will be developed in the following example. Since the solution is rather lengthy, appropriate steps will be identified.

EXAMPLE 1   The Heap-O-Power Calculator Company manufactures two types of calculators: the Basic and the Scientific. Each Basic requires 10 minutes in the electronics prefabrication area; each Scientific requires 20 minutes of electronic prefabrication. For the final assembly and packaging, each Basic requires 6 minutes; each Scientific requires only 4 minutes since they are more completely assembled on the electronics assembly line. The company is able to produce only 500 plastic calculator bodies per day. There are only 150 work-hours available per day in the electronics prefabrication area and only 40 work-hours per day in the final assembly and packaging area. If each Basic yields a profit of $4 and each Scientific yields a profit of $5, how many of each type should the company produce daily so as to maximize its profits?

Solution   Step 1.   *Read the problem carefully.* (This is a word problem!)

Step 2.   As with any word problem, we must *label what we are to find.* Thus we let

$$x = \text{number of Basics produced per day}$$
$$y = \text{number of Scientifics produced per day}$$

Step 3.   *Translate the given information into mathematical statements.* Since each Basic requires 10 minutes and each Scientific requires 20 minutes in the electronics prefabrication area, we shall use daily a total of $10x + 20y$ minutes of labor in the electronics prefabrication room. But there are only 150 hours = 9000 minutes available. This constraint translates into the inequality

$$10x + 20y \le 9000 \tag{1}$$

In a similar fashion, the assembly time requirements and the limitations on available assembly time translate into the inequality

$$6x + 4y \le 2400 \tag{2}$$

since 40 hours = 2400 minutes.

Finally, the plastic production capacity limits the daily production to at most 500 units:

$$x + y \le 500 \tag{3}$$

The only other constraints are

$$x \ge 0 \tag{4}$$
$$y \ge 0 \tag{5}$$

since we cannot produce a negative number of units.

Our objective is to maximize the total profit by choosing to produce the "right" number of each type. Producing x Basics and y Scientifics at profits of $4 and $5, respectively, yields a total profit of

$$P = 4x + 5y \tag{6}$$

We now collect the preceding mathematical statements into the following statement of our problem.

MATHEMATICAL FORMULATION OF EXAMPLE 1

Maximize $P = 4x + 5y$

subject to $\begin{cases} 10x + 20y \leq 9000 \\ 6x + 4y \leq 2400 \\ x + y \leq 500 \\ x \geq 0 \\ y \geq 0 \end{cases}$

The mathematical formulation of Example 1 is that of a typical linear programming
problem. Its essential features are that

1. All expressions are linear: $ax + by$

2. The constraints are inequalities: $\leq$, $\geq$

3. The variables must be nonnegative

4. Something is to be maximized or minimized

In a linear programming problem, that which is to be maximized or minimized
is called the **objective function.** A point that optimizes (maximizes or minimizes
as the case may be) the objective function is called an **optimal solution** to the
problem. A point that satisfies the constraints of the problem but does not necessarily optimize the objective function is called a **feasible solution.**

Step 4.   *Find the feasible solutions.* The techniques of the preceding section can
be used to sketch the set of feasible solutions to this problem as illustrated in
Figure 7–20(a).

Step 5.   *Analyze the values of the objective function* on the set of feasible solutions. In this problem the objective function is profit $P = 4x + 5y$. Lines representing various levels of profit are sketched in Figure 7–20(b).

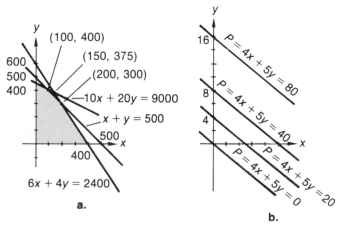

a.

b.

Figure 7–20

Note that the parallel lines $4x + 5y = $ constant each represent a fixed level of profit. As such a "profit" line is moved upward, it represents a greater profit. Step 6. *Find a point that maximizes the profit.* We can find the maximum profit level by moving the "profit" line as high as possible without leaving the set of feasible solutions. This is easily seen to be the line that passes through the point (100, 400) since any higher line would not contain any feasible solutions; any lower line would represent a lower profit (see Figure 7–21). Thus the company should produce 100 Basic and 400 Scientific calculators daily in order to maximize its profit. Its maximum daily profit is then

$$P_{max} = 4 \cdot 100 + 5 \cdot 400$$
$$= \$2400$$

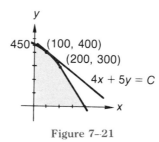

Figure 7–21

Sometimes, when the slope of the profit line is very close to the slope of one of the constraint lines, it may be difficult to ascertain when the profit line leaves the set of feasible solutions on its way upward. However, we need not be especially meticulous about sketching the profit line because, as the profit line is moving upward and is just ready to leave the feasible set, it must be passing through a "vertex" or "corner" of the feasible set.

In case the profit coincides with an edge of the constraint set as it is ready to leave the constraint set, it will be passing through two vertices. In this case, there will be many optimal solutions; nevertheless, at least one of the optimal solutions occurs at a vertex.

> The optimal solution must occur at a vertex of the feasible region. If two vertices give this same optimal value, then all points on the segment joining these two points give the same optimal value.

Thus rather than sketching the profit lines, we need only check the profit at each vertex and choose that vertex that yields the maximum profit. For instance, in Example 1 the vertices of the feasible set are $v_1 = $ (0, 0), $v_2 = $ (0, 450), $v_3 = $ (100, 400), $v_4 = $ (200, 300), and $v_5 = $ (400, 0). The corresponding profits are

$$P_1 = 4 \cdot 0 + 5 \cdot 0 = 0$$
$$P_2 = 4 \cdot 0 + 5 \cdot 450 = 2250$$
$$P_3 = 4 \cdot 100 + 5 \cdot 400 = 2400$$
$$P_4 = 4 \cdot 200 + 5 \cdot 300 = 2300$$
$$P_5 = 4 \cdot 400 + 5 \cdot 0 = 1600$$

The maximum profit of $2400 occurs at $v_3 = (100, 400)$.

Our approach to linear programming problems is summarized as follows.

SOLVING A LINEAR PROGRAMMING PROBLEM

1. *Read the problem* carefully from beginning to end.

2. *Label the quantities* that you are expected to find.

3. *Translate* the given information *into mathematical statements.* This should result in an objective function and a set of constraints described by a system of linear inequalities.

4. *Find the* set of *feasible solutions* by sketching the solution to the system of inequalities.

5. *Evaluate the objective function* at each vertex of the set of feasible solutions.

6. *Choose a vertex* that optimizes the objective function.

The method we have described here generalizes to three variables but requires three-dimensional figures. Since we cannot visualize more than three dimensions, this graphical method breaks down when there are four or more unknowns. There is an abstract algebraic formulation of this procedure known as the **simplex method,** which can then be used. The simplex method is beyond the scope of this book, however.

EXAMPLE 2 In order to treat a certain deficiency, Dr. Quincy prescribes a monthly diet containing at least 60 units of carbohydrates, 40 units of fats, and 45 units of protein. The patient is being treated in Metropolitan Hospital, which is on an austerity budget. Metropolitan seeks to minimize the cost of the food. Providing a variety of foods can be very costly; thus, they purchase only two types of foods: a "Nutritious Mixture" and a "Tasty Feeder." Each unit of Mixture contains 2 units of carbohydrates, 1 unit of fat, 1 unit of protein, and costs $6. Each unit of Feeder contains 2 units of protein, 1 unit of fat, 1 unit of carbohydrates, and costs $9. What combination of Mixture and Feeder should the hospital use to compose a diet of minimum cost that satisfies the requirements of the prescription?

Solution Let

$$x = \text{number of units of Mixture}$$
$$y = \text{number of units of Feeder}$$

to be used monthly in the patient's diet. This combination yields

$$2x + y \quad \text{units of carbohydrates}$$
$$x + 2y \quad \text{units of protein}$$
$$x + y \quad \text{units of fats}$$

at a cost of $C = 6x + 9y$ dollars. Hence we are to

$$\begin{aligned}
\text{minimize} \quad & C = 6x + 9y \\
\text{subject to} \quad & \left\{\begin{array}{l}
2x + y \geq 60 \\
x + 2y \geq 45 \\
x + y \geq 40 \\
x \geq 0 \\
y \geq 0
\end{array}\right.
\end{aligned}$$

The set of feasible solutions is shown in Figure 7–22. This time the lines of the form $6x + 9y = $ constant consist of (x, y)'s that yield the same cost. Since we wish to *minimize* cost, we move the cost line down as far as possible without leaving the feasible set. Clearly, this line will pass through one of the vertices of the feasible set when it has reached this stage. Thus an optimal solution again occurs at one of the vertices. In the chart below we list the vertices and the corresponding costs.

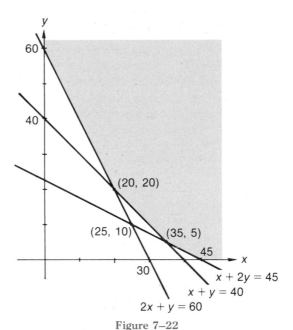

Figure 7–22

Vertex (x, y)	Cost $(6x + 9y)$
(0, 60)	540
(20, 20)	300
(35, 5)	255
(45, 0)	270

Thus the minimum cost occurs at vertex (35, 5). The institution should purchase 35 units of Mixture and 5 units of Feeder in order to minimize its cost at $255 (see Fig. 7–22). ∎

EXERCISES 7.11

1. The Swifty Bicycle Company makes two models, the Alpha and the Beta. Each Alpha requires 1 hour of manufacturing time and 2 hours assembly time and yields a profit of $2. Each Beta requires 2 hours manufacturing time and 1 hour assembly time and yields a profit of $2.50. If 50 bikes can be manufactured simultaneously and both the manufacturing and the assembly areas operate only 12 hours per day, how many of each model should be produced daily so as to maximize profits? What is the maximum profit?

2. Repeat Exercise 1 with the following changes: Each Alpha requires 3 hours of manufacturing time, 1/2 hour of assembly time, and yields $5 profit. Each Beta yields a $2 profit.

3. Change Exercise 2 to have each bicycle yielding the same profit of $3.

4. A farmer is planning to raise oats and corn. Each acre of oats yields a profit of $40; each acre of corn yields a profit of $60. To prepare the soil and plant the crops, he rents three machines M_1, M_2 and M_3 for 70, 40, and 90 hours, respectively. Each acre of oats requires 2 hours of M_1's time and 1 hour each from M_2 and M_3; each acre of corn requires 3 hours of M_3's time and 1 hour each from M_1 and M_2. How many acres of each crop should he plant in order to maximize his profit? What is the maximum profit?

5. A truck manufacturer makes both light duty pickups and heavy duty semis. Each semi requires 5 worker-days labor in the manufacturing area and 4 worker-days in the painting and finishing area and yields a profit of $300. Each pickup requires 3 worker-days in manufacturing and 2 days finishing and yields a profit of $150. Enough men have been hired to provide 210 worker-days labor in manufacturing and 160 worker-days labor in the finishing area per week. How many of each type of truck should be produced in order to maximize the company's profits? What is the maximum profit each week?

6. Do Exercise 5 with the profit figures interchanged.

7. A dietitian must supply a diet including at least 36 units of protein, 24 units of carbohydrates, and 16 units of fats. These requirements can be met with a mixture of two foods F_1 and F_2 costing $20 and $16 per unit, respectively. Each unit of F_1 supplies 9 units of protein, 3 units of carbohydrates, and 1 unit of fat. Each unit of F_2 supplies 2 units of protein, 2 of carbohydrates, and 2 of fat. What mixture of these two foods should be used in order to minimize the cost of supplying the given dietary requirements?

8. Repeat Exercise 7 with F_1 costing $30 per unit and F_2 costing $20 per unit.

9. Repeat Exericse 7 with the cost of F_2 reduced to $12 per unit.

10. The Sunbright Mining Company owns two mines that produce silver and lead. The company has contracted to supply 3200 pounds of silver to the United Ring Company and 5000 pounds of lead to the Acme Plumbing Company. The first mine produces 40 pounds of silver and 100 pounds of lead per day. The second produces 80 pounds of silver and 50 pounds of lead per day. If it costs $150 per day to operate the first mine and $200 per day to operate the second, how many days should each mine be operated in order to fulfill the contract at least expense? What is the total cost of operating the mines?

11. Repeat Exercise 10 with the mining costs reversed.

12. A bank with $10 million in deposits must allocate its resources to loans and bond purchases. Its profit on loans is 6 percent of the amount loaned, whereas its net return on the bonds it purchases is only 4 percent. The Federal Reserve Board requires the bank to keep at least 20 percent of its deposits on hand as a cash reserve; this amount cannot be invested in either loans or bonds. The bank wants to invest at least 70 percent of its deposits with no more than 65 percent of its deposits on loan. What amounts should it invest in bonds and in loans in order to maximize its return? What is its maximum return?

13. A contractor builds houses and apartments. He makes a profit of $6000 on each apartment unit and $8000 on each house he builds. He has enough financial backing to build 240 total residences. However, he has only 320 worker-months of labor available. Each home requires 2 worker-months and each apartment requires 1 worker-month of labor. How many homes and how many apartments should he build so as to

maximize his profits? What is his maximum profit?

14. Repeat Exercise 13 with the profit figures reversed.

15. The Super Suds Soap Company must allocate part of its advertising budget to print and part to radio and television. They have $6 million to spend. Their market analysts have determined that each $1 spent on print advertising eventually returns $30; each $1 spent on radio and television commercials returns $40. The company president wants to spend no more than $4 million on printed advertising and no more than $5 million on radio and television commercials. How should the company distribute its advertising dollars in order to maximize its return? What is its maximum return?

16. Repeat Exercise 15 if the expected returns of $30 and $40 are reversed.

17. Laara likes to mix two kinds of breakfast cereal: Crunchy Bits and Fluffy Flakes. Each unit of Crunchy Bits supplies 12 units of vitamin A, 2 units of vitamin B, and 4 units of B_{12}. Each unit of Fluffy Flakes supplies 4 units of A, 2 units of B, and 20 units of B_{12}. Crunchy Bits cost twice as much per unit as Fluffy Flakes. To supply 48 units of A, 16 units of B, and 80 units of B_{12} at minimal cost, how many units of each cereal should Laara's parents buy?

18. Repeat Exercise 17 with the following changes. Each unit of Crunchy Bits supplies 5, 3, and 5 units of A, B, and B_{12}, respectively, and each unit of Fluffy Flakes supplies 2, 2, and 1 unit of A, B, and B_{12}, respectively. The pediatrician recommends a total of 60, 40, and 35 units of these vitamins in the same order.

19. Hilda's Ice Cream Company manufactures two kinds of ice cream: Plain, which sells for $3/gallon, and Extra Creamy, which sells for $3.30/gallon. Each gallon of Plain is made from 2 quarts of milk and 2 quarts of cream; each gallon of Extra Creamy is made from 4 quarts of cream. Currently on hand there are 1000 gallon containers, 1600 quarts of milk, and 3200 quarts of cream. (a) How many gallons of each type should be produced in order to collect the most money? (b) How much is then collected?

20. A toy manufacturer makes widgits and gadgets by using three machines: a mold, a lathe, and a sprayer. Each widgit requires 18 minutes mold time, 18 minutes lathe time, 6 minutes spraying time and yields $1. Each gadget requires 6 minutes mold time, 9 minutes lathe time, 12 minutes spraying time, and yields a profit of $1.50. The mold is available for 45 hours per week, the lathe for 48 hours, and the sprayer for 40. What production should be scheduled each week in order to maximize revenues? What is the maximum income?

The following are linear programming problems that have already been translated into mathematical statements.

21. Maximize $5x + 3y$

subject to $\begin{cases} 2x + 5y \le 300 \\ x + y \le 90 \\ 0 \le x \le 70 \\ y \ge 0 \end{cases}$

22. Maximize $10x + 5y$

subject to $\begin{cases} 4x + 3y \le 60 \\ x + 3y \le 30 \\ x \ge 0 \\ y \ge 0 \end{cases}$

23. Minimize $5x + 4y$

subject to $\begin{cases} 5x + 6y \ge 90 \\ 5x + 3y \ge 60 \\ x \ge 0 \\ y \ge 0 \end{cases}$

24. Minimize $2x + 3y$

subject to $\begin{cases} x + 3y \ge 6 \\ 3x + y \ge 9 \\ 2x + y \ge 7 \\ x \ge 0 \\ y \ge 0 \end{cases}$

25. Maximize $5x + y$

subject to $\begin{cases} 10x + y \le 175 \\ 5x + 2y \le 125 \\ x + 2y \le 105 \\ x \ge 0 \\ y \ge 9 \end{cases}$

7.12 CHAPTER REVIEW

TERMS AND CONCEPTS

- System of Equations

$$\begin{cases} 2x + y = 3 \\ x - y = 4 \end{cases}$$

  1. Consistent   Exactly one solution.
  2. Inconsistent   No solutions.
  3. Dependent   Infinitely many solutions.

- System of Inequalities

$$\begin{cases} 2x + y \le 3 \\ x - t > 4 \end{cases}$$

- Determinant

$$\begin{vmatrix} a & b \\ c & d \end{vmatrix} = ad - bc$$

- Matrix

$$\begin{bmatrix} 1 & -1 & 2 \\ 2 & 3 & 1 \end{bmatrix}$$

- Identity Matrix

$$I = \begin{bmatrix} 1 & 0 & \cdots & 0 \\ 0 & 1 & \cdots & 0 \\ & \cdot & & \cdot \\ & \cdot & & \cdot \\ & \cdot & & \cdot \\ 0 & \cdots & 0 & 1 \end{bmatrix}, \quad AI = IA = A$$

- Inverse of a Matrix

$$AA^{-1} = A^{-1}A = I$$

RULES AND FORMULAS

- Matrix Algebra
  1. To add or subtract matrices, add or subtract corresponding entries.
  2. To multiply a matrix by a number, multiply each entry of the matrix by that number.
  3. The product of two matrices is illustrated on page 330. AB is not necessarily the same as BA; AB and BA may be equal in special cases.
  4. To find A^{-1}
     a. Write the identity I alongside A.
     b. Try to convert A to I by using the steps of the elimination method on this extended matrix.
     c. If a row of 0's results in the part of the matrix corresponding to A in *(b)*, then A has no inverse.
- Determinants
  1. Transposing a matrix (reflecting through its main diagonal) does not change the value of its determinant.
  2. Interchanging two rows or two columns of a determinant changes its sign.
  3. Multiplying each element of any row or column of a matrix by the same number k multiplies its determinant by k; equivalently, a common factor of each element in a given row or column can be factored out of the determinant.
  4. Adding a multiple of any row (or column) to another row (column, respectively) does not change the value of the determinant.
- Partial Fractions

$$\frac{1}{(x - 1)^2(x^2 + 1)^2} = \frac{A}{x - 1} + \frac{B}{(x - 1)^2} + \frac{Cx + D}{x^2 + 1} + \frac{Ex + F}{(x^2 + 1)^2}$$

SOLUTION TECHNIQUES

- Systems of Equations

  1. Graphing

     Find the point(s) common to the graphs of each equation in the system.

  2. Substitution

     Use one of the equations to express one of the variables in terms of the others and substitute this expression into the other equations.

  3. Elimination

     Perform the following operations to eliminate one or more of the variables from one or several equations:
     a. Interchange equations.
     b. Multiply an equation by a nonzero constant.
     c. Add a nonzero multiple of one equation to another equation.

  4. Matrix (a condensed form of elimination)

     a. Write the variables in the same order in each equation.
     b. Extract the matrix by suppressing the variables and replacing the equal signs with a broken vertical line.
     c. Focus on a given column and use the steps of the elimination method to reduce this column to, at most, one nonzero entry.
     d. Repeat step 3 but focus on another column and take care so as not to lose any zeros in the previously "zeroed" column(s).
     e. When no more columns can be "zeroed" without losing a previously zeroed column, it is time to read off the solution.

  5. Cramer's Rule

     In a system of n linear equations in n unknowns

     $$x_1, \ldots, x_n$$

     a. If $\Delta \neq 0$, the solution is given by

     $$x_j = \frac{\Delta_j}{\Delta} \quad j = 1, \ldots, n$$

     b. If $\Delta = 0$, the system is dependent or inconsistent. Use the elimination or matrix methods for further clarification of the situation.

- Systems of Inequalities in Two Variables
  1. Replace the inequality symbols with equals signs.
  2. Sketch the resulting graphs.
  3. The solution to a given inequality consists of one of the regions determined by the graph of the corresponding equality.
  4. The solution set will consist of region(s) common to all of the regions determined in step 3.
- Linear Programming
  1. *Read the problem* carefully from beginning to end.
  2. *Label the quantities* that you are expected to find.
  3. *Translate* the given information *into mathematical statements*. This should result in an objective function and a set of constraints described by a system of linear inequalities.

4. *Find the* set of *feasible solutions* by sketching the solution to the system of inequalities.
5. *Evaluate the objective function* at each vertex of the set of feasible solutions.
6. *Choose a vertex* that optimizes the objective function.

7.13 SUPPLEMENTARY EXERCISES

Solve the following systems of equations by the method of substitution.

1. $\begin{cases} y + 4z = 11 \\ 5y + 3z = 4 \end{cases}$ **2.** $\begin{cases} w - 3v = 8 \\ w + 2v = 3 \end{cases}$

3. $\begin{cases} 4x - 2y = 1 \\ 10x - 5y = 3 \end{cases}$ **4.** $\begin{cases} u + 2v = 3 \\ u - 2v = 15 \end{cases}$

5. $\begin{cases} 4s + v = 24 \\ 7s - 3v = 4 \end{cases}$ **6.** $\begin{cases} 2r + 3s = 1 \\ 8r + 12s = 4 \end{cases}$

7. $\begin{cases} 3p - 5q = 5 \\ 5p - 3q = 3 \end{cases}$ **8.** $\begin{cases} 5x + 3w = 19 \\ 3x + 4w = 7 \end{cases}$

9. $\begin{cases} 9x + 6y = 3 \\ 12x + 8y = 4 \end{cases}$ **10.** $\begin{cases} x - y^2 = 1 \\ x - 3y = 11 \end{cases}$

Solve the following systems by the elimination method.

11. $\begin{cases} 2x - 3y = 8 \\ x - 2y = 5 \end{cases}$ **12.** $\begin{cases} 2r - 3s = 5 \\ r + s = 5 \end{cases}$

13. $\begin{cases} 3t - 4u = -6 \\ 5t + 3u = 19 \end{cases}$ **14.** $\begin{cases} 12v + 11w = 8 \\ 36v + 33w = 20 \end{cases}$

15. $\begin{cases} z + 3y = 14 \\ 7z + 5y = 2 \end{cases}$

16. $\begin{cases} x - y + 2z = 5 \\ x - 2y + 3z = -1 \end{cases}$

17. $\begin{cases} a + 2b + c = 5 \\ 3a - 4b - 7c = 5 \\ 2a + b - 3c = 5 \end{cases}$

18. $\begin{cases} u - 2v + w = 1 \\ 2u - 4v + 3w = 4 \\ -2u + 8v - w = -1 \end{cases}$

19. $\begin{cases} 3r - s + t = 2 \\ 2r - 4s - t = 8 \\ r + s + t = -2 \end{cases}$

20. $\begin{cases} x + 4y = 11 \\ 5x + 3y = 4 \\ 17x + 34y = 85 \end{cases}$

21. $\begin{cases} 10u + 2v + 3w = -4 \\ -3u - 5v + 2w = 13 \\ 2u - 2v + 5w = 0 \end{cases}$

22. $\begin{cases} t - u + 5v = 2 \\ t + 5u - 7v = -10 \\ 3t - 2u + 3v = 4 \end{cases}$

23. $\begin{cases} 2x + 5z = 1 \\ y + 3z = 1 \\ x + y + 3z = 9 \end{cases}$

Solve the following systems by the matrix method.

24. $\begin{cases} 3r + s + 4t = -4 \\ 5r - s + 3t = 6 \\ r - 2s + t = 4 \end{cases}$

25. $\begin{cases} 2x - 4y - 6z = -1 \\ 2x - 3y + z = 7 \\ x - 2y - 3z = 4 \end{cases}$

26. $\begin{cases} 3u - 2v + 5w = 12 \\ 2u - 5v + 3w = 4 \\ u + 3v + 2w = 8 \end{cases}$

27. $\begin{cases} a - 2b - c = 0 \\ 2a + b - 2c = 0 \\ a - b - c = 0 \end{cases}$

28. $\begin{cases} 9u - 6v + 3w = 12 \\ 6u - 4v + 2w = 8 \\ 3u - 2v + w = 4 \end{cases}$

29. $\begin{cases} r - 2s - 3t = 3 \\ 2r + s - t = 1 \\ r + s + t = -4 \end{cases}$

30. $\begin{cases} x - 2y - 2z = 0 \\ 2x + y + 3z = 0 \\ 2x - y + z = 0 \\ x + y - z = 3 \end{cases}$

31. $\begin{cases} x + y + 4z = 0 \\ x + y + w = 5 \\ x + 2y - 2w = -6 \\ y + 2z + w = 3 \end{cases}$

32. $\begin{cases} t - 3v + w = -14 \\ t + u + v + w = 10 \\ t - 3u - v = -7 \\ 2u + v + w = 11 \end{cases}$

33. $\begin{cases} 2p - q - r + 2s = 8 \\ 4p + q + 5r + 2s = 6 \\ p + q + r + s = 1 \\ 3p + 2q - 2r - s = -3 \end{cases}$

Use Cramer's rule to solve the following systems.

34. $\begin{cases} 2x + y = 4 \\ -x + 3y = 2 \end{cases}$ **35.** $\begin{cases} u + 5v = 1 \\ u - 3v = 7 \end{cases}$

36. $\begin{cases} 2r + s = 6 \\ r + 2s = 0 \end{cases}$ **37.** $\begin{cases} t + 4u = 11 \\ 5t + 3u = 4 \end{cases}$

38. $\begin{cases} 5x + 3y = 17 \\ 3x + 4y = 6 \end{cases}$

39. $\begin{cases} x - 2y + z = 5 \\ 3x + 3y - 2z = -6 \\ 2x - y + 3z = 11 \end{cases}$

40. $\begin{cases} 3p - 4q + r = 3 \\ p + 3q + 2r = -4 \\ 2p - q - r = -1 \end{cases}$

41. $\begin{cases} 8t + 2u + 5v = 4 \\ 3t - u + 2v = 7 \\ t - u - 4v = -5 \end{cases}$

42. $\begin{cases} x - 3y + z = 3 \\ x + y + 3z = 3 \\ 2x + y + 2z = -1 \end{cases}$

43. $\begin{cases} 3u + 4v + 4w + 2x = 7 \\ -2u - 3v + 2w - 2x = -2 \\ 4u + 5v - 2w - 2x = 14 \\ u - v - 4w - 3x = 0 \end{cases}$

Evaluate the following determinants.

44. $\begin{vmatrix} 4 & -9 \\ -6 & 11 \end{vmatrix}$ **45.** $\begin{vmatrix} 3 & 2 & -4 \\ 1 & -3 & 1 \\ 4 & -1 & -3 \end{vmatrix}$

46. $\begin{vmatrix} 3 & 2 & -1 \\ 2 & -1 & 1 \\ 1 & 1 & -2 \end{vmatrix}$ **47.** $\begin{vmatrix} 2 & -2 & -1 \\ -1 & -3 & 2 \\ 2 & 4 & 6 \end{vmatrix}$

48. $\begin{vmatrix} -1 & 1 & -2 \\ -1 & -1 & -1 \\ 2 & 1 & -1 \end{vmatrix}$ **49.** $\begin{vmatrix} 3 & -2 & 2 \\ -1 & -4 & 5 \\ 4 & -2 & 3 \end{vmatrix}$

50. $\begin{vmatrix} 2 & 4 & 1 \\ 4 & 2 & -1 \\ 0 & -1 & -3 \end{vmatrix}$

51. $\begin{vmatrix} 1 & 1 & 0 & 2 \\ 2 & 2 & 1 & -1 \\ -1 & 1 & 2 & 1 \\ 1 & 2 & -1 & 2 \end{vmatrix}$

52. $\begin{vmatrix} 3 & -4 & 4 & -2 \\ 4 & 6 & 1 & 1 \\ -2 & 2 & 3 & 5 \\ 2 & 2 & -2 & 3 \end{vmatrix}$

53. $\begin{vmatrix} 4 & -4 & -1 & -1 \\ 2 & -2 & -1 & 2 \\ 2 & -6 & 2 & -1 \\ 6 & 2 & 1 & 3 \end{vmatrix}$

Sketch and write the equation of the circle passing through the given points. Find its center and radius.

54. $(0, 0)$, $(3, 1)$, and $(7, -1)$

55. $(11, 2)$, $(3, 14)$, and $(-7, -10)$

56. $(-26, -10)$, $(22, 4)$, and $(5, 21)$

Find the partial fractions decomposition of each of the following.

57. $\dfrac{x + 3}{(x + 2)(x + 1)}$

58. $\dfrac{4x - 2}{(x + 1)(2x - 3)}$

59. $\dfrac{2x - 2}{x(x + 1)^2}$

60. $\dfrac{-x^2 + 6x + 6}{(x - 2)(x^2 + 3)}$

61. $\dfrac{2x^4 + 4x^3 + 9x^2 + 7x + 2}{(x + 1)(x^2 + x + 1)^2}$

62. $\dfrac{2x^5 + x^4 - x^3 - x^2 + 7x + 2}{x^3 - x}$

Solve the following nonlinear systems.

63. $\begin{cases} \log |x - 1| + 2y = 1 \\ \log |x + 1| - 2y = -1 \end{cases}$

64. $\begin{cases} x^2 + 3xy + y^2 = 20 \\ x^2 - xy = 0 \end{cases}$

65. $\begin{cases} 4x^2 + 3y^2 = 16 \\ 3x^2 - 2y^2 = -5 \end{cases}$

Let $A = \begin{bmatrix} 1 & 3 & 2 \\ 1 & 0 & 4 \end{bmatrix}$, $B = \begin{bmatrix} 1 & 2 \\ -1 & 3 \\ 4 & 0 \end{bmatrix}$,

$C = \begin{bmatrix} 1 & 4 \\ 3 & 1 \\ 2 & 0 \end{bmatrix}$, $D = \begin{bmatrix} -1 & 2 \\ 3 & 1 \end{bmatrix}$, *and*

$E = \begin{bmatrix} 1 & 0 & 1 \\ 3 & 1 & -2 \\ 1 & 0 & 4 \end{bmatrix}$.

Find

66. $A(B + 2C)$ **67.** ABD

68. BDA **69.** AEC

70. AEB **71.** EBA

72. $ECDA$ **73.** $BDAE$

Find the inverse for each of the following matrices.

74. $\begin{bmatrix} 5 & -2 \\ -3 & 1 \end{bmatrix}$ **75.** $\begin{bmatrix} 2 & -1 \\ 5 & -3 \end{bmatrix}$

76. $\begin{bmatrix} 8 & 3 & -1 \\ 3 & 1 & -1 \\ 6 & 2 & -1 \end{bmatrix}$

77. $\begin{bmatrix} 1 & -3 & 2 \\ -1 & 4 & -2 \\ 1 & -3 & 3 \end{bmatrix}$

Write the following systems in the form AX = B and use A^{-1} to solve the system. A^{-1} will have been found in the preceding exercises.

78. $\begin{cases} 5x - 2y = -1 \\ -3x + y = 1 \end{cases}$

79. $\begin{cases} 2x - y = 1 \\ 5x - 3y = 2 \end{cases}$

80. $\begin{cases} 8x + 3y - z = 0 \\ 3x + y - z = 1 \\ 6x + 2y - z = 3 \end{cases}$

81. $\begin{cases} x - 3y + 2z = -1 \\ -x + 4y - 2z = 0 \\ x - 3y + 3z = -2 \end{cases}$

Graph the solutions for the following inequalities.

82. $|3x - 4y + 6| < 3$

83. $\dfrac{2}{2x + y} \le \dfrac{1}{x + 2y} \le 0$

84. $0 \le \dfrac{1}{x + y} < \dfrac{1}{x - y}$

85. $\begin{cases} x^2 + y^2 < 4 \\ x < y \end{cases}$

86. $\begin{cases} x^2 - 6y + y^2 < 0 \\ y - x \ge 3 \end{cases}$

87. $\begin{cases} x^2 + y^2 - 4y < 0 \\ x^2 + y^2 - 4x \ge 0 \end{cases}$

88. $\begin{cases} x^2 - 6x + y^2 + 5 < 0 \\ x^2 + y^2 - 6x - 4y + 12 < 0 \end{cases}$

89. $\begin{cases} x^2 - 4x + y^2 < 0 \\ y^2 < 3x - 6 \end{cases}$

90. $\begin{cases} y^2 - 5x - 50 \le 0 \\ x^2 + 10x + y^2 < 0 \end{cases}$

91. Find two numbers whose sum is 90 and whose difference is 36.

92. A number is two less than half of another number. The second number is seven less than three times the first. Find the numbers.

93. A 60-mile boat trip upriver takes 12 hours, and the return trip takes only 4 hours. What are the speeds of the boat and of the current?

94. The tens digit of a two-digit number is twice the units digit. If the digits are reversed and the result is doubled, the final result is 9 more than the original. Find the number. (See Exer. 39, Sect. 7.1.)

95. The sum of the digits of a three-digit number is one less than three times its hundreds digit. The number is eight more than 111 times its hundreds digit. The tens digit is three less than the sum of the hundreds and units digits. Find the number. (See Exer. 2.9, Sect. 7.3.)

96. Cindy has 42¢ in pennies, nickels, and dimes. If the number of dimes is one less than the number of nickels, and she has a total of twelve coins, find the number of each type of coin she has.

97. Mrs. Schoen can buy 15 pounds of pork, 15 pounds of fish, and 20 pounds of steak for $102.50. She could also buy $7\frac{1}{2}$ pounds of pork, 30 pounds of fish, and 15 pounds of steak for $97.50. Or, she could buy $22\frac{1}{2}$ pounds of pork, 15 pounds of fish, and $7\frac{1}{2}$ pounds of steak for $82.50. What is the price per pound of each of these foods?

98. Ken, Les, and Roger can paint a garage in 4 hours working together. But Ken and Roger quit after 2 hours; Les finishes the job in $6\frac{1}{2}$ more hours. If Roger had quit after 3 hours, Ken and Les could finish in $4\frac{1}{2}$ more hours. How long would it take each one alone to paint the garage? (See Example 4, Sect. 7.2.)

Solve the following linear programming problems.

99. Maximize $10y + 15x$

subject to $\begin{cases} y \le 2x \\ x + y \le 90 \\ y \ge 10 \end{cases}$

100. Maximize $4x + 8y$

subject to $\begin{cases} 3x + 7y \le 10 \\ 2x + 5y \le 3 \\ x \ge 0 \\ y \ge 0 \end{cases}$

101. Minimize $5x + 10y$

subject to $\begin{cases} x + 2y \ge 80 \\ x + y \ge 50 \\ 5x + 2y \ge 150 \\ x \ge 0 \\ y \ge 0 \end{cases}$

102. Minimize $150x + 200y$

subject to $\begin{cases} 10x + y \ge 10 \\ x \le y + 5 \\ x \ge 0 \\ y \ge 0 \end{cases}$

103. Minimize $6x + 3y$

subject to $\begin{cases} x + 3y \ge 34 \\ x + y \ge 14 \\ x \ge 4 \\ y \ge 0 \end{cases}$

104. Maximize $20x + 32y$

subject to $\begin{cases} x + y \le 210 \\ x + 2y \le 400 \\ 9x \le 180 + 10y \\ x \ge 0 \\ y \ge 0 \end{cases}$

105. Maximize $10x + 20y$

subject to $\begin{cases} 7x + 4y \le 200 \\ 3x + y \le 60 \\ x \ge 0 \\ y \ge 0 \end{cases}$

106. Maximize $20x + 14y$

subject to $\begin{cases} 8x + 40 \ge 7y \\ 2x + y \ge 10 \\ 4x \le 40 + y \\ x + y \le 20 \\ y \ge 0 \end{cases}$

107. Minimize $4y + 2x$

subject to $\begin{cases} 4y + 2 \le x \\ x + 2y \ge 5 \\ y \ge 0 \end{cases}$

108. Minimize $x - y$

subject to $\begin{cases} 8x + 5y \ge 230 \\ 6x \le 160 + 5y \\ 10y \le x + 120 \\ x \ge 0 \\ y \ge 0 \end{cases}$

8 Mathematical Induction, Progressions, and Probability

Have you ever wondered how a banker computes the monthly payment required on a loan or how pension planners compute the premium needed to generate a given retirement income? Each of these computations is made by using formulas established by **mathematical induction.** Mathematical induction is generally applied to situations that progress in a step-by-step fashion. For instance, the amount still owed on a loan is reduced systematically after each payment.

Mathematical induction involves the discovery of a certain pattern in the first several stages of a continuing process. Then this pattern is established for the remaining stages by showing that the validity of the pattern or formula at a given stage forces its validity at the next stage.

We continue to deal with orderly patterns of computation by establishing various counting formulas regarding the number of combinations and rearrangements of a given collection. These formulas are in turn useful in determining the likelihood or **probability** that certain events will happen. Industrial concerns determine probabilities of failure for their products in order to develop proper warranty policies. Police work can sometimes rely on probability, for instance, in that it is very unlikely that two identical cars from different states have the same license number and are on the same street at the same time. At the gambling tables at Las Vegas, Reno, and Atlantic City that attract millions of people each year, the odds in a betting situation are determined by the laws of probability.

8.1 MATHEMATICAL INDUCTION

We have already encountered the need to establish the truth of certain statements for every positive integer n. For instance, in the discussion of exponents we stated that $(xy)^n = x^n y^n$. In the section on compound interest, we found a formula for the value of an investment after n compounded-interest payments. We were able to convince ourselves of the validity of each of these formulas without too much trouble. However, there are situations in which a formula is desired for every positive integer n but its form and its verification are in no ways obvious.

The **principle of mathematical induction** is based on the axioms of the system of natural numbers. We shall only state and use this principle; its proof is not within the scope of this book. However, its statement is so simple and the principle itself seems so natural that we should have no qualms regarding its use.

PRINCIPLE OF MATHEMATICAL INDUCTION

Suppose that for each positive integer n, there is a corresponding statement S_n. If

1. S_1 is true, and

2.   The truth of S_k implies the truth of S_{k+1} for every k

then S_n is true for every n.

If the conditions 1 and 2 are satisfied, then S_1 is true, and S_2 is true because S_1 is true. S_3 is true because S_2 is true, S_4 is true because S_3 is true, S_5 is true because S_4 is true, and so on.

Here we see the resemblance of mathematical induction to the domino principle. For if dominos are lined up so that when any one topples, it topples the next one, the entire line can be toppled simply by pushing the first domino. In the same way, if statements are lined up so that when any one is true the next one is true, the entire list can be verified simply by showing that the first statement is true (see Figure 8–1).

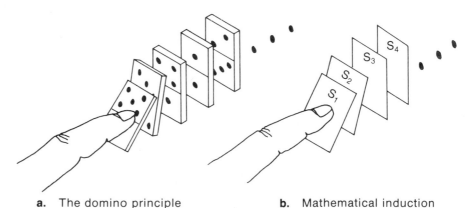

a.   The domino principle   **b.**   Mathematical induction

Figure 8–1

The applications of the principle of mathematical induction are straightforward; we just verify conditions 1 and 2; that is,

1. *Prove* that S_1 is true

2. Show that the truth of S_k implies the truth of S_{k+1}

Even though there are apparently only two things to be done in using the principle of mathematical induction, step 2 sometimes requires a certain amount of skill and ingenuity and can involve very complex algebraic manipulations.

EXAMPLE 1   Use Mathematical Induction to prove that

$$1 + 2 + 3 + \cdots + n = \frac{n(n + 1)}{2}$$

for $n = 1, 2, \ldots$

Solution   Let S_n be the statement $1 + 2 + \cdots + n = \dfrac{n(n + 1)}{2}$ and observe that S_1 is true since

$$1 = \frac{1(1 + 1)}{2}$$

Next we assume the truth of S_k and try to show that then S_{k+1} must be true; that is, we assume that

$$1 + \cdots + k = \frac{k(k + 1)}{2}$$

and try to show that

$$1 + \cdots + k + (k + 1) = \frac{(k + 1)[(k + 1) + 1]}{2}$$

Now

$$\begin{aligned}
1 + \cdots + k + (k + 1) &= (1 + \cdots + k) + (k + 1) \\
&= \frac{k(k + 1)}{2} + (k + 1) \\
&\qquad\qquad \text{by the assumption on } S_k \\
&= \frac{k(k + 1) + 2(k + 1)}{2} \\
&= \frac{(k + 1)(k + 2)}{2} \\
&= \frac{(k + 1)[(k + 1) + 1]}{2}
\end{aligned}$$

as was to be shown. Thus S_n is true for every n.   ■

EXAMPLE 2   Use mathematical induction to prove that

$$1^2 + 2^2 + \cdots + n^2 = \frac{n(n + 1)(2n + 1)}{6}$$

for $n = 1, 2, \ldots$

Solution   Let S_n be the statement in question. Then S_1 is true since

$$1^2 = \frac{1 \cdot 2 \cdot 3}{6}$$

Next, we assume that S_k is true and try to prove S_{k+1}; that is, we assume

$$1^2 + 2^2 + \cdots + k^2 = \frac{k(k + 1)(2k + 1)}{6}$$

Then for S_{k+1},

$$
\begin{aligned}
1^2 + 2^2 + \cdots + k^2 + (k + 1)^2 &= (1^2 + \cdots + k^2) + (k + 1)^2 \\
&= \frac{k(k + 1)(2k + 1)}{6} + (k + 1)^2 \\
&= \frac{k(k + 1)(2k + 1) + 6(k + 1)^2}{6} \\
&= \frac{(k + 1)[k(2k + 1) + 6(k + 1)]}{6} \\
&= \frac{(k + 1)(2k^2 + 7k + 6)}{6} \\
&= \frac{(k + 1)(k + 2)(2k + 3)}{6} \\
&= \frac{(k + 1)[(k + 1) + 1][2(k + 1) + 1]}{6}
\end{aligned}
$$

Hence S_n is true for every n.   ■

EXAMPLE 3   The "Tower of Hanoi," an ancient puzzle, consists of three pegs or "towers" on one of which are stacked discs of decreasing size as shown in Figure 8–2. The problem is to move the discs *one at a time* from one peg to another, never placing a larger disc above a smaller one, in such a way that the entire stack is finally located on another peg. Determine the minimum number of steps that are required to achieve this task if the tower consists of n discs.

Solution   This is a more difficult problem in that we must not only verify the answer by using mathematical induction, but we must also find the answer! In determining the correct answer, we shall use the inductive technique—we shall first study a tower with one disc, then a tower with two discs, and so on until we discern a clue. We shall let M_n be the solution for a tower of n discs.

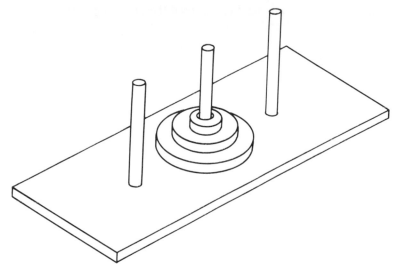

Figure 8–2   The Tower of Hanoi.

For a tower with one disc, it clearly takes only one move to complete the task. Thus $M_1 = 1$. For a tower with two discs, we move the top disc to one vacant peg, the bottom disc to the other vacant peg; finally, place the small disc over the larger for a total of three moves as illustrated in Figure 8–3. Thus $M_2 = 3$.

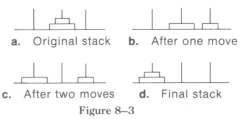

a. Original stack   **b.** After one move

c. After two moves   **d.** Final stack

Figure 8–3

For a tower with three discs, we observe that the bottom disc cannot be removed until the top two are stacked on another peg; that is, we must first work the Hanoi puzzle for $n = 2$. This takes $M_2 = 3$ moves. Then the bottom disc is moved to the vacant peg. Finally, the top two discs must be moved again to sit above the bottom disc; the Hanoi puzzle for $n = 2$ must be solved again for $n = 2$. This involves another M_2 moves. Hence

$$M_3 = M_2 + 1 + M_2 = 3 + 1 + 3 = 7$$

It appears that the discussion for $n = 3$ is perfectly general; that is, to solve the puzzle with $k + 1$ discs, we must

1.   Solve the puzzle with k discs

2.   Move the bottom peg

3.   Solve the puzzle with k discs again

One need only reread the preceding discussion (for a tower of three discs) with k and $k + 1$ in place of 2 and 3, respectively. Hence

$$M_{k+1} = M_k + 1 + M_k$$
$$= 2M_k + 1$$

Since $M_1 = 1$, $M_2 = 3$, $M_3 = 7$, we conjecture $M_n = 2^n - 1$. Letting S_n be the statement: $M_n = 2^n - 1$, we observe that S_1, S_2, and S_3 are true. Assuming S_k:

$$M_k = 2^k - 1$$

we try to prove S_{k+1}. Now

$$M_{k+1} = 2M_k + 1$$
$$= 2(2^k - 1) + 1 \qquad \text{by the assumption on } S_k$$
$$= 2 \cdot 2^k - 2 + 1$$
$$= 2^{k+1} - 1$$

Thus the *least* number of moves required to solve a Hanoi puzzle with n discs is $2^n - 1$. With the currently marketed version having 8 discs, $2^8 - 1 = 255$ moves are required. ■

If care is not taken in showing that the truth of S_k really does imply the truth of S_{k+1} for all k, some silly statements can seemingly be "proved" by using mathematical induction. For instance, Charles the Charlatan who is color-blind says that "all birds are the same color." In attempting to prove this statement, he reasons as follows:

1. In any set of birds consisting of only one bird, all birds (there is only one) are the same color.

2. Assume that in any set of birds consisting of k birds, all birds are the same color. Consider any set of $k + 1$ birds as shown in Figure 8–4. The first k birds are the same color; the last k birds are the same color. Thus all $k + 1$ birds are the same color.

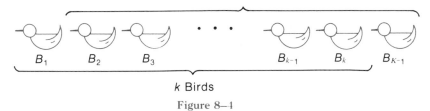

k Birds

Figure 8–4

3. Thus Charles concludes that all birds are the same color.

Evidently something is wrong with Charles' reasoning since the conclusion is obviously false. In step 2, we note that Charles is depending on the overlapping part of the two sets he considered, each of which contains k birds; $B_2, \cdots, B_k$ are in both these sets. Then B_1 and B_{k+1} are both the same color as these; hence they are all the same color. His argument breaks down when

Figure 8–5

$k = 1$ since then the two sets do not overlap (Fig. 8–5). Thus S_1 does not imply S_2 although *if* S_2 were true, then S_3 would be true, and so on.

EXAMPLE 4   Show that $6^n - 1$ is divisible by 5 for $n = 1, 2, 3, \ldots$

Solution   Observe that the statement is true for $n = 1$ since

$$6^1 - 1 = 5$$

If the statement is true for $n = k$, then for $n = k + 1$, we have

$$\begin{aligned} 6^{k+1} - 1 &= (6^k \cdot 6) - 1 \\ &= 6^k(5 + 1) - 1 \\ &= 6^k \cdot 5 + (6^k - 1) \end{aligned}$$

and this is clearly divisible by 5 if $6^k - 1$ is. Thus the truth of the statement for $n = k$ implies its truth for $n = k + 1$. By mathematical induction, the statement is true for every $n = 1, 2, \ldots$ ∎

EXERCISES 8.1

Use Mathematical Induction to establish the following formulas.

1. $1 + 8 + 16 + \cdots + 8(n - 1) = (2n - 1)^2$

2. $2^2 + 4^2 + 6^2 + \cdots + (2n)^2$
$$= \frac{2n(n + 1)(2n + 1)}{3}$$

3. $1^3 + 2^3 + 3^3 + \cdots + n^3 = \left[\dfrac{n(n + 1)}{2}\right]^2$

4. $2^3 + 4^3 + \cdots + (2n)^3 = 2[n(n + 1)]^2$

5. $2 + 4 + \cdots + 2^n = 2(2^n - 1)$

6. $4 + 16 + \cdots + 4^n = \dfrac{4(4^n - 1)}{3}$

7. $\dfrac{1}{2} + \dfrac{1}{4} + \cdots + \dfrac{1}{2^n} = 1 - \dfrac{1}{2^n}$

8. $1 + x + \cdots + x^n = \dfrac{1 - x^{n+1}}{1 - x}, \quad x \neq 1$

9. $1 \cdot 2 + 2 \cdot 3 + \cdots + n(n + 1)$
$$= \frac{n(n + 1)(n + 2)}{3}$$

10. $1 \cdot 3 + 2 \cdot 5 + \cdots + n(2n + 1)$
$$= \frac{n(n + 1)(4n + 5)}{6}$$

11. $1 \cdot 3 + 3 \cdot 5 + \cdots + (2n - 1)(2n + 1)$
$$= \frac{n(4n^2 + 6n - 1)}{3}$$

12. $1 \cdot 2 + 2 \cdot 4 + \cdots + n \cdot 2^n$
$$= (n - 1)2^{n+1} + 2$$

13. $1 \cdot 2 \cdot 3 + 2 \cdot 3 \cdot 4 + \cdots$
$$+ n(n + 1)(n + 2)$$
$$= \frac{n(n + 1)(n + 2)(n + 3)}{4}$$

14. $\dfrac{1}{1 \cdot 2} + \dfrac{1}{2 \cdot 3}$
$$+ \cdots + \frac{1}{n(n + 1)} = \frac{n}{n + 1}$$

15. $\dfrac{1}{1 \cdot 2 \cdot 3} + \dfrac{1}{2 \cdot 3 \cdot 4} + \cdots$
$$+ \frac{1}{n(n + 1)(n + 2)}$$
$$= \frac{n(n + 3)}{4(n + 1)(n + 2)}$$

Establish the following statements using mathematical induction.

16. $4^n - 1$ is divisible by 3

17. $5^n - 1$ is divisible by 4

18. $9^n - 1$ is divisible by 4

19. $2^{2n+1} + 1$ is divisible by 3

20. $2^{[2^n]} - 1$ is divisible by 3

21. $n^3 + 6n^2 + 2n$ is divisible by 3

22. $n(n^2 + 5)$ is divisible by 6

23. $a^n - b^n$ is divisible by $a - b$, $a \neq b$

24. $a^{2n} - b^{2n}$ is divisible by $a + b$, $a \neq -b$

25. $a^{2n-1} + b^{2n-1}$ is divisible by $a + b$, $a \neq -b$

26. For any two real or complex numbers a, b,

show by mathematical induction that for every positive integer n

a. $(ab)^n = a^n b^n$

b. $\left(\dfrac{a}{b}\right)^n = \dfrac{a^n}{b^n}$, when $b \neq 0$

27. Use mathematical induction to establish the formula for the value of an investment of P_0 dollars after n compounding periods:

$$P_n = P_0(1 + \beta)^n$$

Recall from Chapter 6 that $P_n = P_{n-1}(1 + \beta)$.

28. **a.** What is the total number of gifts "my true love gave to me" during the 12 days of Christmas?
   b. How many gifts will "my true love give to me" during the n days of Christmas? (*Hint:* The formulas in Example 1 and Exercise 9 are useful.)

29. Advertising agencies frequently sponsor contests to promote certain products. One such contest requires the participants to count the number of ways in which the message "Our product is best" can be found in the pyramid shown in the figure. Use mathematical induction to find the correct solution for the given pyramid. (*Hint:* Consider "messages" of increasing length T, STS, ESE as indicated.

```
              S     E
              ESTSE

                O
              O U O
            O U R U O
            O U R P R O
          O U R P R P R U O
        O U R P R O R P R U O
        O U R P R O D
        O U R P R O D U
      O U R P R O D U C
                T
                I
                S
               B
                      B E B
 O U R P R O D      B E B
 O U R P R O D U C T I S B E S E B S I T C U D O R P R U O
 O U R P R O D U C T I S B E S T S E B S I T C U D O R P R U O
```

8.2 SEQUENCES AND PROGRESSIONS

We have already considered the numbers 1, 2, 3, . . ., their reciprocals 1, 1/2, 1/3, . . ., and the inverse powers of two: 1, 1/2, 1/4, 1/8, ... Each of these is an example of sequence. Intuitively, a **sequence** is an infinite list of objects or numbers in which there is a first object in the list, a second, and so on. Formally a sequence is a function whose domain is the set of natural numbers. For such a function f, we consider $f(1)$, $f(2)$, $f(3)$, ... In working with sequences however, we usually dispense with functional notation and instead write

$$a_1, a_2, a_3 \ldots$$

a_n is called the nth term of the sequence.

EXAMPLE 1   Write out the first five terms of the sequences whose nth term is

a. $\dfrac{1}{n + 1}$

b. $\dfrac{n^2}{2n + 1}$

Solution   We are to write out the respective terms for $n = 1, \ldots, 5$. These are:

a. $\dfrac{1}{2}, \dfrac{1}{3}, \dfrac{1}{4}, \dfrac{1}{5}, \dfrac{1}{6}$

b. $\dfrac{1}{3}, \dfrac{4}{5}, \dfrac{9}{7}, \dfrac{16}{9}, \dfrac{25}{11}$ ■

Arithmetic Progression

Of particular interest are the sequences that form arithmetic and geometric progressions.

DEFINITION

A sequence is an **arithmetic progression** if the difference between any two successive terms is constant; that is, there is some **common difference** d such that

$$a_{n+1} - a_n = d \quad \text{for } n = 1, 2, \ldots$$

In an arithmetic progression, $a_{n+1} = a_n + d$; an arithmetic progression is built up by adding the same d to any term to get the next term. Thus, if we know any term and the common difference, we can find any other term since

$$
\begin{aligned}
a_2 &= a_1 + d \\
a_3 &= a_2 + d = a_1 + 2d \\
a_4 &= a_3 + d = a_1 + 3d
\end{aligned}
$$

.
.
.

$$a_n = a_{n-1} + d = a_1 + (n-1)d$$

EXAMPLE 2   Write out the first four terms and the twelfth term of the arithmetic progression whose first term is 2 and whose common difference is 3.

Solution   The first four terms are 2, 5, 8, 11. The twelfth term is

$$
\begin{aligned}
a_{12} &= a_1 + 11d \\
&= 2 + 11 \cdot 3 \\
&= 35
\end{aligned}
$$
■

EXAMPLE 3   If the fifteenth term of an arithmetic progression is 59 and its twenty-seventh term is 107, find its tenth term.

Solution   The given information yields the following system of equations.

$$
\begin{aligned}
a_{15} &= a_1 + 14d = 59 \\
a_{27} &= a_1 + 26d = 107
\end{aligned}
$$

Subtracting the first equation from the second, we obtain

$$12d = 48$$
$$d = 4$$

Then

$$a_1 + 14 \cdot 4 = 59$$
$$a_1 = 3$$

Finally,

$$a_{10} = a_1 + 9d$$
$$= 3 + 9 \cdot 4$$
$$a_{10} = 39$$

    ■

The terms of an arithmetic progression that appear between two given terms a and b are called **arithmetic means** of a and b. If there is only one term, it is the arithmetic mean; if there are two, they are called the two arithmetic means; n terms are called n arithmetic means.

EXAMPLE 4   Insert three arithmetic means between 10 and 22.

Solution   We want to form an arithmetic progression of the form . . . 10, $10 + d, 10 + 2d, 10 + 3d, 22, . . .$

Thus

$$22 = 10 + 4d$$

or

$$d = 3$$

The arithmetic means requested are 13, 16, 19.   ■

Geometric Progression

> **DEFINITION**
> A **geometric progression** is a sequence in which each successive term is obtained by multiplying the given term by the same constant; that is, there is some number r called the **common ratio** such that
>
> $$a_{n+1} = r \cdot a_n \qquad \text{for } n = 1, 2, 3, . . .$$

If the first term of a geometric sequence is a and the common ratio is r, the sequence has the form

$$a, ar, ar^2, ar^3, . . .$$

thus

$$a_n = ar^{n-1}$$

EXAMPLE 5 Find the first six terms and the twelfth term of the geometric progression whose first term is 2 and whose common ratio is 1/2.

Solution Any given term is multiplied by 1/2 to get the next term. Thus the first six terms are

$$2, 1, \frac{1}{2}, \frac{1}{4}, \frac{1}{8}, \frac{1}{16}$$

The twelfth term is

$$a_{12} = 2 \cdot \left(\frac{1}{2}\right)^{11} = \frac{1}{2^{10}}$$
$$= \frac{1}{1024}$$ ∎

As with arithmetic progressions, if we know any two terms of a geometric progression we can find the others. The technique is illustrated in the following example.

EXAMPLE 6 If the fourth term of a geometric progression is 24, and the seventh term is 192, find the second term.

Solution Again, the given information yields a system of equations.

$$a_4 = a_1 r^3 = 24$$
$$a_7 = a_1 r^6 = 192$$

Then

$$\frac{192}{24} = \frac{a_1 r^6}{a_1 r^3} = r^3$$

thus

$$r^3 = 8$$
$$r = 2$$

Now,

$$24 = a_1 r^3 = a_1 \cdot 8$$

hence

$$a_1 = 3$$

Finally,

$$a_2 = a_1 r = 3 \cdot 2$$
$$a_2 = 6$$

■

The terms of a geometric progression that appear between two given terms a and b are called **geometric means** of a and b. If there is only one term, it is called the geometric mean of a and b; if there are n terms, they are called the n geometric means.

EXAMPLE 7   Insert three geometric means between 6 and 486,

Solution   We are seeking part of a geometric progression of the form
. . . 6, $6r$, $6r^2$, $6r^3$, 486, ... Thus

$$486 = 6r^4$$
$$r^4 = 81$$
$$r = \pm 3$$

There are two sets of three geometric means between 6 and 486. They correspond to $r = 3$ and $r = -3$. For $r = 3$, the geometric means are

18, 54, 162

for $r = -3$, they are

-18, 54, -162 ■

We have already seen examples of everyday applications of arithmetic and geometric progressions. In Chapter 6 we studied exponential growth and decay. If a given population P_0 doubles every hour, the hourly population figures form a geometric progression with first term P_0 and common ratio $r = 2$. For if the initial population is P_0, than 1 hour later the population is $P_1 = 2P_0$; still another hour later it is $P_2 = 4P_0$. After n hours, the population is $P_n = 2^n P_0$. Listing the population at each hour yields the geometric progression

$$P_0, 2P_0, 4P_0, 8P_0, \ldots$$

We also studied compound interest in Chapter 6. If 5 percent interest is compounded annually on an investment of P_0 dollars, the value of the investment after n years was seen to be $P_n = P_0 \cdot (1.05)^n$. Thus the yearly values of the investment form a geometric progression with first term P_0 and common ratio $r = 1.05$:

$$P_0, P_0(1.05), P_0(1.05)^2, \ldots, P_0(1.05)^n, \ldots$$

On the other hand, if the interest is not compounded, the yearly values of the investment form an arithmetic progression. For after n years, the value is $P_0 + (0.05)P_0 \cdot n$.

$$P_0, P_0 + 0.05P_0, P_0 + 2(0.05P_0), \ldots, P_0 + n(0.05)P_0, \ldots$$

EXAMPLE 8 Mr. Jaworski is enclosing a stairway with poles as indicated in Figure 8–6. The length of the shortest pole is 30 inches; the longest is 8 feet. The span is 12 feet. If he places a pole every 6 inches, how long should each pole be?

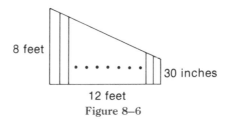

8 feet

30 inches

12 feet

Figure 8–6

Solution The length of the poles increases steadily as we move from right to left. Thus the lengths form an arithmetic progression. In 12 feet there are 24 6-inch steps to be taken. Thus there are 23 arithmetic means to be placed between 30 and 96 ($= 12 \cdot 8$) inches. We form an arithmetic progression

$$30, 30 + d, 30 + 2d, \ldots, 30 + 23d, 30 + 24d = 96$$

From the last relation we obtain

$$30 + 24d = 96$$
$$24d = 66$$
$$d = \frac{66}{24} = \frac{11}{4}$$
$$d = 2\frac{3}{4} \text{ inches}$$

The lengths of the poles are, in inches,

$$30, 32\frac{3}{4}, 35\frac{1}{2}, 38\frac{1}{4}, 41, 43\frac{3}{4}, 46\frac{1}{2}, 49\frac{1}{4}, 52, 54\frac{3}{4}, 57\frac{1}{2}, 60\frac{1}{4}, 63,$$

$$65\frac{3}{4}, 68\frac{1}{2}, 71\frac{1}{4}, 74, 76\frac{3}{4}, 79\frac{1}{2}, 82\frac{1}{4}, 85, 87\frac{3}{4}, 90\frac{1}{2}, 93\frac{1}{4}, 96$$ ■

EXERCISES 8.2

Write out the first five terms and the tenth term of the sequences whose nth term is given.

1. $3n$
2. $2n - 1$
3. $2n + 3$
4. $5n + 3$
5. $2n^2$
6. $3n^2 + 2$
7. $1 + (-1)^n$
8. $1 + 2^n$
9. $(-2)^n$
10. $1 + \left(-\dfrac{1}{2}\right)^n$

Write out the first four terms and the eighth term of the sequences whose nth term is given.

11. $\dfrac{n-1}{n+1}$ **12.** $\dfrac{n}{2n-1}$

13. $\dfrac{n+1}{n+2}$ **14.** $\dfrac{n(n+1)}{2}$

15. $\dfrac{3}{n+5}$ **16.** $\dfrac{1}{n(n+1)}$

17. $\dfrac{2n+3}{2n-3}$ **18.** $\dfrac{1}{2n^2}$

19. $\dfrac{(-1)^n}{n^3}$ **20.** $\dfrac{n}{n^2+1}$

Write out the first five terms of the sequences defined as follows.

21. $a_1 = -2,\ a_n = -a_{n-1}$

22. $a_1 = 6,\ a_n = a_{n-1} + 1$

23. $a_1 = 1,\ a_n = \dfrac{na_{n-1}}{2}$

24. $a_1 = 2,\ a_n = \dfrac{1}{a_{n-1}}$

25. $a_1 = 2,\ a_n = (a_{n-1} - 1)^2$

26. $a_1 = 1,\ a_2 = 1,\ a_n = (a_{n-1} + a_{n-2})$

Write out the first five terms and the fifteenth term of the arithmetic progression with the following data.

27. $a_1 = 2,\ d = 3$ **28.** $a_1 = 3,\ d = -2$

29. $a_1 = -5,\ d = 2$ **30.** $a_4 = 15,\ d = 6$

31. $a_2 = 20,\ d = -2$ **32.** $a_{12} = -40,\ d = -3$

Find the requested term of an arithmetic progression given the following information.

33. $a_4 = -5,\ a_{23} = 52$, find a_{30}

34. $a_{53} = 108,\ a_{86} = 174$, find a_{100}

35. $a_{14} = 0,\ a_{10} = -12$, find a_{21}

36. $a_{16} = 53,\ a_{23} = 74$, find a_6

37. $a_6 = -9,\ a_{14} = -25$, find a_4

38. $a_7 = 0,\ a_{10} = -6$, find a_3

Insert the requested number of arithmetic means between the following numbers.

39. 7 means between 6 and 30

40. 3 means between -6 and 14

41. 6 means between 20 and 55

42. 4 means between -8 and 17

43. 5 means between 13 and 49

44. 2 means between 16 and 31

Write out the first four terms and the eighth term of the geometric series with the following data.

45. $a_1 = 2,\ r = 3$ **46.** $a_1 = 48,\ r = \dfrac{1}{2}$

47. $a_3 = 48,\ r = 4$ **48.** $a_3 = \dfrac{1}{250},\ r = \dfrac{1}{5}$

49. $a_7 = 18,\ r = \sqrt{3}$ **50.** $a_{12} = 512,\ r = 2$

Find the requested term of a geometric progression given the following information.

51. $a_3 = 12,\ a_6 = 96$, find a_{10}

52. $a_5 = 45,\ a_{13} = 3645$, find a_8

53. $a_4 = 81,\ a_8 = 9$, find a_2

54. $a_9 = 48,\ a_3 = 6$, find a_{11}

55. $a_4 = -12,\ a_7 = 96$, find a_2

56. $a_5 = 18,\ a_8 = -486$, find a_3

Insert the requested number of geometric means between the following numbers.

57. 5 terms between 21 and 567

58. 7 terms between 15 and 1215

59. 2 terms between 18 and -486

60. 3 terms between 648 and 8

61. 9 terms between 2048 and 2

62. 4 terms between -24 and 768

63. Bob borrows $1000 interest-free from his father in order to buy a car. He agrees to repay his father $200 in 1 month and $50 per month thereafter. **a.** List the amount still owed to his father after each payment. **b.** Do these amounts form an arithmetic or geometric progression?

64. Sue has been jogging for 1 month and can now run 1 mile comfortably.

a. She wishes to increase her distance by 1/10 of a mile every other day. Write a sequence indicating the distance she runs every day for the next 2 weeks.

b. She wishes to increase her distance by 10 percent every other day. Write a sequence indicating the distance she runs every day for the next 2 weeks.

65. The altitude of an object t seconds after it falls from a height of h_0 feet is $h_0 - 16t^2$ feet.

a. Write a sequence that indicates its altitude after each second.

b. Write a sequence that indicates the distance fallen *during each second.*

c. Show that the sequence in (b) forms an arithmetic progression.

d. How far does an object fall during the tenth second? the twentieth second?

66. A certain ball bounces to three-quarters the height from which it falls. If it is initially dropped from a building 120 meters high, write the altitudes to which it rises on the first four bounces. Do these numbers form an arithmetic or geometric progression?

67. Suppose that an automobile loses 25 percent of its value each year. **a.** List the yearly values of a car that initially sells for $8000. **b.** What is its value after 10 years? **c.** Do these numbers form an arithmetic or geometric progression?

68. Betty wishes to flush her car radiator but cannot completely drain the engine block. The cooling system holds 16 liters, but she can drain only 12 liters from the radiator. She dilutes the old solution by successively filling the radiator with water, running the engine to mix the old solution with the water, and then draining the radiator again. **a.** How much of the old solution remains after each of the first five successive draining operations? **b.** Do these numbers form an arithmetic or geometric progression?

8.3 THE SUMMATION NOTATION

Although it is a simple matter to write down short sums, very long sums can be cumbersome. Lengthy sums appear often enough that it is convenient to have a shorthand notation to describe them. The **summation symbol Σ** is used for this purpose.

DEFINITION

$$\sum_{j=k}^{l} a_j = a_k + a_{k+1} + \cdots + a_l$$

In the preceding definition j is called the **index of summation.** It is a dummy variable in the sense that $\displaystyle\sum_{j=k}^{l} a_j = \sum_{n=k}^{l} a_n$ since both represent $a_k + \cdots + a_l$.

EXAMPLE 1 Evaluate $\displaystyle\sum_{j=2}^{5} (2j + 1)$.

Solution $\displaystyle\sum_{j=2}^{5} (2j + 1) = (2 \cdot 2 + 1) + (2 \cdot 3 + 1) + (2 \cdot 4 + 1) + (2 \cdot 5 + 1)$

$$= 5 + 7 + 9 + 11$$
$$= 32 \qquad\qquad \blacksquare$$

EXAMPLE 2 Write $\displaystyle\sum_{j=10}^{27} 2^j$ in expanded form.

Solution $\displaystyle\sum_{j=10}^{27} 2^j = 2^{10} + 2^{11} + \cdots + 2^{26} + 2^{27} \qquad\qquad \blacksquare$

EXAMPLE 3 Write $(3 + 8 + 13 + 18 + 23)$ in shorthand form by using the summation symbol.

Solution   There are several ways of writing a solution to this problem. Two of these are given here

$$3 + 8 + 13 + 18 + 23 = \sum_{j=0}^{4} (3 + 5j)$$

$$= \sum_{j=1}^{5} (-2 + 5j) \qquad \blacksquare$$

We have already seen several situations in which the summation symbol could have been used to advantage. For instance, the formulas in Examples 1 and 2 of Section 8.1 are rewritten here.

$$1 + 2 + \cdots + n = \frac{n(n + 1)}{2}$$

$$\boxed{\sum_{j=1}^{n} j = \frac{n(n + 1)}{2}} \qquad (1)$$

$$1^2 + 2^2 + \cdots + n^2 = \frac{n(n + 1)(2n + 1)}{6}$$

$$\boxed{\sum_{j=1}^{n} j^2 = \frac{n(n + 1)(2n + 1)}{6}} \qquad (2)$$

The summation symbol is also useful in computing the sum of a finite number of successive terms in an arithmetic or geometric progression. Before discussing these applications, several elementary properties of the Σ symbol should be established.

The symbol $\sum_{j=1}^{n} C$ denotes the addition of C to itself repeatedly from step 1 to step n. Thus

$$\sum_{j=1}^{n} C = \underset{\underset{j=1}{\uparrow}}{C} + \underset{\underset{j=2}{\uparrow}}{C} + \cdots + \underset{\underset{j=n}{\uparrow}}{C} = nC$$

$$\boxed{\sum_{j=1}^{n} C = nC} \qquad (3)$$

In general,

$$\boxed{\sum_{j=k}^{l} C = (l - k + 1)C} \qquad (3')$$

Because of the "rearrangement" properties of addition, we obtain

$$(a_k + b_k) + \cdots + (a_l + b_l) = (a_k + \cdots + a_l) + (b_k + \cdots + b_l)$$

that is,

$$\sum_{j=k}^{l} (a_j + b_j) = \sum_{j=k}^{l} a_j + \sum_{j=k}^{l} b_j \qquad (4)$$

The distributive law becomes the *sliding constant* property. A constant multiplier can slide through the Σ:

$$Ca_k + \cdots + Ca_l = C(a_k + \cdots + a_l)$$

that is,

$$\sum_{j=k}^{l} (Ca_j) = C\left(\sum_{j=k}^{l} a_j\right) \qquad (5)$$

Although "sliding constant" very aptly describes this property, its correct mathematical name is the property of **homogeneity.**

Finally, we observe that two Σ's can be linked into one if their indices match up correctly. For instance,

$$\sum_{j=2}^{5} a_j + \sum_{6}^{10} a_j = (a_2 + \cdots + a_5) + (a_6 + \cdots + a_{10})$$

$$= \sum_{j=2}^{10} a_j$$

In general,

$$\sum_{j=k}^{l} a_j + \sum_{j=l+1}^{m} a_j = \sum_{j=k}^{m} a_j \qquad (6)$$

Now to address the task of adding a finite number of successive terms of an arithmetic progression we consider

$$\cdots \; a_N, \; a_N + d, \; a_N + 2d, \; \ldots, \; a_N + (n-1)d, \; \cdots$$

$$\underbrace{\hphantom{a_N, \; a_N + d, \; a_N + 2d, \; \ldots, \; a_N + (n-1)d}}_{n \text{ terms}}$$

The sum of n successive terms beginning at a_N is

$$a_N + (a_N + d) + \cdots + [a_N + (n-1)d]$$

$$= \sum_{j=0}^{n-1} (a_N + jd)$$

$$= \sum_{j=0}^{n-1} a_N + \sum_{j=0}^{n-1} jd \qquad \text{(formula 4)}$$

$$= \sum_{j=0}^{n-1} a_N + d\sum_{j=0}^{n-1} j \qquad \text{(formula 5)}$$

$$= \sum_{j=0}^{n-1} a_N + d\sum_{j=1}^{n-1} j \qquad (j = 0 \text{ does not add to the sum})$$

$$= na_N + d\frac{(n-1)n}{2} \qquad \begin{array}{l}\text{(Note that the last } j \text{ is } n - 1; \text{ then} \\ \text{use formulas 3' and 1.)}\end{array}$$

$$= \frac{n}{2}[2a_N + (n-1)d]$$

$$= \frac{n}{2}[a_N + \{a_N + (n-1)d\}]$$

This result can be stated as follows.

SUM OF AN ARITHMETIC PROGRESSION

The sum of n successive terms of an arithmetic progression equals $\frac{n}{2}$ times the sum of the first and last terms used:

$$a_1 + \cdots + a_n = \frac{n}{2}[a_1 + a_n] \qquad (7)$$

EXAMPLE 4   Evaluate the sums given in Examples 1 and 3.

Solution   Note that the terms of Example 1 form an arithmetic progression with $d = 2$. Thus applying formula 7,

$$\sum_{j=2}^{5} (2j + 1) = 5 + 7 + 9 + 11$$

$$= \frac{4}{2}(5 + 11) \qquad \text{(There are four terms.)}$$

$$= 32$$

In Example 3, we also have an arithmetic progression, this time with $d = 5$.

$$\underbrace{3 + 8 + 13 + 18 + 23}_{5 \text{ terms}} = \frac{5}{2}(3 + 23)$$
$$= 65$$

EXAMPLE 5   Find the sum of the odd integers between 50 and 490.

Solution   We are to find

$$51 + 53 + \cdots + 487 + 489$$

These terms form an arithmetic progression with $d = 2$, 51 as the first term, and 489 as the last term:

$$a_n = 489 = 51 + (n - 1)2$$

Solving this for n, we obtain $n = 220$. The sum of these 220 terms is then obtained as

$$\frac{220}{2} (51 + 489) = 110 \cdot 540$$
$$= 59,400$$

Knowing the sum of an arithmetic progression can be useful in many everyday situations as illustrated in the following example.

EXAMPLE 6 When booking a performance in an arena, the promoter wants to know the number of seats that are available in a given section. The section is wedge-shaped with 15 seats in the first row and each succeeding row having two additional seats. If there are 20 rows in the section, what is the total number of seats in the section?

Solution There are 15 seats in the first row, each succeeding row has two additional seats. Thus the twentieth row has $(15 + 19 \cdot 2)$ seats. The total number of seats is

$$15 + 17 + 19 + \cdots + (15 + 19 \cdot 2)$$
$$= \frac{20}{2} [15 + (15 + 19 \cdot 2)] \qquad \text{(formula 7)}$$
$$= 10 (30 + 38)$$
$$= 680 \text{ seats}$$

In order to find the sum of the first n terms of a geometric progression with common ratio $r \neq 1$, we write

$$S_n = a + ar + \cdots + ar^{n-1}$$

then

$$rS_n = ar + \cdots + ar^n$$

Subtracting, we get

$$S_n(1 - r) = a - ar^n = a(1 - r^n)$$

and

$$S_n = \frac{a(1 - r^n)}{1 - r}$$

This observation is stated as follows:

SUM OF A GEOMETRIC PROGRESSION

$$a + ar + \cdots + ar^{n-1} = \sum_{j=0}^{n-1} ar^j = a\left(\frac{1 - r^n}{1 - r}\right) \qquad r \neq 1 \qquad (8)$$

EXAMPLE 7 Evaluate the sum in Example 2.

Solution

$$\sum_{j=10}^{27} 2^j = 2^{10} + 2^{11} + \cdots + 2^{27}$$

$$= 2^{10}(1 + 2 + \cdots + 2^{17})$$

$$= 2^{10}\sum_{j=0}^{17} 2^j$$

$$= 2^{10}\left(\frac{1 - 2^{18}}{1 - 2}\right) \qquad \text{(formula 8)}$$

$$= 2^{10}(2^{18} - 1)$$

$$= 268{,}434{,}432 \qquad \text{(from a calculator)} \qquad \blacksquare$$

The formula for the sum of a geometric progression is useful in financial situations such as the study of investments, annuities, insurance, and mortgage payments. This formula is also useful in scientific or engineering situations as indicated in the following example.

EXAMPLE 8 If a ball falls from an altitude of 200 feet and has the property that it rebounds to three quarters of its original altitude after each fall, how far has it traveled by the time it hits the ground for the tenth time?

Solution Figure 8–7 shows the situation. Since the ball initially falls 200 feet, then *rises and falls* nine times, the total distance traveled is

$$200 + 2 \cdot \left(200 \cdot \frac{3}{4}\right) + 2\left[200 \cdot \left(\frac{3}{4}\right)^2\right] + \cdots + 2\left[200 \cdot \left(\frac{3}{4}\right)^8\right] + 2\left[200\left(\frac{3}{4}\right)^9\right]$$

$$= 200 + \sum_{n=1}^{9} 2 \cdot 200 \cdot \left(\frac{3}{4}\right)^n$$

To use (8) we need $\sum_{0}^{9}$. This form is obtained by adding and subtracting the term corresponding to $(3/4)^0$. Then we continue as follows.

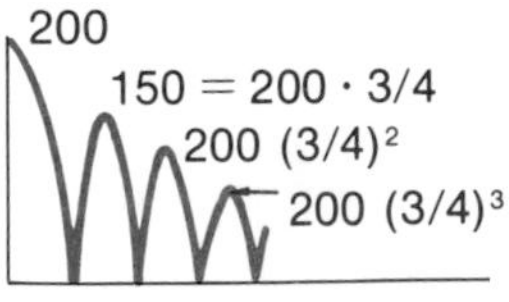

$$= 200 + \sum_{n=0}^{9} 2 \cdot 200\left(\frac{3}{4}\right)^n - 2 \cdot 200$$

$$= -200 + 2 \cdot 200 \sum_{n=0}^{9} \left(\frac{3}{4}\right)^n \qquad \text{(formula 5)}$$

200
150 = 200 · 3/4
200 (3/4)²
200 (3/4)³

Figure 8–7

$$= -200 + 400\,\frac{1 - (3/4)^{10}}{1 - (3/4)} \qquad \text{(formula 8)}$$

$$= -200 + 1600\left[1 - \left(\frac{3}{4}\right)^{10}\right]$$

$$\approx -200 + 1510 \qquad \text{(from a calculator)}$$

$$= 1310 \text{ feet}$$

EXAMPLE 9 Evaluate $\displaystyle\sum_{n=1}^{10} (2n + 3n^2 + 2^n)$.

Solution $\displaystyle\sum_{n=1}^{10} (2n + 3n^2 + 2^n)$

$$= \sum_{n=1}^{10} 2n + \sum_{n=1}^{10} 3n^2 + \sum_{n=1}^{10} 2^n \qquad \text{(formula 4)}$$

$$= 2\sum_{n=1}^{10} n + 3\sum_{n=1}^{10} n^2 + \sum_{n=0}^{10} 2^n - 2^0 \qquad \begin{array}{l}\text{(Use formula 5 twice; then add}\\\text{and subtract the term } 2^0.)\end{array}$$

$$= 2\,\frac{10 \cdot 11}{2} + 3\,\frac{10 \cdot 11 \cdot 21}{6} + \frac{1 - 2^{11}}{1 - 2} - 1 \qquad \text{(Use formulas, 1, 2, and 8.)}$$

$$= 110 + 1155 + 2047 - 1$$

$$= 3311$$

EXERCISES 8.3

Expand and evaluate the following.

1. $\displaystyle\sum_{n=1}^{4} \frac{1}{n}$

2. $\displaystyle\sum_{k=2}^{3} k(-1)^k$

3. $\displaystyle\sum_{n=1}^{3} (2n - 1)(2n)$

4. $\displaystyle\sum_{n=3}^{5} n^2(-1)^n$

5. $\displaystyle\sum_{n=2}^{6} (n^3 - n)$

6. $\displaystyle\sum_{j=2}^{7} j(j + 2)$

7. $\displaystyle\sum_{k=1}^{3} \frac{1}{k(k + 1)}$

8. $\displaystyle\sum_{n=3}^{5} \frac{2^n}{2n}$

Write the following in Σ form.

9. $5 + 7 + 9 + 11 + 13$

10. $1 + 4 + 9 + 16 + \cdots + 81 + 100$

11. $\dfrac{1}{2} + \dfrac{1}{4} + \dfrac{1}{6} + \dfrac{1}{8} + \dfrac{1}{10}$

12. $3 + 9 + 27 + 81 + 243$

13. $20 + 30 + 40 + 50 + 60 + 70$

14. $2 \cdot 3 + 3 \cdot 4 + 4 \cdot 5 + 5 \cdot 6$

15. $2^2 + 3^3 + 4^4 + \cdots + 99^{99}$

16. $1 - 2 + 3 - 4 + 5 - 6 + 7$

Evaluate the following by using the results of this section.

17. $\displaystyle\sum_{k=1}^{10} (3k + 1)$

18. $\displaystyle\sum_{i=0}^{6} (2 + 3i)$

19. $\displaystyle\sum_{k=2}^{80} (1 + 4k)$

20. $\displaystyle\sum_{j=1}^{15} (2j - 1)$

21. $\displaystyle\sum_{n=15}^{25} (3n - 1)$

22. $\displaystyle\sum_{n=1}^{15} (n^2 + 2n)$

23. $\displaystyle\sum_{n=1}^{10} (2n - 1)(2n)$

24. $\displaystyle\sum_{n=1}^{10} (n - 2)(n - 3)$

25. $\displaystyle\sum_{n=1}^{14} (n^2 + 3n + 2)$

26. $\displaystyle\sum_{n=0}^{5} 5 \cdot \left(\frac{1}{2}\right)^n$

27. $\displaystyle\sum_{n=0}^{5} 2 \cdot 3^n$

28. $\displaystyle\sum_{n=1}^{7} 16 \cdot \left(\frac{1}{2}\right)^n$

29. $\displaystyle\sum_{k=1}^{4} 3 \cdot 4^k$

30. $\displaystyle\sum_{n=1}^{10} (n^2 - 4n + 3^n)$

31. $\displaystyle\sum_{k=1}^{5} (3k + 3^k \cdot 4 + 4k^2)$

32. $\displaystyle\sum_{l=1}^{20} [3 \cdot 2^l + l(l + 3) + 4]$

33. A grocer stacks cans in a pyramid with 30 cans on the bottom row, one less can on each higher row until the peak has just one can. How many cans are in the pyramid?

34. Barrels are stacked in a brewery in a pyramid fashion with 50 barrels on the bottom row, one less barrel in each successive row, and a total height of ten rows. How many barrels are there?

35. Karen deposits $1 on May 1, $2 on May 2, . . . $31 on May 31. How many dollars does she deposit in May?

36. Rob deposits 1¢ on May 1, 2¢ on May 2, 4¢ on May 3, and so on, doubling the deposit each day. How much does he deposit in May?

37. The annual premium on a certain insurance policy is 2 percent of the value of an article. If a given article depreciates at the rate of 10 percent per year so that its value at the end of a given year is 90 percent of its value at the beginning of that year, what is the total of the insurance premiums for a period of ten years?

38. Fred receives a chain letter including a list of five names. He is requested to send $1.00 to the name at the top of the list, remove this name, and add his name to the bottom of the list. He is then supposed to mail ten copies of the new letter with the same instructions.
   a. How much money will he eventually receive if the chain is unbroken?
   b. How many steps can this chain take before it will have involved the entire U.S. population of about 250 million people?
   c. How many steps can this chain take before it will have involved the entire world population of about 4 billion people?

39. The late 1960s and early 1970s witnessed the popularity of many pyramid selling schemes. There were many variations on the schemes, but common to all of them was an emphasis on "building a sales organization" rather than actually selling products. One such plan requires an investment of $10,000. Then, for each person you convince to join your organization, you collect $10,000. You keep $500 of this and turn the rest over to your "sponsor." He then keeps $250 and passes the rest up to his sponsor and so on until the chain reaches the company president. The company expects each participant to enlist five people into the sales organization: assume each participant enlists five people.
   a. How many echelons must be built below you before you recoup your initial investment?
   b. Through how many echelons can this pyramid proceed before including the entire available adult U.S. population of about 100 million people?

40. Number the squares on a checkerboard from 1 to 64 and then place one dot on square 1, two dots on square two, four dots on square 3, and so on.
   a. How many dots are required on square 64?
   b. How many dots are required in total?
   c. Answer the preceding questions if three dots are placed on square 3, four dots on square 4, and so on.

8.4 THE BINOMIAL THEOREM

Expanding a binomial such as $(a + b)^{10}$ by repeated multiplication can be a very laborious task and provides many opportunities for error. The binomial theorem provides a straightforward way of writing such expansions directly without repeated multiplications. For a clue regarding a pattern for repeated multiplication of a binomial, consider the following:

$$
\begin{aligned}
(a + b)^0 &= 1 \\
(a + b)^1 &= a + b \\
(a + b)^2 &= a^2 + 2ab + b^2 \\
(a + b)^3 &= a^3 + 3a^2b + 3ab^2 + b^3 \\
(a + b)^4 &= a^4 + 4a^3b + 6a^2b^2 + 4ab^3 + b^4 \\
(a + b)^5 &= a^5 + 5a^4b + 10a^3b^2 + 10a^2b^3 + 5ab^4 + b^5
\end{aligned}
$$

We can make several observations about the expansion of $(a + b)^n$.

1. The first term is $a^n = a^n b^0$.

2. The last term is $b^n = a^0 b^n$.

3. All terms are of the form $a^k b^l$ with $k + l = n$.

4. In moving from left to right to successive terms, the power of a decreases by 1 and the power of b increases by 1.

The coefficients of $a^k b^l$ in the expansion of $(a + b)^n$ are called **binomial coefficients.** If we extract the binomial coefficients from the preceding expansions we obtain **Pascal's triangle** as shown here.

PASCAL'S TRIANGLE

```
                    1
                 1     1
              1     2     1
           1     3     3     1
        1     4     6     4     1
     1     5    10    10     5     1
  1     6    15    20    15     6     1
1     7    21    35    35    21     7     1
   .     .     .     .     .     .     .
```

In Pascal's Triangle the leading entry and the final entry in each row is a 1. Inside each row, a given entry is the sum of the two entries above it. For instance, consider the rows corresponding to $(a + b)^4$ and $(a + b)^5$.

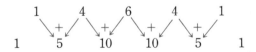

The binomial coefficients appearing in the expansion of $(a + b)^n$ are found in the corresponding row of Pascal's triangle; the second entry in that row will be n.

We can use Pascal's triangle to read off

$$
\begin{aligned}
(a + b)^7 = {}& a^7 + 7a^6b + 21a^5b^2 + 35a^4b^3 + 35a^3b^4 \\
& + 21a^2b^5 + 7ab^6 + b^7
\end{aligned}
$$

EXAMPLE 1   Use Pascal's triangle to expand the following.

a. $(x + 2y)^4$

b. $(x - 2)^5$

Solution   **a.**   According to Pascal's triangle, the coefficients for $n = 4$ are 1, 4, 6, 4, 1 in that order. Thus

$$(x + 2y)^4 = x^4 + 4x^3(2y) + 6x^2(2y)^2 + 4x(2y)^3 + (2y)^4$$
$$= x^4 + 8x^3y + 24x^2y^2 + 32xy^3 + 16y^4$$

b.   The coefficients for $n = 5$ are 1, 5, 10, 10, 5, 1. Thus

$$(x - 2)^5 = [x + (-2)]^5$$
$$= x^5 + 5x^4(-2) + 10x^3(-2)^2 + 10x^2(-2)^3$$
$$+ 5x(-2)^4 + (-2)^5$$
$$= x^5 - 10x^4 + 40x^3 - 80x^2 + 80x - 32$$ ■

Although Pascal's triangle is quite useful for small n, it could be quite a chore to generate 100 rows of the triangle in order to find some coefficient in $(a + b)^{100}$. A formula for the binomial coefficients is certainly desirable. It will be helpful to introduce the following notation:

DEFINITION

$$n! = 1 \cdot 2 \cdot \cdots \cdot n$$
$$0! = 1$$

The ! symbol is read **"factorial."** The expression **0!** is seldom used; its notation is introduced merely to simplify writing certain expressions later. As examples in the use of factorials, we have

$$\frac{5!}{3!} = \frac{\cancel{1} \cdot \cancel{2} \cdot \cancel{3} \cdot 4 \cdot 5}{\cancel{1} \cdot \cancel{2} \cdot \cancel{3}} = 4 \cdot 5 = 20$$

and

$$\frac{5!}{2!(5 - 2)!} = \frac{5!}{2!3!} = \frac{\cancel{1} \cdot \cancel{2} \cdot \cancel{3} \cdot 4 \cdot 5}{(\cancel{1} \cdot \cancel{2})(1 \cdot 2 \cdot \cancel{3})} = \frac{20}{2} = 10$$

Also note that

$$n!(n + 1) = (n + 1)!$$

The factorial notation helps in writing a formula for the binomial coefficients. These coefficients are embodied in the binomial theorem.

THE BINOMIAL THEOREM

$$(a + b)^n = \sum_{l=0}^{n} \frac{n!}{l!(n - l)!} a^{n-l}b^l$$

Proof   It is a simple matter to verify that the binomial theorem holds for small values of n, say, for $n = 1, 2, 3$. We then assume the formula holds for $n = k$ and attempt to establish it for $n = k + 1$. For $n = k$, we have

$$(a + b)^k = \sum_{l=0}^{k} \frac{k!}{l!(k - l)!} a^{k-l}b^l$$

Then $(a + b)^{k+1} = (a + b)(a + b)^k$

$$= (a + b)\left[\sum_{l=0}^{k} \frac{k!}{l!(k - l)!} a^{k-l}b^l \right]$$

$$= a\sum_{l=0}^{k} \frac{k!}{l!(k - l)!} a^{k-l}b^l + b\sum_{l=0}^{k} \frac{k!}{l!(k - l)!} a^{k-l}b^l$$

$$= \sum_{l=0}^{k} \frac{k!}{l!(k - l)!} a^{k+1-l}b^l + \sum_{l=0}^{k} \frac{k!}{l!(k - l)!} a^{k-l}b^{l+1}$$

$$= a^{k+1} + \sum_{l=1}^{k} \frac{k!}{l!(k - l)!} a^{k+1-l}b^l + \sum_{l=0}^{k-1} \frac{k!}{l!(k - l)!} a^{k-l}b^{l+1} + b^{k+1}$$

Replace the (dummy) index of summation l with $l - 1$ in the second summation.

$$= a^{k+1} + \sum_{l=1}^{k} \frac{k!}{l!(k - l)!} a^{k+1-l}b^l + \sum_{l=1}^{k} \frac{k!}{(l - 1)!(k - l + 1)!} a^{k-l+1}b^l + b^{k+1}$$

$$= a^{k+1} + \sum_{l=1}^{k} \left[\frac{k!}{l!(k - l)!} + \frac{k!}{(l - 1)!(k - l + 1)!} \right] a^{k+1-l}b^l + b^{k+1}$$

$$= a^{k+1} + \sum_{l=1}^{k} \left[\frac{k!(k - l + 1)}{l!(k - l)!(k - l + 1)} + \frac{k!l}{(l - 1)!(k - l + 1)!l} \right] a^{k+1-l}b^l + b^{k+1}$$

$$= a^{k+1} + \sum_{l=1}^{k} \frac{k!(k - \cancel{l} + 1 + \cancel{l})}{l!(k + 1 - l)!} a^{k+1-l}b^l + b^{k+1}$$

$$= a^{k+1} + \sum_{l=1}^{k} \frac{(k + 1)!}{l!(k + 1 - l)!} a^{k+1-l}b^l + b^{k+1}$$

$$= \sum_{l=0}^{k+1} \frac{(k + 1)!}{l!(k + 1 - l)!} a^{k+1-l}b^l$$

Thus the truth of the formula for $n = k$ implies its truth for $n = k + 1$. Since the formula holds for $n = 1, 2,$ and 3, mathematical induction then tells us that it holds for all positive integer powers n.   ∎

From the binomial theorem we note that the coefficient of $a^j b^k$ in $(a + b)^n$ is just $n!/j!k!$; of course, $j + k = n$.

EXAMPLE 2 The kth term in the expansion of $(a + b)^n$ is the term involving b^{k-1}; thus, in $(a + b)^2 = a^2 + 2ab + b^2$, a^2 is the first term, $2ab$ is the second term; and b^2 is the third term. Use the binomial theorem to find the following.

  a. The third term of $(x - 3)^{15}$

  b. The fifth term of $(x + 2y)^{10}$

Solution **a.** The third term of $(x - 3)^{15} = [x + (-3)]^{15}$ is the one involving $x^{13}(-3)^2$:

$$\frac{15!}{2!(15 - 2)!} x^{13}(-3)^2 = \frac{15!}{2!13!} x^{13}(-3)^2$$
$$= \frac{14 \cdot 15}{2} \cdot 9x^{13}$$
$$= 945x^{13}$$

  b. The fifth term of $(x + 2y)^{10}$ is the one involving $x^6(2y)^4$.

$$\frac{10!}{4!6!} x^6 \cdot (2y)^4 = \frac{7 \cdot 8 \cdot 9 \cdot 10}{2 \cdot 3 \cdot 4} \cdot 16x^6y^4$$
$$= 3360 \, x^6y^4$$

EXAMPLE 3 Use the binomial theorem to evaluate $(99)^4$.

Solution
$$(99)^4 = (100 - 1)^4 = [100 + (-1)]^4$$
$$= 100^4 + \frac{4!}{1!3!}100^3(-1) + \frac{4!}{2!2!}100^2(-1)^2$$
$$+ \frac{4!}{3!1!}100(-1)^3 + \frac{4!}{4!0!}(-1)^4$$
$$= (100)^4 - 4(100)^3 + 6(100)^2 - 4(100) + 1$$
$$= 100,000,000 - 4,000,000 + 60,000 - 400 + 1$$
$$= 96,059,601$$

EXERCISES 8.4

1. Evaluate the following factorial expressions.

  a. 6! **b.** $\dfrac{9!}{5!}$ **c.** $\dfrac{3!2!}{4}$

  d. $\dfrac{6!}{4!3!}$ **e.** $\dfrac{7! \cdot 8 \cdot 9}{10!}$ **f.** $\dfrac{100!3!}{99!4!}$

Use Pascal's triangle to expand the following.

2. $(t - 3)^5$
3. $(r - s)^{10}$
4. $(2x + y)^6$
5. $(4x - y)^3$
6. $(3a - 5b)^4$
7. $(v - w^2)^5$
8. $(u^2 - 3v)^4$
9. $(2x + 3y^2)^4$

10. $(a^3 - 2b^2)^6$
11. $\left(x + \dfrac{1}{x}\right)^6$
12. $\left(\dfrac{3}{u} - \dfrac{u}{3}\right)^5$
13. $(t - u^{-2})^8$
14. $\left(2p - \dfrac{2}{q^2}\right)^5$

Use the binomial theorem to evaluate the following.

15. 11^5
16. 9^5
17. $(101)^4$
18. $(98)^4$
19. $(1.1)^6$
20. $(1.02)^5$

21. $(0.995)^3$

Use the binomial theorem to find the indicated term of the following.

22. $(2a - b)^7$, fifth term

23. $(x + y)^{17}$, fourth term

24. $(2v + w)^{20}$, eighteenth term

25. $(r + 2s)^{20}$, third term

26. $(x^2 + y)^{15}$, eleventh term

27. $(u^2 - 2v)^{10}$, fourth term

28. $(3a - 2b^2)^5$, fourth term

29. $\left(t + \dfrac{1}{t}\right)^{12}$, eighth term

30. $\left(x - \dfrac{3}{y}\right)^{12}$, seventh term

31. $\left(\dfrac{p^2}{q} - \dfrac{2}{p}\right)^{10}$, fourth term

---

8.5 PERMUTATIONS AND COMBINATIONS

Whenever an object or quantity can assume several different characteristics, in various combinations with one another, it can sometimes be a time-consuming process to enumerate all possible configurations for the system. Accordingly, some kind of systematic counting procedure should be developed to determine the total number of such configurations. Consider, for example, a simple—yet important—situation from everyday life: classification of human blood types.

EXAMPLE 1 Three substances, called antigens, affect the blood donor–recipient relationship. The medical profession calls these antigens A, B, and Rh. Any, all, or none of these may be present in the blood of a given individual. How many blood types can be classified by using these three antigens?

Solution The human population can be divided into two classes: those whose blood contains or does not contain antigen A:

A present
A absent

Each of these groups is further subdivided into two smaller groups: those whose blood does or does not contain antigen B:

A present and B present
A present but B absent

A absent but B present
A absent and B absent

These groups are in turn subdivided into two classes each, by the presence or absence of Rh (usually denoted Rh+ and Rh−, respectively). These classes along with the common designation of blood type are as follows:

	Designation of Blood Type
A present, B present, Rh+	AB+
A present, B present, Rh−	AB or AB−
A present, B absent, Rh+	A+
A present, B absent, Rh−	A or A−

A absent, B present, Rh+   B+
A absent, B present, Rh−   B or B−
A absent, B absent, Rh+   O+
A absent, B absent, Rh−   O or O−

There are eight different blood types. ■

In Example 1 the eight different blood types were found by splitting the human population into two classes determined by the present or absence of antigen A. Then the presence or absence of B split *each* of these classes into two smaller classes yielding $2 \cdot 2 = 4$ classes. Then the Rh factor (+ or −) further split each of these into two parts yielding

$$2 \cdot 2 \cdot 2 = 8$$

classifications by blood type.

The following example differs only slightly from the first.

EXAMPLE 2 A manufacturing concern developing a top secret new product uses three of the letters *A, B, C, D, E* to label the versions it is currently testing. If no letter can appear more than once on a given label, how many different labels are possible?

Solution If only one letter were allowed, then just five labels would be possible. If two letters are to be used, then once the first letter is chosen, it can be followed by any one of the four remaining letters. Since there are five possible choices for the first letter and each of these can then be followed by four possible second letters, there are

$$5 \cdot 4 = 20$$

possible two-letter labels:

AB	BA	CA	DA	EA
AC	BC	CB	DB	EB
AD	BD	CD	DC	EC
AE	BE	CE	DE	ED

For a three-letter label, only three letters remain after the choice of the first two letters. Thus each of the preceding two-letter labels can be followed by any one of three remaining letters; for example, we have

ABC
ABD
ABE

Consequently, there are

$$5 \cdot 4 \cdot 3 = 60$$

possible three-letter labels if no letter can appear more than once on a given label. ■

Had repetitions of letters been permitted in Example 2, there would have been five choices for the second letter. This yields

$$5 \cdot 5 = 25$$

two-letter labels. Then any one of five letters could be used in the third position yielding

$$5 \cdot 5 \cdot 5 = 125$$

three-letter labels if repetitions are permitted.

These observations are embodied in the following principle.

FUNDAMENTAL COUNTING PRINCIPLE

Suppose that n events are to take place and that

1. The first event can result in N_1 different outcomes.

2. After the first event, there remains N_2 possible outcomes for the second event.

.
.
.

n. After the preceding $(n - 1)$ events, there remain N_n possible outcomes for the nth event.

Then the total number of possible outcomes from the sequence of n events is the product

$$N_1 \cdot N_2 \cdot \cdots \cdot N_n$$

In Example 1 the "events" were the determination of the presence or absence of the respective antigens: A, B, Rh; in Example 2 the events were the selection of the first, second, and third letters, respectively, in the labels.

EXAMPLE 3 An automobile license numbering system currently used by several states consists of three letters followed by three digits. Find the total number of distinct license numbers of this type if

a. Repetitions are permitted

b. Repetitions are not permitted

c. Letters may not be repeated, digits may be repeated, but 0 may not be the first digit

Solution **a.** There are 26 choices for each of the three letters and 10 choices for each of the three digits. Thus the total number of distinct license numbers is

$$26 \cdot 26 \cdot 26 \cdot 10 \cdot 10 \cdot 10 = 17{,}576{,}000$$

b. If no repetitions are permitted, then there are

26 choices for the first letter
25 choices for the second letter
24 choices for the third letter
10 choices for the first digit
9 choices for the second digit
8 choices for the third digit

This gives a total of

$$26 \cdot 25 \cdot 24 \cdot 10 \cdot 9 \cdot 8 = 11{,}232{,}000$$

distinct license numbers.

c. The differences between this case and (b) are that only nine choices are permitted for the first digit and ten choices are available for each of the remaining digits. Thus the total is

$$26 \cdot 25 \cdot 24 \cdot 9 \cdot 10 \cdot 10 = 14{,}040{,}000 \qquad \blacksquare$$

Permutations

DEFINITION

Let S be a collection of n objects or symbols. An arrangement of r of these objects ($r \le n$) in which

1. order is important

2. repetitions are not permitted

is called a **permutation** of S. The total number of permutations of n objects taken r at a time is denoted $_nP_r$.

To find the total number of permutations of n objects taken r at a time, observe that there are

n ways to choose the first object
$n - 1$ ways to choose the second object
$n - 2$ ways to choose the third object
.
.
.

$$n - (r - 1) = n - r + 1 \text{ ways to choose the } r\text{th object}$$

Thus by the fundamental counting principle, there are

$$n \cdot (n - 1) \cdot (n - 2) \cdot \cdots \cdot (n - r + 1)$$

such permutations. Another way to write this is to observe that

$$
\begin{aligned}
n(n &- 1)(n - 2) \cdots (n - r + 1) \\
&= \frac{n(n - 1) \cdots (n - r + 1)[(n - r) \cdots 2 \cdot 1]}{[(n - r) \cdots 2 \cdot 1]} \\
&= \frac{n!}{(n - r)!}
\end{aligned}
$$

The total number of permutations of n objects taken r at a time is

$$
\begin{aligned}
_nP_r &= n(n - 1) \cdots (n - r + 1) \\
&= \frac{n!}{(n - r)!}
\end{aligned}
$$

In particular, the number of permutations in which all n objects are used is

$$
\begin{aligned}
_nP_n &= n(n - 1) \cdots 2 \cdot 1 \\
&= n!
\end{aligned}
$$

(Recall from Section 8.4 that $0! = 1$ by definition. Thus $n! = n!/(n - n)!$ in agreement with the first formula.)

EXAMPLE 4 On Competition Day at Southdale Elementary School, one of the events is a softball game between the fifth- and sixth-grade classes. If 15 persons in the sixth grade are interested in playing softball, in how many ways can a nine-person team be formed?

Solution Since position is important and no person can play two positions simultaneously, the number is $_nP_r$ with $n = 15$, $r = 9$:

$$
\begin{aligned}
_{15}P_9 &= \frac{15!}{9!} = \frac{15 \cdot 14 \cdot \cdots \cdot 10 \cdot 9 \cdots 2 \cdot 1}{9 \cdot 8 \cdots 2 \cdot 1} \\
&= 15 \cdot 14 \cdot 13 \cdot 12 \cdot 11 \cdot 10 \\
&= 3,603,600
\end{aligned}
$$

Combinations

In many types of arrangements, the order is immaterial. For instance, in a bridge hand only the "hand" of 13 cards is important. The order in which these 13 cards are received does not matter.

DEFINITION

Let S be a collection of n objects or symbols. A collection of r of these objects in which

1. Order is irrelevant

2. Repetitions are not permitted

is called a **combination** from S. The total number of combinations of n objects taken r at a time is denoted $_nC_r$.

EXAMPLE 5 How many different bridge hands of 13 cards can be dealt from a 52-card deck?

Solution The total number of hands of 13 cards that can be dealt *in a given order* from a 52-card deck is $_{52}P_{13}$; the total number of distinct hands, ignoring the order in which the individual cards are received, is $_{52}C_{13}$. Once the player picks up her hand, she can rearrange the cards any way she chooses. Thus the same hand could be received in $_{13}P_{13} = 13!$ different ways. Hence each hand represents 13! different permutations or ordered hands. Since there are $_{52}C_{13}$ hands and $_{52}P_{13}$ permutations, we have

$$_{52}C_{13} \cdot 13! = \; _{52}P_{13}$$

or

$$
\begin{aligned}
_{52}C_{13} &= \frac{_{52}P_{13}}{13!} \\
&= \frac{52!}{39!13!} \\
&= \frac{52 \cdot 51 \cdot \; \cdots \; \cdot 40 \cdot [39 \;\rule{1.5cm}{0.4pt}\; 2 \cdot 1]}{[39 \cdot 38 \;\rule{1.2cm}{0.4pt}\; 2 \cdot 1] \, [13 \cdot 12 \cdot \; \cdots \; \cdot 2 \cdot 1]} \\
&= \frac{52 \cdot 51 \cdot \; \cdots \; \cdot 41 \cdot 40}{13 \cdot 12 \cdot \; \cdots \; \cdot 2 \cdot 1} \\
&\approx (6.35)10^{11}
\end{aligned}
$$

As with the preceding example, we note that the total number of permutations of n objects taken r at a time is $_nP_r$. If no repetitions are allowed but the order of the listing is immaterial, there are $_nC_r$ combinations of these n objects taken r at a time. But the r objects in a given combination can be arranged in $_rP_r = r!$ ways; that is, each combination represents $r!$ permutations. Hence

$$_nC_r \cdot r! = \; _nP_r$$

or

$$_nC_r = \frac{_nP_r}{r!} = \frac{n!}{(n-r)!r!}$$

> The total number of combinations of n objects taken r at a time is
>
> $$_nC_r = \frac{n!}{(n - r)!r!}$$

Note that the numbers $_nC_r$ have appeared earlier as the binomial coefficients; see the binomial theorem, Section 8.4.

EXAMPLE 6 In panel tests of consumer products, the technique of paired comparisons has seen a certain amount of success. In this technique each product is paired with each of the other products and a direct comparison of the two products is made. With a sample consisting of six different products, how many paired comparisons must be made?

Solution We must find the total number of combinations of six objects taken two at a time (i.e., in pairs). This is just

$$_6C_2 = \frac{6!}{(6 - 2)!2!} = \frac{6!}{4!2!} = \frac{6 \cdot 5 \cdot 4 \cdot 3 \cdot 2 \cdot 1}{(4 \cdot 3 \cdot 2 \cdot 1)(2 \cdot 1)} = \frac{6 \cdot 5}{2} = 15 \quad \blacksquare$$

Counting problems appear in a wide variety of situations. These can range from those that are merely interesting, such as puzzles and games, to very important practical problems that occur in the design of computer systems and telephone-switching networks. A whole branch of mathematics called "combinatorics" is devoted essentially to the development of counting techniques for increasingly complex situations.

In this section we have illustrated a variety of counting problems. The techniques discussed here are by no means exhaustive but they are basic in that many more sophisticated counting techniques rely on these principles of permutations and combinations.

EXERCISES 8.5

1. **a.** How many seven-digit telephone numbers are possible if 0 is not permitted as a first digit?
   b. How many ten-digit telephone numbers (area code and number) are possible if 0 is not permitted as the first digit of either the area code or the number?

2. Compare the possible numbers of five-digit and nine-digit ZIP codes if 0 is permitted as the first digit.

3. In how many ways could you answer a true-false test consisting of ten questions?

4. In how many ways could you answer a multiple choice (*A, B, C, D, E*) test having ten questions?

5. How many different meals can be planned by using three meats, four vegetables, two breads, four fruits, and three drinks if each meal is to contain exactly one item from each group?

6. An automobile can be purchased in three body types, eight colors, one or two of which can be applied to any given car, three engines, two transmissions, and two "rear ends." In how

many different configurations can this automobile be purchased?

7. In selecting skiing equipment, Laara has narrowed her options down to three skis, two bindings, two poles, and three kinds of boots. How many options does she have?

8. **a.** Using at least one and up to ten different ingredients, how many different combination pizzas can be made? **b.** What if three of these ingredients (crust, cheese, sauce) must be used?

9. A builder provides five basic house plans with choice of three carpets in each of five rooms, colonial or contemporary trim, zero to five appliances (two brands each), and two paint colors in each of five rooms. How many options does a buyer have?

10. Computers operate in the binary system—switches are either on or off. How many switches are necessary to distinguish
    a. digits 0 through 9?
    b. letters *a* through *z?*
    c. letters and digits?
    d. both upper and lower case letters?
    e. both upper and lower case letters and digits?
    f. both upper and lower case letters, digits and up to 20 punctuation and mathematics symbols?

11. In how many ways can 12 jurors be seated in the jury box?

12. In how many ways can a president, vice-president, secretary, and treasurer be elected
    a. from a membership of 15 people?
    b. from a membership of 7 males, 8 females if the president must be female and the secretary must be male?

13. In how many ways can six candidates be arranged on a ballot?

14. In how many ways can six candidates for county commission and four candidates for city council be arranged on a ballot if the candidates for a given office must be grouped together?

15. In how many ways can four flags be arranged on a pole if all flags are of a different color?

16. The presidents and sales managers of four companies are to make presentations. In how many ways can these presentations be arranged if
    a. all the presidents must speak before the sales managers?

b. each president is followed by his sales manager?

17. Jane has received invitations to interview six companies for employment. In how many ways can she order these interviews if
    a. no restrictions are placed on the order of the interviews?
    b. a certain two of these interviews must be in succession since they will be in the same city?
    c. the six companies are located in two cities, three in each, and only one trip can be made to each city?

18. How many distinct lines can be drawn through ten points, no three of which fall on the same line?

19. How many distinct circles can be drawn through ten points, no three of which lie on the same line? (*Hint:* Any three noncollinear points determine a unique circle.)

20. Semaphore or flag-signaling is based on holding a flag in each hand and then positioning each flag at one of eight positions around a circle as indicated in the sketch. If identical flags are used and the flags must occupy different positions, how many different signals can be sent?

21. In how many ways can 12 players be split into two teams of 6?

22. How many different "euchre" hands of 5 cards can be dealt from a 24-card deck?

23. How many different (euchre) hands are there if it is recognized that the declaration of one of the four suits as "trump" affects the value of a hand and trump is always declared? (See Exer. 22.)

24–25. Answer Exercises 22–23 for the entire deal of four players.

26. In how many ways can a jury of 12 people be chosen from a pool of 20 people?

27. In how many ways can an extended jury of 13 people (12 jurors and 1 alternate) be chosen from a pool of 20 people? Assume that the order of selection of the 12 jurors is immaterial but that the thirteenth juror selected is designated as the alternate.

28. Rework Exercise 27 under the assumption that 13 jurors are chosen and then any one of these is later designated as the alternate.

29. On a test of ten questions, how many ways are there to choose the questions to answer if
    a. any two can be omitted?
    b. any two of the last five can be omitted?
    c. seven problems must be worked including three of the last five?

30. a. How many different numbers can be formed from the digits in 10632?
    b. How many of these are even numbers?
    c. How many are odd numbers?
    d. How many of the numbers are greater than 5000?
    e. How many are less than 3500?

31. If ten points are placed on a circle and adjacent points are connected to form a polygon, how many distinct diagonals does this polygon have?

32. a. How many different dominoes ranging from double blank to double 6 are there?
    b. How many different sums can be formed on dominoes?

33. How many different combinations can be obtained from rolling
    a. three dice?
    b. three tetrahedra (choose the number on the bottom)?
    c. three dodecahedra?

34. In a family of six girls and five boys, how many different family picture arrangements are possible if the parents alone occupy the first row, the girls occupy the second row, and the boys occupy the last row?

35. In how many ways can three history, six math, four sociology, and five science books be arranged on a shelf if books on the same subject are to be shelved together?

36. In how many ways can 30 cars be arranged in a funeral procession, assuming the first 3 cars are specified in a given order, the next group of 5 is identified but no order within this group is required, three groups of 4 are specified where the groups are to follow in a given order but order within each group is arbitrary?

37. How many different 5-card poker hands can be dealt from a 52-card deck if the first 2 cards are dealt face down (order is immaterial) and the last 3 are dealt face up (order affects the betting)?

38. a. In how many ways can 12 members of a ski club ride in three distinct cars (4 persons each) if seating within each car is ignored? b. What if each owner rides in his own car?

39. In how many ways can 16 people be split into
    a. four bridge "tables" of 4 persons each; ignore the distinction between tables and the division of tables into teams?
    b. eight bridge teams of 2 persons each?

40. In how many ways can union- and management-negotiating teams be seated on opposite sides of a table if each team consists of five members?

41. In how many ways can a group of 12 horses be subdivided into a. four teams of 3? b. one team of 6 and two teams of 3?

42. Answer Exercise 41 if the teams are assigned to distinct jobs.

43–44. Answer Exercises 41–42 if 2 of the horses cannot be made to cooperate on the same team.

45. Three salespeople from the Safer-Alarm Company attend a convention. Twelve presentations are of interest to them but each can attend only four. In how many ways can this be done assuming a. no time conflicts? b. the presentations are given in groups of three but in different rooms?

46. a. How many different four-person committees can be formed from a group of five men and three women?
    b. Answer (a) if at least one committee member must be female.
    c. Answer (a) if the committee must have at least one male and two females.
    d. Answer (a) if the committee must be sexually balanced.
    e. Answer (a) if Lois must be a member of the committee.

8.6 PROBABILITY

Probability is the study of the laws of chance. Probability is used to test the effectiveness of new medications and to test claims of abilities in mental telepathy. Life insurance rates are based on the probability that individuals in certain classifications will live to a certain age. When the Weather Service tells us that the chance of rain is 10 percent, they are making a statement of probability. In this section we shall generally use the decimal or fractional equivalent of percent. Thus we would say that the chance or probability of rain is 0.1 or 1/10. Questions of probability are also of interest in games of chance; games whose progress depends on chance events such as the outcome from tossing a coin, rolling dice or a single die, spinning a wheel, or dealing cards.

A fair coin is one that when tossed is equally likely to turn up heads as tails. (In the unlikely event that the coin is tossed for sharpshooting practice, we shall assume that the marksman misses and that the coin indeed falls to a surface and lands "heads-up" or "tails-up" with equal likelihood.) We say that the probability of turning up heads is 1/2 and the probability of turning up tails is 1/2. This means that if the coin is tossed a great many times, heads would turn up about half the time and tails would turn up about half the time. Heads will not necessarily turn up exactly half the time, though. We are concerned with probability or likelihood here, not precision or exactness.

Similarly, a fair die is one in which no number is more likely than another to turn up when the die is tossed or rolled. If the die is rolled a great many times, each number should turn up about 1/6 of the time. We say that the probability of turning up a 4 on a single roll of the die is 1/6.

Performing a task or succession of tasks whose results depend on chance will be called an **experiment.** The result of an experiment is called an **outcome.** The collection (or set) of all possible outcomes is called the **sample space** S. An **event** is a collection of one or more possible outcomes; it is a subset of S. Thus in the experiment of rolling a die the possible outcomes are 1, 2, 3, 4, 5, 6; we write $S = \{1, 2, 3, 4, 5, 6\}$. Turning up a 4 is an event; likewise, turning up a *4 or a 6* is an event. These events are denoted $E = \{4\}$ and $F = \{4, 6\}$, respectively.

DEFINITION

Assuming that each possible outcome of an experiment is equally likely, the **probability** P of an event E is the number of ways in which E could occur $N(E)$ divided by the total number of possible outcomes T:

$$P(E) = \frac{N(E)}{T}$$

Needless to say, the counting techniques developed in the preceding section should be useful here. Note that the event could occur at most in every outcome; in this case $N(E) = T$ and $P(E) = 1$. This is called the certain event. At worst E could never occur in which case $N(E) = 0$ and $P(E) = 0$. This is called the impossible event. For example, rolling a 7 with one die is impossible;

its probability is 0. On the other hand, we shall certainly turn up one of the numbers 1 to 6 with one die; the probability of rolling a 1, 2, 3, 4, 5, or 6 is 1. Thus

$$0 \leq P(E) \leq 1$$

EXAMPLE 1   Find the probability of rolling a 4 or a 6 with one die.

Solution   Think of rolling the die as a game that is won by rolling a 4 or a 6. The "event" in question is "winning the game." The number of possible outcomes is 6: $S = \{1, 2, 3, 4, 5, 6\}$ and $T = 6$. There are two ways to win the game, by rolling a 4 or a 6: $E = \{4, 6\}$ and $N(E) = 2$. Thus

$$P(E) = \frac{N(E)}{T} = \frac{2}{6} = \frac{1}{3}$$ ■

EXAMPLE 2   Find the probability of rolling a sum of eight with two dice.

Solution   If the dice are fair, it makes no difference whether they are thrown together or one after the other. Identify one as the "first" die and the other as the "second." This is essential to counting the total number of possible outcomes. For instance, a sum of two can occur in only one way: as 1, 1 on the two dice. But a sum of three can occur in two ways: as 1, 2 or as 2, 1 on the first and second dice, respectively. Thus a sum of three is twice as likely to occur as a sum of two.

There are six ways in which the first die could turn up and six ways for the second die. Thus there are $6 \cdot 6 = 36$ possible outcomes from rolling first one die, then a second: $T = 36$. A sum of eight could be obtained from the combinations (2, 6), (3, 5), (4, 4), (5, 3), (6, 2). Thus for the event "sum = 8," $N(E) = 5$. Finally,

$$P(\text{sum} = 8) = \frac{N(E)}{T} = \frac{5}{36}$$ ■

Independent Events

> **DEFINITION**
> Experiments are said to be **independent** if the outcome of one does not influence the outcome of any of the others. Events associated with independent experiments are also said to be independent.

Frequently, an event E can be expressed in terms of two or more independent events. For example, the event E of getting heads on each of two tossed coins can be expressed as

"getting heads on the first coin" (E_1)

and

"getting heads on the second coin" (E_2);

$E = (E_1 \text{ and } E_2)$. Clearly, $P(E_1) = 1/2$ and $P(E_2) = 1/2$. About half the time H will turn up on the first coin. Then for those times when H has turned up on the first coin, H can be expected only half the time on the second coin; $P(HH) = 1/2 \cdot 1/2 = 1/4$. This is also verified by listing the possible outcomes: $S = \{HH, HT, TH, TT\}$.

> For independent events $E_1, \cdots, E_n$
> $$P(E_1 \text{ and } E_2 \text{ and } \cdots \text{ and } E_n) = P(E_1) \cdot P(E_2) \cdot \cdots \cdot P(E_n)$$

EXAMPLE 3   Assume a card is drawn at random from a 52-card deck and returned to the deck. Then the deck is shuffled and again a card is drawn at random.

a.   Find the probability of drawing two hearts.

b.   Find the probability of drawing two cards of the same suit.

Solution   a.   Getting a heart on the first and second draws form independent events E and F. Since hearts comprise 13 of the 52 cards, $P(E) = P(F) = 13/52 = 1/4$. Thus

$$P(\text{two hearts}) = P(E \text{ and } F) = P(E) \cdot P(F)$$
$$= \frac{1}{4} \cdot \frac{1}{4} = \frac{1}{16}$$

b.   Two cards of the same suit are obtained by drawing two spades or two clubs or two diamonds or two hearts. Call these events $E_s,\ E_c,\ E_d,\ E_h$, respectively. As in (a), we find

$$P(E_s) = P(E_c) = P(E_d) = P(E_h) = \frac{1}{16}$$

Thus approximately

$\dfrac{1}{16}$ of the time we shall draw two spades

$\dfrac{1}{16}$ of the time we shall draw two clubs

$\dfrac{1}{16}$ of the time we shall draw two diamonds

$\dfrac{1}{16}$ of the time we shall draw two hearts.

We cannot draw two hearts *and* two spades since only two cards can be drawn, not four. In fact, no two of these events can occur simultaneously. Consequently, we draw two cards of the same suit approximately

$$\frac{1}{16} + \frac{1}{16} + \frac{1}{16} + \frac{1}{16} = \frac{1}{4}$$

of the time:

$$P(\text{two cards of the same suit}) = \frac{1}{4}$$

■

**Mutually
Exclusive
Events**

DEFINITION

Events $E_1, E_2, \cdots, E_n$ are said to be **mutually exclusive** if no two of them can occur simultaneously; that is, if

$$P(E_i \text{ and } E_j) = 0 \qquad \text{when } i \neq j.$$

The four events in Example 3(b) are mutually exclusive. On the other hand, the events $E_1 = $ "drawing two hearts" and $E_2 = $ "drawing a picture card" are not mutually exclusive since both E_1 and E_2 occur when the draw is A♥, K♥; $P(E_1 \text{ and } E_2) > 0$ in this case. The following statement on sums of probabilities has already been illustrated in Example 3(b).

For mutually exclusive events $E_1, E_2, \cdots, E_n$

a. $P(E_i \text{ and } E_j) = 0$ when $i \neq j$

b. $P(E_1 \text{ or } E_2 \text{ or } \cdots \text{ or } E_n) = P(E_1) + P(E_2) + \cdots + P(E_n)$

Probabilities of events that are not mutually exclusive are more complicated. We shall consider only two such events, say, E and F. If the event "E and F" occurs, then certainly E occurs. Thus if we count the number of ways in which E occurs $N(E)$, we shall have included the number of ways in which "E and F" occurs: $N(E \text{ and } F)$. Similarly, $N(F)$ includes $N(E \text{ and } F)$. Thus $N(E) + N(F)$ includes $N(E \text{ and } F)$ twice; consequently,

$$N(E \text{ or } F) = N(E) + N(F) - N(E \text{ and } F) \qquad (1)$$

where the last term is subtracted in order to prevent those occurrences from being counted twice. This relationship is illustrated in Figure 8–8.

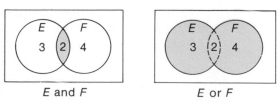

E and F   E or F

Figure 8–8

For example, in Figure 8–8, $N(E \text{ and } F) = 2$, $N(E) = 5 \,(= 3 + 2)$, and $N(F) = 6 \,(= 4 + 2)$. Clearly,

$$N(E \text{ or } F) = 3 + 2 + 4 = (3 + 2) + (4 + 2) - 2$$
$$= N(E) + N(F) - N(E \text{ and } F)$$

Dividing (1) by the total number of possible outcomes T, we have

$$\boxed{P(E \text{ or } F) = P(E) + P(F) - P(E \text{ and } F)}$$

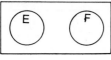

Figure 8–9

This statement also holds for mutually exclusive events since then $P(E \text{ and } F) = 0$. Mutually exclusive events are illustrated in Figure 8–9; E and F cannot occur simultaneously.

EXAMPLE 4 Three cards are drawn at random from a 52-card deck; the cards are not returned to the deck before drawing the next card. What is the probability of drawing at least two aces or at least two clubs?

Solution We may assume that the cards are drawn simultaneously so that the order of the draw is unimportant. Since there are 52 cards, the total number of distinct 3-card hands is $_{52}C_3$. We are interested in drawing either two or three aces. Since there are four aces, there are $_4C_3$ ways to select (or draw) exactly three of them. There are $_4C_2$ ways to draw exactly two aces and for each pair of aces there are 48 possible nonace third cards. Thus there are $48 \cdot \,_4C_2$ two-ace hands and as a result there are

$$_4C_3 + 48 \cdot \,_4C_2$$

distinct hands having two or three aces. Letting E be the event "drawing at least two aces," we have

$$P(E) = \frac{_4C_3 + 48 \cdot \,_4C_2}{_{52}C_3}$$

Since there are 13 clubs, we obtain in a similar fashion

$$P(F) = \frac{_{13}C_3 + 39 \cdot \,_{13}C_2}{_{52}C_3}$$

where F is the event "drawing at least two clubs."

But these events are not mutually exclusive. Both events occur when the cards drawn are A♣, another ace, and another club. There are 3 choices for the other ace and 12 choices for the other club. Thus $N(E \text{ and } F) = 3 \cdot 12 = 36$ and

$$P(E \text{ and } F) = \frac{36}{_{52}C_3}$$

Finally

$$P(2 \text{ clubs or } 2 \text{ aces}) = P(E \text{ or } F) = P(E) + P(F) - P(E \text{ and } F)$$

$$= \frac{{}_4C_3 + 48 \cdot {}_4C_2 + {}_{13}C_3 + 39 \cdot {}_{13}C_2 - 36}{{}_{52}C_3}$$

$$= \frac{\dfrac{4!}{3!} + 48 \cdot \dfrac{4!}{2!2!} + \dfrac{13!}{3!10!} + 39 \cdot \dfrac{13!}{2!11!} - 36}{\dfrac{52!}{3!49!}}$$

$$= \frac{4 + 48 \cdot 3 \cdot 2 + 13 \cdot 2 \cdot 11 + 39 \cdot 13 \cdot 6 - 36}{26 \cdot 17 \cdot 50}$$

$$= \frac{3584}{22100} = \frac{896}{5525}$$

$$\approx 0.162 \qquad \blacksquare$$

Complementary Events

In many situations the computation of probabilities can be very involved but the probability of the event not occurring (called the **complementary event**) may be easier to find. This situation will be illustrated in the following example. Before proceeding with the example, observe that if

$$T = \text{total number of possible outcomes}$$

and

$$N(E) = \text{number of ways in which the event } E \text{ can occur}$$

then the total number of ways in which the event would not occur ($\sim E$) is given by

$$N(\sim E) = T - N(E)$$

Dividing by T, we obtain

$$P(\sim E) = 1 - P(E)$$

EXAMPLE 6 A couple plans to have four children. What is the probability that the family will include at least one boy and at least one girl?

Solution We shall assume that the chances of a male or a female child are equal. Thus for a given birth

$$P(\text{girl}) = P(\text{boy}) = \frac{1}{2}$$

There are several ways to approach this problem. We could list all possible four-child configurations (e.g., BGGB) and count $N(E)$ and T. Or we could calculate

P(exactly one girl)
P(exactly two girls and two boys)
P(exactly one boy)

and then add these since these events are mutually exclusive. This approach also involves counting or a careful analysis to get the correct probabilities. The simplest approach is to consider the complementary event: all boys or all girls. Since successive births are independent events,

$$P(GGGG) = \frac{1}{2} \cdot \frac{1}{2} \cdot \frac{1}{2} \cdot \frac{1}{2} = \frac{1}{16}$$

similarly,

$$P(BBBB) = \frac{1}{16}$$

The events "all boys" and "all girls" are mutually exclusive. Thus

$$P(\text{all girls or all boys}) = \frac{1}{16} + \frac{1}{16} = \frac{1}{8}$$

In other words, about one of every eight four-child families are composed of a single gender. Thus

$$P(\text{mixed gender}) = 1 - P(\text{single gender})$$
$$= 1 - \frac{1}{8} = \frac{7}{8}$$

Note that we consider both independent and mutually exclusive events in the preceding example. It is important not to confuse the two concepts, since probabilities are

1. Multiplied for independent events   [to find $P(E \text{ and } F)$]

2. Added for mutually exclusive events   [to find $P(E \text{ or } F)$]

Also note that two events could be neither independent nor mutually exclusive. Consider, for example, the situation of drawing two cards from a deck and not replacing the first card before the second card is drawn. Let the events be

$$E = \text{the first card is a spade}$$
$$F = \text{the second card is a spade}$$

Clearly, $P(F)$ depends on whether or not the first card was a spade; E and F are not independent. Since we could draw a spade each time, these events are likewise not mutually exclusive.

EXERCISES 8.6

1. In a shell game with a coin placed under one of three shells, what are your chances of guessing the correct shell?

2. Two dice are rolled. Find the probabilities of the following events.
   a. exactly one 1
   b. sum = 7
   c. sum is odd
   d. sum ≤ 9
   e. sum is odd and ≤ 9
   f. sum is odd or ≤ 9
   g. sum ≥ 5
   h. 5 ≤ sum ≤ 9

3. In a set of dominoes containing one of each combination from double blank to double 6, find the probability of drawing each of the following.
   a. double 3
   b. any double
   c. at least one blank
   d. exactly one blank
   e. a sum of 5
   f. a sum ≥ 5.

4. Find the probability that the sum of two distinct numbers chosen at random from the digits 1 to 9 is
   a. odd
   b. even
   c. ≥ 9
   d. ≤ 5

5. With a four-sided die (a tetrahedron), we count the side that is down. With two four-sided dice, find the probability of obtaining a sum
   a. = 4;   b. ≥ 4.

6. The dodecahedron (12-sided solid) can also be used as a gambling die. With two such dice, find the probability of obtaining a sum
   a. = 10;   b. ≥ 10.

7. What are the probabilities of obtaining a double from rolling two dice if the dice are
   a. cubes?   b. tetrahedra?   c. dodecahedra?

8. In Exercise 7, find the probabilities of getting triples from three dice.

9. In a game called "Bye-Bye-Charlie," a red, a white, and a green marble are placed in a container. Each player draws one marble from his container. If the marble is red, it is "Bye-Bye-Charlie"; the player is out of the game. Otherwise the marble is replaced and the procedure is repeated. The game continues until only one player remains. For a given player, what is the probability of being in the game to play the fifth round?

10. Five white marbles and five red marbles are placed in an urn. A marble is drawn out and returned and then the procedure is repeated. What is the probability of drawing
    a. two white marbles?   b. one of each color?

11. Answer Exercise 10 if the first marble is not returned to the urn.

12. A coin is tossed five times. Find the probabilities of the following events.
    a. exactly three successive heads
    b. at least three successive heads
    c. at least three heads or three tails.

13. Five people choose a digit in the range 1 to 9 at random. Find the probability that at least two people choose the same digit.

14. Find the probability that of 30 people, at least 2 have the same birthday. (*Hint:* Compute the complementary probability.)

15. In a telepathy experiment, three people are each to select one of three symbols that are "transmitted" by the subject. What is the probability that at least two could select the correct symbol simply by guessing?

16. In a family of four children, what is the probability of
    a. exactly two boys and two girls?
    b. three children of the same sex?
    c. all children of the same sex?

17. In a family of five girls, what is the probability that the sixth child will be a boy?

18. A single card is drawn from a standard 52-card deck. What is the probability of drawing
    a. an ace?
    b. a diamond?
    c. the ace of diamonds?
    d. an ace or a diamond?
    e. an odd number?
    f. a face card?
    g. a black card?
    h. the ace of diamonds or a heart?
    i. the ace of diamonds or a face card?

19. Two cards are drawn (and not replaced) from a

standard 52-card deck. Find the probability of
a. at least one ace
b. two diamonds
c. the ace of diamonds
d. an ace or a diamond
e. at least one odd number
f. two face cards
g. at least one black card
h. the ace of diamonds or a heart
i. the ace of diamonds and a face card

20. In the game of Tripoley, the 10, J, Q, K and ace of hearts are each potential payoff cards. With four players, what is the probability of a given player being dealt one or more of these payoff cards?

21. In Tripoley, the pair Q, K of hearts and any triple consisting of 8, 9, 10 in one suit are potential bonus payoffs. With four players, what is the probability of a given player being dealt
    a. Q, K of hearts?
    b. 8, 9, 10 of clubs?
    c. Q, K of hearts and 8, 9, 10 of clubs?
    d. Q, K of hearts or 8, 9, 10 of clubs?

22. Answer Exercise 21 (**a** to **c**) for "at least one player" rather than "a given player."

23. What is the probability of exactly one ace in a **a.** 5-card poker hand? **b.** 13-card bridge hand?

In the game of pinochle, a deck of 48 cards is used that consists of two copies of each card in the range 9-Ace. Twelve cards are dealt to each of four players.

24. What is the probability that a given pinochle hand contains no aces and no face cards (only 9's and 10's)?

25. What is the probability that at least one of the four pinochle hands contains no aces and no face cards (only 9's and 10's)?

26. An important combination sought in pinochle is "double pinochle": J♦, J♦, Q♠, Q♠. What is the probability of double pinochle in a given hand?

27. Find the probability of double pinochle if pinochle partners are permitted to pool their hands. (See Exer. 26.)

28. If ten people including Al and Bert are lined up at random, find the probability that **a.** Al precedes Bert; **b.** Bert precedes Al; **c.** Al and Bert are next to one another.

29. What is the probability of guessing ten correct answers on a ten-question true-false test?

30. Of 20,000 patients treated for a given disease with a certain drug, 16,000 recovered. If you contract this disease, what is the probability that you could be cured with this medication?

8.7 CHAPTER REVIEW

TERMS AND CONCEPTS

- Sequences or progressions
  $$a_1, a_2, \ldots$$
- Summation
  $$\sum_{j=1}^{n} aj = a_1 + \cdots + a_n$$
- Factorial
  $$n! = 1 \cdot 2 \cdot \ldots \cdot n$$
- Permutations
  Rearrangements in which order is important and repetitions are not permitted.
- Combinations
  Same as permutations, except that order is irrelevant.
- Probability
  $$P(E) = \frac{\text{number of ways in which } E \text{ could occur}}{\text{total number of possible outcomes}}$$
- Independent Events
  No one of the events influences the outcome of any of the others.

- Mutually Exclusive Events
    No two of the events can occur simultaneously.

RULES AND FORMULAS

- Principle of Mathematical Induction
    If **1.** S_1 is true
    2. The truth of S_k implies the truth of S_{k+1} for every k
    then S_n is true for every n.
- Progressions
    1. Arithmetic

$$a_n = a_{n-1} + d = a_1 + (n - 1)d$$

$$a_1 + \cdots + a_n = \frac{n}{2}[a_1 + a_n]$$

2. Geometric

$$a_{n+1} = r \cdot a_n$$

$$\sum_{j=0}^{n-1} ar^j = a\left(\frac{1 - r^n}{1 - r}\right) \qquad r \neq 1$$

- Binomial Expansions
    1. Binomial Theorem

$$(a + b)^n = \sum_{l=0}^{n} \frac{n!}{l!(n - l)!} a^{n-l}b^l$$

2. Pascal's Triangle

$$
\begin{array}{llll}
(a + b)^0 = & 1 & & 1 \\
(a + b)^1 = & a + b & & 1 \quad 1 \\
(a + b)^2 = & a^2 + 2ab + b^2 & & 1 \quad 2 \quad 1 \\
(a + b)^3 = a^3 + 3a^2b + 3ab^2 + b^3 & & 1 \quad 3 \quad 3 \quad 1 \\
(a + b)^4 & & 1 \quad 4 \quad 6 \quad 4 \quad 1 \\
(a + b)^5 & & 1 \quad 5 \quad 10 \quad 10 \quad 5 \quad 1 \\
(a + b)^6 & & 1 \quad 6 \quad 15 \quad 20 \quad 15 \quad 6 \quad 1 \\
(a + b)^7 & & 1 \quad 7 \quad 21 \quad 35 \quad 35 \quad 21 \quad 7 \quad 1 \\
\end{array}
$$

- Permutations and Combinations
    1. Fundamental counting principle
       If N_k represents the number of possible outcomes for the kth event (after the first $k - 1$ events have taken place), then the total number of possible outcomes from the sequence of n events is $N_1 \cdot N_2 \cdots \cdot N_n$
    2. Permutations
       The total number of permutations of n objects taken r at a time is

$$_nP_r = n(n - 1) \cdots (n - r + 1) = \frac{n!}{(n - r)!}$$

3. Combinations
       The total number of combinations of n objects taken r at a time is $_nC_r = \dfrac{n!}{(n - r)!r!}$

- Probability
    1. $P(E_1 \text{ or } E_2) = P(E_1) + P(E_2) - P(E_1 \text{ and } E_2)$
    2. $P(\sim E) = 1 - P(E)$
    3. Independent events
       $P(E_1 \text{ and } \cdots \text{ and } E_n) = P(E_1) \cdot P(E_2) \cdots \cdot P(E_n)$
    4. Mutually exclusive events
       $P(E_1 \text{ or } \cdots \text{ or } E_n) = P(E_1) + P(E_2) + \cdots + P(E_n)$

8.8 SUPPLEMENTARY EXERCISES

Use mathematical induction to establish the following.

1. $2 + 5 + 10 + \cdots + (n^2 + 1)$
$$= \frac{n(2n^2 + 3n + 7)}{6}$$

2. $1^2 + 3^2 + 5^2 + \cdots + (2n - 1)^2$
$$= \frac{n(2n - 1)(2n + 1)}{3}$$

3. $1^3 + 3^3 + 5^3 + \cdots + (2n - 1)^3$
$$= n^2(2n^2 - 1)$$

4. $3 + 9 + \cdots + 3^n = \dfrac{3(3^n - 1)}{2}$

5. $1 \cdot 3 + 2 \cdot 4 + \cdots + n(n + 2)$
$$= \frac{n(n + 1)(2n + 7)}{6}$$

6. $2 \cdot 5 + 3 \cdot 6 + \cdots + (n + 1)(n + 4)$
$$= \frac{n(n + 4)(n + 5)}{3}$$

7. $1 \cdot 4 + 2 \cdot 9 + \cdots + n(n + 1)^2$
$$= \frac{n(n + 1)(n + 2)(3n + 5)}{12}$$

8. $\dfrac{1}{1 \cdot 3} + \dfrac{1}{3 \cdot 5} + \cdots + \dfrac{1}{(2n - 1)(2n + 1)}$

$$= \frac{n}{2n + 1}$$

9. $\dfrac{1}{2 \cdot 5} + \dfrac{1}{5 \cdot 8} + \cdots + \dfrac{1}{(3n - 1)(3n + 2)}$

$$= \frac{n}{2(3n + 2)}$$

10. $8^n - 1$ is divisible by 7.

11. $4^{2n} - 1$ is divisible by 5.

12. $n(n^2 + 2)$ is divisible by 3.

Write out the first five terms and the twelfth term of the sequence whose nth term is given.

13. $n - 5$ **14.** $\dfrac{3n + 1}{2}$

15. $n^2 + 1$ **16.** $1 + (-2)^n$

17. $2n^2 - 1$

Write out the first five terms and the ninth term of the sequences whose nth term is given.

18. $\dfrac{2n}{3}$ **19.** $\dfrac{n}{n + 1}$

20. $\dfrac{n(n - 1)}{2}$ **21.** $\dfrac{3}{2n - 1}$

22. $\dfrac{n}{n^2 + n}$

Write out the first five terms of the sequences defined here.

23. $a_1 = 3$, $a_n = 2a_{n-1} + 1$

24. $a_1 = 2$, $a_n = 3a_{n-1} - 2$

25. $a_1 = -\sqrt{2}$, $a_n = (a_{n-1})^2(-1)^n$

26. $a_1 = 0$, $a_n = (a_{n-1} + 2)^2$

Write out the first five terms and the tenth term of the arithmetic progression with the following data.

27. $a_1 = 4$, $d = 1$

28. $a_3 = 5$, $d = 4$

29. $a_4 = 16$, $d = 4$

30. $a_{20} = 100$, $d = -5$

Find the requested term of an arithmetic progression given the following information.

31. $a_2 = 7$, $a_6 = 23$, find a_7

32. $a_{10} = 32$, $a_{42} = 128$, find a_{167}

33. $a_8 = 0$, $a_{20} = -24$, find a_3

34. $a_{20} = 14$, $a_{38} = 86$, find a_7

Insert the requested number of arithmetic means between the following numbers.

35. 3 means between 3 and 19

36. 2 means between 67 and 76

37. 3 means between 42 and 58

38. 4 means between -6 and 14

Write out the first four terms and the eighth term of the geometric series with the following data.

39. $a_1 = 1$, $r = 5$

40. $a_1 = 486$, $r = \dfrac{1}{3}$

41. $a_4 = 4$, $r = \sqrt{2}$

42. $a_6 = 250$, $r = 5$

Find the requested term of a geometric progression given the following information.

43. $a_3 = 18$, $a_6 = 486$, find a_8

44. $a_3 = -21$, $a_6 = 63\sqrt{3}$, find a_2

45. $a_6 = 512$, $a_{11} = 16$, find a_3

46. $a_5 = 75$, $a_9 = 1875$, find a_3

Insert the requested number of geometric means between the following numbers.

47. 2 terms between 12 and 96

48. 3 terms between 2 and 162

49. 5 terms between 67 and 536

50. 2 terms between 62 and 496

Write the following in Σ form.

51. $2 + 4 + 6 + 8$

52. $\dfrac{1}{2} + \dfrac{1}{4} + \dfrac{1}{8} + \dfrac{1}{16} + \dfrac{1}{32}$

53. $3 + 7 + 11 + 15 + 19 + 23$

54. $3 - 7 + 11 - 15 + 19 - 23$

55. $2!3 + 3!6 + 4!12 + 5!24 + 6!48 + 7!96$

Expand and evaluate the following.

56. $\displaystyle\sum_{j=1}^{4} (2j - 1)$

57. $\displaystyle\sum_{n=2}^{5} (n - 2)(n - 3)$

58. $\displaystyle\sum_{l=3}^{6} (l + 2)l$

59. $\displaystyle\sum_{m=1}^{3} (2m - 1)(2m + 1)$

60. $\displaystyle\sum_{j=2}^{6} (-1)^j (j + 3)$

61. $\displaystyle\sum_{k=3}^{10} (3k + 1)$

62. $\displaystyle\sum_{j=1}^{16} (2 + j)$

63. $\displaystyle\sum_{n=2}^{12} (n - 2)(n - 3)$

64. $\displaystyle\sum_{n=2}^{10} 3 \cdot 2^n$

65. $\displaystyle\sum_{n=0}^{10} (5 \cdot 2^n + 6n + 2)$

Evaluate the following factorial expressions.

66. $5!$

67. $\dfrac{11!}{9!}$

68. $\dfrac{5!}{4!3!}$

69. $\dfrac{16!5!}{13!}$

Use Pascal's triangle to expand the following.

70. $(a - b)^3$

71. $(a + 2)^9$

72. $(x + 3y)^7$

73. $(3x + 2y^2)^5$

74. $(\sqrt{a} - \sqrt{b})^6$

75. $\left(2\sqrt{x} - \dfrac{1}{x}\right)^6$

76. $\left(2a - \dfrac{1}{b^3}\right)^8$

Use the binomial theorem to find the indicated term of the following.

77. $(x - 2)^{10}$, fourth term

78. $(t - 2)^{20}$, sixth term

79. $(a + b)^{15}$, seventh term

80. $(2u + v)^{12}$, eleventh term

81. $(r^2 - s)^{10}$, sixth term

Use the binomial theorem to evaluate the following.

82. $(102)^4$

83. $(1.02)^3$

84. $(0.98)^6$

85. $(103)^4$

86. How many nine-digit Social Security numbers are possible if zero is not permitted as a first digit?

87. How many pairs of shoes are required to stock a complete inventory of six styles in sizes 6 to 12 including half-sizes and widths A A, A, B, C, D, E, EE, EEE?

88. In selecting backpacking equipment, Cindy has narrowed her options down to three kinds of hiking boots, four sleeping bags, three tents, three backpacks, and two brands of cooking equipment. How many options does she have?

89. A rabbi, priest, minister, moslem, agnostic, and atheist are scheduled to speak on a given evening. In how many ways can their talks be ordered?

90. For a race of 16 dogs, in how many ways can the first three places be won?

91. How many different "sheephead" hands of 6 cards can be dealt from a 24-card deck?

92. Answer Exercise 91 for the entire deal of four players.

93. A student is required to take core courses but must select six of ten electives. In how many ways can this be done if only the list (and not the order) of the six courses is important.

94. An intrastate transit system services seven cities. How many different kinds of tickets must it have printed including both one-way and round-trip tickets?

95. How many different ways are there to form a four-person committee from a group of five men and three women if

  a. the committee members are titled so that order cannot be ignored and

  b. the second position must be filled by a female?

  c. the second position must be filled by a female and the last position must be filled by a male?

  d. the second position must be filled by a female, the last position must be filled by a male, and the committee must be sexually balanced?

  e. Lois must be a member of the committee?

96. Two dice are rolled. Find the probabilities of the following events.

  a. sum = 7 or 11

  b. sum is even

  c. sum is a multiple of 3

d. sum is even and a multiple of 3

e. sum is even or a multiple of 3

97. Two numbers from 1 to 50 are selected at random. Find the probability that the sum is

a. ≥ 50;   **b.** ≤ 10;   **c.** $= 25$.

98. Find the probabilities of obtaining triples when rolling four dice and the dice are

a. cubes;   **b.** tetrahedra;   **c.** dodecahedra

99. Answer Exercise 98 for quadruples.

100. Three white marbles and five red marbles are placed in an urn. A marble is drawn out and returned and then the procedure is repeated. What is the probability of drawing

a. two white marbles?

b. one of each color?

101. Answer Exercise 100 if the first marble is not returned.

102. A coin is tossed four times. Find the probability of

a. exactly two heads;

b. at least two heads;

c. the first toss is heads and at least two heads occur.

103. Five people choose a number 1 to 50 at random. Find the probability that at least two people choose the same number.

104. Of the first 40 U.S. presidents, what is the probability that at least two will share

a. the same birthday?

b. the same day of death?

105. In a telepathy experiment, each of five people is to select one of five symbols "transmitted" by the subject. What is the probability that at least three could select the correct symbol simply by guessing?

106. In a family of six children, what is the probability of

a. exactly three boys and three girls?

b. four children of the same sex?

c. at least four children of the same sex?

107. Two cards are drawn (and not replaced) from a standard 52-card deck. Find the probability of

a. two aces

b. at least one diamond

c. two odd numbers

d. at least one face card

e. two black cards

f. the ace of diamonds and a heart

g. the ace of diamonds or a face card.

108. If a Tripoley player receives six cards, what is the probability of his being dealt one or more payoff cards? (See Exercise 20, Sect. 8.6.)

109. With eight players in Tripoley receiving six cards each, what is the probability of a given player being dealt

a. Q, K of hearts?

b. 8, 9, 10 of diamonds?

c. Q, K of hearts and 8, 9, 10 of diamonds?

d. Q, K of hearts or 8, 9, 10 of diamonds?

110. Answer Exercise 109 (**a** to **c**) for "at least one player" rather than for "a given player."

111. What is the probability that two bridge players are each dealt no aces and no face cards?

112. What is the probability that at least one of the four bridge hands contains no aces and no face cards?

113. If two pinochle players pool their hands, what is the probability that they corner all 12 cards in some suit? (See note preceding Exer. 24, Sect. 8.6.)

114. Of 1500 walnut seedlings planted on a tree farm, 1325 survived the first year. What is the probability of an individual plant surviving?

Appendix
Tables: Exponents and Logarithms

For a detailed explanation regarding the use of tables, see Section 6.4 (page 264).

Table 1 Powers of e

x	0 e^x	0 e^{-x}	1 e^x	1 e^{-x}	2 e^x	2 e^{-x}	3 e^x	3 e^{-x}	4 e^x	4 e^{-x}
0.0	1.000	1.000	1.010	.990	1.020	.980	1.031	.970	1.041	.961
0.1	1.105	.905	1.116	.896	1.127	.887	1.139	.878	1.150	.869
0.2	1.221	.819	1.234	.811	1.246	.803	1.259	.795	1.271	.787
0.3	1.350	.741	1.363	.733	1.377	.726	1.391	.719	1.405	.712
0.4	1.492	.670	1.507	.664	1.522	.657	1.537	.651	1.533	.644
0.5	1.649	.607	1.665	.600	1.682	.595	1.699	.589	1.716	.583
0.6	1.822	.549	1.840	.543	1.859	.538	1.878	.533	1.896	.527
0.7	2.014	.497	2.034	.492	2.054	.487	2.075	.482	2.096	.477
0.8	2.226	.449	2.248	.445	2.270	.440	2.293	.436	2.316	.432
0.9	2.460	.407	2.484	.403	2.509	.399	2.535	.395	2.560	.391
1.0	2.718	.368	2.746	.364	2.773	.361	2.801	.357	2.829	.353
1.1	3.004	.333	3.034	.330	3.065	.326	3.096	.323	3.127	.320
1.2	3.320	.301	3.353	.298	3.387	.295	3.421	.292	3.456	.289
1.3	3.669	.273	3.706	.270	3.743	.267	3.781	.264	3.819	.262
1.4	4.055	.247	4.096	.244	4.137	.242	4.179	.239	4.221	.237
1.5	4.482	.223	4.527	.221	4.572	.219	4.618	.217	4.665	.214
1.6	4.953	.202	5.003	.200	5.053	.198	5.104	.196	5.155	.194
1.7	5.474	.183	5.529	.181	5.585	.179	5.641	.177	5.697	.176
1.8	6.050	.165	6.110	.164	6.172	.162	6.234	.160	6.297	.159
1.9	6.686	.150	6.753	.148	6.821	.147	6.890	.145	6.959	.144
2.0	7.389	.135	7.463	.134	7.538	.133	7.614	.131	7.691	.130
2.1	8.166	.122	8.248	.121	8.331	.120	8.415	.119	8.499	.118
2.2	9.025	.111	9.116	.110	9.207	.109	9.300	.108	9.393	.106
2.3	9.974	.100	10.074	.099	10.176	.098	10.278	.097	10.381	.096

x	5 e^x	5 e^{-x}	6 e^x	6 e^{-x}	7 e^x	7 e^{-x}	8 e^x	8 e^{-x}	9 e^x	9 e^{-x}
0.0	1.051	.951	1.062	.942	1.073	.932	1.083	.923	1.094	.914
0.1	1.162	.861	1.174	.852	1.185	.844	1.197	.835	1.209	.827
0.2	1.284	.779	1.297	.771	1.310	.763	1.323	.756	1.336	.748
0.3	1.419	.705	1.433	.698	1.448	.691	1.462	.684	1.477	.677
0.4	1.568	.638	1.584	.631	1.600	.625	1.616	.619	1.632	.613
0.5	1.733	.577	1.751	.571	1.768	.566	1.786	.560	1.804	.554
0.6	1.916	.522	1.935	.517	1.954	.512	1.974	.507	1.994	.502
0.7	2.117	.472	2.138	.468	2.160	.463	2.182	.458	2.203	.454
0.8	2.340	.427	2.363	.423	2.387	.419	2.411	.415	2.435	.411
0.9	2.586	.387	2.612	.383	2.638	.379	2.664	.375	2.691	.372
1.0	2.858	.350	2.886	.346	2.915	.343	2.945	.340	2.974	.336
1.1	3.158	.317	3.190	.313	3.222	.310	3.254	.307	3.287	.304
1.2	3.490	.287	3.525	.284	3.561	.281	3.597	.278	3.633	.275
1.3	3.857	.259	3.896	.257	3.935	.254	3.975	.252	4.015	.249
1.4	4.263	.235	4.306	.232	4.349	.230	4.393	.228	4.437	.225
1.5	4.712	.212	4.759	.210	4.807	.208	4.855	.206	4.904	.204
1.6	5.207	.192	5.259	.190	5.312	.188	5.366	.186	5.420	.185
1.7	5.755	.174	5.812	.172	5.871	.170	5.930	.169	5.989	.167
1.8	6.360	.157	6.424	.156	6.488	.154	6.553	.153	6.619	.151
1.9	7.029	.142	7.099	.141	7.171	.139	7.243	.138	7.316	.137
2.0	7.768	.129	7.846	.127	7.925	.126	8.004	.125	8.085	.124
2.1	8.585	.116	8.671	.115	8.758	.114	8.846	.113	8.935	.112
2.2	9.488	.105	9.583	.104	9.679	.103	9.777	.102	9.875	.101
2.3	10.486	.095	10.591	.094	10.697	.093	10.805	.093	10.913	.092

For exponents not in this range use $e^{2.303} \approx 10$, $e^{4.605} \approx 100$, $e^{6.901} \approx 1000$, $e^{9.210} \approx 10000$ and the laws of exponents. For example,

$$e^{3.52} \approx e^{2.303 + 1.117} = e^{2.303}e^{1.117} \approx 10 \cdot (3.058) = 30.58$$

and

$$e^{-5.17} \approx e^{-4.605 - 0.565} = e^{-4.605}e^{-0.565} \approx (0.01)(0.568) = (0.00568)$$

Table 2 Natural Logarithms (ln *x*)

x	0	1	2	3	4	5	6	7	8	9
1.0	0.0000	0.0100	0.0198	0.0296	0.0392	0.0488	0.0583	0.0677	0.0770	0.0862
1.1	0.0953	0.1044	0.1133	0.1222	0.1310	0.1398	0.1484	0.1570	0.1655	0.1740
1.2	0.1823	0.1906	0.1989	0.2070	0.2151	0.2231	0.2311	0.2390	0.2469	0.2546
1.3	0.2624	0.2700	0.2776	0.2852	0.2927	0.3001	0.3075	0.3148	0.3221	0.3293
1.4	0.3365	0.3436	0.3507	0.3577	0.3646	0.3716	0.3784	0.3853	0.3920	0.3988
1.5	0.4055	0.4121	0.4187	0.4253	0.4318	0.4383	0.4447	0.4511	0.4574	0.4637
1.6	0.4700	0.4762	0.4824	0.4886	0.4947	0.5008	0.5068	0.5128	0.5188	0.5247
1.7	0.5306	0.5365	0.5423	0.5481	0.5539	0.5596	0.5653	0.5710	0.5766	0.5822
1.8	0.5878	0.5933	0.5988	0.6043	0.6098	0.6152	0.6206	0.6259	0.6313	0.6366
1.9	0.6419	0.6471	0.6523	0.6575	0.6627	0.6678	0.6729	0.6780	0.6831	0.6881
2.0	0.6931	0.6981	0.7031	0.7080	0.7130	0.7178	0.7227	0.7275	0.7324	0.7372
2.1	0.7419	0.7467	0.7514	0.7561	0.7608	0.7655	0.7701	0.7747	0.7793	0.7839
2.2	0.7885	0.7930	0.7975	0.8020	0.8065	0.8109	0.8154	0.8198	0.8242	0.8286
2.3	0.8329	0.8372	0.8416	0.8459	0.8502	0.8544	0.8587	0.8629	0.8671	0.8713
2.4	0.8755	0.8796	0.8838	0.8879	0.8920	0.8961	0.9002	0.9042	0.9083	0.9123
2.5	0.9163	0.9203	0.9243	0.9282	0.9322	0.9361	0.9400	0.9439	0.9478	0.9517
2.6	0.9555	0.9594	0.9632	0.9670	0.9708	0.9746	0.9783	0.9821	0.9858	0.9895
2.7	0.9933	0.9969	1.0006	1.0043	1.0080	1.0116	1.0152	1.0188	1.0255	1.0260
2.8	1.0296	1.0332	1.0367	1.0403	1.0438	1.0473	1.0508	1.0543	1.0578	1.0613
2.9	1.0647	1.0682	1.0716	1.0750	1.0784	1.0818	1.0852	1.0886	1.0919	1.0953
3.0	1.0986	1.1019	1.1053	1.1086	1.1119	1.1151	1.1184	1.1217	1.1249	1.1282
3.1	1.1314	1.1346	1.1378	1.1410	1.1442	1.1474	1.1506	1.1537	1.1569	1.1600
3.2	1.1632	1.1663	1.1694	1.1725	1.1756	1.1787	1.1817	1.1848	1.1878	1.1909
3.3	1.1939	1.1970	1.2000	1.2030	1.2060	1.2090	1.2119	1.2149	1.2179	1.2208
3.4	1.2238	1.2267	1.2296	1.2326	1.2355	1.2384	1.2413	1.2442	1.2470	1.2499
3.5	1.2528	1.2556	1.2585	1.2613	1.2641	1.2669	1.2698	1.2726	1.2754	1.2782
3.6	1.2809	1.2837	1.2865	1.2892	1.2920	1.2947	1.2975	1.3002	1.3029	1.3056
3.7	1.3083	1.3110	1.3137	1.3164	1.3191	1.3218	1.3244	1.3271	1.3297	1.3324
3.8	1.3350	1.3376	1.3403	1.3429	1.3455	1.3481	1.3507	1.3533	1.3558	1.3584
3.9	1.3610	1.3635	1.3661	1.3686	1.3712	1.3737	1.3762	1.3788	1.3813	1.3838
4.0	1.3863	1.3888	1.3913	1.3938	1.3962	1.3987	1.4012	1.4036	1.4061	1.4085
4.1	1.4110	1.4134	1.4159	1.4183	1.4207	1.4231	1.4255	1.4279	1.4303	1.4327
4.2	1.4351	1.4375	1.4398	1.4422	1.4446	1.4469	1.4493	1.4516	1.4540	1.4563
4.3	1.4586	1.4609	1.4633	1.4656	1.4679	1.4702	1.4725	1.4748	1.4770	1.4793
4.4	1.4816	1.4839	1.4861	1.4884	1.4907	1.4929	1.4952	1.4974	1.4996	1.5019
4.5	1.5041	1.5063	1.5085	1.5107	1.5129	1.5151	1.5173	1.5195	1.5217	1.5239
4.6	1.5261	1.5282	1.5304	1.5326	1.5347	1.5369	1.5390	1.5412	1.5433	1.5454
4.7	1.5476	1.5497	1.5518	1.5539	1.5560	1.5581	1.5602	1.5623	1.5644	1.5665
4.8	1.5686	1.5707	1.5728	1.5748	1.5769	1.5790	1.5810	1.5831	1.5851	1.5872
4.9	1.5892	1.5913	1.5933	1.5953	1.5974	1.5994	1.6014	1.6034	1.6054	1.6074
5.0	1.6094	1.6114	1.6134	1.6154	1.6174	1.6194	1.6214	1.6233	1.6253	1.6273
5.1	1.6292	1.6312	1.6332	1.6351	1.6371	1.6390	1.6409	1.6429	1.6448	1.6467
5.2	1.6487	1.6506	1.6525	1.6544	1.6563	1.6582	1.6601	1.6620	1.6639	1.6658
5.3	1.6677	1.6696	1.6715	1.6734	1.6752	1.6771	1.6790	1.6808	1.6827	1.6845
5.4	1.6864	1.6882	1.6901	1.6919	1.6938	1.6956	1.6974	1.6993	1.7011	1.7029
5.5	1.7047	1.7066	1.7084	1.7102	1.7120	1.7138	1.7156	1.7174	1.7192	1.7210
5.6	1.7228	1.7246	1.7263	1.7281	1.7299	1.7317	1.7334	1.7352	1.7370	1.7387
5.7	1.7405	1.7422	1.7440	1.7457	1.7475	1.7492	1.7509	1.7527	1.7544	1.7561
5.8	1.7579	1.7596	1.7613	1.7630	1.7647	1.7664	1.7682	1.7699	1.7716	1.7733
5.9	1.7750	1.7766	1.7783	1.7800	1.7817	1.7834	1.7851	1.7867	1.7884	1.7901

Table 2 (continued)

x	0	1	2	3	4	5	6	7	8	9
6.0	1.7918	1.7934	1.7951	1.7967	1.7984	1.8001	1.8017	1.8034	1.8050	1.8066
6.1	1.8083	1.8099	1.8116	1.8132	1.8148	1.8165	1.8181	1.8197	1.8213	1.8229
6.2	1.8245	1.8262	1.8278	1.8294	1.8310	1.8326	1.8342	1.8358	1.8374	1.8390
6.3	1.8406	1.8421	1.8437	1.8453	1.8469	1.8485	1.8500	1.8516	1.8532	1.8547
6.4	1.8563	1.8579	1.8594	1.8610	1.8625	1.8641	1.8656	1.8672	1.8687	1.8703
6.5	1.8718	1.8733	1.8749	1.8764	1.8799	1.8795	1.8810	1.8825	1.8840	1.8856
6.6	1.8871	1.8886	1.8901	1.8916	1.8931	1.8946	1.8961	1.8976	1.8991	1.9006
6.7	1.9021	1.9036	1.9051	1.9066	1.9081	1.9095	1.9110	1.9125	1.9140	1.9155
6.8	1.9169	1.9184	1.9199	1.9213	1.9228	1.9242	1.9257	1.9272	1.9286	1.9301
6.9	1.9315	1.9330	1.9344	1.9359	1.9373	1.9387	1.9402	1.9416	1.9430	1.9445
7.0	1.9459	1.9473	1.9488	1.9502	1.9516	1.9530	1.9544	1.9559	1.9573	1.9587
7.1	1.9601	1.9615	1.9629	1.9643	1.9657	1.9671	1.9685	1.9699	1.9713	1.9727
7.2	1.9741	1.9755	1.9769	1.9782	1.9796	1.9810	1.9824	1.9838	1.9851	1.9865
7.3	1.9879	1.9892	1.9906	1.9920	1.9933	1.9947	1.9961	1.9974	1.9988	2.0001
7.4	2.0015	2.0028	2.0042	2.0055	2.0069	2.0082	2.0096	2.0109	2.0122	2.0136
7.5	2.0149	2.0162	2.0176	2.0189	2.0202	2.0215	2.0229	2.0242	2.0255	2.0268
7.6	2.0282	2.0295	2.0308	2.0321	2.0334	2.0347	2.0360	2.0373	2.0386	2.0399
7.7	2.0412	2.0425	2.0438	2.0451	2.0464	2.0477	2.0490	2.0503	2.0516	2.0528
7.8	2.0541	2.0554	2.0567	2.0580	2.0592	2.0605	2.0618	2.0631	2.0643	2.0656
7.9	2.0669	2.0681	2.0694	2.0707	2.0719	2.0732	2.0744	2.0757	2.0769	2.0782
8.0	2.0794	2.0807	2.0819	2.0832	2.0844	2.0857	2.0869	2.0882	2.0894	2.0906
8.1	2.0919	2.0931	2.0943	2.0956	2.0968	2.0980	2.0992	2.1005	2.1017	2.1029
8.2	2.1041	2.1054	2.1066	2.1078	2.1090	2.1102	2.1114	2.1126	2.1138	2.1150
8.3	2.1163	2.1175	2.1187	2.1199	2.1211	2.1223	2.1235	2.1247	2.1258	2.1270
8.4	2.1282	2.1294	2.1306	2.1318	2.1330	2.1342	2.1353	2.1365	2.1377	2.1389
8.5	2.1401	2.1412	2.1424	2.1436	2.1448	2.1459	2.1471	2.1483	2.1494	2.1506
8.6	2.1518	2.1529	2.1541	2.1552	2.1564	2.1576	2.1587	2.1599	2.1610	2.1622
8.7	2.1633	2.1645	2.1656	2.1668	2.1679	2.1691	2.1702	2.1713	2.1725	2.1736
8.8	2.1748	2.1759	2.1770	2.1782	2.1793	2.1804	2.1815	2.1827	2.1838	2.1849
8.9	2.1861	2.1872	2.1883	2.1894	2.1905	2.1917	2.1928	2.1939	2.1950	2.1961
9.0	2.1972	2.1983	2.1994	2.2006	2.2017	2.2028	2.2039	2.2050	2.2061	2.2072
9.1	2.2083	2.2094	2.2105	2.2116	2.2127	2.2138	2.2148	2.2159	2.2170	2.2181
9.2	2.2192	2.2203	2.2214	2.2225	2.2235	2.2246	2.2257	2.2268	2.2279	2.2289
9.3	2.2300	2.2311	2.2322	2.2332	2.2343	2.2354	2.2364	2.2375	2.2386	2.2396
9.4	2.2407	2.2418	2.2428	2.2439	2.2450	2.2460	2.2471	2.2481	2.2492	2.2502
9.5	2.2513	2.2523	2.2534	2.2544	2.2555	2.2565	2.2576	2.2586	2.2597	2.2607
9.6	2.2618	2.2628	2.2638	2.2649	2.2659	2.2670	2.2680	2.2690	2.2701	2.2711
9.7	2.2721	2.2732	2.2742	2.2752	2.2762	2.2773	2.2783	2.2793	2.2803	2.2814
9.8	2.2824	2.2834	2.2844	2.2854	2.2865	2.2875	2.2885	2.2895	2.2905	2.2915
9.9	2.2925	2.2935	2.2946	2.2956	2.2966	2.2976	2.2986	2.2996	2.3006	2.3016

For x not in the range listed in this table, use $\ln 10 \approx 2.3026$ and the laws of logarithms. For example,

$$\ln(0.05) = \ln(5 \cdot 10^{-2}) = \ln 5 + \ln 10^{-2} = \ln 5 - 2 \ln 10$$
$$\approx 1.6094 - 2(2.3026) \approx -2.9958$$

and

$$\ln 412 = \ln(4.12 \cdot 10^2) = \ln(4.12) + \ln 10^2 = \ln(4.12) + 2 \ln 10$$
$$\approx 1.4159 + 2(2.3026) \approx 6.0211$$

Table 3   Common Logarithms (log₁₀ *x*)

x	0	1	2	3	4	5	6	7	8	9
1.0	.0000	.0043	.0086	.0128	.0170	.0212	.0253	.0294	.0334	.0374
1.1	.0414	.0453	.0492	.0531	.0569	.0607	.0645	.0682	.0719	.0755
1.2	.0792	.0828	.0864	.0899	.0934	.0969	.1004	.1038	.1072	.1106
1.3	.1139	.1173	.1206	.1239	.1271	.1303	.1335	.1367	.1399	.1430
1.4	.1461	.1492	.1523	.1553	.1584	.1614	.1644	.1673	.1703	.1732
1.5	.1761	.1790	.1818	.1847	.1875	.1903	.1931	.1959	.1987	.2014
1.6	.2041	.2068	.2095	.2122	.2148	.2175	.2201	.2227	.2253	.2279
1.7	.2304	.2330	.2355	.2380	.2405	.2430	.2455	.2480	.2504	.2529
1.8	.2553	.2577	.2601	.2625	.2648	.2672	.2695	.2718	.2742	.2765
1.9	.2788	.2810	.2833	.2856	.2878	.2900	.2923	.2945	.2967	.2989
2.0	.3010	.3032	.3054	.3075	.3096	.3118	.3139	.3160	.3181	.3201
2.1	.3222	.3243	.3263	.3284	.3304	.3324	.3345	.3365	.3385	.3404
2.2	.3424	.3444	.3464	.3483	.3502	.3522	.3541	.3560	.3579	.3598
2.3	.3617	.3636	.3655	.3674	.3692	.3711	.3729	.3747	.3766	.3784
2.4	.3802	.3820	.3838	.3856	.3874	.3892.	.3909	.3927	.3945	.3962
2.5	.3979	.3997	.4014	.4031	.4048	.4065	.4082	.4099	.4116	.4133
2.6	.4150	.4166	.4183	.4200	.4216	.4232	.4249	.4265	.4281	.4298
2.7	.4314	.4330	.4346	.4362	.4378	.4393	.4409	.4425	.4440	.4456
2.8	.4472	.4487	.4502	.4518	.4533	.4548	.4564	.4579	.4594	.4609
2.9	.4624	.4639	.4654	.4669	.4683	.4698	.4713	.4728	.4742	.4757
3.0	.4771	.4786	.4800	.4814	.4829	.4843	.4857	.4871	.4886	.4900
3.1	.4914	.4928	.4942	.4955	.4969	.4983	.4997	.5011	.5024	.5038
3.2	.5051	.5065	.5079	.5092	.5105	.5119	.5132	.5145	.5159	.5172
3.3	.5185	.5198	.5211	.5224	.5237	.5250	.5263	.5276	.5289	.5302
3.4	.5315	.5328	.5340	.5353	.5366	.5378	.5391	.5403	.5416	.5428
3.5	.5441	.5453	.5465	.5478	.5490	.5502	.5514	.5527	.5539	.5551
3.6	.5563	.5575	.5587	.5599	.5611	.5623	.5635	.5647	.5658	.5670
3.7	.5682	.5694	.5705	.5717	.5729	.5740	.5752	.5763	.5775	.5786
3.8	.5798	.5809	.5821	.5832	.5843	.5855	.5866	.5877	.5888	.5899
3.9	.5911	.5922	.5933	.5944	.5955	.5966	.5977	.5988	.5999	.6010
4.0	.6021	.6031	.6042	.6053	.6064	.6075	.6085	.6096	.6107	.6117
4.1	.6128	.6138	.6149	.6160	.6170	.6180	.6191	.6201	.6212	.6222
4.2	.6232	.6243	.6253	.6263	.6274	.6284	.6294	.6304	.6314	.6325
4.3	.6335	.6345	.6355	.6365	.6375	.6385	.6395	.6405	.6415	.6425
4.4	.6435	.6444	.6454	.6464	.6474	.6484	.6493	.6503	.6513	.6522
4.5	.6532	.6542	.6551	.6561	.6571	.6580	.6590	.6599	.6609	.6618
4.6	.6628	.6637	.6646	.6656	.6665	.6675	.6684	.6693	.6702	.6712
4.7	.6721	.6730	.6739	.6749	.6758	.6767	.6776	.6785	.6794	.6803
4.8	.6812	.6821	.6830	.6839	.6848	.6857	.6866	.6875	.6884	.6893
4.9	.6902	.6911	.6920	.6928	.6937	.6946	.6955	.6964	.6972	.6981
5.0	.6990	.6998	.7007	.7016	.7024	.7033	.7042	.7050	.7059	.7067
5.1	.7076	.7084	.7093	.7101	.7110	.7118	.7126	.7135	.7143	.7152
5.2	.7160	.7168	.7177	.7185	.7193	.7202	.7210	.7218	.7226	.7235
5.3	.7243	.7251	.7259	.7267	.7275	.7284	.7292	.7300	.7308	.7316
5.4	.7324	.7332	.7340	.7348	.7356	.7364	.7372	.7380	.7388	.7396

Table 3 (continued)

x	0	1	2	3	4	5	6	7	8	9
5.5	.7404	.7412	.7419	.7427	.7435	.7443	.7451	.7459	.7466	.7474
5.6	.7482	.7490	.7497	.7505	.7513	.7520	.7528	.7536	.7543	.7551
5.7	.7559	.7566	.7574	.7582	.7589	.7597	.7604	.7612	.7619	.7627
5.8	.7634	.7642	.7649	.7657	.7664	.7672	.7679	.7686	.7694	.7701
5.9	.7709	.7716	.7723	.7731	.7738	.7745	.7752	.7760	.7767	.7774
6.0	.7782	.7789	.7796	.7803	.7810	.7818	.7825	.7832	.7839	.7846
6.1	.7853	.7860	.7868	.7875	.7882	.7889	.7896	.7903	.7910	.7917
6.2	.7924	.7931	.7938	.7945	.7952	.7959	.7966	.7973	.7980	.7987
6.3	.7993	.8000	.8007	.8014	.8021	.8028	.8035	.8041	.8048	.8055
6.4	.8062	.8069	.8075	.8082	.8089	.8096	.8102	.8109	.8116	.8122
6.5	.8129	.8136	.8142	.8149	.8156	.8162	.8169	.8176	.8182	.8189
6.6	.8195	.8202	.8209	.8215	.8222	.8228	.8235	.8241	.8248	.8254
6.7	.8261	.8267	.8274	.8280	.8287	.8293	.8299	.8306	.8312	.8319
6.8	.8325	.8331	.8338	.8344	.8351	.8357	.8363	.8370	.8376	.8382
6.9	.8388	.8395	.8401	.8407	.8414	.8420	.8426	.8432	.8439	.8445
7.0	.8451	.8457	.8463	.8470	.8476	.8482	.8488	.8494	.8500	.8506
7.1	.8513	.8519	.8525	.8531	.8537	.8543	.8549	.8555	.8561	.8567
7.2	.8573	.8579	.8585	.8591	.8597	.8603	.8609	.8615	.8621	.8627
7.3	.8633	.8639	.8645	.8651	.8657	.8663	.8669	.8675	.8681	.8686
7.4	.8692	.8698	.8704	.8710	.8716	.8722	.8727	.8733	.8739	.8745
7.5	.8751	.8756	.8762	.8768	.8774	.8779	.8785	.8791	.8797	.8802
7.6	.8808	.8814	.8820	.8825	.8831	.8837	.8842	.8848	.8854	.8859
7.7	.8865	.8871	.8876	.8882	.8887	.8893	.8899	.8904	.8910	.8915
7.8	.8921	.8927	.8932	.8938	.8943	.8949	.8954	.8960	.8965	.8971
7.9	.8976	.8982	.8987	.8993	.8998	.9004	.9009	.9015	.9020	.9025
8.0	.9031	.9036	.9042	.9047	.9053	.9058	.9063	.9069	.9074	.9079
8.1	.9085	.9090	.9096	.9101	.9106	.9112	.9117	.9122	.9128	.9133
8.2	.9138	.9143	.9149	.9154	.9159	.9165	.9170	.9175	.9180	.9186
8.3	.9191	.9196	.9201	.9206	.9212	.9217	.9222	.9227	.9232	.9238
8.4	.9243	.9248	.9253	.9258	.9263	.9269	.9274	.9279	.9284	.9289
8.5	.9294	.9299	.9304	.9309	.9315	.9320	.9325	.9330	.9335	.9340
8.6	.9345	.9350	.9355	.9360	.9365	.9370	.9375	.9380	.9385	.9390
8.7	.9395	.9400	.9405	.9410	.9415	.9420	.9425	.9430	.9435	.9440
8.8	.9445	.9450	.9455	.9460	.9465	.9469	.9474	.9479	.9484	.9489
8.9	.9494	.9499	.9504	.9509	.9513	.9518	.9523	.9528	.9533	.9538
9.0	.9542	.9547	.9552	.9557	.9562	.9566	.9571	.9576	.9581	.9586
9.1	.9590	.9595	.9600	.9605	.9609	.9614	.9619	.9624	.9628	.9633
9.2	.9638	.9643	.9647	.9652	.9657	.9661	.9666	.9671	.9675	.9680
9.3	.9685	.9689	.9694	.9699	.9703	.9708	.9713	.9717	.9722	.9727
9.4	.9731	.9736	.9741	.9745	.9750	.9754	.9759	.9763	.9768	.9773
9.5	.9777	.9782	.9786	.9791	.9795	.9800	.9805	.9809	.9814	.9818
9.6	.9823	.9827	.9832	.9836	.9841	.9845	.9850	.9854	.9859	.9863
9.7	.9868	.9872	.9877	.9881	.9886	.9890	.9894	.9899	.9903	.9908
9.8	.9912	.9917	.9921	.9926	.9930	.9934	.9939	.9943	.9948	.9952
9.9	.9956	.9961	.9965	.9969	.9974	.9978	.9983	.9987	.9991	.9996

Solutions

CHAPTER 1

Section 1.1

1. $2 \cdot \boxed{2} \cdot 2 \cdot 3$ **3.** $2 \cdot 2 \cdot 3 \cdot 3 \cdot 11$ **5.** $3 \cdot 3 \cdot 7 \cdot 13$ **7.** LCM = 225, GCD = 15   **9.** LCM = 645,
GCD = 43   **11.** LCM = 210, GCD = 3

13.

```
       7.3
 +-+-+-+-+-+-+-+-•+-+
 -1 0           8
```

15. -3.2

```
 +-•+-+-+-+-+
         0
```

17. 0.666 . . .

```
 +-+•+-+-+-+
 0
```

19. 5/9   **21.** 3/5   **23.** 7/8   **25.** $-22/15$ **27.** -6 **29.** 2   **31.** 48   **33.** 1   **35.** -52
37. -58 **39.** 22   **41.** 24   **43.** 5/8   **45.** 1/6   **47.** 1/12   **49.** 10/9   **51.** $-3/16$ **53.** 78
55. 1/9   **57.** 5/2   **59.** 14/5   **61.** 5/2   **63.** $0.\overline{857142}$ **65.** $2.\overline{09}$ **67.** $0.70\overline{83}$ **69.** 1/2
71. 282/55   **73.** 2238911/3330   **75.** [(63) .857, (65) 2.091, (67) 0.708, (69) 0.500, (71) 5.127, (73) 672.346]
77. $(20 - 1)40 = 800 - 40 = 760$ **79.** $(25 - 1) \cdot 7 = 25 \cdot 7 - 7 = 175 - 7 = 168$ **81.** $8(20 - 1) =$
$160 - 8 = 152$ **83.** 5, 5.00   **85.** 3/1000, 0.003   **87.** 1.25, 125%   **89.** 75%   **91.** 40%   **93.** 0.18
95. No. Example: $2 - 3 \neq 3 - 2$; No. Example: $2 - (3 - 4) \neq (2 - 3) - 4$ **97.** No. Example: $4 \div (2 + 2) \neq$
$(4 \div 2) + (4 \div 2)$ **99.** 160 bolts   **101.** 13.2 mpg   **103.** 5.1875; the first loaf   **105.** 2250
107. \$263.50, \$400   **109.** \$250

Section 1.2

1. $2 < 7$ **3.** $-2 > -6$ **5.** $2 > -6$ **7.** $-100 < 17$ **9.** $-2 > -7$ **11.** $-4 > -12$
13. $-4 < 12$ **15.** [(9) 1, (11) 2, (13) 2]

17.

```
 ◄────+-+-+-+-)
      0 1 2
```

19.

```
 ─────(-+-]─────
     -2   0
```

21.

```
 ─────+-+-+-+-(-+-)
      0       4  6
```

23.

```
 ─────+-+-+-(──────►
      0     3
```

25. **a)** $a \le b$ if and only if $b - a \ge 0$ **b)** If $a \le b$ and $b \le c$, then $a \le c$ **c)** If $a \le b$, then $(a + b) \le (b + c)$ **d)** If $a \le b$, then
$-a \ge -b$ **e)** If $a \le b$, and $x \ge 0$, then $ax \le bx$ **f)** If $a \le b$ and $x \le 0$, then $ax \ge bx$ **g)** If $0 < a \le b$, then $0 < 1/b \le 1/a$
27. $7/11 > 19/37$ **29.** $6/7 > 54/93$ **31.** $0.0\overline{23} < 0.0\overline{23}$ **33.** $7.3\overline{54} > 7.35\overline{4}$ **35.** $4.8 < 4.87 < 4.8\overline{7} <$
$4.\overline{87} < 4.\overline{8}$ **37.** 5/8 will do   **39.** 7.3545 will do   **41.** The 10 oz package @ 12.9¢/oz   **43.** The 3-1/2 lb box
@ 6.41¢/oz   **45.** $12.42 < \text{mpg} < 12.50$ **47.** To within $\pm \, 0.07002 \text{ cm}^2$

Section 1.3

1. 3^5 **3.** 128 **5.** 1/144 **7.** 1,953,125 **9.** 4 **11.** -32 **13.** 64 **15.** 64 **17.** y^{17}
19. x^{30} **21.** a^5 **23.** y^2/x **25.** y^3/x^3 **27.** $12x^4y^3z^3$ **29.** $-14b^2/a^4$ **31.** $1/x^{12}$ **33.** $x^7y^8z^7$
35. $27/x^2$ **37.** a^9b^6 **39.** y^{16} **41.** x^{12}/y^8 **43.** z/x^7y^4 **45.** $s^3t^3/8$ **47.** s^4 **49.** $c^{18}/a^{30}b^{24}$
51. $x^{18}z^8/y^{22}$ **53.** $2ab^2 + a^2b$ **55.** $(p + q)^2/pq$ **57.** $27 - 8 \neq 1$ **59.** $9 + 9 \neq 36$
61. $(1/25) \neq 1/4 + 1/9$ **63.** $10^{100}, 10^{(10^{100})}$ **65.** 0.000,000,000,000,000,000,000,000,000,911
67. 30,000,000,000 **69.** $2.463(10^3)$ **71.** $(1.4317)(10^2)$ **73.** $1.32(10^{-3})$ **75.** $7.12(10^{-14})$
77. $1.26263(10^{11})$ acres, $5.11(10^{10})$ hectares **79.** $3 \cdot 10^5 = 3000$ **81.** $2 \cdot 10^5 = 200,000$ **83.** $2^{(N-1959)}$
where N = current year **85.** 1024 billion, 64 billion **87.** 67.5 trillion, 278 billion, 833 billion **89.** 89,280,000 mi
91. $4.40(10^{18})$ mi

Section 1.4

1. 7 **3.** $6\sqrt{2}$ **5.** 5 **7.** -2 **9.** 4 **11.** $\sqrt{2}/2$ **13.** 9 **15.** 3 **17.** 4 **19.** 4
21. 3 **23.** 8 **25.** $-1/3$ **27.** 1/9 **29.** 9 **31.** 9 **33.** 27 **35.** 12 **37.** $3r^2|s|$
39. $2ab\sqrt[3]{2a^2b^2}$ **41.** $25|xy|\sqrt{|y|}$ **43.** $-8x^3y^2$ **45.** $u^2v^4w^{10}$ **47.** $3/x$ **49.** $p^{1/4}r^4/q^{1/2}$ **51.** $a^{16}z^2$
53. $xz^{1/2}/y^{1/4}$ **55.** $x^4y^6z^3$ **57.** $1/(u^5v^4w^4)$ **59.** $13\sqrt{3}$ **61.** $2\sqrt[3]{3} - 2\sqrt{2}$ **63.** -7 **65.** 45
67. $37 + 2\sqrt{3} + 15\sqrt{2}$ **69.** 1 **71.** 7 **73.** $\sqrt[3]{9} \neq 1 + 2$ **75.** $\sqrt{8} \neq 4$ **77.** Elmo ≈ 37;
Wiener ≈ 171 **79.** \$10.82 trillion; \$1.20 trillion **81.** 33,636 **83.** 1047°C **85.** 4.76 mg; 11.31 mg

Section 1.5

1. $6x^2 + 6x - 8$ **3.** $3a^4 + 4a^3 - 3a^2 + 9$ **5.** $14t^7 - 37t^6 + 12t^5 - 5t^3 + 4t^2 + 16$ **7.** $12y^2 + 2y - 2$
9. $t^3 - 7t + 6$ **11.** $2s^5 + 11s^3 - 2s^2 + 12s - 8$ **13.** Carry out the indicated multiplications.
15. $16a^2 - 16ab + 4b^2$ **17.** $27x^3 + 135x^2y + 225xy^2 + 125y^3$ **19.** $r^2 - 2rs + s^2 - t^2$ **21.** $u^4 - 16$
23. $25(1 - 2x)$ **25.** $2rs^2(2r^2 - s^6 + 5r^3s^2 - 6r)$ **27.** $(6 + s)(6 - s)$ **29.** $(6x - 7y)(6x + 7y)$
31. $(2 - u)(4 + 2u + u^2)$ **33.** $(4 + v)(16 - 4v + v^2)$ **35.** $(y - 5)(y - 4)$ **37.** $(p - 5)(p + 4)$
39. $(2a - 3)(2a - 5)$ **41.** $(2r - 7s)^2$ **43.** $(a + 1)^2(a - 1)^2$ **45.** $(r + 3)(r - 3)(r^2 + 2)$
47. $(\sqrt{2}x - \sqrt{3})^2(\sqrt{2}x + \sqrt{3})^2(2x^2 + 3)^2$ **49.** $(2r + s)^3$
51. $(\sqrt{5}p - 2q)(\sqrt{5}p + 2q)(5p^2 - 2\sqrt{5}pq + 4q^2)(5p^2 + 2\sqrt{5}pq + 4q^2)$ **53.** $(u^4 + 2u^2 + 2)(u^4 - 2u^2 + 2)$
55. $(t^2 + 7t + 16)(t^2 - 7t + 16)$ **57.** $(p - 5)(p - 1)$ **59.** $(y^2 + 1)(x + 2)(x - 2)$
61. $(x - y)(x^2 + xy + y^2 + x + y)$ **63.** $(y + 3)(y - 3)(x + 5)(x + 1)$ **65. a)** $120x + 75y + 95z$
b) $78x + 43y + 55z$ **c)** $42x + 32y + 40z$

Section 1.6

1. $x^2 - 3x + 1$ **3.** $u + 7$ **5.** $z - 1$ **7.** $2r/(2 - r)$ **9.** $-2/x(x - 3)$ **11.** $8/(p + q)$
13. $w(4w - 3)/(w^2 - 3w + 9)$ **15.** $(9 - t^2)/3t$ **17.** $ab^2(a - b)$ **19.** $(v - 5)/(v - 4)$ **21.** -1
23. $(x + 5)/(x - 3)$ **25.** $1/(x - y)$ **27.** $a^2(a + 2)/(a - 2)^2$ **29.** $(s + 3)/s$ **31.** $-(2a + 1)/(3a - 2)$
33. $(1 - 2w)/(1 + 2w)$ **35.** $6a^2 - 9a + 14$, rem $= -6a - 14$ **37.** $2y - 3$, rem $= 25y - 16$
39. $x^3 - 13x + 12$ **41.** $2\sqrt{6} - 5$ **43.** $10 - 3\sqrt{10}$ **45.** $\sqrt{2} + \sqrt{3} - 1$ **47.** $3 + \sqrt{6} - \sqrt{15}$
49. $2 + \sqrt[3]{2} - \sqrt[3]{4}$ **51.** $2\sqrt[4]{2} + \sqrt[4]{8}$ **53.** $(\sqrt[3]{49} - \sqrt[3]{9})/4$ **55.** $(a + b - 2\sqrt{ab})/(a - b)$
57. $r + \sqrt{r^2 - 1}$ **59.** $v\sqrt[3]{w} + w\sqrt[3]{v}$ **61.** $y\sqrt{2} + 2y^2\sqrt[4]{2}$ **63.** $(7N + 11)/3$ **65.** $7.41(10^{-11})$ on m_1,
$8.34(10^{-11})$ on m_2, $2.41(10^{-11})$ on m_3 **67. a)** $xyz/(xy + yz + xz)$; **b)** 4

Section 1.7

1. $3 - 2i$ **3.** $5 - 4i$ **5.** $4 + 7i$ **7.** $-2 + 2i$ **9.** $-16 + 11i$ **11.** $8 - 12i$ **13.** $-1 + 5i$
15. $-1 - 13i$ **17.** 25 **19.** $3 + 4i$ **21.** $-11 - 2i$ **23.** 0 **25.** 1 **27.** $1 + 2i$ **29.** $8i$
31. $(0 + i)^2 = (0 - 1) + 0i = -1 + 0i$ **33.** $2 + 5i$ **35.** $1 - \sqrt{2}i$ **37.** $-5 + 3i$ **39.** $-2 - 6i$
41. $-3 - 2i$ **43.** $(-1 + 3i)/5$ **45.** $(-3 + 6i)/5$ **47.** -1 **49.** $-i$ **51.** i **53.** $(7 + 11i)/5$
55. $(3 + 4i)/25$ **57.** $(-7 + 6i)/5$

Section 1.9

1. $2 \cdot 2 \cdot 19$ **3.** $2 \cdot 2 \cdot 2 \cdot 2 \cdot 2 \cdot 3 \cdot 5 \cdot 11$ **5.** GCF = 15, LCM = 19950 **7.** 7/22 **9.** $-5/2$
11. -15 **13.** -25 **15.** 3/2 **17.** 23/2 **19.** 119/30 **21.** 3 **23.** 14/27 **25.** 7
27. ![number line with -3.63 marked between -4 and 0] **29.** ![number line from -3 to 6] **31.** ![number line near 0 1]

33. 0.888 . . .   **35.** 137/111   **37.** 152050/9999   **39. a)** 20 **b)** 0.0125 **c)** 300%   **41.** $(7/11) < (2/3) < (5/7)$
$< (3/4)$ **43.** $13.2\overline{45} > 13.2\overline{45} > 13.2\overline{45}$ **45.** 13.81 will do   **47.** 64   **49.** 1/2   **51.** $4|a|$ **53.** 10^{24}
55. 4   **57.** 13   **59.** $3 + 2\sqrt{2}$ **61.** 5   **63.** t^6 **65.** $1/s^{12}$ **67.** $8x^6$ **69.** $x^{48}y^{20}$
71. $1/x^7$ **73.** $12x^2yz$ **75.** $u/(v^8w^7)$ **77.** $x^2y^2/[(y^2 - x^2)(x - y)^2]$ **79.** $4|a^7|b^4$ **81.** $a^3/y^{3/2}z^{5/2}$
83. $x\sqrt{yz}/y$ **85.** $8 + 64 \neq 216$ **87.** $1/4 - 1/9 \neq 1$ **89.** 6,500,000,000
91. 0.000,000,000,000,000,000,000,000,000,000,663   **93.** $2.372(10^{-1})$ **95.** $2.60014(10^1)$ **97.** $3.985(10^{18})$
99. $6r^2 + 11r - 35$ **101.** $(2 - x)/(x^2 + 2x + 4)$ **103.** $(y + 9)/(y + 5)$ **105.** $4(4 + 3a)$

107. $(x - 5)(x + 4)$ **109.** $(3 - a)(3 + a)(9 + a^2)$ **111.** $(x - \sqrt[4]{2})(x + \sqrt[4]{2})(x^2 + \sqrt{2})(x^4 + 2)$
113. $(2x + 3y)^2$ **115.** $(25 + z^2 - 5z)(25 + z^2 + 5z)$ **117.** $(2 - 3y)^2$ **119.** $(2x - y)(2x + y)(y - 1)(y + 1)$
121. $2r^2 - r + 4$, rem $= 1$ **123.** $-2 + 3i$ **125.** $2 + 5i$ **127.** i **129.** 0   **131.** $(-1 - 17i)/10$
133. 1   **135.** $6 - 5i$ **137.** i **139.** 10,383 ft   **141.** $26.83   **143.** $2.54(10^{13})$ mi   **145.** 6.25,
200   **147.** 4000   **149.** 29.2%

CHAPTER 2

Section 2.1

1. 5   **3.** -1 **5.** 18/5   **7.** 6   **9.** 0   **11.** all t **13.** 16   **15.** 26   **17.** 0   **19.** 4
21. 0   **23.** all s **25.** 2   **27.** -2 **29.** 2/13   **31.** all w **33.** -4 **35.** 1   **37.** 2
39. -3 **41.** $4i$ **43.** $-2i$ **45.** $(6 - 7i)/5$ **47.** $x = 5, y = 0$ **49.** $x = 5, y = 5$ **51. a.** 25°
b. $-40°$ **c.** $C = (5/9)(F - 32)$ **53. a.** \$700 **b.** $P = I/rt, r = I/Pt, t = I/Pr$ **55.** $y = rx(x - r)$ **57. a.** $r = \sqrt{s/\pi}/2$ **b.** $r = \sqrt[3]{3V/4\pi}$ **59. a.** $l = A/w = P/2 - w$ **b.** $w = A/l = P/2 - l$

Section 2.2

1. ± 2 **3.** $\pm 2i$ **5.** 2, 3   **7.** -2 **9.** 2, -1 **11.** $-3/2, 2/3$ **13.** $-2/3$ **15.** $1/2, -5$
17. 1/5   **19.** $-3/2, 5$ **21.** $-7, 3$ **23.** 4, -22 **25.** $-2 \pm\sqrt{2}$ **27.** $-1 \pm\sqrt{3}/\sqrt{2}$ **29.** 9,
-8 **31.** 6, -2 **33.** $1 \pm \sqrt{5}/\sqrt{2}$ **35.** 1/2   **37.** $3/2, -1/2$ **39.** 1, -4 **41.** $2/3, -1/3$
43. 3/2   **45.** $-1, -2/3$ **47.** $(1 \pm i\sqrt{3})/2$ **49.** $3 \pm 5i$ **51.** 2, $-2/3$ **53.** 2, $-7/5$
55. $1 \pm i\sqrt{3}$ **57.** $-1 \pm i$ **59.** $-4 \pm 3i$ **61.** $(3 \pm i\sqrt{7})$ **63.** 4, 2/3   **65.** $\sqrt{2}/2 \pm i$
67. $-1 \pm 2i$ **69.** $-2 \pm 3i$ **71.** $1 \pm i$ **73.** $-3 \pm i$ **75.** 2   **77.** 1   **79.** 1   **81.** 2
83. 2   **85.** $(-3/2)i, -i$ **87.** $(-i \pm 5)/2$ **89.** 20 sec after firing.   **91. a.** 410 km, **b.** 418 km
93. 70 kg or about 154 lb   **95.** 8 mph

Section 2.3

1. $x = \pm\sqrt{5}, \pm 1$ **3.** $\pm\sqrt{2}/2$ **5.** $\pm\sqrt{7}, \pm 1$ **7.** $\sqrt[3]{5}, 1$ **9.** $-\sqrt[3]{3}$ **11.** $-\sqrt[3]{5}, \sqrt[3]{2}$
13. $\pm\sqrt{2}, \pm\sqrt[4]{2}$ **15.** $\pm\sqrt[4]{2/3}, \pm 1$ **17.** $\sqrt[5]{3}, -\sqrt[5]{2}$ **19.** $\pm\sqrt[6]{3}$ **21.** 9   **23.** 4
25. 1/4   **27.** 1/4   **29.** 256   **31.** 4   **33.** 14   **35.** 8   **37.** 7   **39.** no solutions   **41.** 13, 5
43. $-3, -2$ **45.** 3   **47.** $\pm 2, \pm 1/2$ **49.** $\pm 3, \pm 1$ **51.** -7 **53.** $-4, -2$ **55.** 0, 2/11
57. 2, 13   **59.** 2/3, 1/2   **61.** $\pm\sqrt{13}, \pm\sqrt{2}$ **63.** $-1, -2\sqrt[3]{2/3}$ **65.** 1/2 cm

67. $c = \sqrt{(-p^2 + \sqrt{(p^4 + 4m^2E^2)})}/m\sqrt{2}$

Section 2.4

1. \$18.00   **3.** \$120   **5.** \$45   **7.** \$5 and \$6   **9.** \$13   **11.** 16 cm $\times$ 13 cm   **13.** 5 ft   **15.** 3, 6
17. 1 ft and 4 ft   **19.** 6 mph, 10 hr and 5 hr   **21.** Bill 6 hr, Jane 5 hr, 300 mi   **23.** 24 hr   **25.** 5 liters
27. 18   **29.** \$12,000   **31.** 40   **33.** 1 hr, 12 min   **35.** Doug 12 days, Todd 6 days   **37.** $5\frac{1}{3}$ hr
39. Through **d.** the computation yields $(2x + 4)/2 = x + 2$; **e.** $x + 2 - x = 2$ **41.** The corners are $x, x + 2, x + 14$, and $x + 16$. The sum is $4x + 32 = 4(x + 8)$. The mentalist divides by 4, then subtracts 8 to get the upper left corner entry. The rest is easy.   **43.** The numbers are $x, x + 1, x + 2, (x + 7), (x + 7) + 1, (x + 7) + 2, (x + 14), (x + 14) + 1$, and $(x + 14) + 2$. The sum is $9x + 3 \cdot 7 + 3 \cdot 14 + 3 \cdot 1 + 3 \cdot 2 = 9x + 72 = 9(x + 8)$. From this formula the sum is obtained quickly for any given x.   **45.** Each eats 8/3 sandwiches. The hungry man pays 80¢ for 8/3 sandwiches or 30¢ per sandwich. He buys 1/3 and 7/3 sandwiches, respectively, from his friends. Thus he pays them 10¢ and 70¢, respectively.
47. 84   **49.** 40   **51.** 6, 8   **53.** 3, 5, 7   **55.** 5 in^2 **57.** 22, 24   **59.** 6, 12

Section 2.5

1. **a.** tripled **b.** $y = x/5$ **c.** 6   3. **a.** reduced to 1/8 original **b.** $M = 4v^3$ **c.** 256   5. **a.** $s^* = s/12$ **b.** $s = 3t^2/\sqrt{uv^2}$ **c.** 1   7. **a.** $y^* = 48y$ **b.** $y = 20w\sqrt{x}/z^3$ **c.** 900   9. $t = 240/r$ 11. $F^* = 27F/2$ 13. **a.** $L = 10wd^2/l$ **b.** $w^* = 20$ in, $d^* = 24$ in **c.** deeper   15. 9 times   17. 625   19. **a.** directly *wrt* T, inversely *wrt* P **b.** directly *wrt* T, inversely *wrt* V **c.** doubled   21. 9 times   23. 2/3 of original area   25. **a.** $m = k_1 S = k_1(k_2 h^2)$; $m = k_3 o = k_3(k_4 Vr) = k_3 k_4(k_5 h^3)r$; Equating these two expressions for m and solving for r, we find $r = k_1 k_2/k_3 k_4 k_5 h = K/h$ **b.** $r(3/2) = 2K/3$, $r(3) = K/3 = (\frac{1}{2}r)(3/2)$, $r(6) = K/6 = (\frac{1}{2}r)(3) = (\frac{1}{4}r)(3/2)$

Section 2.6

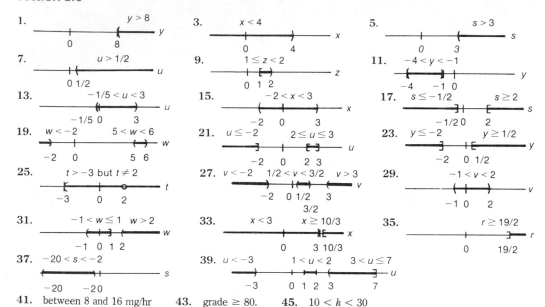

41. between 8 and 16 mg/hr   43. grade ≥ 80.   45. $10 < h < 30$

Section 2.7

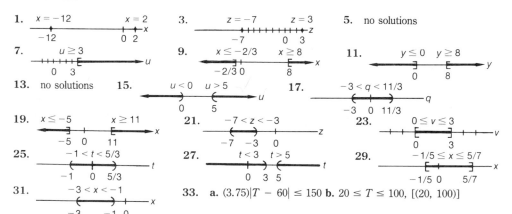

33. **a.** $(3.75)|T - 60| \leq 150$ **b.** $20 \leq T \leq 100$, $[(20, 100)]$

Section 2.9

1. -10 3. 16   5. 4   7. 6   9. 2   11. $x = 4, y = -2$ 13. 3   15. $-1/3, -6$
17. no real solutions   19. $6, -2$ 21. $2/5, -4/3$ 23. $1/3, 3$ 25. $(-1 \pm 3i)/2$ 27. $(3 \pm 3i)/2$
29. $2 \pm i$ 31. $-3 \pm 2i$ 33. $-4 \pm 3i$ 35. $\pm\sqrt{2}, \pm i\sqrt{6}$ 37. $\pm\sqrt{2}, \pm\sqrt[4]{2}, \pm i\sqrt{2}, \pm i\sqrt[4]{2}$
39. The real roots are ± 1 41. 1, 16   43. 9/2, 2   45. 5   47. 1/2   49. 7   51. $\pm\sqrt{7}$
53. -1 55. $-2, -6$ 57. $-2 - i/2$ 59. $i \pm 2$ 61. $2i, -i/3$

63.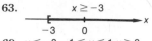

$x \geq -3$

65. $z < -5 \quad z > 1$

67. $-2 < x < 3$

69. $u \leq -3 \; -1 \leq u \leq 1 \; u \geq 2$

71. $-4 \leq s \leq 0$

73. $r > 5$

75. $p \leq -6 \quad p \geq -2$

77. $-5/3 < t < 1$

79. $x \leq -5 \quad x > 11$

81. $7/5 < v < 17$

83. $-3 < p < -1$ but $p \neq -2$

85. **a.** doubled **b.** $r = 10p^3 s / \sqrt{q}$ **c.** 40 **87.** 43.3¢ per mile **89.** **a.** 300 m, **b.** 500 m **91.** \$16,000
93. 15 **95.** **a.** 16 mph, **b.** 3/4 hr, $1\frac{1}{4}$ hr **97.** \$4600 **99.** 10 hr, 8 hr **101.** 2 furlongs by 3 furlongs
103. The entries are n, $n + 1$, $n + 7$, and $(n + 7) + 1$. The sum of these is $4n + 16 = 4(n + 4)$. She divides by 4 and subtracts 4 to get the first entry. The rest is easy. **105.** 17

CHAPTER 3

Section 3.1

1. $A = (2, 1) \; B = (-2, 3) \; C = (-1, -4) \; D = (6, -3) \; E = (4, 10) \; F = (-12, 2) \; G = (-12, -7) \; H = (3, -9) \; I = (12, 11) \; J = (-9, 10) \; K = (-8, -1) \; L = (10, -7) \; M = (11, 4) \; N = (-7, 6) \; O = (-5, -6)$
$3 - 15$

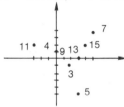

17. Burkettsville—F9, Cassella—L17, Celina—Q15, Chickasaw—O17, Coldwater—O11, Convoy—T12, Cranberry—J14, Ft. Recovery—N7, Maria Stein—L23, Montezuma—P18, Padua—P8, Philothea—N9, Sharpsburg—I7, St. Henry—M11, Van Wert—U14 **19.** (2, 0) **21.** (1/2, 1) **23.** (0, 2) **25.** $\sqrt{82}$ **27.** $\sqrt{122}$ **29.** $\sqrt{2}$ **31.** 3
33. 8 **35.** $3\sqrt{2}$ **37.** 6 1/2 miles **39.** $d[(-2, 4), (0, -1)] = d[(-2, 4), (3, 6)] = \sqrt{29}$ and the points are not collinear **41.** $d[(-2, 1), (0, 4)] = d[(2, 2), (4, 5)] = \sqrt{13} \; d[(-2, 1), (2, 2)] = d[(0, 4), (4, 5)] = \sqrt{17}$
43. Yes. $d[(-1, -4), (2, 2)] + d[(2, 2), (7, 12)] = 3\sqrt{5} + 5\sqrt{5} = 8\sqrt{5} = d[(-1, -4), (7, 12)]$ **45.** If so, (8, 2) must be in the middle. But $d[(2, -2), (8, 2)] + d[(8, 2), (12, 5)] = \sqrt{52} + 5 \approx 12.211 > d[(2, -2), (12, 5)] = \sqrt{149} \approx 12.206$; No
47. The midpoint of each diagonal is $((a + b)/2, c/2)$.

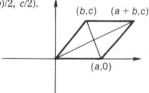

49.

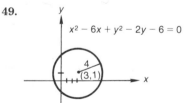

51.

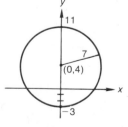

53.

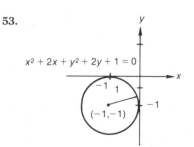

55.

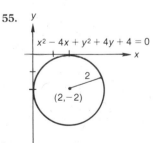

57.

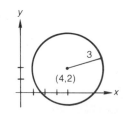

$x^2 - 8x + y^2 - 4y + 11 = 0$

59.

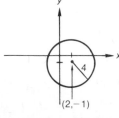

$x^2 - 4x + y^2 + 2y - 11 = 0$

61.

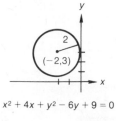

$x^2 + 4x + y^2 - 6y + 9 = 0$

63.

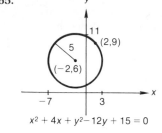

$x^2 - 4x + y^2 + 8y - 205 = 0$

65.

$x^2 + 4x + y^2 - 12y + 15 = 0$

67.

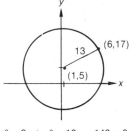

$x^2 - 2x + y^2 - 10y - 143 = 0$

69.

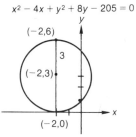

$x^2 + 4x + y^2 - 6y + 4 = 0$

71.

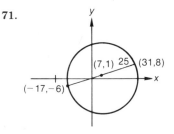

$x^2 - 14x + y^2 - 2y - 575 = 0$

73. Circle: center $(3, -1)$, $r = 5$ **75.** Circle: center $(-2, -4)$, $r = 7$ **77.** Circle: center $(5, -3)$, $r = 2$
79. Circle: center $(5, -5)$, $r = 5$ **81.** Single point $(3, -1)$ **83.** Circle: center $(-1, 2)$, $r = 3$ **85.** Circle:
center $(2, -3)$, $r = 4$

Section 3.2

1.

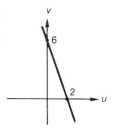

3.

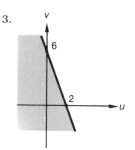

5.

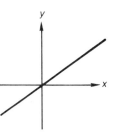

7.

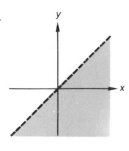

9.

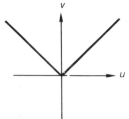

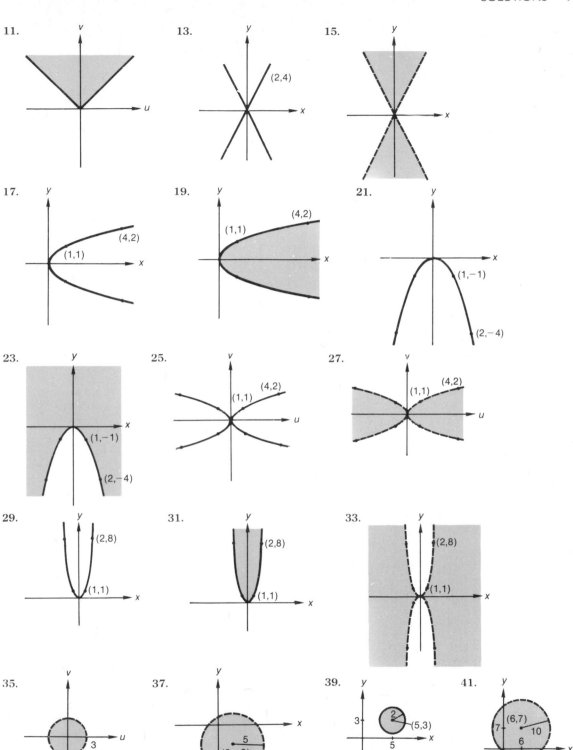

43. 71.5¢/dozen   **45.** In mid-1986, the increase in scrapping costs will exceed the savings in production costs.
47. Roadblocks should be placed on all roads 50 mi from Placerville.   **49.** $y = x$

Section 3.3

1. $3x - y + 6 = 0$ **3.** $2x - y - 12 = 0$ **5.** $10x + y + 4 = 0$ **7.** $5x + y - 4 = 0$
9. $x + 12 - 3y = 0$, $(0, 4)$, $(-12, 0)$ **11.** $x + y = 2$, $(0, 2)$, $(2, 0)$ **13.** $y = -1$, $(0, -1)$, no x-intercept
15. $x + 2y + 12 = 0$, $(0, -6)$, $(-12, 0)$ **17.** $y = x + 5$ **19.** $4x + y + 6 = 0$ **21.** $2x + y = 6$
23. $x - 2y - 4 = 0$ **25.** $y = x + 9$ **27.** $y = 7x - 14$

29. 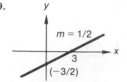   **31.** 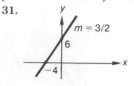   **33.**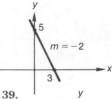

35. 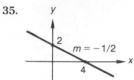   **37.** 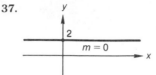   **39.**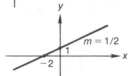

41. perpendicular   **43.** neither   **45.** parallel   **47.** perpendicular   **49.** neither   **51.** yes
53. no   **55.** yes   **57.** $1/a\, x + 1/b\, y - 1 = 0$ is the general form of a line; the line has intercepts $(a, 0)$ and $(0, b)$.
59. $5x - 3y - 15 = 0$, $m = 5/3$ **61.** $2x + 3y + 6 = 0$, $m = -2/3$ **63.** $x + 2y - 2 = 0$, $m = -1/2$
65. Either $0 = 0$ and every (x, y) satisfies the equation, or $c \neq 0$ and no (x, y) satisfies the equation
67. $x = -1$, $y = 3$ **69.** $x + 3y + 7 = 0$; $3x - y + 1 = 0$ **71.** $2x + y + 3 = 0$; $x - 2y - 6 = 0$
73. $y = 2$; $x = 6$ **75.** $x + y = 29$; $x - y = -31$ **77.** $4x - 3y + 25 = 0$ **79.** slope $= 9/27 = 1/3 =$
$4/12$; it is called a 4–12 pitch

81. 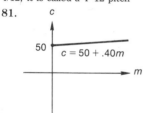   **83.** **a.** 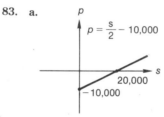   **b.** \$20,000 **c.** \$30,000

85. Solution outlined within exercise.

Section 3.4

1. 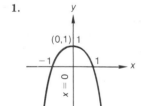   **3.** 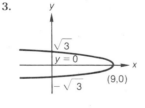   **5.**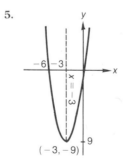

7. 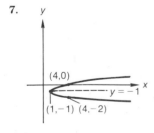   **9.** 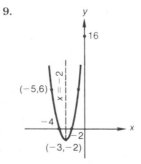   **11.**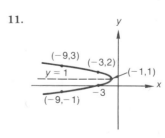

13.
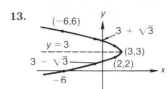

15.
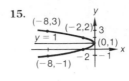

17.
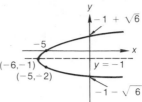

19.
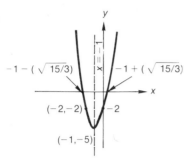

21.
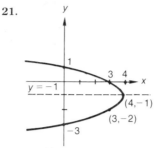

23.
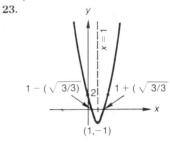

25. a. 6400 ft; **b.** 20 sec; **c.** 40 sec   **27. a.** 10,000 ft; **b.** 20 sec; **c.** 45 sec   **29. a.** 3600 ft; **b.** 10 sec; **c.** 25 sec
31. $-1 \le x \le 1/2$ **33.** $1/2 < x < 5/2$ **35.** $1/3 \le x \le 2$ **37.** all x **39.** $-1 < x < 4$

41. $-3 \le x \le 4$ **43.** $-\sqrt[3]{12} \le x \le \sqrt[3]{2}$ **45.** $x > 9$ **47.** Best selling price $= \$375$ **49.** $\$10.50$ per
gallon   **51.** At 150 units sold per month   **53.** 15 cars

Section 3.5

1.
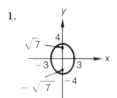

3.
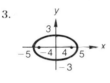

5.
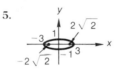

7.
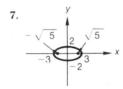

9.
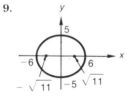

11.
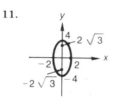

13.
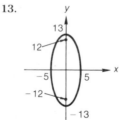

15.
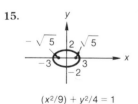

$(x^2/9) + y^2/4 = 1$

17.
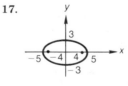

$(x^2/25) + y^2/9 = 1$

19.
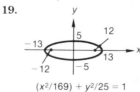

$(x^2/169) + y^2/25 = 1$

21.
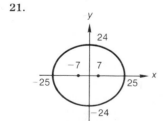

$(x^2/625) + y^2/576 = 1$

23.
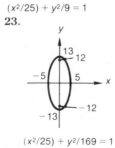

$(x^2/25) + y^2/169 = 1$

25.
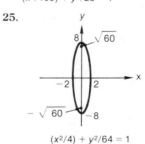

$(x^2/4) + y^2/64 = 1$

27.

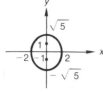

$(x^2/4) + y^2/5 = 1$

29. $(y^2/16) - (x^2/9) = 1$

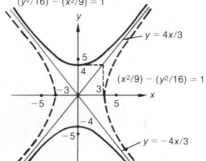

31.

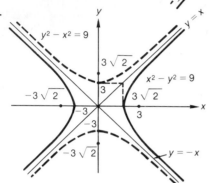

33.

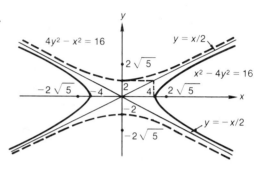

35.

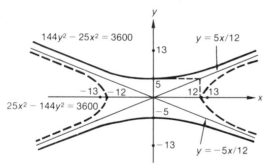

37. $y^2 - 4x^2 = 12$

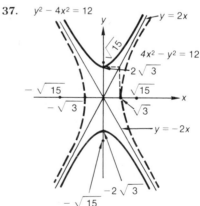

39. $4y^2 - 9x^2 = 36$

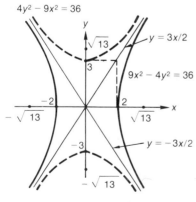

41. $(y^2/64) - (x^2/25) = 1$

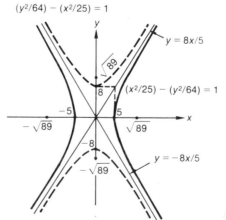

43.

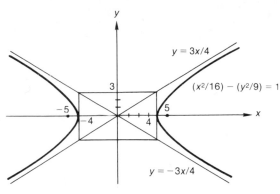

$y = 3x/4$

$(x^2/16) - (y^2/9) = 1$

$y = -3x/4$

45.

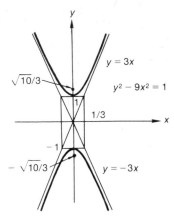

$y = 3x$

$y^2 - 9x^2 = 1$

$\sqrt{10}/3$

$-\sqrt{10}/3$

$y = -3x$

47.

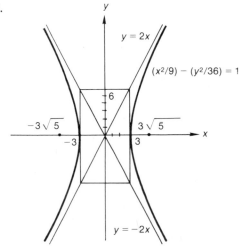

$y = 2x$

$(x^2/9) - (y^2/36) = 1$

$y = -2x$

49.

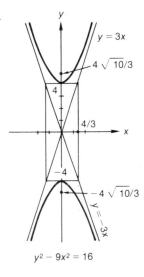

$y = 3x$

$4\sqrt{10}/3$

$-4\sqrt{10}/3$

$y = -3x$

$y^2 - 9x^2 = 16$

51.

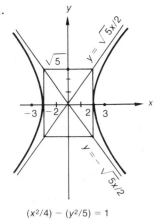

$y = \sqrt{5}x/2$

$\sqrt{5}$

$y = -\sqrt{5}x/2$

$(x^2/4) - (y^2/5) = 1$

53.

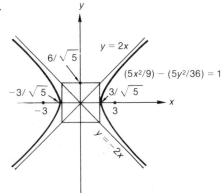

$y = 2x$

$6/\sqrt{5}$

$(5x^2/9) - (5y^2/36) = 1$

$-3/\sqrt{5}$

$3/\sqrt{5}$

$y = -2x$

55.

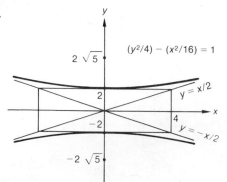

$(y^2/4) - (x^2/16) = 1$

$2\sqrt{5}$

$y = x/2$

$y = -x/2$

$-2\sqrt{5}$

57.

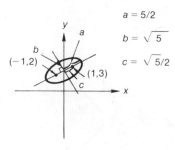

$(-1,2)$

$(1,3)$

$a = 5/2$

$b = \sqrt{5}$

$c = \sqrt{5}/2$

$21x^2 - 4xy + 24y^2 + 10x - 120y + 25 = 0$

59.

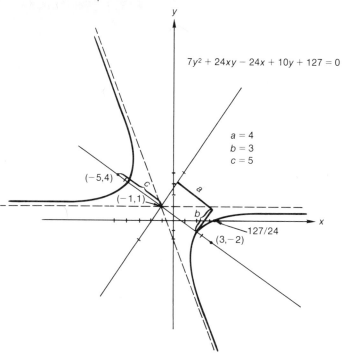

$7y^2 + 24xy - 24x + 10y + 127 = 0$

$a = 4$
$b = 3$
$c = 5$

$(-5,4)$

$(-1,1)$

$127/24$

$(3,-2)$

61.

15

25 20 20 25

50 m of chain
Stakes located as shown.

Section 3.6

1.

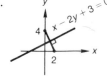

$x - 2y + 3 = 0$

4

2

3.

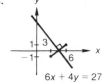

1 3

-1 6

$6x + 4y = 27$

5.

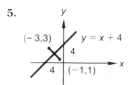

$(-3,3)$

$y = x + 4$

4

$(-1,1)$

7.

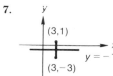

$(3,1)$

$y = -1$

$(3,-3)$

9.

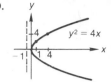

$y^2 = 4x$

11.

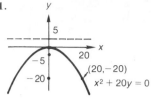

$(20,-20)$
$x^2 + 20y = 0$

13.

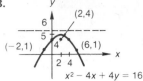

$(2,4)$
$(-2,1)$ $(6,1)$
$x^2 - 4x + 4y = 16$

15.

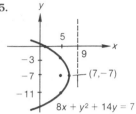

$(7,-7)$
$8x + y^2 + 14y = 7$

17.

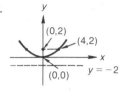

$(0,2)$ $(4,2)$
$(0,0)$ $y = -2$

19.

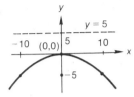

$y = 5$
$(0,0)$

21.

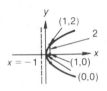

$(1,2)$
$x = -1$ $(1,0)$
$(0,0)$

23.

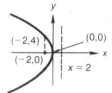

$(-2,4)$ $(0,0)$
$(-2,0)$ $x = 2$

25.

$(1,2)$
$(-1,2)$
$8(x - 1) = -(y - 2)^2$

27.

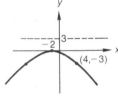

$(4,-3)$
$12y = -(x + 2)^2$

29.

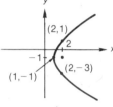

$(2,1)$
$(1,-1)$ $(2,-3)$
$4(x - 1) = (y + 1)^2$

31.

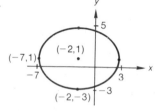

$(-2,1)$
$(-7,1)$ $(-2,-3)$
$(x + 2)^2/25 + (y - 1)^2/16 = 1$

33.

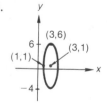

$(3,6)$
$(1,1)$ $(3,1)$
$(x - 3)^2/4 + (y - 1)^2/25 = 1$

35.

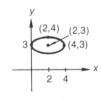

$(2,4)$ $(2,3)$
$(4,3)$
$[(x - 2)^2/4] + (y - 3)^2 = 1$

37.

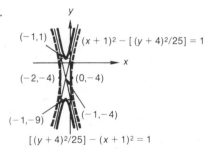

$(-1,1)$ $(x + 1)^2 - [(y + 4)^2/25] = 1$
$(-2,-4)$ $(0,-4)$
$(-1,-9)$ $(-1,-4)$
$[(y + 4)^2/25] - (x + 1)^2 = 1$

39.

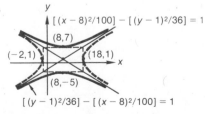

$[(x - 8)^2/100] - [(y - 1)^2/36] = 1$
$(8,7)$
$(-2,1)$ $(18,1)$
$(8,-5)$
$[(y - 1)^2/36] - [(x - 8)^2/100] = 1$

41.

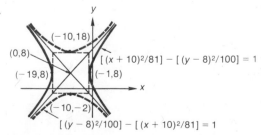

$(-10,18)$
$(0,8)$
$[(x + 10)^2/81] - [(y - 8)^2/100] = 1$
$(-19,8)$ $(-1,8)$
$(-10,-2)$
$[(y - 8)^2/100] - [(x + 10)^2/81] = 1$

43.

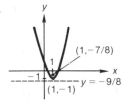

$(1,-7/8)$

$(1,-1)$ $y = -9/8$

$(y + 1) = 2(x - 1)^2$

45.

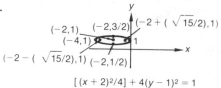

$(-2,1)$ $(-2,3/2)$ $(-2 + (\sqrt{15}/2),1)$

$(-4,1)$ 1

$(-2 - (\sqrt{15}/2),1)$ $(-2,1/2)$

$[(x + 2)^2/4] + 4(y - 1)^2 = 1$

47.

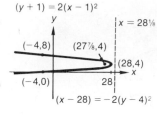

$x = 28\frac{1}{8}$

$(-4,8)$ $(27\frac{7}{8},4)$

$(28,4)$

$(-4,0)$ 28

$(x - 28) = -2(y - 4)^2$

49.

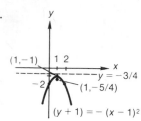

$(1,-1)$ 1 2

$y = -3/4$

-2 $(1,-5/4)$

$(y + 1) = -(x - 1)^2$

51. $(-2,-1 + 2\sqrt{2})$

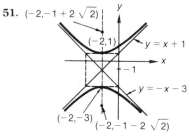

$(-2,1)$ $y = x + 1$

-1

$y = -x - 3$

$(-2,-3)$

$(-2,-1 - 2\sqrt{2})$

$[(y + 1)^2/4] - [(x + 2)^2/4] = 1$

53. $(-2,3 + \sqrt{29})$

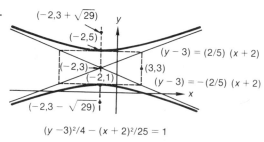

$(-2,5)$

$(y - 3) = (2/5)(x + 2)$

$(-2,3)$ $(3,3)$

$(-2,1)$ $(y - 3) = -(2/5)(x + 2)$

$(-2,3 - \sqrt{29})$

$(y - 3)^2/4 - (x + 2)^2/25 = 1$

55.

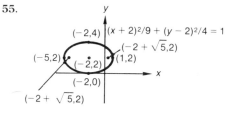

$(-2,4)$ $(x + 2)^2/9 + (y - 2)^2/4 = 1$

$(-2 + \sqrt{5},2)$

$(-5,2)$ $(-2,2)$ $(1,2)$

$(-2,0)$

$(-2 + \sqrt{5},2)$

57.

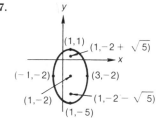

$(1,1)$ $(1,-2 + \sqrt{5})$

$(-1,-2)$ $(3,-2)$

$(1,-2)$ $(1,-2 - \sqrt{5})$

$(1,-5)$

$[(x - 1)^2/4] + [(y + 2)^2/9 = 1$

59.

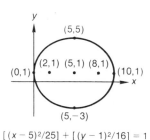

$(5,5)$

$(2,1)$ $(5,1)$ $(8,1)$

$(0,1)$ $(10,1)$

$(5,-3)$

$[(x - 5)^2/25] + [(y - 1)^2/16] = 1$

Section 3.8

1. a.

$(-7,0)$

$(-3,0)$

b. 4 **c.** $(-5, 0)$ **d.** $y = 0$ **e.** $x = -5$

3. a. 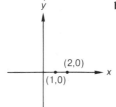 **b.** $\sqrt{13}$ **c.** (5, 7/2) **d.** $2y = 3x - 8$ **e.** $4x + 6y = 41$

5. a. 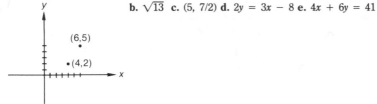 **b.** $\sqrt{5}$ **c.** (1/2, 1) **d.** $y = -2x + 2$ **e.** $4y = 2x + 3$

11. $x^2 + 2x + y^2 + 4y - 15 = 0$

7. $x^2 + 6x + y^2 - 8y = 0$ **9.** $x^2 - 8x + y^2 - 2y + 12 = 0$

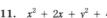

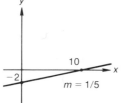   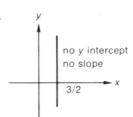   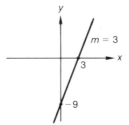

13. Yes   **15.** Yes   **17.** $6x + y = 3$ **19.** $y = 2$ **21.** $x + 2y = -14$ **23.** $4x - 3y + 12 = 0$
25. $3x + 4y = 25$ **27.** parallel   **29.** parallel   **31.** perpendicular   **33.** $y = 5; x = -1$
35. $x = 0; y = 7$

37.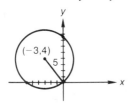

39.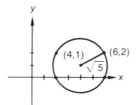

41.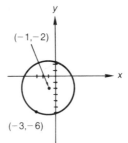

43.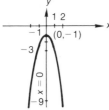

45.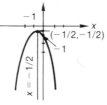

47.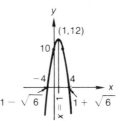

49.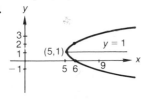

51.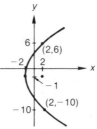

$16 (x + 2) = (y + 1)^2$

53. $-2 \leq x \leq 3$ **55.** $x < -5/3, x > 4$ **57.** $-\sqrt{3} < x < \sqrt{3}$

59.

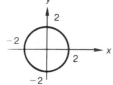

61.

63.

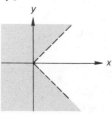

65.

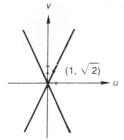

67.

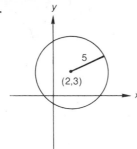

69.

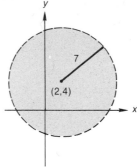

71.

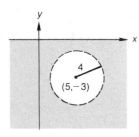

73.

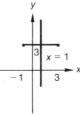

75.

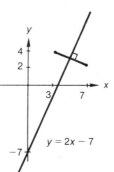

77.

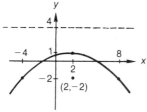

$12(y - 1) = -(x - 2)^2$

79.

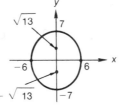

$(x^2/36) + y^2/49 = 1$

81.

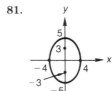

$25x^2 + 16y^2 = 400$

83.

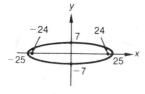

$49x^2 + 625y^2 = 30,625$

85.

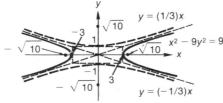

$9y^2 - x^2 = 9$

87.

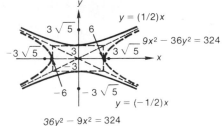

$9x^2 - 36y^2 = 324$

$36y^2 - 9x^2 = 324$

89.

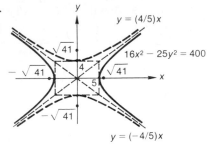

$16x^2 - 25y^2 = 400$

$25y^2 - 16x^2 = 400$

91.

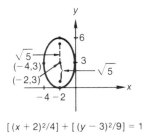

$[(x + 2)^2/4] + [(y - 3)^2/9] = 1$

93.

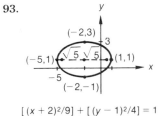

$[(x + 2)^2/9] + [(y - 1)^2/4] = 1$

95.

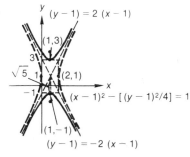

$(y - 1) = 2(x - 1)$

$(x - 1)^2 - [(y - 1)^2/4] = 1$

$(y - 1) = -2(x - 1)$

$[(y - 1)^2/4] - (x - 1)^2 = 1$

97.

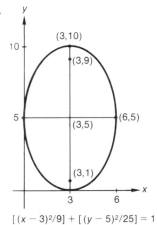

$[(x - 3)^2/9] + [(y - 5)^2/25] = 1$

99.

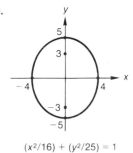

$(x^2/16) + (y^2/25) = 1$

101.

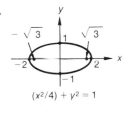

$(x^2/4) + y^2 = 1$

103.

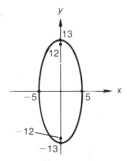

$(x^2/25) + (y^2/169) = 1$

105.

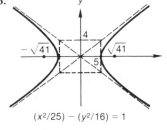

$(x^2/25) - (y^2/16) = 1$

107.

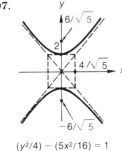

$(y^2/4) - (5x^2/16) = 1$

109.

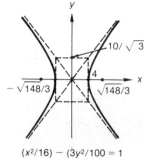

$(x^2/16) - (3y^2/100 = 1$

111.

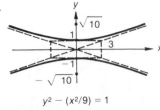

$y^2 - (x^2/9) = 1$

113.

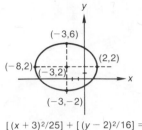

$[(x + 3)^2/25] + [(y - 2)^2/16] = 1$

115.

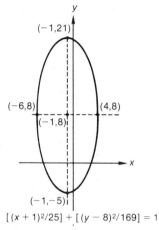

$[(x + 1)^2/25] + [(y - 8)^2/169] = 1$

117.

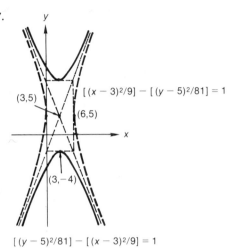

$[(y - 5)^2/81] - [(x - 3)^2/9] = 1$

119.

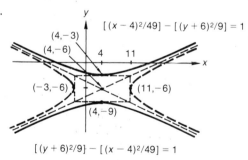

$[(y + 6)^2/9] - [(x - 4)^2/49] = 1$

121. 144 feet, $t = 1$, $t = 4$

CHAPTER 4

Section 4.1

1. y is a function of x **3.** Neither is a function of the other.   **5.** x is a function of y **7.** Each is a function of the other.   **9.** x is a function of y **11.** 4, 9, -21 **13.** $-5, -5, 25$ **15.** 0, 21, 25   **17.** 1, 0, -3
19. 3, 4, 1   **21.** domain: $x \geq 1$; range: $f(x) \geq 0$ **23.** domain: $v \neq 2$; range: $F(v) > 0$ **25.** domain: $|r| < 5$,
$r \neq -2$, range: $-\infty < g(r) < \infty$ **27.** domain: $-3 < x \leq -2$ or $2 \leq x < 3$; range: $G(x) > 0$ **29.** domain: $-6 \leq x \leq 6$, $x \neq -1,2$

31.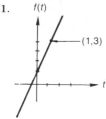

33.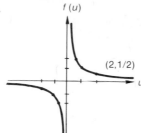

35.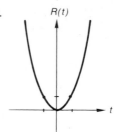

37.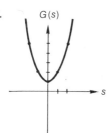

39.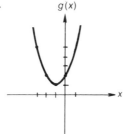

41.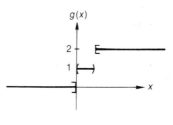

43.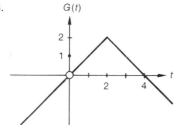

45. **a.** neither **b.** neither **c.** $y = f(x)$ **d.** both **e.** $y = f(x)$ **f.** neither **g.** both **h.** both **i.** both **j.** $y = f(x)$ **k.** $y = f(x)$ **l.** both **m.** neither **n.** $y = f(x)$ **o.** neither **47.** **a.** $d = 55t$ **b.** 137.5 mi **c.** $D: t \geq 0$; $R: d \geq 0$
49. **a.** $C(n) = 2n$ $0 \leq n \leq 9$; $1.75n$, $10 \leq n \leq 99$; $1.5n$, $n \geq 100$ **b.** No. **c.** If $n \geq 86$, he should order 100 cases
d. $D:$ integers $n \geq 0$; $R:$ 2, 4, . . ., 18, 17.5, 19.25, . . ., 173.25, 150, 151.50, . . . **51.** **a.** $C = (\$3.75) \cdot |T - 60|$
b. \$75 **c.** \$150 **d.** $D: T \geq$ absolute 0; $R: C \geq 0$ **53.** **a.** $f(x) = 0$, $x = 0$; 0.75, $0 < x \leq 50$; $0.015x$, $50 < x \leq 600$; 9 +
$(0.01) \cdot (x - 600)$, $x > 600$ **b.** \$10.50 **c.** 75¢ **d.** $D: x \geq 0$; $R: f(x) \geq 0$ in multiples of 0.01 **55.** **a.** $h(x) = 120 - 5x/2$
b. 20 mg **c.** $0 \leq x < 44$ **d.** $R: 10 \leq h(x) \leq 120$

Section 4.2

1. **a.**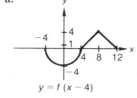
$y = f(x - 4)$

b.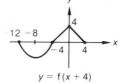
$y = f(x + 4)$

c.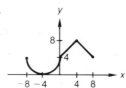
$y = f(x) + 4$

d.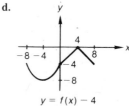
$y = f(x) - 4$

e.

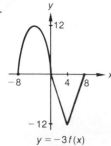

$$y = -3f(x)$$

f.

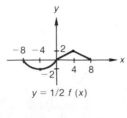

$$y = 1/2\, f(x)$$

g.

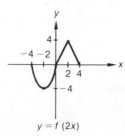

$$y = f(2x)$$

h.

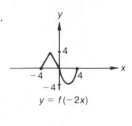

$$y = f(-2x)$$

i.

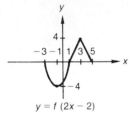

$$y = f(2x - 2)$$

j.

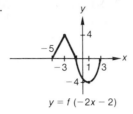

$$y = f(-2x - 2)$$

k.

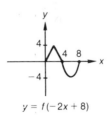

$$y = f(-2x + 8)$$

l.

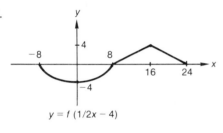

$$y = f(1/2x - 4)$$

m.

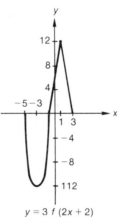

$$y = 3\, f(2x + 2)$$

n.

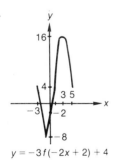

$$y = -3f(-2x + 2) + 4$$

o.

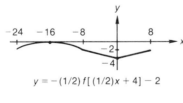

$$y = -(1/2)\, f[\,(1/2)x + 4\,] - 2$$

p.

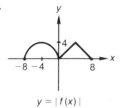

$$y = |f(x)|$$

q.

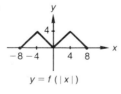

$$y = f(|x|)$$

3. **a–c.**

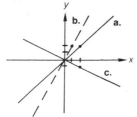

d–f.

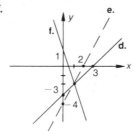

g–h.

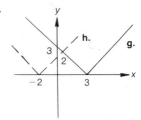

i–j.

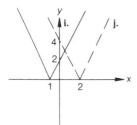

k.

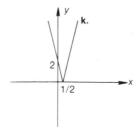

l.

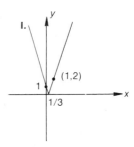

m.

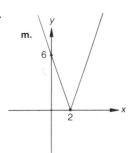

5. **a.**

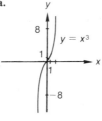

b.

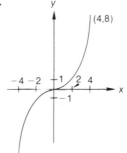

c.

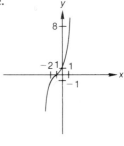

d.

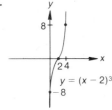

e.

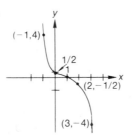

f.

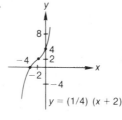

g.

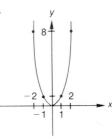

h.

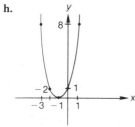

i.

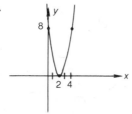

j.

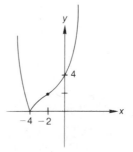

k.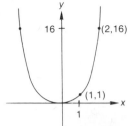

l.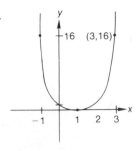

m.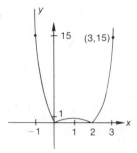

7.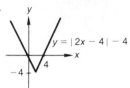

9.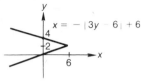

11.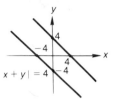

13.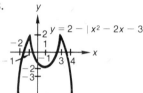

15.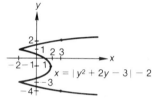

17.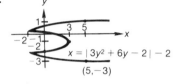

19.
a.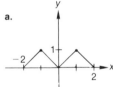
b.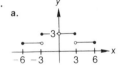

21.
a.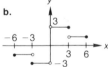
b.

23. even   **25.** odd   **27.** neither   **29.** neither   **31.** odd   **33.** even   **35.** both even: $f(-x)g(-x) = f(x)g(x)$; both odd: $f(-x)g(-x) = [-f(x)] \cdot [-g(x)] = f(x)g(x)$

37.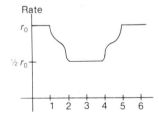

39.

Section 4.3

1. $f(x) + g(x) = 2x^2 + x - 1$, $f(x) - g(x) = -x + 11$, $f(x) \cdot g(x) = x^4 + x^3 - x^2 + 5x - 30$, $f(x)/g(x) = (x^2 + 5)/(x^2 + x - 6)$, $2f(x) = 2x^2 + 10$ **3.** $f(x) + g(x) = x^3 + 2x^2 - 2x - 6$, $f(x) - g(x) = x^3 - 2x^2 - 4x + 6$, $f(x) \cdot g(x) = 2x^5 + x^4 - 12x^3 - 3x^2 + 18x$, $f(x)/g(x) = (x^3 - 3x)/(2x^2 + x - 6)$, $3f(x) = 3x^3 - 9x$ **5.** $f(x) + g(x) = x^3 - x$, $f(x) - g(x) = x^3 - 2x^2 - x + 2$, $f(x) \cdot g(x) = x^5 - x^4 - 2x^3 + 2x^2 + x - 1$, $f(x)/g(x) = x - 1$, $4f(x) = 4x^3 - 4x^2 - 4x + 4$ **7.** $f(1/x) = (1/x) + x$, $f(x^2) = x^2 + (1/x^2)$, $f(2x) = 2x + (1/2x)$ **9.** $g(1/t) = (3/t^2) - (1/t) + 2$, $g(-x) = 3x^2 + x + 2$, $g(t + 2) = 3t^2 + 11t + 12$ **11.** $G(y - 2) = y^2 - 5y + 6$, $G(x) = (x + 2)^2 - 5(x + 2) + 6$, $G(0) = G(2 - 2) = 2^2 - 5 \cdot 2 + 6 = 0$, $G(2) = G(4 - 2) = 4^2 - 5 \cdot 4 + 6 = 2$ **13.** $P(1) = 1, P(2) =$

4. $P(a + 1) = a^2 + 2a + 1, P(a) = P[(a - 1) + 1] = (a - 1)^2 + 2(a - 1) + 1 = a^2$ **15.** $h(4) = 21, h(t) = t^2 + 2t - 3, h(2x + 1) = 4x^2 + 8x, h(y - 3) = y^2 - 4y$ **17.** $f \circ g(u) = 12u^2 + 4u + 1, f \circ g(0) = 1, g \circ f(t) = 6t^2 + 4t + 2, g \circ f(0) = 2$ **19.** $g \circ f(r) = r^2 - 4r + 3, g \circ f(-1) = 8, f \circ g(r) = r^2 - 3, f \circ g(-1) = -2$ **21.** $f(t) = t - 1, g(u) = u^2; g \circ f(t) = (t - 1)^2$ **23.** $f(n) = n + 1, g(x) = 1/x^2; g \circ f(n) = 1/(n + 1)^2$ **25.** $f(u) = u^3 + 2,$ $g(t) = 3t^2 + 2t + 4; g \circ f(u) = 3(u^3 + 2)^2 + 2(u^3 + 2) + 4$ **27.** **a.** $L(2^2) = L(4) = 8, [L(2)]^2 = 4$ **b.** $L(\sqrt{4}) = L(2) = 2, \sqrt{L(4)} = \sqrt{8} = 2\sqrt{2}$ **c.** $L(-1) = -7, -L(1) = 1$ **29.** **a.** $F(1 + 1) = F(2) = 5, F(1) + F(1) = 4$ **b.** $F(1 - 1) = F(0) = 1, F(1) - F(1) = 0$ **c.** $F(1 \cdot 1) = 2, F(1) \cdot F(1) = 4$ **31.** 0 **33.** -3 **35.** $4x + 2h$ **37.** $2x + 2 + h$ **39.** $-10x + 4 - 5h$ **41.** $4x^3 + 6x^2h + 4xh^2 + h^3$ **43.** $f(x) = (0.05)x, g(x) = (0.07)x,$ $R(x) = f(x) + g(x) = (0.12)x$ **45.** $e(q) = 3q^2 + 2q, t(q) = 2q + 1, d(q) = e(q) \cdot t(q) = 6q^3 + 7q^2 + 2q$ **47.** $C(s) = (0.5)s; B(C) = 1000 + (0.1)C;$ Bonus $= B(C(s)) = 1000 + (0.1)(0.5)s;$ Income = bonus + commission = $1000 + (0.05)s + (0.5)s = 1000 + (0.55)s$ **49.** $V(r) = 20 \cdot 2000r = 40,000r; R(V) = V/200; F(R) = (0.2)R =$ $F(R(V(r))) = (0.2)V/200 = (0.2) \cdot 40,000r/200 = 40r$

Section 4.4

1. $f^{-1}(x) = (1 - x)/4$ **3.** $g^{-1}(v) = \sqrt[5]{v}/2$ **5.** $P^{-1}(s) = \sqrt[4]{s}/2, s \geq 0$ **7.** no inverse **9.** no inverse **11.** $s^{-1}(u) = 2 + \sqrt{u + 2}, u \geq -2$ **13.** $s^{-1}(t) = (t^2 + 3)/2, t \geq 0$ **15.** $F^{-1}(x) = F(x)$ **17.** $p^{-1}(t) = p(t)$ **19. through 27.** Insert the expressions for $u = f(x)$ and $v = g(x)$ into $g[u], f[v]$, respectively

29. **(a and d)**

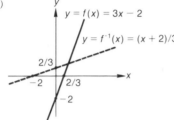

$y = f(x) = 3x - 2$
$y = f^{-1}(x) = (x + 2)/3$

31. **(a and d)**

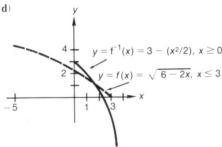

$y = f(x) = x^2 - 4x + 3, x \geq 2$
$y = f^{-1}(x) = 2 + \sqrt{x + 1}, x \geq -1$
$(-1, 2)$

33. **(a and d)**

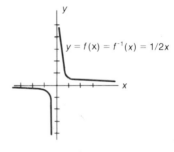

$y = f^{-1}(x) = 3 - (x^2/2), x \geq 0$
$y = f(x) = \sqrt{6 - 2x}, x \leq 3$

35. **(a and d)**

$y = f(x) = f^{-1}(x) = 1/2x$

37. **(a and d)**

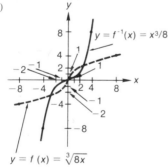

$y = f^{-1}(x) = x^3/8$
$y = f(x) = \sqrt[3]{8x}$

39. **a.** Increasing on $(-3, -2)$ and on $(0, 4)$; decreasing on $(-\infty, -3)$ and on $(4, 7)$ **b.** Increasing on $(0, 2)$; decreasing on $(-4, 0)$ and on $(2, \infty)$ **c.** Increasing on $(-\infty, -4)$ and on $(-2, 8)$; decreasing on $(-4, -2)$ and on $(8, 10)$ **d.** Increasing on $(2, \infty)$; decreasing on $(-\infty, 2)$ **e.** Increasing on $(-\infty, -5)$; decreasing on $(-5, 5)$ and on $(5, \infty)$ **f.** Increasing on $(-\infty, -8)$ and on $(-2, 0)$; decreasing on $(-8, -2)$ and on $(0, 6)$ **g.** Increasing on $(0, 5)$ and on $(8, \infty)$; decreasing on $(-15, 0)$ and on $(5, 8)$ **41.** Increasing on $(-\infty, \infty)$ **43.** Increasing on $(-\infty, 0)$ decreasing on $(0, \infty)$ **45.** Decreasing on $(-\infty, 2)$; increasing on $(2, \infty)$ **47.** Constant $(= 0)$ on $(-\infty, 0)$; increasing on $(0, \infty)$ **49.** $T = F(C) = \frac{9}{5}C + 32; C = \frac{5}{9}(T - 32) = F^{-1}(T)$ **51.** $I = (0.05)P = f(P); P = I/0.05 = 20I = f^{-1}(I)$ **53.** No, yes

Section 4.6

1. $x = f(y)$ **3.** Both **5. a.** both **b.** $x = f(y)$ **c.** $y = f(x)$ **d.** neither **7.** 36, 9, 1 **9.** $-1, 5/7, 2$
11. $4x^2 + 4x + 3$, $x^4 + 2x^2 + 3$, $(1 + 2x + 3x^2)/x^2$ **13.** 50, 38, $t^2 - 13t + 50$, $t^2 - 15 + 64$ **15. a.** 16
and 100; **b.** 10 and 4; **c.** $6x + 4$ and $6x + 8$; **d.** -2 and -10 **17.** $w \geq -1$, $w \neq 1$ **19.** no v

21.

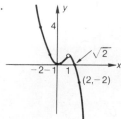

23.

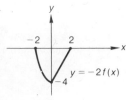

25.

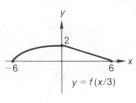

27.

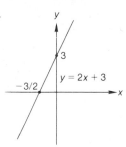

29.

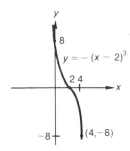

31.

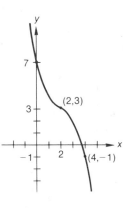

33.

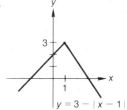

35.

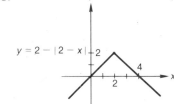

37.

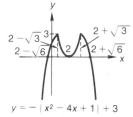

39.

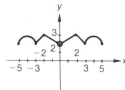

41. odd **43.** 0 **45.** $8x - 2 + 4h$ **47.** $-(2x + h)/x^2(x + h)^2$ **49.** $f(x) + g(x) = x + 3$; $f(x) - g(x) = 5x - 3$; $f(x) \cdot g(x) = 9x - 6x^2$; $f(x)/g(x) = 3x/(3 - 2x)$, $x \neq 2/3$; $f \circ g(x) = 9 - 6x$; $g \circ f(x) = 3 - 6x$; $2f(x) = 6x$; $f \circ g(2) = -3$; $g \circ f(2) = -9$ **51.** $f(x) + g(x) = x^2 + x$; $f(x) - g(x) = x^2 - 3x - 2$; $f(x) \cdot g(x) = 2x^3 - x^2 - 3x - 1$; $f(x)/g(x) = (x^2 - x - 1)/(2x + 1)$, $x \neq -1/2$; $f \circ g(x) = 4x^2 + 2x - 1$; $g \circ f(x) = 2x^2 - 2x - 1$; $-3f(x) = -3x^2 + 3x + 3$; $f \circ g(-3) = 29$; $g \circ f(-3) = 23$ **53.** $f(t) = (t + 1)/(t - 1)$; $g(u) = u^2$; $g \circ f(t) = [(t + 1)/(t - 1)]^2$ **55.** $f(u) = u^2 + 2u + 5$; $g(v) = v^{2/3}$; $g \circ f(u) = (u^2 + 2u + 5)^{2/3}$ **57.** No
59. $p^{-1}(t) = t^7/128$ **61.** $g^{-1}(u) = (u - 1)/(u + 1)$ **63. and 65.** Insert the expressions for $u = f(x)$ and $v = g(x)$ into $g[u]$, $f[v]$, respectively.

67.

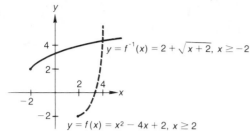

$y = f^{-1}(x) = 2 + \sqrt{x + 2},\ x \geq -2$

$y = f(x) = x^2 - 4x + 2,\ x \geq 2$

69.

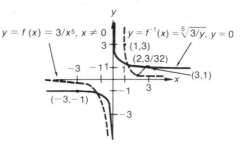

$y = f(x) = 3/x^5,\ x \neq 0$ $\qquad y = f^{-1}(x) = \sqrt[5]{3/y},\ y = 0$

(1,3)

(2,3/32)

(3,1)

(−3,−1)

71. Increasing on $(-5, -3)$ and on $(0, 1)$; decreasing on $(-3, -1)$, on $(-1, 0)$, and on $(3, 4)$; constant on $[1, 3]$
73. Decreasing on $(-\infty, -1)$ and on $(-1, \infty)$ **75.** Decreasing on $(-\infty, -1)$ and on $(1, 3)$; increasing on $(-1, 1)$ and on $(3, \infty)$

CHAPTER 5

Section 5.1

1. $r = -2$ **3.** $r = 1$ **5.** $r = -2$ **7.** $r = 41$ **9.** $r = 20$ **11.** $r = 7$ **13.** $r = 0$
15. $r = 1$ **17.** no   **19.** not a factor   **21.** $(b - 2)(b^3 - 4b^2 - 2b - 2)$
23. $(x - 2)(x^4 + 2x^3 + 4x^2 + 8x + 16)$ **25.** $(w + 2)(w^4 - 2w^3 - 6w^2 + 12w - 18)$
27. $(y + 2)(y^5 - 2y^4 + 2y^3 - 4y^2 + 8y - 6)$
29. $(p + 2)(p^5 - 2p^4 + 6p^3 - 10p^2 + 20p - 10)$

Section 5.2

1. $v = 1, -2, 4$ **3.** $s = 1/2, -3/2, -1/3$ **5.** $u = -1/3, -1$ **7.** $t = \pm 2$ **9.** 1 positive root, 0 or 2 negative roots   **11.** 0 or 2 positive roots, 1 or 3 negative roots   **13.** 0 or 2 positive roots, 1 negative root
15. 0, 2, or 4 positive roots, 1 negative root   **17.** $(-4, 2)$ **19.** $(-1, 1)$ **21.** $(-1, 2)$ **23.** $(-3, 1)$
25. $u \approx -2.2$ **27.** $y \approx -0.7$ **29.** $y \approx -1.2$ **31.** $s \approx -0.9$ **33.** $x = 3, x \approx 3.2, x \approx -1.2$ ($x = -1 \pm \sqrt{5}$)   **35.** $x = -1/2, 5/2, -7/2$ **37.** $x \approx -2.3$ **39.** $y = 1/2, y \approx -3.1, y \approx -1.7$ **41.** $x = 1,$
$x \approx \pm 1.7, x \approx 1.4$ ($x = \pm\sqrt{3}, \pm\sqrt{2}$)   **43.** $s \approx 2.4, s \approx 2.7, s \approx -2.0$ **45.** 18

Section 5.3

1. $x^3 - x^2 + 4x - 4$ **3.** $x^4 + 4$ **5.** $x^4 - 3x^3 + 3x^2 - 3x + 2$ **7.** $x^5 - 5x^4 + 10x^3 - 10x^2 + 9x - 5$
9. $x^6 - 6x^5 + 16x^4 - 24x^3 + 25x^2 - 18x + 10$ **11.** $\pm 5i, -3;$ $(t^2 + 25)(t + 3)$ **13.** $1 \pm i, -1 \pm i;$
$(x^2 - 2x + 2)(x^2 + 2x + 2)$ **15.** $\pm i, 1/2;$ $(w^2 + 1)(2w - 1)$ **17.** $x = 1, (-1 \pm \sqrt{3}i)/2;$ $(x - 1)(x^2 + x + 1)$
19. $(x + 1)(4x^2 - 4x + 2);$ $-1, (1 \pm i)/2$ **21.** $(t - 3)(t^2 + 2t + 2);$ $3, -1 \pm i$ **23.** $2 \pm 3i, \pm i, -2;$
$(y^2 - 4y + 13)(y^2 + 1)(y + 2)$ **25.** $s = \pm\sqrt{2}, 1 \pm \sqrt{5}i;$ $(s - \sqrt{2})(s + \sqrt{2})(s^2 - 2s + 6)$

Section 5.4

1.

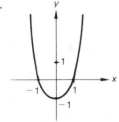

3.

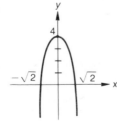

5.

$y = 1 - x^3$

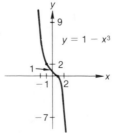

7.

$y = (x - 2)^4 - 3$

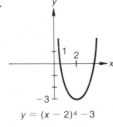

9.

11.

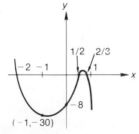

13.

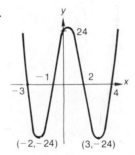

15.

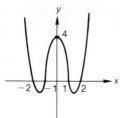

17.

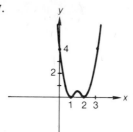

19.

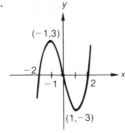

21.

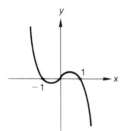

23.

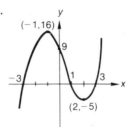

25.

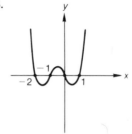

27.

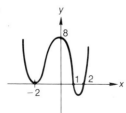

29.

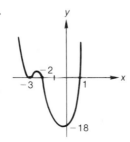

Section 5.5

1.

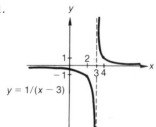

$y = 1/(x - 3)$

3.

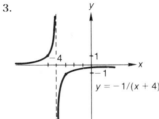

$y = -1/(x + 4)$

5.

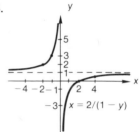

$x = 2/(1 - y)$

7.

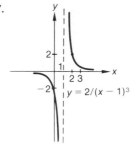

$y = 2/(x-1)^3$

9.

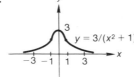

$y = 3/(x^2 + 1)$

11.

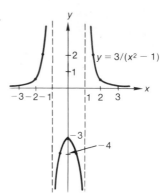

$y = 3/(x^2 - 1)$

13.

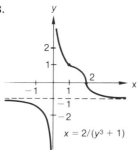

$x = 2/(y^3 + 1)$

15.

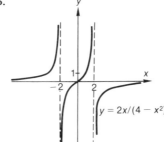

$y = 2x/(4 - x^2)$

17.

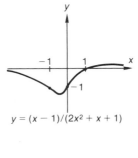

$y = (x - 1)/(2x^2 + x + 1)$

19.

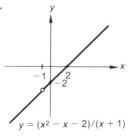

$y = (x^2 - x - 2)/(x + 1)$

21.

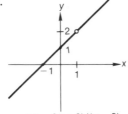

$x = (y^2 - 3y + 2)/(y - 2)$

23.

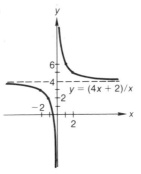

$y = (4x + 2)/x$

25.

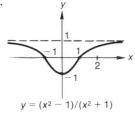

$y = (x^2 - 1)/(x^2 + 1)$

27.

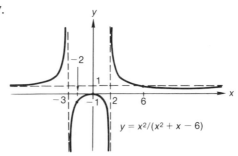

$y = x^2/(x^2 + x - 6)$

29.

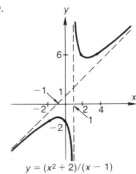

$y = (x^2 + 2)/(x - 1)$

31.

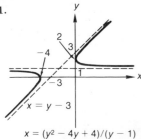

$$x = (y^2 - 4y + 4)/(y - 1)$$

33.

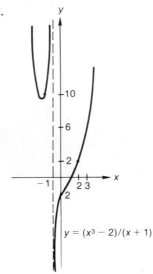

$$y = (x^3 - 2)/(x + 1)$$

35.

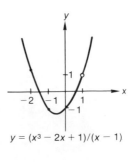

$$y = (x^3 - 2x + 1)/(x - 1)$$

37.

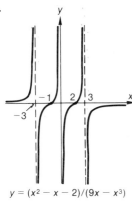

$$y = (x^2 - x - 2)/(9x - x^3)$$

39.

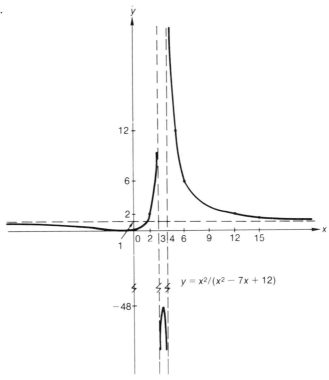

$$y = x^2/(x^2 - 7x + 12)$$

Section 5.7

1. $r = -6$ **3.** $r = -6$ **5.** not a factor **7.** $(s + 1)(s^4 + 3s^3 - 2s^2 + 2s - 1)$ **9.** $x = 2, -1/3$
11. no rational roots **13.** 1 or 3 positive roots; no negative roots **15.** 1, 3, or 5 positive roots; no negative roots
17. $(-2, 1)$ **19.** $(-1, 3)$ **21.** $r \approx -0.8$ **23.** $s \approx 1.2$ **25.** $x \approx 0.8, -2.2, -0.6$ **27.** $u \approx -1.9,$
$0.3, 1.3$ **29.** $x^3 - 4x^2 - 2x + 20$ **31.** $x^4 - 2x^3 + x^2 + 2x - 2$ **33.** $z = 2, \pm 2i$ **35.** $t = 0, 1, \pm i$
37. $w = -1 \pm i, (-1 \pm \sqrt{5})/2$ **39.** $(w - 4)(2w^2 - 4w - 3)$ **41.** $(v - 1)(v + 1)(v^2 + 2v + 2)$

43.

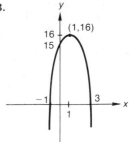

45.

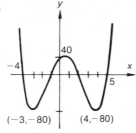

47.

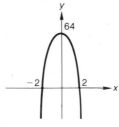

49.

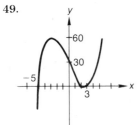

51.

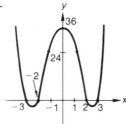

53.

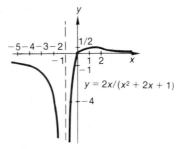

$y = 2x/(x^2 + 2x + 1)$

55.

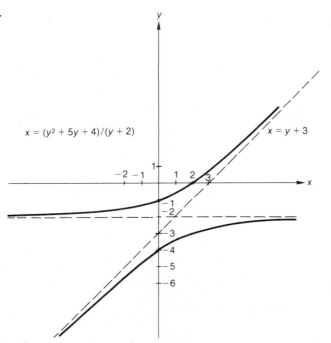

$x = (y^2 + 5y + 4)/(y + 2)$

$x = y + 3$

57.

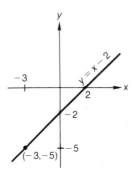

$y = x - 2$

$(-3, -5)$

CHAPTER 6

Section 6.1

1. 3.322   **3.** 365.443   **5.** 16.552

7.
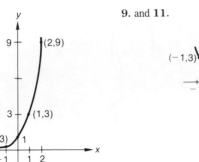

9. and 11.
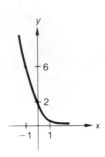

13.

15.
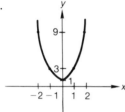

17.
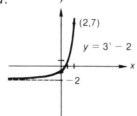

19.
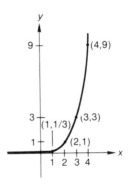

21.
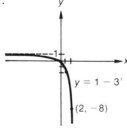

23.
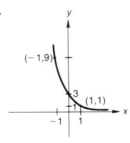

25.
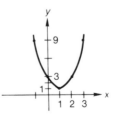

27.
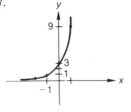

29.
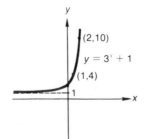

31.
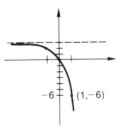

33.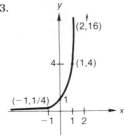

35.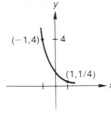

37.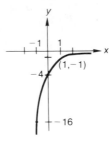

39.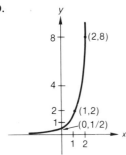

41.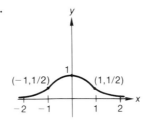

43.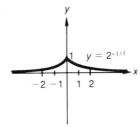

45.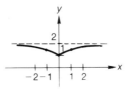

47.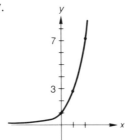

49.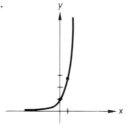

51.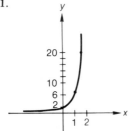

53.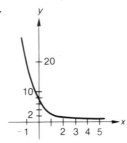

55.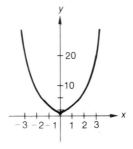

57.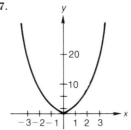

59. $x = 4$ **61.** $v = 2$ **63.** $w = -2$ **65.** $x = 0$ **67.** $z = 2$ **69.** $x = 2$ **71.** $z = -1$
73. $r = 2$ **75.** $t = \pm 2$ **77.** $v = \pm 4$ **79.** $s = 4$ **81.** **a.** $\sim 55.78°$F, $\sim 53.36°$F **b.** 1 ft: 80°F and
$-2.5°$F; 3 ft: 61.25°F and 41.875°F **c.**

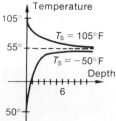

83. **a.** N*(q)

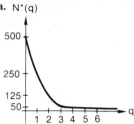

b. almost immediately **c.** less than 1/2 year

Section 6.2

1. $\log_{10} 100{,}000 = 5$ **3.** $\log_5(1/125) = -3$ **5.** $\log_{0.001} 1 = 0$ **7.** $4^{3/2} = 8$ **9.** $(1/6)^{-2} = 36$ **11.** 16
13. 27   **15.** -8 **17.** 2   **19.** 8   **21.** 4   **23.** 2   **25.** 3/2   **27.** -2 **29.** 3/2
31. 8/3   **33.** $-2/3$

35.

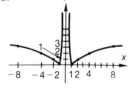

37.

39.

41.

43.

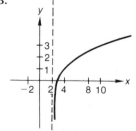

45.

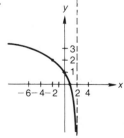

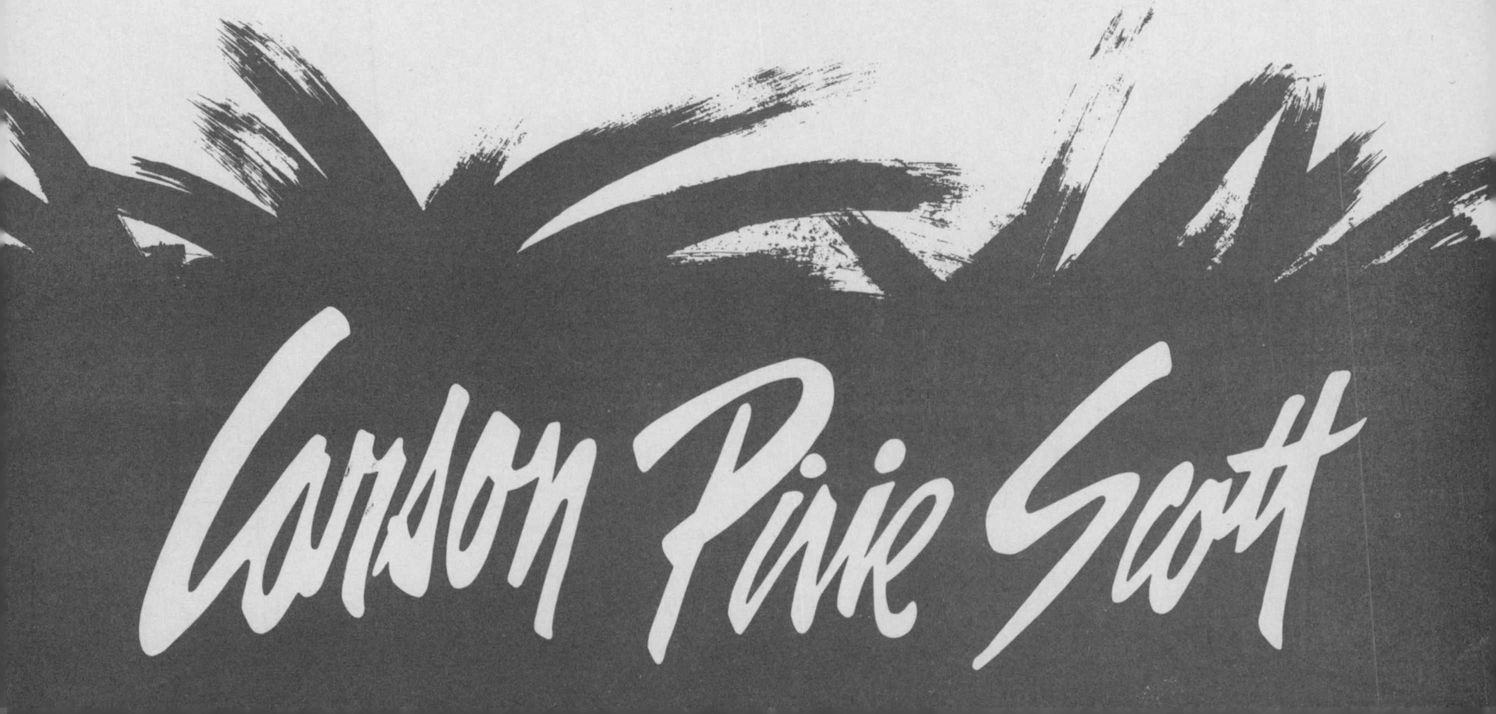
Carson Pirie Scott

47.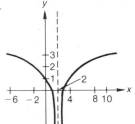

49.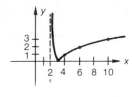

51.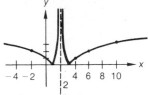

53.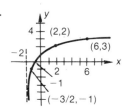

55.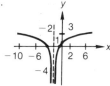

57.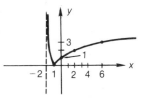

59.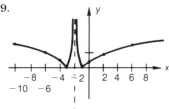

61.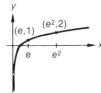

63.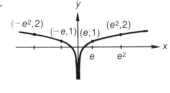

65. 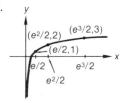 **67. a.** 302 hr, 2 hr **b.**

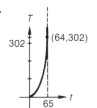

c. $t = 65 - 2^{(302 - T)/50}$

Section 6.3

1. 1.602   **3.** 1.155   **5.** −7.331   **7.** 0.5187   **9.** −0.7570   **11.** 1.7712   **13.** 3.8928
15. 3.0391   **17.** 0.2519   **19.** 1.2851   **21.** 0   **23.** $-\log_b 8 = -3\log_b 2$ **25.** 0   **27.** $\log_e (u + 1)$
29. 5   **31.** $\sqrt{3}$ **33.** 2   **35.** 1/125   **37.** −1/32   **39.** ±5   **41.** 81   **43.** 1/10   **45.** 8
47. 8   **49.** ±3   **51.** no solutions   **53.** 2   **55.** $(2\log 3 - 3\log 10)/(\log 3 - \log 10) \approx 3.9125$
57. $(5\log 4 + \log 3)/(2\log 3 - \log 4) \approx 9.9023$ **59.** $(2\log 3 - \log 2)/(\log 2 + \log 3) \approx 0.8394$
61. $(\log 7 - \log 2)/(\log 7 + \log 2) \approx 0.4747$ **63.** $(2\log 13 + \log 11)/(\log 11 - \log 13) \approx -45.062$ **65.** Let $x = \log_b u$.
Then $u = b^x$ so $u^r = (b^x)^r = b^{xr}$ and $\log_b u^r = xr = r\log_b u$ **67.** $b^1 = b$, thus $1 = \log_b b$

69. 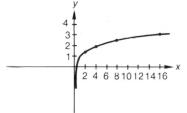   **71.**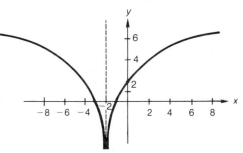

73.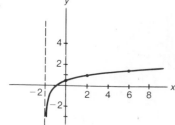

75.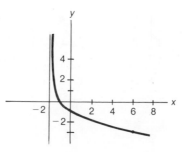

77.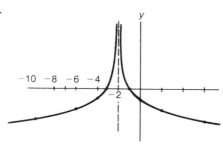

79.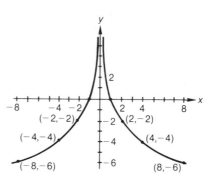

81.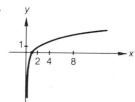

83. **a.** $5^5 \cdot \$10 = \$31{,}250$ **b.** 12 **c.** 14   **85.** 20 times as loud; 100 times as intense   **87.** **a.** $10^{1.5} \approx 31.6$ **b.** 100 times more motion, $(10^{1.5})^2 = 10^3 = 1000$ times more energy **c.** The Alaska quake experienced $10^{0.2} \approx 1.58$ times as much motion and released $(31.6)^{0.2} \approx 2$ times as much energy as the San Francisco quake. The SF quake experienced $10^{1.6} \approx 40$ times as much motion and released $(31.6)^{1.6} \approx 250$ times as much energy as the Yellowstone quake. **d.** $E = 10^{(1.5M + 11.4)} = 10^{1.5M} \cdot 10^{11.4} \approx (2.51) \cdot 10^{11} \cdot (31.6)^M$

Section 6.4

1. 2.0899 4.0899 -2.9101 **3.** -0.2629 5.8452 -35.6904 **5.** 1.2246 -1.2246 -0.5580
7. -6.3750 -5.6250 1.0938   **9.** 14.5593 -0.3823 -0.9510 **11.** 5210 52.1 5,210,000
13. 0.00603 0.000603 0.00000603   **15.** 805 8,050,000 8050   **17.** 0.0911 0.000911 9110
19. 268 26.8 2.68   **21.** 0.4202   **23.** 2.8304   **25.** 4.4973   **27.** -3.1888 **29.** 1.671
31. 0.4539   **33.** 0.0001751   **35.** 0.00002078   **37.** 4,096,000   **39.** 0.009278

Section 6.5

1. **a.** 46,009 **b.** 12 yr 8 mos; **c.** 34 yr 7 mo   **3.** $131,250   **5.** 40%   **7.** 129,904   **9.** **a.** 2115 yr ago **b.** in 5315 yr **c.** 1957 units   **11.** **a.** 5.8 hr **b.** 6.6 hr ago **c.** 673   **13.** **a.** in 3.2 min **b.** 2.63 min ago **c.** 43.5 **d.** 70.7
15. **a.** in 13.5 min **b.** 45 min from now **c.** 6.56 min ago **d.** 561.167   **17.** **a.** 5.614% **b.** 8.16% **c.** 7.25% **d.** 7.25%
e. 7.79%   **19.** **a** $13,498.60 **b.** $8869.20   **21.** **a.** $\approx$14 yrs **b.** $\approx$11 3/4 yr **c.** $\approx$10 yr   **23.** 6.93%
25. **a.** $7871.71 **b.** $7830.69   **27.** **a.** $1487.11 **b.** $1481.64   **29.** 4.6 billion yr   **31.** **a.** 9.12 yr **b.** 25 yr **c.** 50 yr
33. 2.59 . . ., 2.705 . . ., 2.717 . . ., 2.7181 . . ., 2.718268 . . ., 2.718280 . . .

Section 6.7

1. $\log_5 125 = 3$ **3.** $\log_{16} 1 = 0$ **5.** $64^{1/3} = 4$ **7.** 132   **9.** 25   **11.** -6 **13.** 2   **15.** 1/3
17. 7/2   **19.** 18   **21.** $\log_2 2 = 1$ **23.** 1   **25.** 11.73   **27.** -0.491 **29.** 1.994   **31.** 3.2619

33.
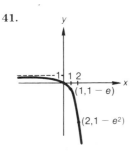

35.

37.
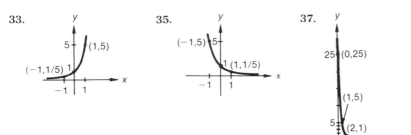

39.
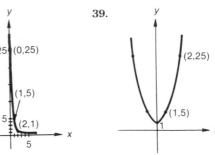

41.
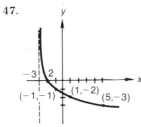

43.

45.
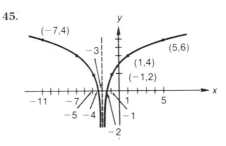

47.
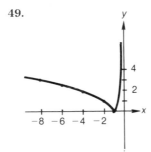

49.

51. 4   **53.** 4/3   **55.** 1   **57.** 0, 1   **59.** 81   **61.** 4   **63.** 2   **65.** 15   **67.** 3   **69.** 8
71. $(2 \log 4 - \log 5)/(2 \log 5 - \log 4) \approx 0.635$ **73.** $(2 \log 12 + \log 15)/(\log 15 - \log 12) \approx 34.408$ **75.** **a.** 1066
gal **b.** 0.925 gal **c.** 34.64 gal **d.** 0.63 hr   **77.** **a.** $600; 13,150.13; 13,956.39; 14,388,41; 14,836.23; 14,841.32 **b.** $1600;
1790.85; 1806.11; 1814.02; 1822.03; 1822.12 **c.** $2800; 2938.66; 2960.49; 2971.89; 2983.52; 2983.65 **d.** $5906.25; 6093.37;
6121.42; 6135.93; 6150.61; 6150.77 **e.** $6875; 7178.15; 7225.22; 7249.74; 7274.68; 7274.96
79. **a.** 46,300,000   4630   463 **b.** 2,160,000   0.000216   0.00216 **c.** 0.0656   0.656   0.000656 **d.** 9,610,000   9610   0.00961
e. 983   0.0983   0.00983   **81.** **a.** 9.142 **b.** 89,440,000 **c.** 0.0007013 **d.** 5695 **e.** 0.003001

CHAPTER 7

Section 7.1

1. **b.** consistent
d. $(x,y) = (1,-1)$ **a.** and **c.**

3. **b.** consistent
d. $(p,q) = (2,-1)$ **a.** and **c.**

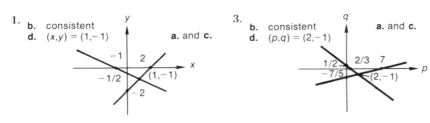

5. b. consistent
  d. $(t,u) = (4,-5)$

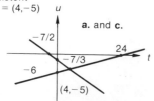

7.
  b. Inconsistent

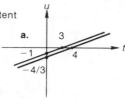

9.
  b. dependent: $z = 1 - 2y$

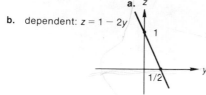

11.
  b. consistent
  d. $(x,w) = (-3,5)$

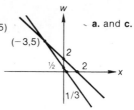

13. b. consistent
  d. $(u,v) = (4,-2)$

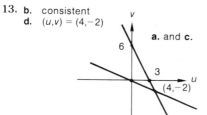

15. b. dependent: $w = (3 - 5v)/7$

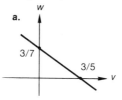

17. b. consistent
  d. $(u,v) = (1,-1)$

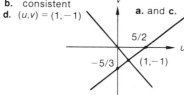

19. b. consistent
  d. $(w,z) = (1,-2)$

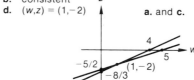

21.
  b. Inconsistent

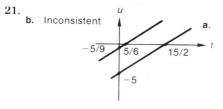

23. b. consistent
  d. $(x,y) = (2,1)$

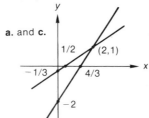

25. b. consistent
d. $(v,r) = (3,2)$

a. and c.

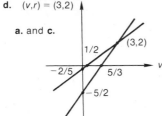

27. $(x, y) = (1, 2)$　　**29.** $(x, y) = (2, 1)$　　**31.** $(x, y) = (4, 11)$

33.

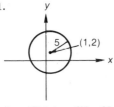

35. a. $\begin{cases} d = 8/5 + 1/5t \\ d = 5t \end{cases}$ **b.**

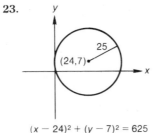

c. $t = 1/3$ hr $= 20$ min; $d = 1\frac{2}{3}$ mi

37. $V = 14, M = 36$　　**39.** 52

Section 7.2

1. $(t, u) = (-1, 1)$　　**3.** $(w, z) = (1, -1)$　　**5.** $(x, y) = (2, -1)$　　**7.** $(x, z) = (2, 3)$　　**9.** dependent; $z = 3 - 3y/2$　　**11.** $(t, v) = (5, 3)$　　**13.** $(r, s) = (-4, -5)$　　**15.** $(z, t) = (3, -1)$　　**17.** inconsistent
19. $(w, y) = (-3/4, 1/4)$　　**21.** $(x, y) = (3/2, 1/2)$　　**23.** $(r, s) = (7, 5)$　　**25.** $(p, q) = (2, 1)$　　**27.** 86
29. fudge 90¢, mints 16¢　　**31.** \$5200 at 7%, \$4400 at 6%　　**33.** 6 and 34　　**35.** first pipe 36 hr, second pipe 45 hr　　**37.** 1200 in²

Section 7.3

1. $x = -5 - 7y, z = 9 + 11y$ for any real number y　　**3.** $r = 2t + 1, s = 1 - t$ for any real number t
5. $(x, y, z) = (-2, -1, -2)$　　**7.** inconsistent　　**9.** $x = 2z - 3, y = z + 2$ for any real number z
11. $(a, b, c) = (2, 0, -3)$　　**13.** $(y, z, w) = (-1, -1, 4)$　　**15.** $(x, y, z) = (-4, 3, 6)$　　**17.** $(r, s, t) = (-2, -1, 2)$　　**19.** $(r, s, t) = (19s - 16, s, 12 - 11s)$　　**21.** $(u, v, w) = (0, 0, 0)$　　**23.** inconsistent
25. $(p, q, r) = (1, -1, -1)$　　**27.** No 20 percent solution, 1000 lb of the 30 percent solution and 1000 lbs of the 70 percent solution　　**29.** 273　　**31.** 263　　**33.** 154　　**35.** 6 nickels, 4 dimes, 3 quarters

Section 7.4

1. $(x, y, z) = (6, 9, 1)$　　**3.** $(r, s, t) = (0, 0, 0)$　　**5.** $r = t + 1, s = 2 - t$ for any real number t　　**7.** $(p, q, r) = (3, -2, 1)$　　**9.** $(a, b) = (4, -2)$　　**11.** $(r, s, t, u) = (-1, 2, 1, 1)$　　**13.** $(r, s, t, u) = (8/15, -2/3, -11/15, 0)$　　**15.** $(a, b, c, d) = (1, -2, 1, 1)$　　**17.** $(r, s, t, u) = (1, 3, 2, 1)$
19. $(u, v, w, x, y) = (-1, 1, -3, 2, 1)$

21.

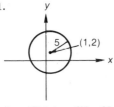

$(x - 1)^2 + (y - 2)^2 = 25$

23.

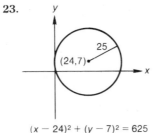

$(x - 24)^2 + (y - 7)^2 = 625$

25.

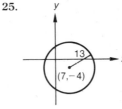

$(x - 7)^2 + (y + 4)^2 = 169$

27. 320 acres @ $80, 240 acres @ $60, 400 acres @ $50   **29.** 5 parts Super-Feeder, 2 parts Weed-Go, 1 part Bulk
31. 2, -7, -6

Section 7.5

1. 5   **3.** -20 **5.** -13 **7.** -5 **9.** 60   **11.** -10 **13.** 63   **15.** 7   **17.** -8
19. 450   **21.** -6 **23.** -6 **25.** 832   **27.** 0   **29.** -18

Section 7.6

1. $(x, y) = (-5, -8)$ **3.** $(t, u) = (5/2, -5/4)$ **5.** $(u, v) = (11, 7)$ **7.** $(w, z) = (1, -1)$ **9.** $(t, y) =$
$(1, -2)$ **11.** $(x, y, z) = (-11/9, 26/9, 1/3)$ **13.** $(r, s, t) = (6, 9, 1)$ **15.** $(x, y, z) = (3/4, -1/6, -11/12)$
17. $(w, z, u) = (1/2, 0, -3/2)$ **19.** $a = -1, b = c = d = 1$ **21.** 4 and 13   **23.** $5000 @ 5%,
$11,000 @ 8%   **25.** Gideon is 5, father is 26

Section 7.7

1. $3/(x - 3) + 5/(x + 2)$ **3.** $(-1/3)/(x + 1) + (4/3)/(x - 2)$ **5.** $2 + 1/(x - 3) - 3/(x + 2)$
7. $(9/2)/(x + 2) + (1 - 9x/2)/(x^2 + 2x + 2)$ **9.** $1/x + (-x + 1)/(x^2 - 2)$ **11.** $1/(x + 2) + (3x + 1)/(x^2 + 2)$
13. $(1/2)/x - (1/2)/(x + 2) - 1/(x + 2)^2$ **15.** $4/(x - 2) - 1/(x + 1) + 3/(x + 1)^2$
17. $2x + 1 - 3/x + 4/x^2 + 3/(x - 3)$ **19.** $1/(x - 1) + (2x + 1)/(x^2 + 1)^2$ **21.** $-1/(x - 1) +$
$(x - 1)/(x^2 + 2) + (x + 1)/(x^2 + 2)^2$ **23.** $1/(x - 2) + 3/(x^2 + 2) - (x + 10)/(x^2 + 2)^2$ **25.** $2x + 1/(x + 2) +$
$1/(x^2 + 2)^2$

Section 7.8

1. $(x, y) = (3, 2)$ and $(4, 3/2)$ **3.** $(x, y) = (2/5, 16/5)$ and $(2, 0)$ **5.** $(x, y) = (3, 2), (3, -2), (-3, 2)$, and
$(-3, -2)$ **7.** $(x, y) = (\pm 3\sqrt{2}, \pm 2)$ **9.** $(x, y) = (-1/3, 1), (1, -3), (2/9, 1/9), (2, 1)$ **11.** $(\pm \sqrt{2}, 0)$
13. $(x, y) = (1, 0)$ **15.** $(x, y) = (4, 2)$ **17.** $(p, q) = (2, 0)$ or $(3, \pm \sqrt{\log_2 5 - 2})$ **19.** $(x, y) = (1, 2)$ and
$(0, 1)$ **21.** $(u, v) = (6, 1)$ **23.** $x = 2$ **25.** $x = 3$ **27.** $(x, y) = (2, 3)$ and $(1, 2)$ **29.** $(x, y) =$
$(0, \pm 3), (-1, 1), (4/3, 1)$ **31.** $(1, 7)$ and $(2, 14)$ **33.** 12 units of either chemical and 14 units of the other
35. 8, 9   **37.** **a.** $v_1 + v_1' = v_2 + v_2'$ **b.** $v_2' = -2, v_1' = -11$ **c.** yes, object 1; $1 - 11 = -10 = -8 - 2$

Section 7.9

1. $\begin{bmatrix} 1 & 3 \\ 3 & 2 \end{bmatrix}$ **3.** $\begin{bmatrix} 4 & 10 \\ 4 & 10 \end{bmatrix}$ **5.** $\begin{bmatrix} 8 & -1 \\ 12 & -4 \end{bmatrix}$ **7.** $\begin{bmatrix} 8 & 0 \\ 17 & 10 \end{bmatrix}$ **9.** $\begin{bmatrix} 1 & 1 \\ -19 & -51 \end{bmatrix}$ **11.** $\begin{bmatrix} 17 & 43 \\ 28 & 72 \end{bmatrix}$

13. $\begin{bmatrix} 3 & 1 \\ 4 & 1 \\ -1 & 2 \end{bmatrix}$ **15.** $\begin{bmatrix} 0 & 7 \\ 10 & 3 \end{bmatrix}$ **17.** $\begin{bmatrix} 5 & 4 & -1 \\ 10 & 2 & 2 \\ 4 & 8 & -4 \end{bmatrix}$ **19.** $\begin{bmatrix} -12 & 13 \\ 12 & 1 \end{bmatrix}$ **21.** $\begin{bmatrix} 7 & 2 & 1 \\ 29 & -14 & 19 \\ 5 & 46 & -29 \end{bmatrix}$

23. $\begin{bmatrix} -4 & -6 & -3 & 9 \\ 4 & 5 & 20 & -5 \end{bmatrix}$ **25.** $\begin{bmatrix} 5 & -1 & -1 & -6 \\ -6 & 1 & 0 & -10 \end{bmatrix}$ **27.** $\begin{bmatrix} -47 & -27 & 6 & 60 \\ -22 & -2 & 55 & 74 \end{bmatrix}$

29. $\begin{bmatrix} -6 & -1 & -3 & 14 \\ 1 & 4 & 20 & 0 \\ 9 & -3 & -1 & -13 \\ -18 & -8 & -1 & 18 \end{bmatrix}$ **31.** $\begin{bmatrix} 3 & -5 \\ -1 & 2 \end{bmatrix}$ **33.** $\begin{bmatrix} 7 & 1 \\ 6 & 1 \end{bmatrix}$ **35.** no inverse

37. $\begin{bmatrix} -1 & -2 & 3 \\ 2 & 3 & -5 \\ 3 & 8 & -10 \end{bmatrix}$ **39.** $\begin{bmatrix} 1 & -1 & 0 \\ -2 & 5 & 1 \\ -1 & 3 & 1 \end{bmatrix}$ **41.** $(x, y) = (-9, 4)$ **43.** $(x, y) = (-5, -13)$

45. $(x, y, z) = (4, -6, -15)$ **47.** $(x, y, z) = (-5, 20, 12)$

49. $\begin{bmatrix} 2 & 1 & 2 \\ 4 & 2 & 1 \\ 1 & 3 & 1 \\ 1 & 2 & 3 \end{bmatrix} \begin{bmatrix} 10 \\ 5 \\ 20 \end{bmatrix} = \begin{bmatrix} 65 \\ 70 \\ 45 \\ 80 \end{bmatrix};$ stamper 65 hr
press 70 hr
welder 45 hr
painter 80 hr

51. $(0.6)(20 \quad 15 \quad 25 \quad 10)\begin{bmatrix} 2 & 1 & 2 \\ 4 & 2 & 1 \\ 1 & 3 & 1 \\ 1 & 2 & 3 \end{bmatrix} + (0.4)(10 \quad 12 \quad 20 \quad 10)\begin{bmatrix} 2 & 1 & 2 \\ 4 & 2 & 1 \\ 1 & 3 & 1 \\ 1 & 2 & 3 \end{bmatrix} = (120.20, \, 132.60, \, 98.80);$

bicycle \$120.20, moped \$132.60, tricycle \$98.80

Section 7.10.

1.

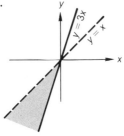

3.

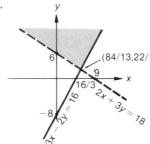

5.

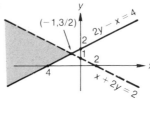

7.

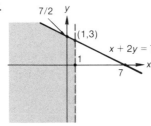

9.

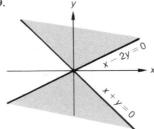

11.

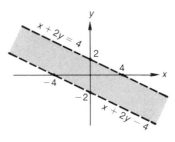

13.

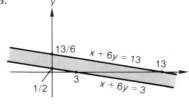

15.

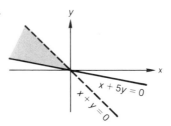

17.

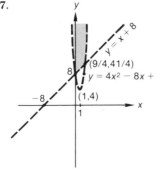

19.

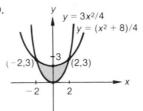

21.

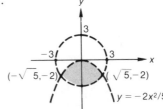

23.

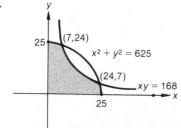

25.

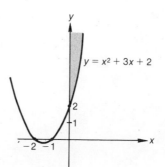

Section 7.11

1. 200 of each, \$900   **3.** 600 alphas, no betas, \$1800   **5.** Any combination of S semis and P pickups with $30 \le S \le 40$, $20 \ge P \ge 0$ and $2S + P = 80$; \$12,000   **7.** 4 units F_1, 6 units F_2; \$176   **9.** 2 units F_1, 9 units F_2, \$148
11. Mine No. 1 40 days, mine No. 2 20 days; \$11,000   **13.** 80 homes, 160 apartment units; \$1,600,000   **15.** \$1 million on print and \$5 million on radio and TV; \$230 million   **17.** 2 units of Crunchy Bits, 6 units of Fluffy Flakes
19. a. 400 gal Plain, 600 gal Extra Creamy **b.** \$3180   **21.** maximum = 410 @ $(x, y) = (70, 20)$ **23.** minimum = 70 @ $(x, y) = (6, 10)$ **25.** maximum = 100 @ $(x, y) = (15, 25)$

Section 7.13

1. $(y, z) = (-1, 3)$ **3.** inconsistent   **5.** $(s, v) = (4, 8)$ **7.** $(p, q) = (0, -1)$ **9.** dependent: $y = (1 - 3x)/2$ **11.** $(x, y) = (1, -2)$ **13.** $(t, u) = (2, 3)$ **15.** $(z, y) = (-4, 6)$ **17.** $(a, b, c) = (4, 0, 1)$
19. $(r, s, t) = (-1, -3, 2)$ **21.** $(u, v, w) = (1, -4, -2)$ **23.** $(x, y, z) = (8, 10, -3)$ **25.** inconsistent
27. dependent: $b = 0$, $a = c$ **29.** $(r, s, t) = (-3, 3, -4)$ **31.** $(x, y, z, w) = (4/3, 0, -1/3, 11/3)$
33. $(p, q, r, s) = (1, -2, 0, 2)$ **35.** $(u, v) = (19/4, -3/4)$ **37.** $(t, u) = (-1, 3)$ **39.** $(x, y, z) = (2/3, -2/3, 3)$ **41.** $(t, u, v) = (0, -3, 2)$ **43.** $(u, v, w, x) = (-1, 3, 1/2, -2)$ **45.** 0   **47.** -74
49. -16 **51.** 23   **53.** -328

55.

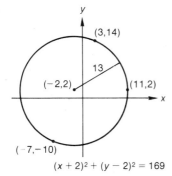

57. $-1/(x + 2) + 2/(x + 1)$ **59.** $-2/x + 2/(x + 1) + 4/(x + 1)^2$ **61.** $2/(x + 1) + 3x/(x^2 + x + 1)^2$
63. $x = \sqrt{2}$, $y = [1 + \log(\sqrt{2} + 1)]/2$; $x = -\sqrt{2}$, $y = [1 - \log(\sqrt{2} + 1)]/2$; $x = 0$, $y = 1/2$ **65.** $(x, y) = (1, \pm 2)$ and $(-1, \pm 2)$

67. $\begin{bmatrix} 27 & 23 \\ -11 & 36 \end{bmatrix}$ **69.** $\begin{bmatrix} 27 & 51 \\ 39 & 20 \end{bmatrix}$ **71.** $\begin{bmatrix} 7 & 15 & 18 \\ 3 & -18 & 24 \\ 19 & 51 & 42 \end{bmatrix}$ **73.** $\begin{bmatrix} 80 & 15 & 83 \\ 125 & 30 & 47 \\ -8 & -12 & 124 \end{bmatrix}$

75. $\begin{bmatrix} 3 & -1 \\ 5 & -2 \end{bmatrix}$ **77.** $\begin{bmatrix} 6 & 3 & -2 \\ 1 & 1 & 0 \\ -1 & 0 & 1 \end{bmatrix}$ **79.** $(x, y) = (1, 1)$ **81.** $(x, y, z) = (-2, -1, -1)$

83.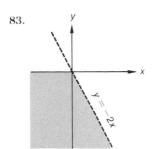

85.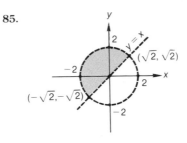

87.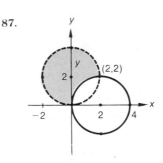

89.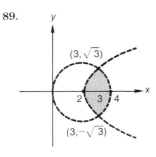

91. 63 and 27   **93.** Boat = 10 mph   river = 5 mph   **95.** 674   **97.** Steak = \$2.50, fish = \$2.00, pork = \$1.50   **99.** maximum = 1300 @ $(x, y) = (80, 10)$ **101.** any combination of x and y satisfying $20 \le x \le 80$, $30 \ge y \ge 0$, $x + 2y = 80$ for a minimum of 400   **103.** minimum = 54 @ $(x, y) = (4, 10)$ **105.** maximum = 1000 @ $(x, y) = (0, 50)$ **107.** any combination of x and y satisfying $4 \le x \le 5$, $1/2 \ge y \ge 0$, $x + 2y = 5$ for a minimum of 10

CHAPTER 8

Section 8.1

1. Must eventually show that $(2n - 1)^2 + 8[(n + 1) - 1] = [2(n + 1) - 1]^2$ **3.** Must show that $[n(n + 1)/2]^2 + (n + 1)^3 = \{(n + 1)[(n + 1) + 1]/2\}^2$ **5.** Show that $2(2^n - 1) + 2^{n+1} = 2(2^{n+1} - 1)$ **7.** Show that $(1 - 1/2^n) + 1/2^{n+1} = 1 - 1/2^{n+1}$ **9.** Show that $n[(n + 1)(n + 2)]/3 + (n + 1)[(n + 1) + 1] = (n + 1)[(n + 1) + 1][(n + 1) + 2]/3$ **11.** Show that $[n(4n^2 + 6n - 1)/3] + [2(n + 1) - 1][2(n + 1) + 1] = (n + 1)[4(n + 1)^2 + 6(n + 1) - 1]/3$ **13.** Show that $[n(n + 1)(n + 2)(n + 3)/4] + (n + 1)[(n + 1) + 1][(n + 1) + 2] = \{(n + 1)[(n + 1) + 1][(n + 1) + 2][(n + 1) + 3]\}/4$ **15.** Show that $n(n + 3)/4(n + 1)(n + 2) + 1/n(n + 1)(n + 2) = (n + 1)[(n + 1) + 3]/4[(n + 1) + 1][(n + 1) + 2]$ **17.** $5^{n+1} - 1 = 5^n \cdot 5 - 1 = (5^n - 1) \cdot 5 + (5 - 1)$
19. $2^{2(n+1)+1} + 1 = 2^{2n+1} \cdot 4 + 1 = (2^{2n+1} + 1)4 - 4 + 1$ **21.** $(n + 1)^3 + 6(n + 1)^2 + 2(n + 1) = n^3 + 9n^2 + 17n + 9 = (n^3 + 6n^2 + 2n) + 3(n^2 + 5n + 3)$ **23.** $a^{n+1} - b^{n+1} = a^n a - b^n b = (a^n - b^n)a + (a - b)b^n$ **25.** $a^{2(n+1)-1} + b^{2(n+1)-1} = (a^{2n-1})a^2 + (b^{2n-1})b^2 = (a^{2n-1})(a^2 - b^2) + b^2(a^{2n-1} + b^{2n-1})$ **27.** $P_{n+1} = P_n(1 + \beta) = [P_0(1 + \beta)^n](1 + \beta) = P_0(1 + \beta)^{n+1}$
29. $2^{16} - 1 = 65,531$

Section 8.2

1. 3, 6, 9, 12, 15; 30   **3.** 5, 7, 9, 11, 13; 23   **5.** 2, 8, 18, 32, 50; 200   **7.** 0, 2, 0, 2, 0; 2   **9.** -2, 4, -8, 16, -32; 1024   **11.** 0, 1/3, 1/2, 3/5; 7/9   **13.** 2/3, 3/4, 4/5, 5/6; 9/10   **15.** 1/2, 3/7, 3/8, 1/3; 3/13   **17.** -5, 7, 3, 11/5; 19/13   **19.** -1, 1/8, $-1/27$, 1/64; 1/512   **21.** -2, 2, -2, 2, -2 **23.** 1, 1, 3/2, 3, 15/2   **25.** 2, 1, 0, 1, 0   **27.** 2, 5, 8, 11, 14; 44   **29.** -5, -3, -1, 1, 3; 23   **31.** 22, 20, 18, 16, 14; -6 **33.** 73   **35.** 21   **37.** -5 **39.** 9, 12, 15, 18, 21, 24, 27   **41.** 25, 30, 35, 40, 45, 50   **43.** 19, 25, 31, 37, 43   **45.** 2, 6, 18, 54; 4374   **47.** 3, 12, 48, 192; 49,152   **49.** 2/3, $2\sqrt{3}/3$, 2, $2\sqrt{3}$, 6; $18\sqrt{3}$ **51.** 1536   **53.** 243   **55.** -3 **57.** For $r = \sqrt{3}$: $21\sqrt{3}$, 63, $63\sqrt{3}$, 189, $189\sqrt{3}$ For $r = -\sqrt{3}$: $-21\sqrt{3}$, 63, $-63\sqrt{3}$, 189, $-189\sqrt{3}$ **59.** -54, 162   **61.** 1024, 512, 256, 128, 64, 32, 16, 8, 4
63. **a.** \$800, \$750, \$700, \$650, \$600, \$550, \$500, \$450, \$400, \$350, \$300, \$250, \$200, \$150, \$100, \$50, \$0; **b.** arithmetic
65. **a.** $a_n = h_0 - 16n^2$ **b.** 16, 48, 80, . . . **c.** $16(n + 1)^2 - 16n^2 = 16 + 32n$ **d.** 304 ft, 624 ft   **67.** \$6000, \$4500, \$3375, . . ., \$6000(3/4)^n, . . .; **b.** \$337.88; **c.** geometric

Section 8.3

1. 25/12   **3.** 44   **5.** 420   **7.** 3/4   **9.** $\sum_{n=0}^{4} (5 + 2n)$ **11.** $\sum_{n=1}^{5} (1/2n)$ **13.** $\sum_{n=2}^{7} 10n$ **15.** $\sum_{n=2}^{99} n^n$
17. 175   **19.** 13,035   **21.** 649   **23.** 1430   **25.** 1358   **27.** 728   **29.** 1020   **31.** 1717
33. 465   **35.** \$496   **37.** 13% of the original value   **39. a.** 3 echelons will more than suffice; two will not
b. 12 echelons will more than suffice; 11 will not

Section 8.4

1. a. 720   **b.** 3024   **c.** 3   **d.** 5   **e.** 1/10   **f.** 25   **3.** $r^{10} - 10r^9s + 45r^8s^2 - 120r^7s^3 + 210r^6s^4 - 252r^5s^5 +$
$210r^4s^6 - 120r^3s^7 + 45r^2s^8 - 10rs^9 + s^{10}$ **5.** $64x^3 - 48x^2y + 12xy^2 - y^3$ **7.** $v^5 - 5v^4w^2 + 10v^3w^4 - 10v^2w^6 +$
$5vw^8 - w^{10}$ **9.** $16x^4 + 96x^3y^2 + 216x^2y^4 + 216xy^6 + 81y^8$ **11.** $x^6 + 6x^4 + 15x^2 + 20 + 15/x^2 + 6/x^4 + 1/x^6$
13. $t^8 - 8t^7/u^2 + 28t^6/u^4 - 56t^5/u^6 + 70t^4/u^8 - 56t^3/u^{10} + 28t^2/u^{12} - 8t/u^{14} + 1/u^{16}$ **15.** 161,051
17. 104060401   **19.** 1.771561   **21.** 0.985074875   **23.** $680x^{14}y^3$ **25.** $760r^{18}s^2$ **27.** $-960u^{14}v^3$
29. $792/t^2$ **31.** $-960p^{11}/q^7$

Section 8.5

1. a. 9,000,000   **b.** 8,100,000,000   **3.** 1024   **5.** 288   **7.** 36   **9.** $5 \cdot 3^{10} \cdot 2^6 = 18,895,680$
11. 12! = 479,001,600   **13.** 720   **15.** 24   **17. a.** 720   **b.** 240   **c.** 72   **19.** 120   **21.** 462
23. 170,016   **25.** $(_{24}C_5)(_{19}C_5)(_{14}C_5)(_9C_5)4 \approx 4.987(10^{14})$ **27.** 1,007,760   **29. a.** 45   **b.** 10   **c.** 110
31. 35   **33. a.** 56   **b.** 20   **c.** 364   **35.** 298,598,400   **37.** 155,937,600   **39. a.** 63,063,000   **b.** 2,027,025
41. a. 15,400   **b.** 9240   **43. a.** 12,600   **b.** 6300   **45. a.** 34,650   **b.** 1296

Section 8.6

1. 1/3   **3. a.** 1/28   **b.** 1/4   **c.** 1/4   **d.** 3/14   **e.** 3/28   **f.** 19/28   **5. a.** 3/16   **b.** 13/16   **7. a.** 1/6   **b.** 1/4
c. 1/12   **9.** 1/16   **11. a.** 2/9   **b.** 5/9   **13.** $1627/2187 \approx 0.74$ **15.** 7/27   **17.** 1/2
19. a. $1 - [(48/52)(47/51)] \approx 0.149$ **b.** 1/17   **c.** 1/26   **d.** $1 - [(36/52)(35/51)] \approx 0.525$ **e.** $116/221 \approx 0.525$ **f.** $11/221 \approx$
0.050   **g.** $1001/1326 \approx 0.755$ **h.** $623/1326 \approx 0.47$ **i.** $2/221 \approx 0.009$ **21. a.** 1/17   **b.** $11/850 \approx 0.013$ **c.** $33/66640 \approx$
0.0005   **d.** $23747/333200 \approx 0.071$ **23. a.** $4(_{48}C_4)/_{52}C_5 = 3243/10829 \approx 0.30$ **b.** $4(_{48}C_{12})/(_{52}C_{13}) = 9139/20825 \approx 0.44$
25. $(_{16}C_{12}/_{48}C_{12}) \cdot 4 = (1820/69,668,534,460) \cdot 4 \approx 0.000000104$ **27.** $(_{44}C_{20})/(_{48}C_{24}) \approx 0.0546$ **29.** 1/1024

Section 8.8

1. Must eventually show that $[n(2n^2 + 3n + 7)/6] + [(n + 1)^2 + 1] = \{(n + 1)[2(n + 1)^2 + 3(n + 1) + 7]\}/6$
3. Must show that $n^2(2n^2 - 1) + [2(n + 1) - 1]^3 = (n + 1)^2[2(n + 1)^2 - 1]$
5. Show that $n(n + 1)(2n + 7) + (n + 1)[(n + 1) + 2] = (n + 1)[(n + 1) + 1][2(n + 1) + 7]/6$
7. Show that $[n(n + 1)(n + 2)(3n + 5)/12] + (n + 1)[(n + 1) + 1]^2 = (n + 1)[(n + 1) + 1][(n + 1) + 2][3(n + 1) + 5]/12$
9. Show that $n/2(3n + 2) + 1/(3[n + 1] - 1)(3[n + 1] + 2) = [n + 1]/2(3[n + 1] + 2)$ **11.** $4^{2(n+1)} - 1 =$
$4^{2n} \cdot 16 - 16 + 16 - 1 = 16(4^{2n} - 1) + (16 - 1)$ **13.** $-4, -3, -2, -1, 0;$ 7   **15.** 2, 5, 10, 17, 26;   145
17. 1, 7, 17, 31, 49;   287   **19.** 1/2, 2/3, 3/4, 4/5, 5/6;   9/10   **21.** 3, 1, 3/5, 3/7, 1/3;   3/17   **23.** 3, 7, 15, 31,
63   **25.** $-\sqrt{2}, 2, -4, 16, -256$ **27.** 4, 5, 6, 7, 8;   13   **29.** 4, 8, 12, 16, 20;   40   **31.** 27   **33.** 10
35. 7, 11, 15   **37.** 46, 50, 54   **39.** 1, 5, 25, 125;   78,125   **41.** $\sqrt{2}, 2, 2\sqrt{2}, 4;$ 16   **43.** 4374
45. 4096   **47.** 24, 48   **49.** For $r = \sqrt{2}$: $67\sqrt{2}, 134, 134\sqrt{2}, 268, 268\sqrt{2}$

For $r = -\sqrt{2}$: $-67\sqrt{2}, 134, -134\sqrt{2}, 268, -268\sqrt{2}$ **51.** $\sum_{1}^{4} 2n$ **53.** $\sum_{0}^{5} (3 + 4n)$

55. $\sum_{1}^{6} (n + 1)! \cdot 3 \cdot 2^{n-1}$ **57.** 8   **59.** 53   **61.** 164   **63.** 330   **65.** 10,587   **67.** 110

69. 403,200   **71.** $a^9 + 18a^8 + 144a^7 + 672a^6 + 2016a^5 + 4032a^4 + 5376a^3 + 4608a^2 + 2304a + 512$
73. $243x^5 + 810x^4y^2 + 1080x^3y^4 + 720x^2y^6 + 240xy^8 + 32y^{10}$ **75.** $64x^3 - 192x^{3/2} + 240 - 160x^{-3/2} + 60/x^3 -$
$12x^{-9/2} + 1/x^6$ **77.** $-960x^7$ **79.** $5005a^9b^6$ **81.** $-252r^{10}s^5$ **83.** 1.061208   **85.** 112,550,881
87. 624   **89.** 720   **91.** $_{24}C_6 = 134,596$ **93.** 210   **95. a.** 1680   **b.** 630   **c.** 450   **d.** 240   **e.** 840
97. a. $649/1225 \approx 0.53$ **b.** $4/245 \approx 0.016$ **c.** $12/1225 \approx 0.0098$ **99. a.** 1/216   **b.** 1/64   **c)** 1/1728
101. a. 3/28   **b.** 15/28   **103.** $72811/390625 \approx 0.19$ **105.** $181/3125 \approx 0.06$ **107. a.** 1/221   **b.** 15/34
c. 20/221   **d.** 91/221   **e.** 25/102   **f.** 1/102   **g.** 195/442   **109. a.** $5/442 \approx 0.01$ **b.** $1/1105 \approx 0.0009$ **c.** 1/433160
d. $407/33320 \approx 0.012$ **111.** $(_{36}C_{13})/(_{52}C_{13}) \approx 0.00364$ **113.** $4(_{36}C_{12})/(_{48}C_{24}) - (_4C_2/_{48}C_{24}) \approx 0.00016$

Index